267133

ELECTROCHEMISTRY OF
BIOLOGICAL MOLECULES

ELECTROCHEMISTRY OF
BIOLOGICAL MOLECULES

GLENN DRYHURST

Department of Chemistry
University of Oklahoma
Norman, Oklahoma

ACADEMIC PRESS   New York   San Francisco   London   1977

A Subsidiary of Harcourt Brace Jovanovich, Publishers

C28 99077✓

CHEMISTRY

ACADEMIC PRESS, INC.
111 Fifth Avenue, New York, New York 10003

United Kingdom Edition published by
ACADEMIC PRESS, INC. (LONDON) LTD.
24/28 Oval Road, London NW1

Library of Congress Cataloging in Publication Data

Dryhurst, Glenn,   Date
      Electrochemistry of biological molecules.

      Includes bibliographies and indexes.
      1.   Organonitrogen compounds—Analysis.   2.   Hetero-
cyclic compounds—Analysis   3.   Electrochemical analysis.
I.   Title
QP801.N55D78   547'.593   75-40608
ISBN 0−12−222650−X

PRINTED IN THE UNITED STATES OF AMERICA

To June, Claire, Tory, and Greg

Contents

Preface

Nitrogen heterocyclic molecules are found extensively in biological systems, and have been chosen by nature to be involved in many, if not most, of the fundamental reactions of living organisms. Most of the nitrogen heterocyclic molecules found in nature contain extensive delocalized or mobile (π) electron systems. A result of this property is that these compounds are generally quite good electron donors and/or acceptors. This in turn allows very many biologically important nitrogen heterocyclic molecules to be studied by various electrochemical techniques, that is, their electrochemical oxidation and/or reduction and related chemical processes may be studied. The purpose of this book is to present a fairly complete summary of the electrochemistry of the more important groups of nitrogen heterocyclic molecules including purines and pyrimidines and their nucleosides and nucleotides, polynucleotides and nucleic acids, pteridines, flavins, pyrroles, porphyrins, and pyridines.

The treatment of the material tends to be, on occasion, somewhat encyclopedic. However, this is necessary to present a reasonably complete summary of the available information. Although this work will be of greatest use to electrochemists working on nitrogen heterocyclic and related systems, it is hoped that biochemists and biologists will also find the information summarized to be of use. Because of the involvement of these molecules in biological electron transfer processes, it is likely that many of the electrochemical reaction routes and intermediates will be similar to the biological electron-transfer processes.

I would like to acknowledge the extensive assistance provided by the Faculty Research Committee of the University of Oklahoma in the preparation of this book. I wish to express my appreciation to Dr. Bruce Laube, Dr. Jean P. Pinson, and Dr. David L. McAllister for reading many sections of the original manuscript, and particularly to Dr. C. LeRoy Blank who very carefully read the final manuscript and made many valuable suggestions. However, any errors in the book, of course, must be attributed to me. I would also like to express my deep appreciation to Professor Philip J. Elving, who first gave me the opportunity to study the electrochemistry of nitrogen heterocycles. It will be obvious from this book that Professor Elving, probably more than anyone else, was responsible for opening up the modern era of electrochemistry of purines and pyrimidines, and he continues to be a major contributor to the field. I am especially grateful to my wife, June Diane Dryhurst, for her constant support and encouragement throughout the years.

Glenn Dryhurst

1

Electrochemistry and Biological Processes

Studies of the electron-transfer reactions of atoms, ions, or molecules by polarographic or voltammetric techniques* can provide an extraordinary amount of information about such reactions. Thus, a characteristic potential at which the reaction occurs can be measured. It is generally referred to as the half-wave potential in DC polarography and some related techniques, or as the half-peak potential or peak potential in voltammetric techniques at stationary electrodes. These potentials characterize the electrode process to some degree and in certain instances may have direct thermodynamic significance.

By use of modern electrochemical techniques it is possible to decide precisely how many electrons are involved in the electron-transfer reaction at a particular potential. It is also possible to detect very unstable intermediates or products produced after the transfer of electrons has occurred. For example, many electrode (and biological) reactions proceed by one-electron-transfer reactions to give a free radical, a radical cation, or a radical anion species. There are examples in which the radicals so produced are extremely unstable such that electron spin resonance spectroscopy, for example, is generally incapable of detecting them, whereas they are detectable by electrochemical techniques such as fast sweep cyclic voltammetry.

It is also quite easy to examine electron-transfer reactions electrochemically under an extraordinarily large number of conditions in which most other chemical or biological studies often cannot be performed satisfactorily. Thus, it is not unusual to study an electrode reaction over a very wide pH range, in the

* Throughout this book polarography refers to electrochemical processes studied at the dropping mercury electrode. Voltammetry refers to electrochemical reactions occurring at any other type of solid or liquid electrode.

1

presence of quite different buffer systems, to employ a variety of temperatures and solvents, and to study the reaction in the presence and almost total absence of protons. Often, as a result of such studies, rather subtle changes in mechanism occur, or intermediates or unstable products can be observed more clearly. This potentially allows a much more detailed reaction mechanism for the electron-transfer process and related reactions to be deduced.

When a dropping mercury electrode or other microelectrodes are used, only extremely small amounts of electroactive material are involved in the electron-transfer reaction, and only minute amounts of products are formed. However, it is possible, and indeed quite common, to scale up the polarographic or voltammetric experiments several hundreds or thousands of times so that appreciable amounts of products can be isolated and identified. In relation to this it is now becoming appreciated that electrochemical methods can be utilized for some rather unique synthetic applications.

Some electrochemical techniques, in particular pulse polarography, have an extraordinary sensitivity so that they can be occasionally utilized for determining electroactive material from the 10^{-2} to 10^{-8} M concentration level. There are few techniques that can rival this range of utility or lower level of analytical detection. Indeed, although pulse polarography has not been widely employed for the solution analysis of organic compounds, it is, in fact, a most attractive analytical technique.

Thus, in essence, modern electrochemical techniques can be employed to study the electron-transfer reaction mechanisms and products of atoms, ions, and molecules, and certain techniques can provide valuable synthetic or analytical tools. The pertinent question to be asked is, "What can electrochemical studies tell one about biological electron-transfer and related processes?" First of all, there is a definite set of similarities between electrochemical and biological (e.g., enzymatic) reactions:

1. Electrochemical and biological electron-transfer, or oxidation–reduction, reactions both involve essentially heterogeneous electron-transfer processes. Electrochemically, this process occurs at the electrode–solution interface; biologically, it occurs at an enzyme–solution interface.

2. Both electrochemical and biological reactions can take place at similar pH and in the presence of similar ionic strengths of inert electrolyte.

3. Both types of processes can occur effectively under nonaqueous conditions.

4. Both types of reactions normally occur at very similar temperatures.

5. Both at an electrode and at the active site of an enzyme it is likely that the substrate molecule has to be oriented in a rather specific fashion before the electron transfer can occur.

These statements are not meant to imply that the unique selectivity often associated with an enzyme can in any way be duplicated by an electrode. On the other hand, an enzyme cannot cause a thermodynamically impossible reaction to proceed; all of the laws and principles of chemistry are still applicable to an enzyme-catalyzed reaction. There is, however, sufficient superficial similarity between the electrochemical and biological reactions, which is not duplicated in other chemical systems, to warrant extensive study of the electrochemistry of biologically important molecules. Such studies should yield an enormous amount of evidence regarding the mechanisms of biological electron-transfer processes.

Normally, when one considers electron-transfer processes in biological systems, one tends to think of the electron-transport or respiratory chain, where an organic substrate is oxidized and usually oxygen is ultimately reduced. However, this reaction does not proceed by direct interaction of the substrate and oxygen but via a series of enzyme-catalyzed electron-transfer reactions, first between the substrate and, for example, a pyridine nucleotide, then through a large number of other transfers, until ultimately a reduced cytochrome is oxidized by oxygen and the oxygen is correspondingly reduced. There are, however, many more electron-transfer or oxidation–reduction type of reactions. Typical of these are those involved in catabolism of many biologically important organic compounds. A group of compounds of this type consists of the purines, in which the enzymatic and electrochemical oxidations are very similar. The electrochemical studies, however, reveal a great deal more information about the fine detail of the reaction than is available from enzymatic studies.

It is a reasonable contention, therefore, that biological and electrochemical redox processes are sufficiently similar that extensive studies of the electrochemistry of biologically important molecules should shed considerable light on the fundamentals of the biological reaction mechanisms.

Having made these somewhat sweeping and highly optimistic statements, it is only proper to point out that not a great deal of electrochemical research has been directed at really understanding the electron-transfer reactions of biologically important molecules, particularly in relation to biological processes. A very large number of biologically important molecules have been studied, generally polarographically, but more often than not these studies unfortunately boil down to a tabulation of half-wave potentials, measurement of the limiting current at a variety of pH values, an approximate guess at the number of electrons involved in the reaction, and postulation of a rather superficial mechanism. It is all too often found that products have not been isolated, identified, or subjected to structure elucidation. It is still common for only a single electrochemical technique to be utilized. Since the early to mid-1960s, however, there has been a marked upsurge in the development of the theory of modern electrochemical techniques and reactions. In addition, the advent of the operational amplifier has led to some spectacular advances in electrochemical instrumentation. The results of these two areas of advancement have led to the rather unusual situation that both theory and instrument design are considerably ahead of their experimental application, particularly to biologically important molecules.

In this book a fairly complete summary of the state of knowledge regarding the electrochemistry of a number of families of *N*-heterocyclic compounds is presented. Not all of the members of these families are normally found in biological systems. However, in order to understand the complete picture of the electrochemistry of the biologically important members of a family of compounds, a review of the pertinent literature regarding all the members of the family has been presented.

There are two unifying features that relate nearly all of the compounds discussed. The first, of course, is that they all contain heterocyclic nitrogen atoms. The second is that the compounds are characterized by molecular systems with mobile electrons. Nitrogen heterocycles containing delocalized or mobile (π) electrons have been chosen by nature to perform or to be involved in many, if not most, of the fundamental reactions of living organisms. For example, the most important constituents of nucleic acids are the purines and pyrimidines. The information stored in the sequences of the purines and pyrimidines in the nucleic acids directs protein synthesis and transmits genetic information. The energy-rich compounds, such as adenosine triphosphate, contain purines and are vital reactants in intermediary metabolism. Although there are hundreds of enzymes, most of these can exert their catalytic effects only in the presence of a suitable coenzyme. There are only a few coenzymes and most of these are conjugated *N*-heterocycles such as the pyridine nucleotides (NAD^+ and $NADP^+$), the flavin nucleotides (FAD and FMN), and the porphyrins, which are the heme prosthetic groups of the cytochromes. The latter are principally redox coenzymes. Folic acid, pyridoxal, and thiamine are part of the vitamin B complex, yet all contain conjugated *N*-heterocyclic groups. Apart from their involvement in folic acid, the pteridines are widely dispersed in nature, yet present considerable mystery as to their exact biological function. They are now suspected of being involved in reactions of sight and in electron transport in the photosynthetic process.

Most of the compounds just mentioned are characterized as being theoretically and experimentally quite good electron donors and/or acceptors. There are, therefore, valid reasons for electrochemically studying the electron-transfer reactions of these compounds and the consequences of such transfers.

It might also be pointed out that a very large number of drugs capable of acting on living cells are, at least in part, conjugated *N*-heterocyclic compounds.

The nitrogen heterocycles which are biologically important and which will be discussed subsequently are synthesized biologically at considerable expense. Since "nature does not indulge in luxuries"[1] there are necessarily deep and fundamental reasons why these compounds are used biologically and why they can so readily accept and/or donate electrons. It is therefore reasonable to expect that electrochemical studies, in conjunction with all of the other experimental and theoretical tools of chemistry, should be able to contribute significantly to the ultimate understanding of many biological electron-transfer reactions.

The material presented in subsequent chapters deals primarily with electron-transfer reactions of biologically important and related compounds. This does not imply, however, that the sole application of electrochemistry to biology is the study of such reactions. Thus, electrochemists are studying modes of ion transport, membrane and surface phenomena, effects of potential and current on tissue healing and regeneration, and many other problems. Other authoritative texts and literature reports deal with these studies.

REFERENCE

1. A. Szent-Györgi, "Introduction to Submolecular Biology." Academic Press, New York, 1960.

2

Theory and Instrumentation

I. INTRODUCTION

No attempt will be made in this chapter to give an extensive account of the theory and instrumentation of electrochemistry. Rather, a summary of the usual working equations, their significance and utility, and electrochemical jargon will be presented. In addition, the principles and circuits of some, but by no means all, electrochemical instrumentation will be discussed.

II. POTENTIOMETRY

Most electrochemical techniques of the type to be discussed can, to a greater or lesser extent, be regarded as derived from potentiometry. Accordingly, it is worth reviewing this technique, even though very little mention is made of it in subsequent chapters.

If a platinum wire electrode is immersed in a solution containing two components of a redox couple (Eq. 1), a potential develops across the

$$Ox + ne \rightleftharpoons Red \tag{1}$$

electrode—solution interface. Unfortunately, it is not possible to measure such a single-electrode potential directly. For this reason, the potential of the platinum electrode must be measured against a second electrode having a constant potential, i.e., a reference electrode. Potentials determined potentiometrically are normally expressed relative to the normal hydrogen electrode (NHE), which again does not have a known absolute potential, but by convention is assigned a potential of 0.0000 V. In order to measure the voltage between the reference electrode and the indicating or platinum electrode, a salt bridge is placed between the two electrode solutions to maintain electrolytic contact, and an apparatus of the type shown in Fig. 2-1 is employed. The DC voltage supply and the electrochemical cell are connected so that their potentials are in opposition. The sliding contact C is moved along the slidewire AB until upon momentarily closing the tapping key the galvanometer (G) shows no deflection. This implies that the voltage across AC is equal to and of the same polarity as the voltage across the cell and no current flows through the galvanometer. The voltage across AC and hence of the cell is then read from the voltmeter (V). Under these conditions the potential E across the cell is given by the Nernst equation (Eq. 2).

$$E = E^0 - \frac{RT}{nF} \ln \frac{A_{Red}}{A_{Ox}} \qquad (2)$$

In this expression E^0 is the standard potential for the reaction shown in Eq. 1 versus the same reference electrode, which, in other words, means the value of E when the activity of Ox and Red are equal (i.e., $A_{Red} = A_{Ox}$); R is the ideal gas

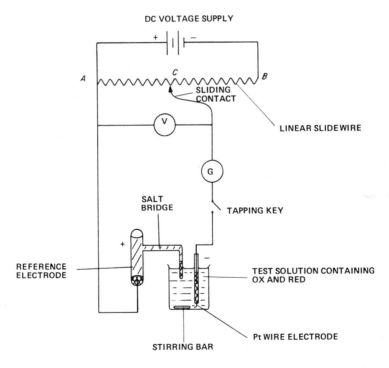

FIG. 2-1. Schematic diagram of simple apparatus for potentiometry. A,B,C: Linear slidewire; V: voltmeter; G: galvanometer.

constant (8.312 international joules), T is the absolute temperature, F is the faraday (96,500 C), and n is the number of electrons involved in the reaction as shown in Eq. 1.

The potentials determined potentiometrically for a redox system are of direct thermodynamic significance. Unfortunately, only a rather limited number of redox systems can be examined potentiometrically. In fact, only those reactions that exhibit electrochemical reversibility can be studied by direct potentiometry of the type described above. In a practical sense, electrochemical reversibility implies that both the oxidized and reduced forms of the redox couple are stable, and that the reaction can be carried out rapidly in either direction so that the Nernst equation is applicable. One can often find redox couples referred to as "sluggish," which implies that the reactions are not rapid and take a long time to reach equilibrium under potentiometric conditions.

The great difference between the potentiometrically determined electrode potential and that for the majority of techniques to be discussed subsequently is that in potentiometry the potential is measured when the voltage from the slidewire across AC is exactly equal to and of the same polarity as the voltage across the electrochemical cell so that the measurement is taken at zero current flow.

III. MEASUREMENTS WITH NET CURRENT FLOW

In many of the techniques to be discussed, a slightly modified form of experimental arrangement is employed. Instead of measuring the zero current potential of an electrode (versus a suitable reference electrode), a potential is applied to the electrode and the resultant flow of current is measured. The most elementary equipment that could be employed is shown in Fig. 2-2. The tapping key shown in Fig. 2-1 has been removed, and the galvanometer has been replaced by an ammeter (A). Normally, in experiments with a net current flow only a low concentration of Ox and/or Red is employed, i.e., in the vicinity of 0.01–10 mM. Because current is going to flow through the circuit, provision has to be made for decreasing the solution resistance. This is done by adding a very large excess of an indifferent or supporting electrolyte such as 1 M KCl or a suitable buffer system. There are other reasons for adding this supporting electrolyte which will be discussed subsequently.

Assuming that the concentration of Ox and Red are equal, one could use the equipment in Fig. 2-2 to plot a manual or point-by-point current–voltage curve. The shape of the curve would be that shown in Fig. 2-3. The applied voltage at which the current is zero would be very close to the value one would measure potentiometrically using the apparatus shown in Fig. 2-1, and since the concentrations of Ox and Red are equal, it would be close to the formal

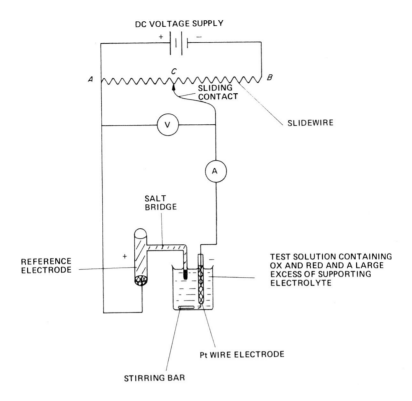

FIG. 2-2. Schematic diagram for measurement of a voltammetric current–voltage curve. *A,B,C*: Linear slidewire; *V*: voltmeter; *A*: ammeter.

potential $(E^{0'})$ of the system. As the applied voltage is made more positive or more negative than the zero current potential, the current clearly increases up to a limiting value, not surprisingly called the anodic $(i_L)_a$ or cathodic $(i_L)_c$ limiting current plateau. The reason for this increase in current is quite simple. If, for example, the applied voltage is made more negative than the zero current potential, then the Nernst relationship demands that the concentration of Ox decreases or, alternatively, that the concentration of Red increases. The only conceivable way in which this can happen is by electrons flowing down the platinum electrode, called the *working electrode* in voltammetry, across the electrode–solution interface and reducing Ox to Red according to Eq. 3. Hence, momentarily the concentration of Ox decreases to the desired level and the Nernst equation is obeyed. It is not necessary that the total solution undergo this concentration change, merely the solution in the immediate vicinity of the electrode surface. Accordingly, under these conditions the Nernst equation

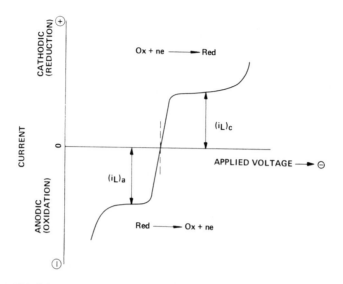

FIG. 2-3.   Voltammogram for equal concentrations of Ox and Red.

could be better written by Eq. 4, where $[Red]_0$ and $[Ox]_0$ refer to the concentrations of the Red and Ox species at the electrode surface.

$$Ox + ne \longrightarrow Red \tag{3}$$

$$E_{Pt} = E^{0'} - \frac{RT}{nF} \ln \frac{[Red]_0}{[Ox]_0} \tag{4}$$

As a result of the decrease in concentration of Ox at the electrode surface, a concentration gradient is immediately established between the surface and bulk concentrations of Ox, and diffusive and convective (i.e., stirred solution) forces supply more Ox to the surface. Therefore, in order to maintain the concentration of $[Ox]_0$ at the necessary level, a continuous current flows, the magnitude of which depends on the rate of arrival of Ox at the electrode surface, i.e., on the forces of *mass transport*. As the applied voltage is made more negative, the surface concentration of Ox decreases, and, accordingly, a more pronounced concentration gradient is obtained and the current increases. Finally, at sufficiently negative applied voltages the surface concentration of Ox is essentially decreased to zero. The greatest possible concentration gradient now exists between the electrode surface and the bulk solution, and the current reaches a limiting value, called the cathodic (reduction) limiting current. Exactly the same qualitative interpretation explains the formation of the anodic (oxidation) limiting current at applied voltages more positive than the zero current value.

The rise in current at voltages much more negative and positive than the limiting currents is due to electrochemical reduction and oxidation, respectively, of either the solvent or some component of the supporting electrolyte. Because both of these are present in such high concentrations, they do not normally give rise to any limiting current plateau.

All polarographic and voltammetric techniques are based on the experiment just described. Normally, an automatic instrument is used which applies some voltage waveform across an electrochemical cell, and the current that results is recorded. A linear waveform (or DC ramp voltage or linear voltage sweep) might be applied; e.g., from an initial voltage of 0.0 V versus the reference electrode, a slowly increasing negative voltage might be applied at a rate of -5 mV/sec, up to -2.0 V, as is typical in DC polarography. Other much more complex waveforms or faster voltage scan rates can also be employed. These will be discussed briefly later in Section IX.

IV. MASS TRANSPORT PROCESSES

In electrochemical experiments there are potentially three types of mass transport processes, i.e., processes that can deliver electroactive material to the working electrode surface. These are convection, migration, and diffusion.

Convection is simply brought about by any type of mechanical or thermal disturbance in the solution. Convection is important when a rotating disc electrode in a quiet solution is employed or when the solution is stirred. Only in the former case has the quantitative interpretation of the processes been evaluated. In many electrochemical techniques convection is not important.

Diffusion is simply the process of mass transfer that occurs whenever there is a concentration gradient between the electrode surface and the bulk solution. Clearly, diffusion is important whenever any electrode reaction occurs.

Migration is an effect that can occur when the electroactive species is ionic. If, for example, the electroactive species is a cation and it is electrochemically reduced at a negatively charged working electrode, there will be a mutual electrostatic attraction between the ion and the electrode. Under these conditions ions will arrive at the electrode more rapidly than if the electrode and the ion had a like charge. It is not possible to quantitatively evaluate these migration effects. Accordingly, they are effectively eliminated by adding to the solution a very large excess of indifferent or supporting electrolyte, usually of at least 100-fold higher concentration than that of the electroactive material. In effect, the addition of the excess of supporting electrolyte virtually eliminates the current-carrying capacity of the electroactive ion. This is because the migration current of an ion depends on its transport number. The transport number (t_+) of a cation is given by Eq. 5, where the C terms refer to concentration in equivalents per liter, and the λ terms refer to the equivalent

$$t_+ = C_+\lambda_+/\Sigma\ C_i\lambda_i \tag{5}$$

ionic conductance. Thus, as the value of C_i, the total concentration of ions in solution, is made a hundred times or more greater than C_+, the value of t_+ becomes very small. Accordingly, the migration of electroactive ions is effectively eliminated. The supporting electrolyte that is added is selected so that it is not itself oxidized or reduced in the voltage regions where the sample under investigation is electroactive. Often, the supporting electrolyte is made up of several different salts forming a buffer system of some desired pH.

V. DIRECT CURRENT POLAROGRAPHY

Undoubtedly, the most widely employed electrochemical technique is direct current (DC) polarography using a dropping mercury electrode (**DME**). Numerous texts have appeared describing the theory and application of DC

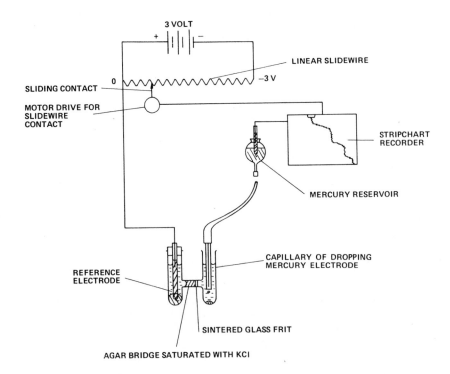

FIG. 2-4.   Schematic diagram of a DC polarograph.

polarography.[1−6] The texts of Meites,[2] Kolthoff and Lingane,[1] and Heyrovský and Kůta[3] are highly recommended.

In modern DC polarography a continuously changing voltage is applied across a DME and a suitable reference electrode. The rate of change of applied voltage is very slow (2−5 mV/sec), so that during the lifetime of a single drop (2−5 sec) the applied voltage remains almost constant. The current that flows as the voltage is varied is recorded continuously. A schematic apparatus is shown in Fig. 2-4. A typical polarogram obtained for the reduction of 1 mM Cd^{2+} in 0.1 M KCl is shown in Fig. 2-5. The oscillations on the polarogram are due to the growth and fall of the individual mercury drops. Normally, the complete current excursions for each drop are not recorded, but rather the current recording device is sufficiently damped so that the oscillations occur about the average current during the drop life. The average current over the whole polarogram is then obtained simply by measuring the midpoints of all the oscillations.

Before proceeding with a discussion of the information that can be obtained from a polarogram, it is worthwhile considering a further utility of the supporting electrolyte normally employed in polarography. In a two-electrode cell of the type shown in Fig. 2-4, current that flows in the external circuit must also flow through the solution by ionic migration. If the cell resistance (R) is about 10,000 Ω and, for example, -2.0 V is applied across the cell and the

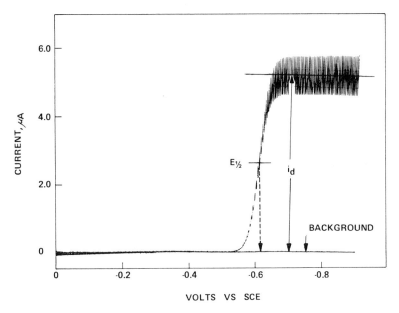

FIG. 2-5. A direct current polarogram of 1 mM Cd^{2+} in 0.1 M KCl. (SCE, Saturated calomel electrode.)

resultant current (i) is 10 μA, then the actual voltage applied between the DME and the reference electrode will be less than −2.0 V by the product

$$iR = 10^{-5} \times 10^4 = 10^{-1} \text{ V}$$

In other words, the real potential between the reference electrode and the DME is −1.90 V, and 0.1 V of the total applied voltage is used to overcome the solution resistance. In most commercial polarographs the recorder chart paper is fed out at some rate proportional to the rate of change of the total applied voltage across the cell, so that the chart division can be calibrated readily in terms of applied voltage. However, if because of iR (ohmic) losses the real voltage between the DME and the reference electrode is not the same as, or very close to, the total applied voltage, then the apparent voltages on the recorded polarograms are in error. In the case of aqueous solutions the addition of indifferent electrolyte up to the 0.1−1 M level reduces the resistance of an aqueous solution to a few hundred ohms, which for normal purposes introduces a negligible iR loss. If nonaqueous solutions are employed, however, the resistance can often not be reduced to the very low levels necessary in two-electrode polarography. In these cases three-electrode instruments must be employed; these will be discussed in Sections XIV, F and G.

Interpretation of DC Polarograms

1. The Limiting Current

The total height of a DC polarographic wave is called the limiting current. Often, in the absence of any complicating factors, the limiting current is made up of three separate contributions: diffusion current, impurity current, and charging current, the latter two often being described as the residual current. The diffusion current is simply due to the diffusion of electroactive species to the electrode and is the quantity that should be measured. The impurity current is due to reduction or oxidation of traces of electroactive impurities in the solvent and supporting electrolyte. The charging current is a small nonfaradaic (i.e., electrons do not cross the electrode−solution interface) current necessary to charge the electrical double layer of each individual mercury drop. Simply by recording a polarogram of the solvent and supporting electrolyte alone, and by subtracting it from the total limiting current, one can obtain the diffusion current.

The Ilkovič equation (Eq. 6) describes the dependence of the average diffusion current i_d, in microamperes, on the DME characteristics and the

$$i_d = 607nD^{1/2}m^{2/3}t^{1/6}C \tag{6}$$

properties of the electroactive species. In this equation n describes the number of electrons involved in the electrode reaction, D is the diffusion coefficient of the electroactive species in cm^2 per second, and C is the bulk concentration of the electroactive species in millimoles per liter. The m and t terms describe the flow rate of mercury through the capillary in milligrams per second and the lifetime of a drop in seconds at the potential corresponding to the diffusion current, respectively. Often, Eq. 6 is rearranged with all the variables on one side and all the constant terms on the other side:

$$i_d/Cm^{2/3}t^{1/6} = 607nD^{1/2} = I \tag{7}$$

The term I is called the diffusion current constant and clearly depends on the n and D values for the electroactive species in question. By evaluating the left-hand side of the equation, which contains all experimentally measurable quantities, the value of I can be computed. It often turns out that the value of I is somewhere between 1.5 and 2.0 x n; i.e., a value of I between 3 and 4 might indicate a two-electron polarographic reaction. Unfortunately, the values of D have a very great influence on the value of I, yet they are seldom known. Coupled with the fact that in very many instances diffusion may not be the sole mode of mass transport, considerable errors in the estimation of n values by the Ilkovič equation can result. Under no circumstances should values of n so calculated be employed without alternative methods of verifying the value.

2. The Half-Wave Potential

The polarographic half-wave potential is the potential at which the polarographic current is one-half of the diffusion current or on occasion the limiting current. In the case of a reversible electrode reaction the half-wave potential $E_{1/2}$ is closely related to the standard electrode potential E^0 by Eq. 8. Here, f refers to activity coefficient and D to diffusion coefficient. More

$$E_{1/2} = E^0 - \frac{RT}{nF} \ln \left(\frac{f_{Red}D_{Ox}^{1/2}}{f_{Ox}D_{Red}^{1/2}} \right) \tag{8}$$

often than not the f and D values of the oxidized and reduced species are quite close so that indeed $E_{1/2}$ is very closely related to E^0.

In many electrode reactions not only electrons but also protons are involved in the reaction (Eq. 9). Under these conditions the half-wave potential becomes pH dependent (Eq. 10).

$$Ox + ne + pH^+ \rightleftharpoons Red\, H_p \tag{9}$$

$$E_{1/2} = E^0 - \frac{RT}{nF} \ln \left(\frac{f_{Red}D_{Ox}^{1/2}}{f_{Ox}D_{Red}^{1/2}} \right) - 2.303 \frac{RT}{nF} p \cdot pH \tag{10}$$

Since at $25°C$ $2.303RT/F$ becomes 0.05915, the shift of $E_{1/2}$ with pH is given by

$$\frac{dE_{1/2}}{d(\text{pH})} = \frac{-0.05915}{n}p \tag{11}$$

If the value of n is known, clearly p, the number of protons involved in the electrode reaction, can be determined. It must be stressed, however, that these equations apply only to reversible electrode reactions.

The majority of electrode reactions are irreversible. This simply means that, for a reaction of the type shown in Eq. 12, the rate constants for the

$$\text{Ox} + n_a e \underset{k_{b,h}}{\overset{k_{f,h}}{\rightleftharpoons}} \text{Red} \tag{12}$$

electron-transfer reaction $k_{f,h}$ and $k_{b,h}$ are so small that, as the potential applied to an electrode is shifted from its equilibrium value, the surface concentrations do not alter sufficiently rapidly, so the Nernst equation is not obeyed. Accordingly, the current and half-wave potential are not governed by the Nernst equation and hence by thermodynamic parameters, but rather by the kinetics of the electrode reaction.

The current for an irreversible cathodic (reduction) polarographic reaction can be calculated from absolute rate theory by way of Eq. 13.[2] In this

$$i_c = nFAC_{Ox}^0 k_{f,h}^0 \exp\left[\frac{-\alpha n_a F}{RT}(E + 0.2412)\right] \tag{13}$$

expression i_c is the cathodic current in microamperes, A is the area of the electrode in cm^2, C_{Ox}^0 is the surface concentration of Ox in millimoles per $1000\ cm^3$, $k_{f,h}^0$ is the formal heterogeneous rate constant in centimeters per second at $0.00\ V$ versus NHE, α is the electron-transfer coefficient, n_a is the number of electrons involved in the potential controlling reaction, and E is the applied potential in volts versus the saturated calomel electrode (SCE). This equation is adapted with respect to the SCE because this electrode is the most widely employed reference electrode in electrochemistry. The value of n is the total number of electrons involved in the electrode reaction, which may be larger than or equal to n_a.

The half-wave potential for an irreversible reaction at $25°C$ is given by Eq. 14,

$$E_{1/2} = -0.2412 + \frac{0.05915}{\alpha n_a} \log \frac{1.349 k_{f,h}^0 t^{1/2}}{D^{1/2}} \tag{14}$$

where it can be seen that $E_{1/2}$ (here again referred to SCE, which is $0.2412\ V$ versus NHE) is dependent not only on the kinetics of the electrode reaction, but also on the drop time of the DME. In the case of an irreversible electrode reaction involving protons, Eq. 15 describes the half-wave potential. Since $E_{1/2}$

$$E_{1/2} = -0.2412 + \frac{0.05915}{\alpha n_a} \log \frac{1.349 k_{f,h}^0 t^{1/2}}{D^{1/2}} - \frac{0.05915}{\alpha n_a} p \cdot \text{pH} \quad (15)$$

for an irreversible electrode reaction is drop time dependent, Meites[2] has proposed that a corrected value of the half-wave potential, $E_{1/2}^0$, be employed, where

$$E_{1/2}^0 = E_{1/2} - \frac{0.02958}{\alpha n_a} \log t_{1/2} \quad (16)$$

In this expression $t_{1/2}$ is the drop time at $E_{1/2}$. Since $E_{1/2}^0$ is independent of the drop time, it is clearly of much greater value for correlation of data from one worker to another. Accordingly, Eq. 14 becomes Eq. 17, and Eq. 15 becomes Eq. 18.

$$E_{1/2}^0 = -0.2412 + \frac{0.05915}{\alpha n_a} \log \frac{1.349 k_{f,h}^0}{D^{1/2}} \quad (17)$$

$$E_{1/2}^0 = -0.2412 + \frac{0.05915}{\alpha n_a} \log \frac{1.349 k_{f,h}^0}{D^{1/2}} - \frac{0.05915}{\alpha n_a} p \cdot \text{pH} \quad (18)$$

From Eq. 18 it is clear that $E_{1/2}^0$ is pH dependent (Eq. 19).

$$\frac{dE_{1/2}^0}{d(\text{pH})} = \frac{-0.05915}{\alpha n_a} p \quad (19)$$

These equations are not meant to be comprehensive, but simply indicate that in the case of a reversible reaction (Nernstian reaction) the half-wave potential is a well-defined, thermodynamically significant moiety. In the case of an irreversible reaction (non-Nernstian) the half-wave potential is a complex parameter influenced primarily by the kinetics of the electrode reactions.

3. Equations of the Polarographic Wave

For a reversible electrode reaction in which both the reactant and product are soluble, the equation for a simple cathodic polarographic wave at $25°C$ is

$$E_{\text{DME}} = E_{1/2} - \frac{0.05915}{n} \log \frac{i_c}{(i_d)_c - i_c} \quad (20)$$

where i_c is the average cathodic current at any potential, E_{DME}, applied to the dropping mercury electrode, and $(i_d)_c$ is the average cathodic diffusion current. In the case of a reversible anodic (oxidation) wave at 25°C a similar equation results

$$E_{DME} = E_{1/2} - \frac{0.05915}{n} \log \frac{(i_d)_a - i_a}{i_a} \tag{21}$$

When both the oxidized and reduced forms of a reversible redox couple are present in solution, polarography gives a composite anodic–cathodic wave, the equation for which is

$$E_{DME} = E_{1/2} - \frac{0.05915}{n} \log \frac{i - (i_d)_a}{(i_d)_c - i} \tag{22}$$

Equations 20, 21, and 22 lead to the very widely applied test for a reversible reaction. A plot of E_{DME} versus $\log i_c/[(i_d)_c - i_c]$ (for a cathodic reaction) should give a straight line of slope $-0.05915/n$ V at 25°C. Unfortunately, this test is far from infallible, and it is not unlikely that irreversible reactions can also give slopes very close to the values expected for a reversible reaction. Thus, while the slope of a plot of E_{DME} versus the logarithmic terms of Eqs. 20–22 might give a value of $-0.05915/n$ V at 25°C, this should be regarded only as an indication of reversibility, *not* as proof. Thus, a slope of -60 mV is often taken as proof of the electrode reaction being a one-electron reversible process. It could equally well be a totally irreversible multielectron reaction.

In general, therefore, it is probably advisable to use the slope of the log plot to decide whether the process is totally irreversible, i.e., if the slope appreciably exceeds 59 mV, but not to decide either the electron number or the reversibility. There are far better ways to determine the latter two factors.

4. Kinetic and Catalytic Currents in Polarography

Apart from diffusion-controlled waves obtained in polarography, there are other types of current control that are observed.

A *kinetic current* is a polarographic current that is limited by the rate of a chemical reaction that precedes the electrode reaction proper. An example is found with formaldehyde, which exists predominantly in aqueous solution in its hydrated form (I, Eq.,23). However, only the dehydrated species (II, Eq. 23) is polarographically reducible (Eq. 24). Hence, the limiting current observed polarographically depends on the kinetics of the homogeneous chemical dehydration reaction. If the rate constant, k_1, is very small, then the polarographic limiting current is also much smaller than normal. This current is called a kinetic current.

$$\underset{(I)}{HC\overset{\displaystyle OH}{\underset{\displaystyle OH}{\Big\langle}}} \quad \underset{k_2}{\overset{k_1}{\rightleftharpoons}} \quad \underset{(II)}{HCHO + H_2O} \tag{23}$$

$$HCHO + 2H^+ + 2e \longrightarrow CH_3OH \tag{24}$$

A *catalytic current* is generally much larger than the current observed for a diffusion-controlled wave. In organic polarography one of the most common catalytic waves is the catalytic hydrogen wave that is frequently observed with *N*-heterocyclic compounds, thiols, proteins, and similar compounds. The following mechanism generally applies:

$$B + HA \rightleftharpoons BH^+ + A^- \tag{25a}$$
$$BH^+ + e \longrightarrow BH \tag{25b}$$
$$2BH \longrightarrow 2B + H_2 \tag{25c}$$

where HA is a proton donor, usually in aqueous solution the hydronium ion. The presence of B, the organic catalyst, reduces the activation energy for reduction of hydrogen ions. Clearly, the current observed for the catalytic wave depends on the concentration of hydrogen ions and on the kinetics of the reactions shown in Eqs. 25a and 25b. In many polarographic catalytic processes the organic catalyst B is electrochemically reduced and catalyzes hydrogen ion reduction as well. Often, electrochemical reduction of an organic species gives a product that catalyzes the reduction of hydrogen ion.

5. Adsorption Waves

If either the electroactive species or the product of the electrode reaction is adsorbed at the DME surface, an adsorption wave *might* be observed. At very low concentrations of electroactive species only a single wave is observed that behaves like a diffusion-controlled wave. Above concentrations at which the adsorbed material forms a monolayer on the electrode surface, the polarographic wave shows two steps (Fig. 2-6). If the electroactive species is adsorbed, an adsorption postwave is produced. If the product is adsorbed, an adsorption prewave is observed. The adsorption pre- or postwaves reach a limiting height corresponding to formation of the complete surface coverage of the electrode. The total polarographic wave height increases with bulk concentration as predicted by the Ilkovič equation.

It should be noted, however, that in many polarographic (or voltammetric) processes the product or reactant of an electrode reaction may be adsorbed, yet no prewave or postwave is observed. This effect is noticed for irreversible electrode reactions or in the case of a reversible reaction in which reactant and product are equally strongly adsorbed.

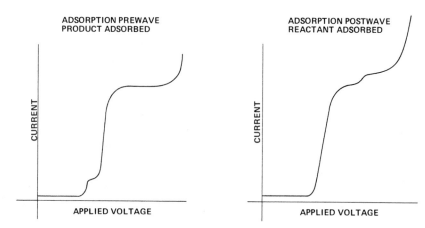

FIG. 2-6.   Appearance of polarographic adsorption prewaves and postwaves.

6. Differentiation among Diffusion-, Kinetic-, Catalytic-, and Adsorption-Controlled Polarographic Waves

It is beyond the scope of this chapter to discuss the details of each of the above types of DC polarographic waves. Often, however, a wave might be partly under two types of control. As a guide to the methods that might be employed to differentiate polarographically among the various types of polarographic waves, a tabulation of some of their properties is presented in Table 2-1. For a detailed discussion of the theory of all these polarographic processes, authoritative texts should be consulted.[1-3]

VI. ALTERNATING CURRENT POLAROGRAPHY

In alternating current (AC) polarography a slow, linear potential sweep (DC ramp) is applied to an electrode exactly as in DC polarography, but, in addition, a small sinusoidal voltage is superimposed on the DC voltage (Fig. 2-7). Usually, the alternating voltage has a frequency of about 100 Hz and an amplitude of 10–20 mV. The waveform that is actually applied to the working electrode, or that the working electrode "sees," is an alternating potential which oscillates about an average value, the DC ramp voltage at any time. A polarogram is then recorded of the resultant alternating current versus the applied DC voltage.

The relationship between DC and AC polarography is shown in Fig. 2-8 for a perfectly reversible, uncomplicated electrode reaction (see Eq. 1). Thus, if the DC voltage corresponds to the DC polarographic half-wave potential, then as the superimposed alternating voltage shifts the voltage more negative, the Nernst

TABLE 2-1   Summary of Some Important Diagnostic Criteria to Decide the Control Mechanisms for Polarographic Limiting Currents

Variable	Principal current controlling process			
	Diffusion	Kinetic	Catalytic	Adsorption
Concentration of electroactive species, C	αC and predicted from Ilkovič equation	αC but smaller than predicted from Ilkovič equation	Generally αC and also $[Z]^{1/2}$ where $[Z]$ is the catalyst concentration (not universally true)	αC up to a limit
Corrected height of mercury column, h_{corr}[a]	$\alpha h_{corr}^{1/2}$	Independent	Independent	αh_{corr}
Temperature	αT	αT	αT	$\alpha(1/T)$
Temperature coefficient[b]	1.3–1.6% per °C	8–12% per °C[c]	5–10% per °C[c]	Variable, often negative[c]
Current–time curves[d]	$(i_d)_\tau \alpha \tau^{1/6}$			$(i_1)_\tau \alpha \tau^{-1/3}$

[a] h_{corr} in centimeters is equal to $h_{Hg} - [3.1/(mt)^{1/3}]$ when h_{Hg} is the height of the mercury column in centimeters, m is the flow rate of mercury through the DME capillary in milligrams per second, and t is the drop time in seconds.

[b] Temperature coefficient is equal to $2.303/(T_2 - T_1) \log i_2/i_1$, where i_2 and i_1 are the average limiting currents at temperatures T_2 and T_1, respectively.

[c] These are typical values but large variations are possible.

[d] The instantaneous current τ is recorded versus time for single mercury drops. Usually an oscilloscope is required for these measurements.

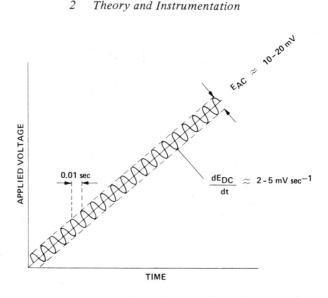

FIG. 2-7.   Voltage signal applied to a DME in AC polarography.

equation demands a decrease in the surface concentration of Ox; hence, an increase in cathodic current results. Similarly, as the alternating voltage shifts the applied voltage more positive than $E_{1/2}$, an increase in anodic current results. Since the alternating voltage is sinusoidal, at $E_{1/2}$ a large sinusoidal alternating current results. It can readily be seen that at DC potentials displaced from the $E_{1/2}$ the resultant alternating current is considerably less. On the residual current or limiting current plateau of the DC polarogram no faradaic alternating current flows. The alternating current observed in these regions is due to the sinusoidal charging of the electrical double layer. Thus, an AC polarogram shows a peak, the summit potential (E_s) of which is at the same potential as the half-wave potential, $E_{1/2}$. The latter statement is true only for a perfectly reversible electrode reaction.

When the formal heterogeneous rate constant for the electrode reaction is large compared to the angular frequently of the applied alternating voltage, the peak or maximum alternating current (Δ_{i_s}) is given by

$$\Delta_{i_s} = \frac{n^2 F^2 A \; \Delta E \omega^{1/2} D^{1/2} C}{4RT} \tag{26}$$

Here, Δ_{i_s} is the amplitude of the alternating current (amperes, peak to peak) at E_s, A is the electrode area (cm^2) $(0.0085m^{2/3}t^{2/3})$, C is the concentration of the electroactive species in the bulk solution (moles per milliliter), ΔE is the amplitude of the applied alternating voltage (volts, peak to

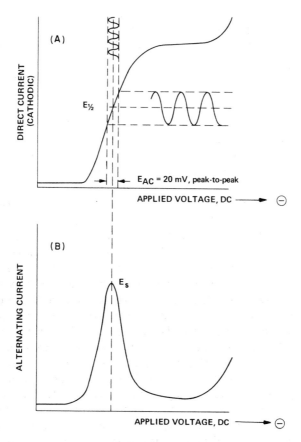

FIG. 2-8. Diagrammatic representation of relationship between (A) direct current and (B) alternating current polarography.

peak), ω is the angular frequency of the applied alternating voltage, i.e., $2\pi f$ where f is the frequency in hertz, D is the diffusion coefficient of the electroactive species (cm^2 per second), R is the gas constant (8.312 J), and T is the absolute temperature. The values of n and F have been defined earlier. When the rate constant for the electrode reaction becomes small compared to the angular frequency [actually compared to $(\omega D)^{1/2}$], which is the case if the reaction tends to be irreversible or if the applied angular frequency is very high, then the peak alternating current is given by

$$\Delta_{i_s} = \frac{n^2 F^2 A \ \Delta E C \, k_{s,h}(1-\alpha)^{1-\alpha}\alpha^{\alpha}(D_{Red}/D_{Ox})^{-\alpha/2}}{RT} \tag{27}$$

where $k_{s,h}$ is the value of the formal heterogeneous rate constant for the redox couple at its formal potential $(E^{0'})$ in centimeters per second and α is the electron-transfer coefficient; D_{Red} and D_{Ox} are the diffusion coefficients of the reduced and oxidized species, respectively.

Clearly, the more irreversible an electrode reaction, i.e., the smaller $k_{s,h}$, the smaller the alternating current. A totally irreversible electrode reaction generally gives a very small AC polarographic peak. Therefore, AC polarography is a much more reliable technique than DC polarography to quickly decide whether an electrode reaction is reversible.

The theory of AC polarography and its various modifications is too complex to discuss here; the excellent summaries by Breyer and Bauer[7] and Smith[8] should be consulted.

A problem that arises in AC polarography is in the measurement of the experimental peak current, because it is not strictly valid to run the sample in the appropriate background and subtract the current observed in background solution alone. This is because the faradaic current and the double-layer charging current, called the *base current* in AC polarography, are not in phase. In other words, the base current arises from charging the electrical double layer, which is equivalent to applying an alternating voltage across a small capacitor. The applied alternating voltage and the resultant alternating base current are 90° out of phase. In fact, the voltage across the capacitor lags the current by $\frac{1}{2}\pi$ rad or 90° (Fig. 2-9). The faradaic alternating current, i.e., that due to the electrode process, is 45° out of phase with the applied alternating voltage, at least for a perfectly reversible, uncomplicated electrode reaction. The phase relationships between the applied alternating voltage, the faradaic alternating current, and the base or charging current are shown in Fig. 2-9. In other words, a vector diagram relating the total current, faradaic current, base current, alternating voltage, and the phase relations between them must be prepared in order to compute the true faradaic current. Methods for doing this have been outlined[7,8] but are rather tedious. A simpler way is to electronically suppress the base current. This can be done by using tuned phase selective amplifiers. Phase selective amplifiers can be made to reject the signal that is 90° out of phase with the applied alternating voltage (i.e., the alternating base current). Signals with phase differences other than 90° give responses dependent on their phase angle. Thus, these amplifiers do not sense the base current at all but do respond to the faradaic current that is only 45° or less out of phase with the applied alternating voltage. However, only a fraction of, rather than the total, faradaic current is recorded, the exact amount depending on the phase angle. This technique is known as phase selective AC polarography.

It should perhaps be mentioned that the scalar difference between the base

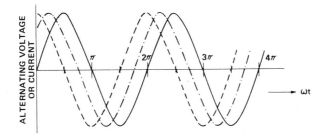

FIG. 2-9. Theoretical phase relationships among the applied alternating voltage (dashed-line), the faradaic alternating current (dot-dashed line), and the base current (solid line) in AC polarography.

current and the faradaic peak current is sufficiently linearly related to the bulk concentration of the electroactive species to be used for most analytical purposes.

A. Adsorption and AC Polarography

The base current in AC polarography is directly proportional to the capacity of the electrical double layer and consequently can be used to follow changes in the structure of the double layer under various conditions. When an uncharged organic molecule is adsorbed at the DME surface, it is usually most strongly adsorbed in the region where the electrode is uncharged. The potential on the electrode when the electrode is uncharged is called the electrocapillary maximum or sometimes the potential of zero charge. The electrocapillary maximum potential is dependent to a large extent on the anions of the supporting electrolyte. For example, in 0.1 M KCl solution the electrocapillary maximum potential is -0.461 V versus SCE, while in 0.1 M KBr it is -0.535 V and in 0.1 M KI it is -0.693 V. The increasingly negative values of the electrocapillary maximum potential in the presence of the latter salts are related to the increasingly strong adsorption of the anions at the DME. A more detailed discussion of the electrocapillary maximum is presented elsewhere.[1-3] Owing to the displacement of ions from the electrical double layer upon adsorption of an uncharged organic molecule at the electrode surface, which generally decreases the dielectric constant of the material between the ions of the double layer and the electrode surface, the double-layer capacity decreases and consequently the AC base current decreases. Accordingly, comparison of the base current in the

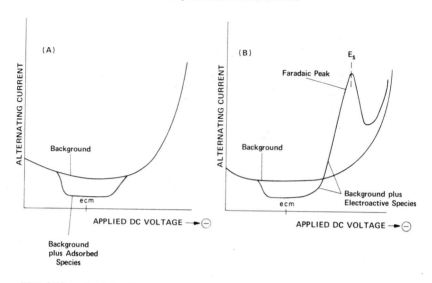

FIG. 2-10. An AC polarogram in the presence of (A) an uncharged, electro-inactive adsorbed molecule, and (B) an uncharged, electroactive adsorbed molecule (ecm, electro-capillary maximum).

presence and absence of the organic molecule can reveal whether it is adsorbed and over what potential regions it is adsorbed. The adsorbed organic molecule does not have to be electroactive to show this base current depression (Fig. 2-10). One would expect *a priori* that charged molecules would be preferentially adsorbed when the electrode carries an opposite charge. This, however, does not always appear to be the case. Generally if the reactant in an electrode reaction is adsorbed, the base current is depressed before the AC peak (Fig. 2-11A), while if the product of the reaction is adsorbed, the base current after the AC peak is depressed (Fig. 2-11B). If both the reactant and product are adsorbed, the base current is depressed on both sides of the AC peak (Fig. 2-11C).

B. Tensammetry

Many molecules are adsorbed at the DME so that by AC polarography the base current is depressed. With certain compounds, at potentials sufficiently removed from the region where they are most strongly adsorbed, they are sharply desorbed. If the adsorption–desorption process occurs over a very narrow range of potentials, periodic changes in the double-layer capacity may occur at the same frequency as the applied alternating voltage. The result is that sharp AC peaks are produced which, not being due to a faradaic electron-transfer

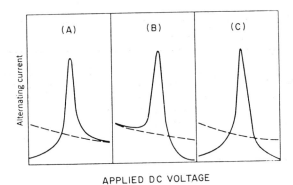

FIG. 2-11. Three AC polarograms: (A) with reactant adsorbed; (B) with product adsorbed; (C) with both reactant and product adsorbed. Dashed line represents the alternating base current.

process, are called tensammetric peaks. For certain uncharged surface-active materials a tensammetric peak is observed at potentials both positive and negative of the electrocapillary maximum (Fig. 2-12). These very pronounced tensammetric peaks are observed relatively rarely. In fact, throughout Chapters 3–10 no example of this type is mentioned. There are instances, however, where small, usually single tensammetric peaks are observed which are often referred to

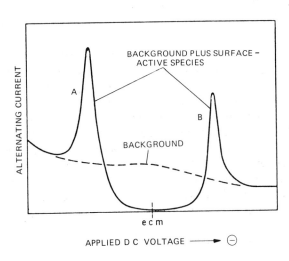

FIG. 2-12. Schematic tensammetric polarogram. A, Positive tensammetric peak; B, negative tensammetric peak (ecm, electrocapillary maximum).

TABLE 2-2   Diagnostic Criteria to Differentiate between Faradaic AC Peaks and Tensammetric Peaks

	Response	
Variable	Faradaic AC peak	Tensammetric AC peak
Concentration of electroactive species, C	(a) E_s generally almost independent (b) $\Delta_{i_s} \alpha C$	(a) E_s shifts linearly with $\log[C]$ (b) Δ_{i_s} nonlinear function of C
Temperature increase	Δ_{i_s} increases	Δ_{i_s} decreases
Temperature coefficient	+1.0—1.3% per $°$C	Negative
DC polarography	Gives DC wave	No DC wave
Current—time curves	Sinusoidal, undistorted waveform at small amplitude (10—20 mV) alternating voltages	Very distorted waveforms

as capacitive or nonfaradaic peaks. These are usually proposed to be associated with adsorption—desorption effects. The ways of differentiating a tensammetric peak from a normal faradaic AC peak have been outlined in Table 2-2.

VII. VOLTAMMETRY AT A ROTATING DISC ELECTRODE

The simplest type of equipment required for voltammetry at a rotating disc electrode is the same as that shown in Fig. 2-4, except that the DME is replaced by a solid disc electrode (e.g., Fig. 2-13), which is rotated at a known and constant speed. The solution is not otherwise stirred. Electroactive material is supplied to a rotating disc electrode by convection and diffusion. Adams[9] has discussed the theory and practice of these electrodes at some length.

The current—voltage curve obtained at a rotating disc electrode looks very much like a polarogram at the DME except, of course, there are no oscillations (Fig. 2-14). A plateau or limiting current is obtained, and a half-wave potential can be measured. Levich[10] has calculated the value of the limiting current i_L, in microamperes, obtained at a rotating disc electrode:

$$i_L = 1.50 \times 10^5 nAD^{2/3}v^{-1/6}N^{1/2}C \qquad (28)$$

where n is the number of electrons involved in the electrode reaction, A is the area of the electrode in cm^2, D is the diffusion coefficient of the electroactive species in cm^2 per second, v is the kinematic viscosity of the solution in cm^2 per

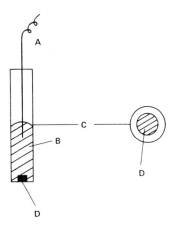

FIG. 2-13.   Practical form of rotated disc electrode. A, External connector; B, mercury pool for electrical contact with solid electrode material; C, glass tube; D, solid electrode material. Only the end, which is flat and disc-shaped, is exposed to the test solution.

second, N is the rotation rate of the electrode in revolutions per second, and C is the bulk concentration of the electroactive species in millimoles per liter.

The equations of the current-voltage curve for a reversible electrode reaction have been written,[11] but generally interpretation of the voltammograms has not yet been attempted.

The rotating disc electrode finds its principal application in analysis and in measurements of the forward and backward heterogeneous rate constants for

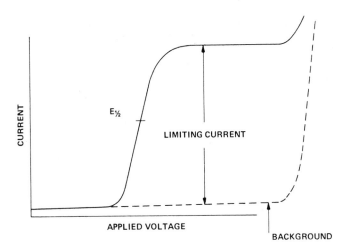

FIG. 2-14.   Schematic voltammogram with a rotating disc electrode.

redox reactions having fairly large values of $k_{s,h}$. A fairly detailed summary of the application of the rotating disc electrode to electrode kinetics and mechanisms has been presented by Adams.[9]

VIII. LINEAR SWEEP VOLTAMMETRY AT STATIONARY ELECTRODES

If a stationary solid or liquid microelectrode is immersed in a quiet, unstirred solution of electroactive species containing a large excess of supporting electrolyte, then a voltammogram (current–voltage curve) can be obtained in an apparatus similar to that in Fig. 2-4, the DME being replaced by the stationary electrode. The voltammogram would normally show a peak (Fig. 2-15) rather than a wave with a plateau.

The peak current i_p is given for a reversible electrode reaction by the Randles–Ševčík[12,13] equation (Eq. 29), where i_p is the peak current in

$$i_p = 2.687 \times 10^5 n^{3/2} A D^{1/2} C v^{1/2} \tag{29}$$

microamperes, C is the bulk concentration of the electroactive species in millimoles per liter, A is the electrode area in (centimeters)^2, n is the electron number, and v is the scan rate of the applied linear voltage sweep in volts per second. Clearly, unlike earlier expressions, the current depends on the scan rate of the applied voltage sweep.

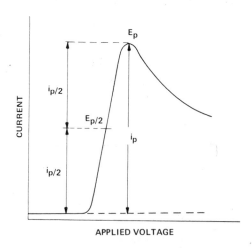

FIG. 2-15.   Schematic voltammogram obtained at a stationary microelectrode in quiet solution.

The voltammogram is characterized by the peak potential E_p or the half-peak potential $E_{p/2}$. If a stationary mercury electrode such as the hanging mercury drop electrode is employed for the voltammogram, then for a reversible system the voltammetric peak potential and the polarographic half-wave potential (DME) are related by the expression for a reduction reaction,[14]

$$E_p = E_{1/2} - \frac{0.0285}{n} \qquad \text{at } 25°C \qquad (30)$$

or

$$E_{p/2} = E_{1/2} + \frac{0.028}{n} \qquad \text{at } 25°C \qquad (31)$$

In other words, for a reduction E_p is $28.5/n$ mV more negative than $E_{1/2}$, while $E_{p/2}$ is $28/n$ mV more positive than $E_{1/2}$. Conversely, for an oxidation E_p is $28.5/n$ mV more positive than $E_{1/2}$, while $E_{p/2}$ is $28/n$ mV more negative than $E_{1/2}$. Based on this information, it is easy to show that for a reversible electrode reaction

$$(E_{p/2})_c - (E_{p/2})_a = 56/n \text{ mV} \qquad \text{at } 25°C \qquad (32)$$

where $(E_{p/2})_c$ refers to the cathodic half-peak potential for reduction of the oxidized form of the reversible couple, and $(E_{p/2})_a$ is the anodic half-peak potential for oxidation of the reduced form of the couple.

For a totally irreversible electrode reaction the peak current i_p, in microamperes, is given by[14]

$$i_p = 2.985 \times 10^5 n(\alpha n_a)^{1/2} A D^{1/2} C v^{1/2} \qquad (33)$$

where all terms have been defined previously.

The peak potential E_p is given by the expression

$$E_p = E^0 - \frac{0.05915}{\alpha n_a} \left[0.4565 + \log \frac{(\alpha n_a D v)^{1/2}}{k_{s,h}} \right] \qquad (34)$$

and the half-peak potential is given by

$$E_{p/2} = E_p + \frac{0.048}{\alpha n_a} \qquad (35)$$

The term $k_{s,h}$ (centimeters per second) is the heterogeneous rate constant for the electrode reaction at the formal potential of the couple.

It might at first sight appear that with increasing scan rate the peak currents would increase and an extraordinary sensitivity might be attained at very high scan rate. However, the charging current increases linearly with scan rate (v), while the faradaic peak current increases only with $v^{1/2}$. Accordingly, one can increase the scan rate only up to a limited value without the charging current becoming large in comparison to the faradaic peak current.

Although Eqs. 30–32 can be employed to decide if a particular electrode reaction is reversible, in fact there is a far superior method. Nicholson and Shain[14] have solved theoretical equations for linear sweep voltammograms at stationary electrodes in quiet solution and have presented tables of numerical functions for calculating theoretical voltammograms. By comparing the theoretical and experimental voltammograms, it is an easy matter to decide whether the electrode reaction being examined is obeying the theoretical criteria of reversibility. Adams[9] has presented a simplified summary of the Nicholson–Shain methods both for simple electrode reactions and for electrode reactions that are complicated by coupled chemical reactions.

IX. CYCLIC VOLTAMMETRY

As the name implies, in cyclic voltammetry a cyclic or triangular voltage waveform is applied to a stationary microelectrode. Typically, an isosceles triangle waveform is applied having a voltage scan rate ranging from perhaps 5 mV sec^{-1} to 1000 V sec^{-1}. A typical applied voltage waveform and resultant cyclic voltammogram for a 1 mM solution of Cd^{2+} in 0.1 M KCl is shown in Fig. 2-16. The solution used for Fig. 2-16 originally contained only Cd^{2+}. As the potential is scanned toward negative potential, the well-formed cathodic peak $(E_p)_c = -0.625$ V versus SCE is observed due to reduction of Cd^{2+} according to Eq. 36. The peak current is described by the Randles–Ševčík equation (Eq. 29).

$$Cd^{2+} + 2e \; \underset{}{\overset{Hg}{\rightleftharpoons}} \; Cd(Hg) \qquad (36)$$

Once the peak has been scanned, the voltage sweep reverses direction and scans toward positive potential. Provided that the scan rate is sufficiently rapid, it is clear that some of the Cd(Hg) produced on the initial cathodic sweep will still be in the vicinity of the electrode surface and accordingly can be reoxidized to Cd^{2+}, giving rise to the anodic peak $(E_p)_a = -0.585$ V versus SCE. The $Cd^{2+}/Cd(Hg)$ system under the conditions shown in Fig. 2-16 is almost reversible. The anodic and cathodic peak potentials for such a system do not occur at the same potential but for a completely reversible system are separated by the potential increment:

$$(E_p)_a - (E_p)_c = 0.058/n \text{ V} \qquad (37)$$

Equation 37 holds when the rate constant for the electrode reaction is large compared to the voltage scan rate. As the scan rate is increased, so that by comparison the rate constant is small, the anodic and cathodic peaks begin to

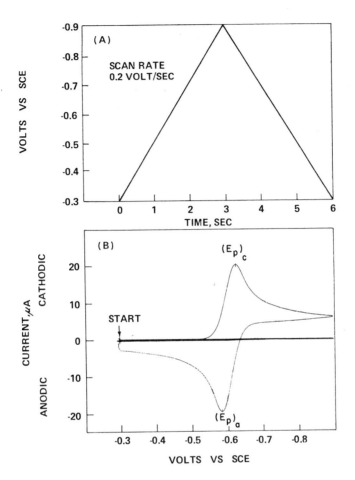

FIG. 2-16. Voltage wave form (A) applied to a hanging mercury drop electrode to obtain a cyclic voltammogram (B) of 1 mM Cd^{2+} in 0.1 M KCl. Scan rate: 0.2 V/sec; area of hanging mercury drop electrode: 2.22 mm^2.

separate by a larger amount. In fact, this phenomenon is the basis for a very simple determination of heterogeneous rate constants ($k_{s,h}$) for electrode reactions using[9,15] cyclic voltammetry. If the electrode reaction is *quasi-reversible* (i.e., intermediate between reversible and totally irreversible), then even at low scan rates the peak potentials will be separated by more than expected from Eq. 37. For a totally irreversible system one peak may not even appear, or the cathodic and anodic peaks might be separated by several tenths of a volt.

By use of very fast sweep cyclic voltammetry it is often possible to detect unstable but electroactive intermediates or products in an electrode reaction.

Several examples of this are shown in Chapter 3. There are also many instances in which the primary product of an electrode reaction undergoes a homogeneous reaction to give further electroactive materials. By studying the electrode reaction with cyclic voltammetry, it is often possible to obtain a rather complete qualitative picture of the overall electrode and related chemical processes. Often the homogeneous rate constants for follow-up chemical reactions can be calculated.[9] In practical cyclic voltammetry it is not necessary to apply just a single voltage cycle. In fact, in order to observe follow-up reactions, it is often necessary to record several voltage cycles.

X. OSCILLOPOLAROGRAPHY

Oscillopolarography is a technique developed by Czechoslovakian workers in the 1940s. The technique is widely employed in that country, but with a few exceptions rarely elsewhere.

Essentially, the technique involves applying a constant alternating current of the order of 100–1000 μA across the electrochemical cell containing a DME. Then, by means of a suitable differentiator circuit, the rate of change of potential with time, dE/dt, is recorded versus potential E. At the potentials where an electrode process occurs, dE/dt decreases, passes through a minimum, and then increases again. The net result is that the transitions appear as incisions or indentations on the dE/dt versus $f(E)$ trace (Fig. 2-17). The potential at the peak of the incision is close to the DC polarographic half-wave potential. For a reversible redox system the anodic and cathodic incisions are close together. In the case of quasi-reversible or totally irreversible processes the cathodic and anodic incisions are farther apart, or one may be totally absent. The location of the incision is defined by the Q value where

$$Q = \frac{\text{linear distance of incision peak from potential of anodic mercury dissolution}}{\text{linear distance from potential of anodic mercury dissolution and that of cathodic background electrolyte discharge}}$$

Although theoretical equations for oscillopolarography have been derived, they are infrequently employed for elucidating electrode mechanisms. The depth of the incisions in the dE/dt versus $f(E)$ are not linearly proportional to concentration, and hence the use of oscillopolarography for analytical work requires very carefully prepared calibration curves. The precision of measurement of the incision peak potential, which is recorded on an oscilloscope, is also modest, at best. The detailed theory and practice of oscillopolarography have been reviewed by Heyrovský and Kalvoda[16] and Kalvoda.[17]

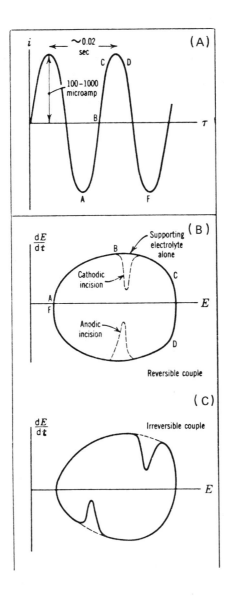

FIG. 2-17. (A) Typical alternating current waveform applied in oscillopolarography. (B) Oscillopolarogram of supporting electrolyte and supporting electrolyte plus a reversible electroactive couple. (C) With an irreversible electroactive couple. (Adapted from Meites[2] with permission of Interscience Publishers, New York.)

XI. PULSE POLAROGRAPHY

As the name implies, pulse polarography utilizes a dropping mercury electrode. In normal pulse polarography a single square-wave voltage impulse having a duration of about 40 msec is applied to each mercury drop at a predetermined instant after its birth. The amplitude of the successive pulses on successive drops is increased linearly with time, as is shown schematically in Fig. 2-18. Typically, the square-wave pulse might be applied 2 sec after the birth of the drop. Now, at a DME the charging current at the first instant that a voltage is applied to the electrode is large but decays relatively rapidly with time. The faradaic current, however, due to the electron-transfer reaction between the electrode and the electroactive species, decays much more slowly. Thus, approaching the end of the 40 msec voltage pulse, the major part of the current flowing at the electrode is faradaic, the charging current having decayed away (Fig. 2-19). It is at this latter time that the current is measured; typically, the current is sampled over the last 20 msec or less of the applied voltage pulse. Thus, a pulse polarogram appears as a series of steps but is of the same qualitative shape as a regular DC polarogram (Fig. 2-20).

The particular beauty of pulse polarography is that the current is sampled when the charging current has almost completely decayed. The net result is that much lower concentrations of electroactive species can be determined. Thus, in normal DC polarography, where the faradaic and charging currents are measured simultaneously, the lower level of concentrations of electroactive species that can be determined is about 10^{-5} M. In the case of pulse polarography the lower limit may be as low as 10^{-7} M.

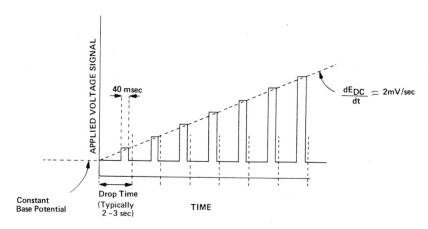

FIG. 2-18.   Schematic diagram of signal waveform applied to the DME in pulse polarography.

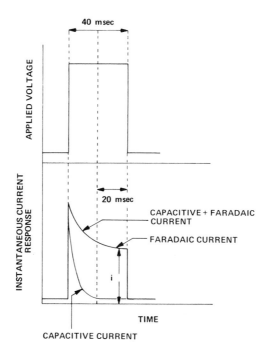

FIG. 2-19. Schematic diagram of applied voltage signal and current response in pulse polarography.

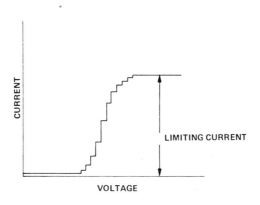

FIG. 2-20. Typical pulse polarogram.

Differential pulse polarography is a technique in which again a square-wave voltage pulse is applied to a mercury drop at some predetermined time after its birth. However, the voltage pulses (40–60 msec) all have a constant amplitude of, typically, between 5 and 100 mV and are superimposed on a relatively slow linear voltage sweep (Fig. 2-21). The voltage pulse occurs at the end of the drop life. The total drop time would normally be controlled by an electronically activated mechanical drop dislodger at 1 or 2 sec. In the differential pulse technique the current flowing at the working electrode (DME) is sampled twice during the lifetime of each drop. The first current sample is taken immediately before application of the voltage pulse, as shown in Fig. 2-21. Then, the voltage pulse is applied and this may have a duration of 40–60 msec. The application of this change in potential produces a resultant current, which can be derived from two sources. The first is the current required to charge the electrical double layer to the newly applied potential (capacitive current). Capacitive current normally decays very rapidly at a rate governed by the magnitude of the capacitance and series resistance of the system. Simultaneously, additional faradaic current may flow if the new potential is such that the equilibrium between the oxidized and reduced forms of the electroactive species at the electrode surface is shifted.

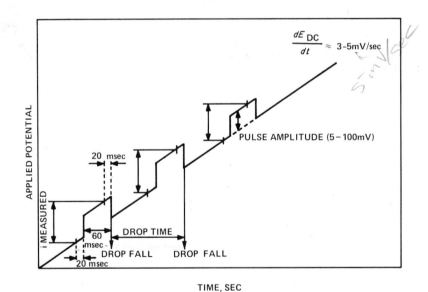

FIG. 2-21.   Schematic diagram of signal waveform applied to the DME in differential pulse polarography.

In order to illustrate the processes involved consider a situation in which an electroactive species Ox gives a normal DC polarogram of the type shown in Fig. 2-22A. Now consider the processes occurring in a differential pulse polarogram, for example when a new drop starts to grow and the initial potential is −0.420 V, i.e., a potential corresponding to the rising portion of the DC polarogram. Assuming that the linear voltage sweep rate is 5 mV sec^{-1} and a 1.0 sec drop time is employed, then as the drop grows the capacitive or charging current continuously decreases. Correspondingly, the faradaic current increases (Fig. 2-22C). Immediately before application of the 5 mV amplitude pulse, the current flowing is sampled (usually over a 10−20 msec time period), i.e., i_1 in Fig. 2-22C. The voltage pulse is then applied and a current increase is noted (Fig. 2-22C). The capacitive current decays rapidly but the faradaic current decays much more slowly.* The duration of the voltage pulse is usually 40−60 msec. Over the last 10−20 msec of the pulse the current is clearly almost all faradaic in nature and again the current is sampled over this time period, i.e., i_2 in Fig. 2-22C. The difference between i_2 and i_1 is the recorded or measured current in differential pulse polarography. These two currents are almost entirely faradaic; any small, residual capacitive currents measured at i_1 and i_2 virtually cancel out. At potentials where no faradaic current flows, e.g., 0 to −0.3 V (Fig. 2-22A), similar current sampling techniques before and at the end of the applied voltage pulse will give values of i_1 and i_2 that are very small, almost identical, and entirely capacitive in nature, i.e., $i_2 - i_1$ will be zero. At potentials corresponding to the plateau of the DC polarographic wave (−0.6 to −1.0 V) the current sampled before and at the end of the voltage pulse will be almost entirely faradaic in nature but i_2 will be equal to i_1; e.g., shifting the electrode potential from −0.800 to −0.805 V (Fig. 2-22A) would not result in a significant increase in the faradaic current, and hence $i_2 - i_1$ will be zero again. Accordingly, significant differences between i_2 and i_1 will be noticed only at potentials corresponding to the rising portion of the DC polarographic wave. At potentials prior to the wave and on the plateau of the DC wave the difference between i_2 and i_1, i.e., the current recorded in differential pulse polarography, will be zero. Thus, a differential pulse polarogram has the appearance of a peak (Fig. 2-23).

Differential pulse polarography is used primarily for analytical work and can be used for analyses at the $10^{-6}-10^{-8}$ M concentration levels. The theoretical aspects of pulse polarographic techniques have been studied by Barker and Gardner[18] and Boduy and co-workers.[19]

---

* Actually, because the duration of the pulse is so short, the area of the drop can be regarded as constant and the faradaic current decays as $t^{-1/2}$ as predicted by the Cottrell equation (Ref. 9), $i_t = nFAD^{1/2}C/\pi^{1/2}t^{1/2}$, where all terms have been defined previously.

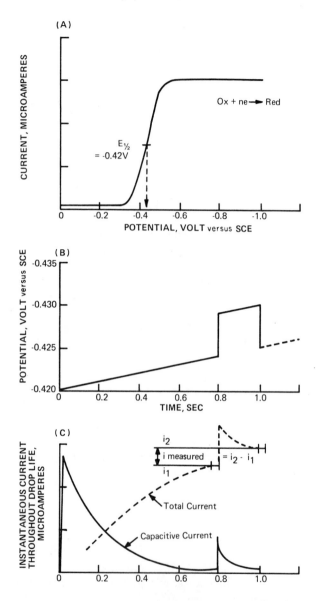

FIG. 2-22. (A) DC polarogram for the polarographic reduction of Ox to Red. (B) Voltage waveform applied to DME during differential pulse polarography; potentials correspond to rising portions of DC polarogram. Linear voltage sweep: 5 mV/sec; voltage pulse: 5 mV; drop time: 1.0 sec. (C) Instantaneous capacitive and total current responses during drop lifetime on application of voltage waveform shown in (B). Note that the pulse lifetime shown in (B) and (C) is not drawn to scale.

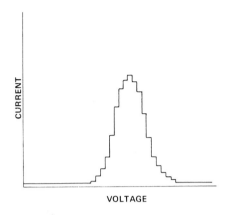

CURRENT

VOLTAGE

FIG. 2-23. Schematic differential pulse polarogram.

XII. CONTROLLED POTENTIAL ELECTROLYSIS AND COULOMETRY

Controlled potential electrolysis is normally employed for two principal reasons in electrochemistry: first, to prepare sufficient products of the electrode reaction to allow their isolation and/or identification and, second, to determine the number of electrons involved in an electrode process.

Generally, a three-electrode system should be employed for such large-scale electrolyses. There are three principal reasons for employing three-electrode systems in electrochemistry: first, to provide a means of overcoming the iR drop in the electrolysis cell; second, to maintain the desired potential on the working electrode versus a suitable reference electrode; and third, to prevent excessive current flowing through the reference electrode.

A typical three-electrode cell is shown in Fig. 2-24. The reference electrode is more often than not a saturated calomel electrode (SCE). The agar plugs, saturated with KCl usually, and the fine sintered glass frits are used to minimize diffusion of solutions between the compartments. The mercury pool electrode is the working electrode, i.e., the electrode at which the reaction of interest is to take place. The solution in the working electrode compartment contains the electroactive species in a solution of suitable supporting electrolyte. The working electrode could alternatively be a series of graphite rods or plates, gold, or platinum gauze. The third electrode is called the counterelectrode (or auxiliary electrode) and is normally immersed in a solution of supporting electrolyte.

The instrument used for controlled potential electrolysis is called a potentiostat. A schematic diagram of a manual potentiostat is shown in

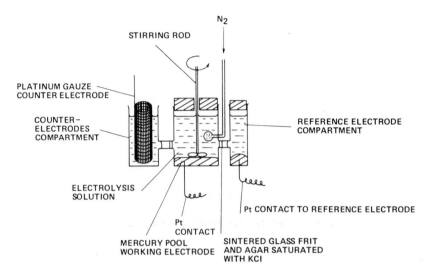

FIG. 2-24.   Typical three-electrode cell for controlled potential electrolysis and coulometry.

Fig. 2-25. In this cell the working electrode (W) is the cathode, and the counterelectrode (C) is the anode. All of the electrolysis current flows between these two electrodes and is recorded on the ammeter. The potential of the working electrode is measured versus the reference electrode (R) in a second circuit by means of a voltage measuring device, such as a potentiometer or perhaps a high-impedance vacuum tube voltmeter. The total voltage applied between the counterelectrode and working electrode is manually adjusted so that the working electrode is at the desired potential versus the reference electrode. As the electrolysis proceeds, current, indicated on the ammeter, decreases (*vide infra*). This necessitates continuous readjustment of the total applied voltage because the *iR* drop through the cell decreases, causing a consequent change in the potential of the working electrode.

As the electrochemical reaction of interest takes place at the working electrode, i.e., a reduction in Fig. 2-25, an anodic reaction takes place at the counterelectrode which is usually oxidation of the solvent and/or supporting electrolyte. Automatic potentiostats are now available and are discussed in Section XIV.

In controlled potential electrolysis the electrolysis is allowed to proceed generally until the current has decayed to a very low level and all of the initial electroactive species has disappeared. Suitable examination and isolation of the reaction products can then be undertaken.

In controlled potential coulometry, however, one is interested in determining

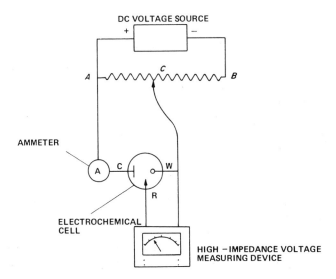

FIG. 2-25. Circuitry for a manual potentiostat. *A, B, C:* Linear slidewire; A: ammeter; W: working electrode; R: reference electrode, C: counterelectrode.

the total amount of electricity that has been used in completely oxidizing or reducing the electroactive species. If the electrode reaction is written

$$A \pm ne \longrightarrow B \qquad (38)$$

then it is observed experimentally for controlled potential electrolysis in stirred solution without any complicating factors that the current decays exponentially toward zero as the electrolysis proceeds.

Lingane[20] has shown that the current at time *t*, in seconds, after initiation of the electrolysis (i_t) is related to the initial current (i_0) by Eq. 39, where *k* is a

$$i_t = i_0 10^{-kt} \qquad (39)$$

constant. The total coulombs of electricity (Q) passed from the beginning of the electrolysis to time *t* is given by

$$Q = \int_0^t i_t \; dt \qquad (40)$$

There are several ways to determine Q. The least accurate but simplest is to make use of Eqs. 39 and 40 to derive Eq. 41. As *t* gets larger, 10^{-kt} gets smaller,

$$Q = \frac{i_0}{2.303k} (1 - 10^{-kt}) \qquad (41)$$

and when kt is greater than about 3, then Q reaches a limiting value,

$$Q_{total} = i_0/2.303k \tag{42}$$

Thus, by recording the electrolysis current at known time intervals and plotting log i_t against time, from Eq. 39 a straight line can be expected having a slope of $-k$ and an intercept log i_0 at $t = 0$. By substituting the values of i_0 and k into Eq. 42, the value of Q_{total} in coulombs can be calculated. If the initial concentration of electroactive species is known, the number of electrons n involved in the electrode reaction can be calculated from

$$Q_{total} = nFN \tag{43}$$

where N is the number of moles of electroactive species initially present in the solution, and F is the faraday (96,500 C).

Because of instrumental problems and complicating secondary chemical reactions that often occur under conditions of controlled potential electrolysis, the above method is rarely satisfactory for determining n. Accordingly, other methods have been devised for measuring the quantity of electricity involved in an electrode reaction under controlled potential electrolysis conditions. These methods utilize a coulometer or current integrator.

There are two types of coulometer: chemical and electronic. The latter have almost completely displaced chemical coulometers. However, in order to perform occasional coulometric experiments, the chemical coulometer provides an inexpensive, moderately accurate, and simple method. Lingane[20] has discussed chemical and other coulometers at length. The chemical coulometer we have found most convenient is the titration coulometer (Fig. 2-26). This consists of a helical silver wire (area ca. 70–80 cm^2) and a platinum gauze cathode (of about the same size) immersed in 70–80 ml of 0.03 M KBr and 0.2 M K$_2$SO$_4$. The latter salt is used to provide reasonable electrical conductivity in the coulometer. The electrolyte is stirred and deaerated with nitrogen, and we have found it expedient to wrap the vessel with aluminum foil to keep light out of the solution. The coulometer is connected in the counterelectrode loop of the potentiostat circuit (in series with the counterelectrode) with the platinum electrode connected as the cathode and the silver wire as the anode. As soon as any current flows through the electrolysis cell, it also passes through the coulometer and the following reactions occur:

$$Ag + Br^- \longrightarrow AgBr + e \qquad \text{(silver anode)} \tag{44a}$$
$$H_2O + e \longrightarrow \tfrac{1}{2}H_2 + OH^- \qquad \text{(platinum cathode)} \tag{44b}$$

Thus, 1 mole of hydroxide ion is produced at the cathode for each faraday of electricity that flows through the coulometer. This is titrated with standard 0.01 M HCl to pH 7 using the glass electrode to monitor the pH. It can be

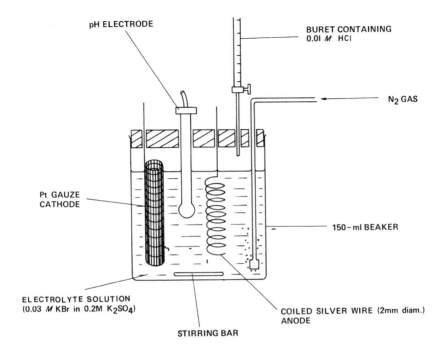

pH ELECTRODE

BURET CONTAINING
0.01 *M* HCl

N₂ GAS

Pt GAUZE
CATHODE

150-ml BEAKER

ELECTROLYTE SOLUTION
(0.03 *M* KBr in 0.2M K₂SO₄)

COILED SILVER WIRE (2mm diam.)
ANODE

STIRRING BAR

FIG. 2-26. Schematic representation of a titration coulometer.

readily shown that 1.00 ml of 0.01 *M* HCl corresponds to 0.9650 C. The solution should be titrated continuously for best results. Lingane[20] has discussed practical aspects of the titration coulometer in some detail. Electronic integrators will be discussed in Section XIV, J.

XIII. CHRONOPOTENTIOMETRY

Chronopotentiometry involves the measurement of the potential of a working electrode as a function of time as a constant current is passed through an electrochemical cell between the working electrode and a suitable auxiliary electrode or counterelectrode. A typical apparatus for such studies is shown in Fig. 2-27.

The constant-current source (50–1000 μA) can be a simple battery connected in series with a large resistor and the electrochemical cell. The resistance of the series resistor is large compared to that of the cell so that small changes in the cell resistance have no appreciable effect on the current. The current flows

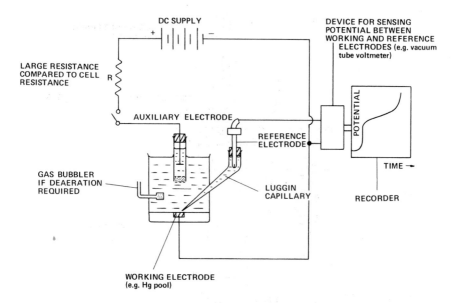

FIG. 2-27. Simple apparatus required for chronopotentiometry.

between the auxiliary and working electrodes. The former electrode is usually isolated from the test solution by means of a fritted tube. The potential of the working electrode is monitored with respect to a suitable reference electrode by means of a vacuum tube voltmeter or similar device. The output of the vacuum tube voltmeter is fed to a suitable recorder and the potential is displayed as a function of time.

A typical chronopotentiogram, which is obtained for the electrolysis of a 10 mM solution of potassium ferricyanide in 1 M potassium chloride using a platinum working electrode in unstirred solution, is shown in Fig. 2-28. The potential of the platinum electrode in this solution without current flowing is of little importance for the present discussion. However, after a few minutes the potential reaches a steady value as shown by the line AB. As soon as the electrolysis current is turned on, the electrode potential shifts very rapidly to more negative values, i.e., C, where ferricyanide begins to be reduced to ferrocyanide. As the electrolysis proceeds, the ferricyanide ion concentration at the electrode surface decreases and, correspondingly, the ferrocyanide ion concentration increases. Since this is a reversible system, the Nernst equation demands that the electrode potential shift to a more negative value, which it is seen to do between C and D. At D the surface concentration of ferricyanide at the electrode is essentially zero, and at this point the supply of ferricyanide

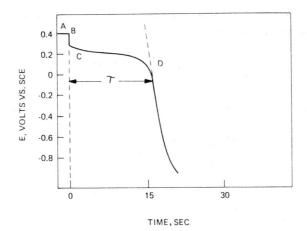

FIG. 2-28. Chronopotentiogram of 10 m*M* potassium ferricyanide in 1 *M* KCl at a platinum working electrode. For label definitions, see text.

diffusing to the electrode becomes insufficient to maintain the current flow. The potential of the platinum electrode thus shifts to more negative potential, to a value where the next most easily reduced solution species is reduced, in this case the potassium ions of the background electrolyte. The interval from *C* to *D*, measured as shown in Fig. 2-28, is called the *transition time τ*.

For an unstirred solution of an electroactive species electrolyzed in this way, the transition time is described by Eq. 45, the Sand[21] equation, where *τ* is the

$$\tau^{1/2} = \frac{\pi^{1/2} n F D^{1/2} A C^{\text{b}}}{2i} \tag{45}$$

transition time in seconds, *D* is the diffusion coefficient of the electroactive species in cm² per second, *A* is the electrode area in cm², *C* is the bulk concentration of the electroactive species in millimoles per liter, and *i* is the constant current in microamperes.

For the straightforward reversible reduction of an oxidized species (Ox) to a soluble reduced species (Red) according to Eq. 1, the equation of the chronopotentiometric potential–time curve is given by Eq. 46,[5] where *E* is the

$$E = E_{\tau/4} - \frac{0.05915}{n} \log \frac{t^{1/2}}{\tau^{1/2} - t^{1/2}} \qquad \text{at } 25°\text{C} \tag{46}$$

potential of the working electrode at some time *t* after the initiation of the electrolysis but before *τ*, and $E_{\tau/4}$ is the quarter transition-time potential (i.e., *E* at *t* = *τ*/4 sec). The chronopotentiometric $E_{\tau/4}$ and the DC polarographic half-wave potential $(E_{1/2})$ are identical for a reversible reaction of the type described above, and hence

$$E_{1/2} = E_{\tau/4} = E^{0\prime} - \frac{RT}{nF} \ln\left(\frac{f_{Red}D_{Ox}^{1/2}}{f_{Ox}D_{Red}^{1/2}}\right) \tag{47}$$

For reversible electrode reactions involving protons, $E_{\tau/4}$ is pH dependent:

$$\frac{dE_{\tau/4}}{d(\text{pH})} = \frac{-0.05915}{n}p \qquad \text{at } 25°\text{C} \tag{48}$$

Here, p is the number of protons involved in the electrode reaction. Similar equations can be written for oxidation reactions.[5]

In the case of a totally irreversible cathodic electrode reaction the equation of the potential–time curve is

$$E = -0.2412 + \frac{0.05915}{\alpha n_a} \log \frac{2k_{f,h}^0}{\pi^{1/2}D^{1/2}} + \frac{0.05915}{\alpha n_a} \log(\tau^{1/2} - t^{1/2}) \tag{49}$$

where E is the electrode potential in volts versus SCE, and $k_{f,h}^0$ is the heterogeneous rate constant at 0.0 V versus the normal hydrogen electrode. The other terms have been described previously. The $E_{\tau/4}$ value for a totally irreversible couple at $25°\text{C}$ is described by Eq. 50.[5,22,23]

$$E_{\tau/4} = -0.2412 + \frac{0.5915}{\alpha n_a} \log \frac{0.564 k_{f,h}^0 \tau^{1/2}}{D^{1/2}} \tag{50}$$

Although Eqs. 46–50 can be used for determining the kinetics and mechanism of electrode reactions, the most important and useful equation in chronopotentiometry is the Sand equation (Eq. 45). This predicts that $\tau^{1/2}$ is directly proportional to the bulk concentration of electroactive species. The Sand equation also predicts that for a normal, uncomplicated, diffusion controlled electrode reaction the product $i\tau^{1/2}$ should be a constant (at a fixed concentration of the electroactive species), i.e.,

$$i\tau^{1/2} = \frac{\pi^{1/2}nFAD^{1/2}C^b}{2} = \text{constant} \tag{51}$$

and independent of i or τ. Deviations from these relationships can often be used to elucidate electrode reactions. Kinetic, catalytic, and adsorption processes can also be observed chronopotentiometrically. For details on these processes other literature should be consulted.[2,5,9,24]

XIV. INSTRUMENTATION

In most areas of electrochemistry there is continuous development in instrument design. In fact, unlike most other chemists, electrochemists are often primarily equipped with instrumentation designed and constructed in their own laboratories. This is due principally to an instrumental concept that appeared 10–15 years ago, namely, the operational amplifier, which is particularly adaptable to electrochemical instrumentation. Virtually all electrochemical techniques are accessible utilizing operational-amplifier-based instruments. This section is not intended as an in-depth report of electronic instrumentation, but merely as a guide to those uninitiated in such instrumentation.

A. The Operational Amplifier

Several scattered reviews of the basic principles of operational amplifiers have appeared.[8,25–30] An operational amplifier is a high-gain, wide-band amplifier. It is characterized by a very high input impedance so that it has an almost negligible current drain on the input source.

An operational amplifier is normally represented in circuit diagrams as a large triangle (Fig. 2-29). For most purposes it can be regarded as having two inputs and one output. The two input voltages e_a and e_b (Fig. 2-29A) and the output voltage e_o are referred to a common ground, which is often the power supply ground. The relationship between the input and output voltages is given by

$$e_o = A(e_b - e_a) \tag{52}$$

where A is the open-loop gain of the amplifier. This is typically very large and of the order of 10^4–10^6. The amplifier shown in Fig. 2-29A therefore responds to the potential difference between its two inputs and is accordingly said to have differential input. In many instances the input b is grounded (Fig. 2-29B), in which case the amplifier is said to be single ended and

$$e_o = -Ae_a = -Ae_i \tag{53}$$

It is seen that the amplifier inverts the input signal; i.e., the polarity of e_o is opposite to that of e_i. For this reason the input designated a is called the inverting input. It will be seen that a signal applied to the input b is not inverted, and therefore this input is called the noninverting input. The inverting and noninverting inputs are usually designated − and +, respectively, although in many circuit diagrams even these symbols are omitted (Fig. 2-29C).

Since the open-loop gain A of an operational amplifier is so high, it is clear that the application of a very small input signal will result in a large output signal. Typically, for a solid-state operational amplifier the maximum output voltage is about ±10 V, hence, the maximum input signal in the open-loop mode

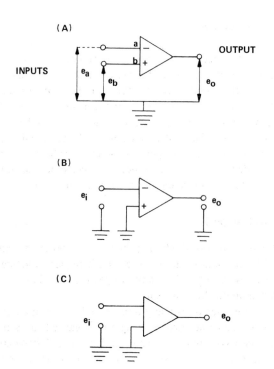

FIG. 2-29.   Schematic representation of operational amplifiers.

is $\pm 10^{-5}$ V, using an amplifier in the configuration shown in Fig. 2-29B, C having an open-loop gain of 10^6.

Before passing on to some typical circuits, it is worth pointing out that, if the input signal is not DC, then above a certain frequency the open-loop gain begins to decrease with increasing frequency. Without going into details, this means that the amplifier has only a limited frequency response. Thus, in AC polarography the upper frequency level at which the amplifier can be used is about 1 kHz in many instances, although amplifiers with somewhat higher frequency responses are now available.

B. Inverting Circuits

In order to utilize reasonable input signals and to perform certain mathematical operations, the operational amplifier is usually programmed by some type of external circuitry. The simplest circuit is that shown in Fig. 2-30. Thus, the signal source e_i is connected through the input resistor R_i to the inverting input, and a feedback resistor R_f is connected between the amplifier

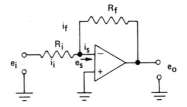

FIG. 2-30. Typical operational amplifier configuration as voltage multiplier. For label definitions, see text.

output and the inverting input. Recalling that the output voltage e_o is of opposite polarity to the input voltage e_i it is clear that feedback through R_f decreases the signal at the input of the amplifier. Since the gain of the amplifier A is large, the voltage at the amplifier inverting input e_s must be $-e_o/A$. Accordingly, because A is so large, the voltage at e_s must be very small or, in other words, is at virtual ground. Kirchoff's current law states in effect that the sum of all currents entering a junction is equal to the sum of all currents leaving a junction. Applying this law to the junction of the input and feedback resistors,

$$i_i = i_s + i_f \tag{54}$$

but because of the very high impedance of the amplifier, i_s is negligibly small, and therefore

$$i_i = i_f \tag{55}$$

Since both R_i and R_f are connected to virtual ground (e_s), Ohm's law can be applied:

$$e_i/R_i = -e_o/R_f \tag{56}$$

or

$$e_o = -R_f e_i/R_i \tag{57}$$

In other words, the circuit shown in Fig. 2-30 amplifies the input signal by $-R_f/R_i$.

Several input signals can be summed and then amplified, as shown in Fig. 2-31. The amplifier input e_s is at virtual ground so there is no interaction between the sources. Under these conditions

$$e_o = -\left(\frac{e_1 R_f}{R_1} + \frac{e_2 R_f}{R_2} + \frac{e_3 R_f}{R_3} + \cdots \right) \tag{58}$$

and when all of the resistors R_1, R_2, R_3, R_f are equal,

$$e_o = -(e_1 + e_2 + e_3) \tag{59}$$

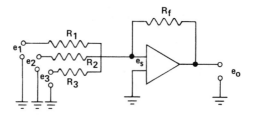

FIG. 2-31.   Summing amplifier. For label definitions, see text.

Hence, the amplifier is algebraically summing all of the input signals. For this reason the amplifier is called a summing amplifier. The amplifier input voltage e_s is for the same reason called the summing point and is always at virtual ground potential. If R_f is not equal to the input resistors, then the summed input signals, $e_1 + e_2 + e_3 + \cdots$, are multiplied by a constant factor. A typical multiplication performed by an operational amplifier is shown in Fig. 2-32. Similarly, operational amplifiers may be used as a subtractor (Fig. 2-33).

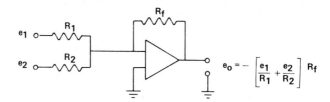

FIG. 2-32.   Operational amplifier as a multiplier. For label definitions, see text.

The insertion of the resistors into the input terminal of an operational amplifier gives the clue that the amplifier is really a current sensing device. Accordingly, it can be employed as a current measuring amplifier (Fig. 2-34). In the configuration shown in Fig. 2-34A

$$e_o = -i_i R_f \tag{60}$$

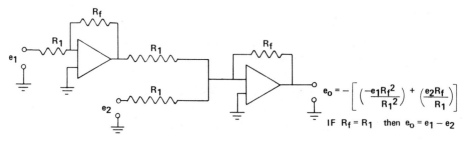

FIG. 2-33.   Operational amplifier as a subtractor. For label definitions see text.

(A)

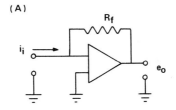

(B)

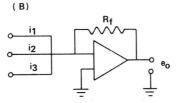

FIG. 2-34. (A) Current amplifier; (B) current summing amplifier. For label definitions, see text.

while in Fig. 2-34B a current summing amplifier is presented, where

$$e_o = -(i_1 + i_2 + i_3)R_f \qquad (61)$$

By placing a capacitor in the feedback loop instead of a resistor, an operational amplifier may be made into an integrator (Fig. 2-35). Recalling that the input and the feedback currents are identical (Eq. 55) and that the potential at the summing point e_s is virtual ground, the charge on the capacitor Q is given by

$$Q = Ce_o \qquad (62)$$

where C is the capacitance in farads of the capacitor. But

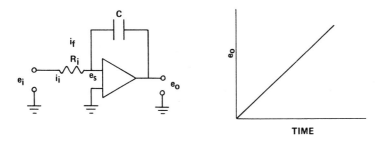

FIG. 2-35. Operational amplifier as an integrator. For label definitions, see text.

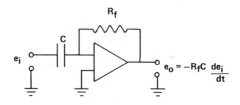

FIG. 2-36.   Operational amplifier as a differentiator. For label definitions, see text.

$$Q = \int_0^t i_f \, dt = \int_0^t i_i \, dt \qquad (63)$$

Therefore,

$$e_o = \frac{1}{C} \int_0^t i_i \, dt \qquad (64)$$

and since

$$e_i = i_i R_i \qquad (65)$$

then

$$e_o = -\frac{1}{R_i C} \int_0^t e_i \, dt \qquad (66)$$

The negative sign appears because of the inverting nature of the amplifier input. Accordingly, if a constant DC voltage is applied as the amplifier input signal, a DC ramp output signal voltage is produced, as shown in Fig. 2-35. The slope of the ramp depends on e_i, R_i, and C. Simply by reversing the positions of the capacitor and resistor, a differentiator may be produced (Fig. 2-36).

C. Noninverting Amplifiers

All of the preceding operational amplifier circuits utilize the inverting input with consequent change in polarity between the input and output signals. Noninverting circuits are far less common, but one noninverting circuit, the *voltage follower* or simply *follower* (Fig. 2-37), is widely employed in electrochemical instruments. The follower circuit employs the noninverting (+) input of the amplifier. Negative feedback is provided for by connecting the

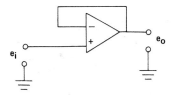

FIG. 2-37. The voltage follower. For label definitions, see text.

output directly to the inverting input. In this circuit

$$e_o = A(e_i - e_o) \tag{67}$$

or

$$e_o = \frac{A}{1 + A} e_i \tag{68}$$

Since $A \approx 1 + A$ because A is very large,

$$e_o = e_i \tag{69}$$

Thus, with the follower amplifier the input and output signals have the same sign and magnitude.

The particular utility of this amplifier is that it has a very high input impedance and a very low output impedance. In a practical sense this means that the amplifier draws a very small current from the source voltage, yet a large current may be drawn from the output. In effect, it isolates the signal and measuring circuits. Operational amplifiers can also be used for transforming signals to their logarithmic analog and vice versa.[29-31]

D. Constant-Current and Constant-Voltage Sources

There are two other basic operational amplifier circuits that are often useful. A constant-current source is conveniently obtained by placing a voltage source, e.g., a Weston cell, into the feedback loop of the amplifier, as shown in Fig. 2-38. The amplifier simply injects sufficient current acrosss the variable resistor (or load resistance, R_{load}) so that the potential at A is equal to that of the voltage source, E_s; i.e., the input to the amplifier is at virtual ground. Thus, a constant voltage is applied between point A and ground. Depending on the value of R_{std} selected, any constant current within the capability of the amplifier can be selected. A constant-voltage source is shown in Fig. 2-39.

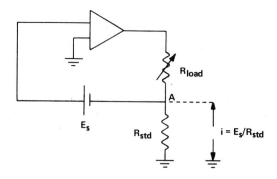

FIG. 2-38.   Constant-current supply. For label definitions, see text.

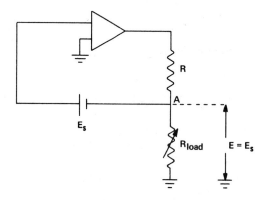

FIG. 2-39.   Constant-voltage source. For label definitions, see text.

E. Properties of Typical Operational Amplifiers

There are a very large number of operational amplifiers commercially available. There are, however, two principal types: vacuum tube and solid-state amplifiers. The former are rarely used today except when high power output requirements are important.

The properties of particular interest in an operational amplifier usually are voltage output, current output, open-loop DC gain, differential input, single-ended input, input impedance, slew rate or rate limit, settling time, voltage offset, bias current, chopper stabilized.

Normally, solid-state amplifiers have a maximum voltage output of ±10 V, while for vacuum tube amplifiers it is ±100 V. The output currents normally range from 2 to 20 mA. In the event that a greater effective output voltage or output current is required, booster or power amplifiers can be obtained which are simply wired in series with the operational amplifiers. Typically, the open-loop gain A of an amplifier is around 10^4-10^6. Many operational amplifiers (chopper stabilized, principally) are not designed to operate with both inputs active and are called single-ended amplifiers. If both inputs can be active, the amplifier is said to have differential input. However, with a differential input amplifier both inputs do not have to be active; i.e., more often than not the noninverting input is grounded.

The input impedance decides effectively how much current the amplifier will draw from the source. Once all other requirements have been satisfied, it is a good rule to select an amplifier with the highest input impedance. Values of input impedance range from about 10^5 to 10^{12} Ω or even greater.

The slew rate or rate limit describes how fast the amplifier reacts to a signal and is usually expressed in volts per microsecond. It is in effect a measure of the ability of the amplifier to produce large, rapid changes in output voltages. The settling time indicates how long it takes the output of the amplifier to reach its correct value within some specified limits, i.e., ±0.1%. In other words, the amplifier output usually slightly overshoots the theoretically correct value and for a very short time oscillates somewhat before settling to the correct value. If the instrument design calls for a very fast response, amplifiers should be selected with fast slew rates and very short settling times.

From what has already been said about operational amplifiers, it would seem that, if both inputs were grounded, the output should be zero. In fact, it is usually observed that an output voltage is found under these conditions. Dividing this voltage by the amplifier gain gives an error voltage referred to the input. This error is termed the *voltage offset* Many amplifiers have a built-in means of adjusting this offset to zero, or a simple external circuit is built. Since the offset voltage is temperature variable, it is clear that it is very difficult to maintain this offset at zero. For high-precision work amplifiers having very low offset voltages should be employed.

The bias current is that current which flows in or out of either terminal under zero signal conditions. For many purposes this is negligible, although external circuits can be employed to reduce it. These are particularly desirable when integrating circuits are constructed. Such circuits are abundant in the technical literature of operational amplifier manufacturers.

In a chopper-stabilized amplifier the voltage offset is reduced by means of internal devices. Chopper-stabilized amplifiers are always single ended.

Finally, if one plans to use the operational amplifier for non-DC applications, e.g., AC polarography, it is wise to check the frequency response of the amplifier. It is beyond the scope of this chapter to discuss the details of the characteristics of operational amplifiers. However, these are readily found by consulting the manufacturers' technical literature[32,33] or standard reference texts.[30,31,34]

F. The Potentiostat

The heart of all electrochemical instruments which require that a known potential be applied between the working electrode and a reference electrode is a potentiostat. There are three basic requirements for a potentiostat: (a) a known potential, which may or may not be time variant, is maintained between the working and reference electrodes; (b) only a negligible current flows through the reference electrode; and (c) the current flowing through the working electrode can be measured.

All modern potentiostats employ three electrodes: a reference electrode (R), a working electrode (W) (e.g., dropping mercury electrode, mercury pool, or rotating platinum disc), and a counterelectrode (C). (See Fig. 2-40.) All the electrolysis current flows between the working electrode and counterelectrode.

For purposes of illustration a one-amplifier potentiostat circuit will first be considered (Fig. 2-40). The first thing to note about this circuit is that the working electrode is connected directly to ground. In order to understand the operation of this circuit, it is of value to consider the cell as a resistive network, as shown in the dummy cell in Fig. 2-40. With the dummy cell in circuit let us assume we require that the working electrode have a potential of -1.0 V versus the reference electrode. In order to achieve this situation, -1.0 V (versus ground) is applied from the DC voltage supply to the input of the control amplifier CA across the 100 K input resistor. This implies that momentarily the voltage at the summing point of the amplifier is above virtual ground. In order to offset this voltage, clearly $+1.0$ V has to be applied across the second 100 K input resistor. This is done quite simply by the control amplifier injecting sufficient positive voltage across R_1 and R_2 so that the voltage tapped off between W and R (equivalent to the working and reference electrodes, respectively, in a real cell) is $+1.0$ V. Under these conditions the summing point of the control amplifier reaches virtual ground [i.e., $+1.0$ V $+ (-1.0$ V$) = 0$]. Now, the particular beauty of this circuit is that if the value of $R_1 + R_2$, which is equivalent to the resistance of the counterelectrode and the electrolysis solution between the counterelectrode and working electrode, is changed, the voltage output of the control amplifier also changes to maintain the necessary potential between the working and reference electrodes. In other words, provided that the required total voltage across R_1 and R_2 does not exceed the maximum voltage

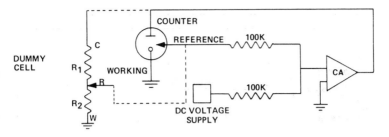

FIG. 2-40. Single operational amplifier potentiostat. CA: control amplifier. For label definitions, see text.

output of the control amplifier, the solution resistance (R_1) is automatically compensated.

If the electrochemical cell contains some electroactive species, then at an appropriate applied potential between the working and reference electrodes the electrochemistry of the system demands that a current flows. The control amplifier accordingly injects sufficient voltage between the counterelectrode and working electrode to maintain the necessary potential between the working and reference electrodes. Sufficient current flows between the counterelectrode and working electrode to satisfy the electrochemistry of the system.

The potentiostat shown in Fig. 2-40 does not fulfill all the requirements of a practical potentiostat since there is no provision for measuring the current flowing through the working electrode. If the resistance of the reference electrode is low, there is also a possibility of appreciable current flowing through that loop of the circuit. Accordingly, a typical practical potentiostat circuit is shown in Fig. 2-41. Again, it should be noted that the working electrode is connected directly to the summing point of the current amplifier (CF) so that in effect it is connected to ground.

The variable DC voltage supply could be obtained from the circuits shown in Fig. 2-42A or 2-42B. Generally, the power supply voltages for operational amplifiers (±15 V for solid-state amplifiers) are very stable, so that a variable DC voltage supply may be derived from the supply using just a single operational amplifier (Fig. 2-42A). Alternatively, a standard Weston cell may be employed (Fig. 2-42B).

All of the electrolysis current flows between the counterelectrode and working electrode. The current flowing is measured by the current amplifier (CF), which simply gives a voltage proportional to the current. When very large currents are flowing, i.e., when current booster amplifiers are employed in the counter electrode loop, the working electrode is normally connected directly to ground and some sort of ammeter is inserted between the output of the control amplifier/booster amplifier system and the counterelectrode.

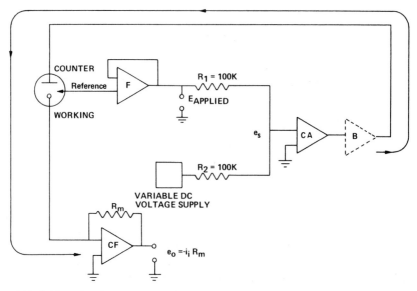

FIG. 2-41. Simple potentiostat circuit. F, Voltage follower; CA, control amplifier; B, booster amplifier (optional); CF, current amplifier. See text for discussion.

G. Direct Current Polarograph

In order to convert the potentiostatic circuit shown in Fig. 2-40 into a DC polarograph, all that is required is a source of slowly varying DC potential (a DC ramp) and a means of recording the applied voltage and the resultant current at the working electrode. This is shown schematically in Fig. 2-43. The initial voltage is set at the appropriate level with the amplifier V by appropriately adjusting the 0–100 K variable resistor. The DC voltage ramp is derived from the amplifier I, which is wired as an integrator. By appropriate choice of the value of R and C and application of Eq. 66, the desired scan rate can be achieved. By altering the polarity of the input voltage, the scan can be made to sweep in a positive or negative direction. The switch S is a shorting switch that allows the capacitor C to be discharged at the end of the voltage sweep. It simply resets the polarograph to its initial state.

For normal polarographic work currents of the order of 0–100 μA are usual. By placing a 1 K resistor in feedback of the current amplifier (CF), each 1 μA of current at its input gives 1 mV output, which is a very useful conversion with most $X - Y$ recorders.

The voltage at the output of the voltage follower, F, is equal, but of opposite sign, to the voltage of the working electrode versus the reference electrode. This is usually fed to the X axis of the $X - Y$ recorder.

(A)

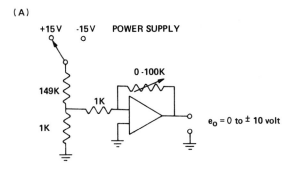

+15V -15V   POWER SUPPLY

0-100K

149K

1K

1K

e_o = 0 to ± 10 volt

(B)

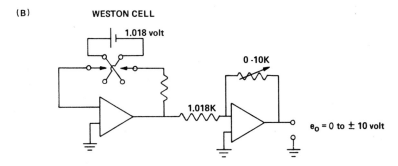

WESTON CELL

1.018 volt

0-10K

1.018K

e_o = 0 to ± 10 volt

FIG. 2-42. Variable DC voltage sources. (A) Single operational amplifier and ±15 V power supply. (B) Two operational amplifiers and standard Weston cell.

H. Cyclic Voltammetry

Cyclic voltammetry can be carried out with the circuit shown in Fig. 2-43. The scan rate, the direction of scan, and the potential span can all be selected using the integrator circuit shown. For scan rates above about 500 mV sec^{-1}, manual control of the integrator device, better known as a function generator, is difficult. Very elegant circuits for programming the scan rate, initial direction and potential limits, and number of scans have been described by Myers and Shain.[35] For fast sweep cyclic voltammetry an oscilloscope, preferably of the storage type, must be used for recording voltammograms.

I. Alternating Current Polarography

Normal, or fundamental, AC polarography can be performed with the circuit shown in Fig. 2-44. Again, the heart of the instrument is a potentiostat, but provision is made for a DC ramp and a sinusoidal oscillator in the input signal circuitry. The alternating current flowing in the working electrode is amplified

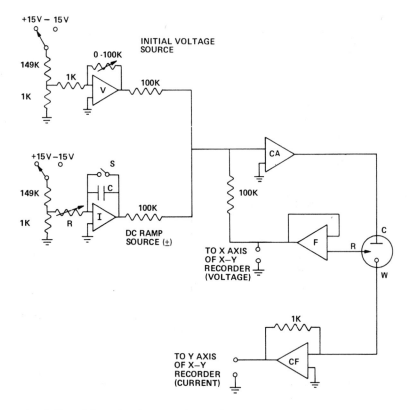

FIG. 2-43. Schematic of a DC polarograph using operational amplifier circuitry. For label definitions, see text.

and converted to an alternating voltage by the current amplifier, CF. In order to block the direct current and any AC noise, a tuned amplifier is inserted immediately after the current amplifier. The tuned amplifier is tuned to the frequency of the sinusoidal oscillator and will pass only signals of that frequency. Since most recorders do not accept alternating voltages, a rectifier is employed. Smith[8] has reviewed in some detail the instrumentation necessary for AC polarography.

J. Electronic Coulometers

There are two types of electronic coulometers. The first is an analog device and generally has only limited applicability. In effect, when one is carrying out, for example, a controlled potential electrolysis, the current—time integral can be determined by placing an operational amplifier wired as an integrator in the

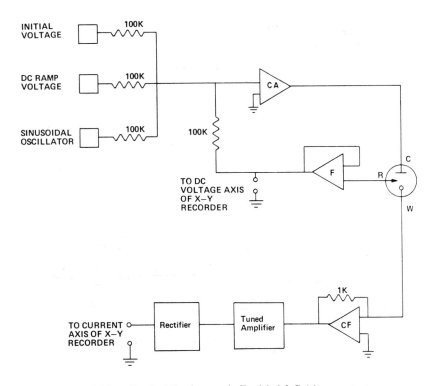

FIG. 2-44. Simple AC polarograph. For label definition, see text.

working electrode loop (Fig. 2-45). The current flowing in the working electrode is amplified and converted to a proportionate voltage by the current amplifier, CF. This voltage is then integrated with respect to time by the operational amplifier I wired as an integrator. The voltage readout is directly proportional to the total coulombs passed during the electrolysis. Clearly, this type of coulometer is useful only up to the point when the amplifier I reaches its maximum output voltage. It is therefore useful only for very short electrolysis or for electrolysis that involves only very small currents.

A more useful electronic coulometer is a digital device (Fig. 2-46). The principle of the method is that a small precision resistor, typically 1 Ω, is placed in the counterelectrode loop of the potentiostat. The electrolysis current passes through this resistor and a voltage develops. This voltage is fed to a voltage-to-frequency converter. This device simply emits a definite number of voltage pulses per second proportional to the input voltage. Typically, for a 1 V input (equivalent to 1 A electrolysis current) the voltage-to-frequency converter puts out 100,000 pulses/sec; for 1 mV input, 100 pulses/sec, etc. These voltage

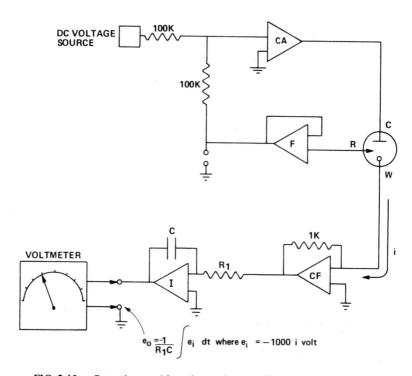

FIG. 2-45.   Potentiostat with analog coulometer. For label definitions, see text.

pulses are counted on an electronic counter. Knowing the relationship between the current and the counts accumulated by the counter, it is an easy matter to compute the total coulombs passed during the electrolysis.

K. Chronopotentiometry

It will be recalled that in chronopotentiometry a constant current is passed between the working electrode and an auxiliary electrode, and the resultant change in the potential of the working electrode versus a suitable reference electrode is recorded.

A typical chronopotentiometric circuit using operational amplifiers is shown in Fig. 2-47. It is seen that the potentiostatic circuitry described for earlier instruments is not present in the chronopotentiometric circuit. The heart of the circuit comprises the control amplifier CA, the DC voltage source, the auxiliary and working electrodes in the cell, and the resistor R_m. The components are analogous to those of the constant-current source shown in Fig. 2-38. The magnitude of the constant current flowing between the auxiliary and working

DIGITAL PULSE COUNTER AND TIMER

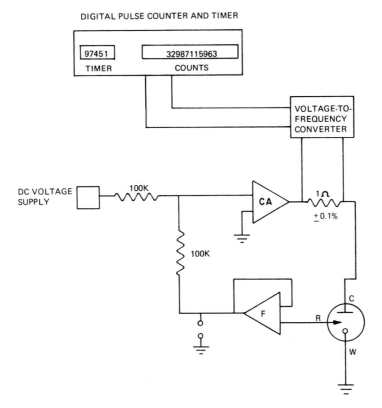

FIG. 2-46. Potentiostat with a digital coulometer. For label definitions, see text.

electrodes is decided by the magnitude of the DC voltage at P and the resistance R_m. Consider the case if P is set at 1 V and R_m at 100 Ω. Momentarily, the summing point of amplifier CA will be above virtual ground. The only way in which this voltage at the summing point can be eliminated is by an equal and opposite voltage being drawn from the output of follower F_2. Since the input and output of the follower must be equal and of the same polarity, clearly, the potential at point X must be -1 V. This can be achieved only by the control amplifier CA injecting sufficient current between the auxiliary and working electrodes and R_m and hence to ground, so that the potential at X is -1 V (i.e., $E_X = iR_m$). Thus, for the condition outlined above, the constant current through the cell will be 10 mA. The variation in potential between the working and reference electrodes is measured by a differential technique with the two followers F_1 and F_2. The voltage difference between the outputs of these two followers is fed to a recorder and its value versus time is traced; i.e., a chronopotentiogram is recorded.

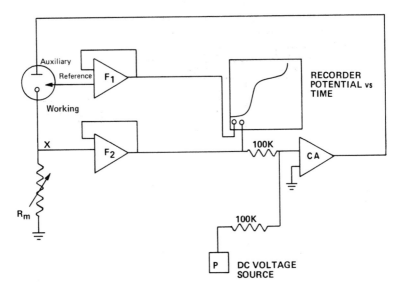

FIG. 2-47.   Typical operational amplifier circuit for chronopotentiometry. For label definitions, see text.

L. Other Instrumentation

The circuitry required for pulse polarography is rather too complex to discuss here. In general, the circuitry and techniques discussed in the preceding section reflect the methods employed by workers whose material is summarized in the ensuing chapters. The information gathered here is not intended to reflect all electrochemical techniques. For the purposes of giving the interested reader a guide to other electrochemical techniques and instrumentation, Supplementary Readings are presented at the end of the chapter.

REFERENCES

1. I. M. Kolthoff and J. J. Lingane, "Polarography," Vols. I and II. Wiley (Interscience), New York, 1952.
2. L. Meites, "Polarographic Techniques," 2nd ed. Wiley (Interscience), New York, 1965.
3. J. Heyrovský and J. Kůta, "Principles of Polarography." Academic Press, New York, 1966.
4. M. Brezina and P. Zuman, "Polarography in Medicine, Biochemistry and Pharmacy." Wiley (Interscience), New York, 1958.
5. P. Delahay, "New Instrumental Methods in Electrochemistry." Wiley (Interscience), New York, 1954.

6. P. Zuman, "Organic Polarographic Analysis." Pergamon, Oxford, 1963.
7. B. Breyer and H. H. Bauer, "Alternating Current Polarography and Tensammetry." Wiley (Interscience), New York, 1963.
8. D. E. Smith, *Electroanal. Chem.* 1, 1 (1966).
9. R. N. Adams, "Electrochemistry at Solid Electrodes." Dekker, New York, 1969.
10. V. G. Levich, "Physicochemical Hydrodynamics." Prentice-Hall, Englewood Cliffs, New Jersey, 1962.
11. I. Fried and P. J. Elving, *Anal. Chem.* 37, 464 and 803 (1965).
12. J. E. B. Randles, *Trans. Faraday Soc.* 44, 327 (1948).
13. A. Ševčík, *Collect Czech. Chem. Commun.* 13, 349 (1948).
14. R. S. Nicholson and I. Shain, *Anal. Chem.* 36, 706 (1964); 37, 178 and 191 (1965).
15. R. S. Nicholson, *Anal. Chem.* 37, 1351 (1965).
16. J. Heyrovský and R. Kalvoda, "Oszillographische Polarographie mit Wechselstrom." Akademie-Verlag, Berlin, 1960.
17. R. Kalvoda, "Techniques of Oscillographic Polarography," 2nd ed. Elsevier, Amsterdam, 1965.
18. G. C. Barker and A. W. Gardner, *Z. Anal. Chem.* 173, 79 (1960).
19. V. I. Boduy, I. V. Kotlova, and U. S. Lyapikov, *Zavod. Lab.* 28, 1042 (1963).
20. J. J. Lingane, "Electroanalytical Chemistry," 2nd ed. Wiley (Interscience), New York, 1966.
21. H. J. S. Sand, *Philos. Mag.* 1, 45 (1901).
22. P. Delahay and T. Berzins, *J. Am. Chem. Soc.* 75, 2486 (1953).
23. C. D. Russel and J. M. Peterson, *J. Electroanal. Chem.* 5, 467 (1963).
24. D. G. Davis, *Electroanal. Chem.* 1, 157 (1966).
25. C. N. Reilley, *J. Chem. Educ.* 39, A853 (1962).
26. C. N. Reilley, *J. Chem. Educ.* 39, A933 (1962).
27. G. L. Booman and W. B. Holbrook, *Anal. Chem.* 35, 1793 (1963).
28. W. M. Schwarz and I. Shain, *Anal. Chem.* 35, 1770 (1963).
29. M. J. D. Brand and B. Fleet, *Chem. Br.* 5, 557 (1969).
30. C. F. Morrison, "Generalized Instrumentation for Research and Teaching." Washington State University, Seattle, 1965.
31. H. V. Malmstadt, C. G. Enke, and E. C. Toren, "Electronics for Scientists." Benjamin, New York, 1963.
32. "Applications Manual for Computing Amplifiers." Philbrick Nexus Research, Inc., Dedham, Massachusetts.
33. "Handbook of Operational Amplifier Applications." Burr-Brown Research Corporation, International Airport Industrial Park, Tucson, Arizona.
34. L. P. Morgenthaler, "Basic Operational Amplifier Circuits for Analytical Chemical Instrumentation." McKee, Pedersen Instruments, Danville, California, 1967.
35. R. L. Myers and I. Shain, *Chem. Instrum.* 2, 203 (1969).

SUPPLEMENTARY READINGS

Potentiometry

W. M. Clarke, "Oxidation-Reduction Potentials of Organic Compounds." Williams & Wilkins, Baltimore, Maryland, 1960.
I. M. Kolthoff and N. H. Furman, "Potentiometric Titrations," 2nd ed. Wiley, New York, 1932.

I. M. Kolthoff and H. A. Laitinen, "pH and Electro-Titrations." Wiley, New York, 1941.
J. J. Lingane, "Electroanalytical Chemistry," 2nd ed. Wiley (Interscience), New York, 1966.
H. Rossotti, "Chemical Applications of Potentiometry." Van Nostrand-Reinhold, Princeton, New Jersey, 1969.

DC Polarography

M. Brezina and P. Zuman, "Polarography in Medicine, Biochemistry and Pharmacy." Wiley (Interscience), New York, 1958.
P. Delahay, "New Instrumental Methods in Electrochemistry." Wiley (Interscience), New York, 1954.
P. J. Elving, *in* "Organic Analysis" (J. Mitchell, ed.), Vol. II. Wiley (Interscience), New York, 1954.
C. A. Hample, "Encyclopedia of Electrochemistry." Van Nostrand-Reinhold, Princeton, New Jersey, 1964.
I. M. Kolthoff and J. J. Lingane, "Polarography," 2nd ed., 2 vols. Wiley (Interscience), New York, 1952.
I. S. Longmuir, ed., "Advances in Polarography," 3 vols., Proceedings of Second International Congress on Polarography, Cambridge, 1959. Pergamon, Oxford, 1960.
S. Mairanovskii, "Catalytic and Kinetic Waves in Polarography." Plenum, New York, 1968.
L. Meites, "Polarographic Techniques." Wiley (Interscience), New York, 1965.
G. W. C. Milner, "The Principles and Applications of Polarography and Other Electroanalytical Processes." Longmans, Green, New York, 1957.
O. H. Müller, "The Polarographic Method of Analysis." Chem. Educ. Publ. Co., Easton, Pennsylvania, 1951.
O. H. Müller, *in* "Physical Methods of Organic Chemistry" (A. Weissberger, ed.), 3rd ed., Part IV, pp. 3155–3279. Wiley (Interscience), New York, 1960.
P. Zuman, ed., "Progress in Polarography," 2 vols. Wiley (Interscience). New York, 1962.
P. Zuman, "Organic Polarographic Analysis." Pergamon, Oxford, 1963.
P. Zuman, "Substituent Effects in Organic Polarography." Plenum, New York, 1967.

AC Polarography

B. Breyer and H. H. Bauer, "Alternating Current Polarography and Tensammetry." Wiley (Interscience), New York, 1963.
P. Delahay, *Adv. Electrochem. Electrochem. Eng.* 1, 233 (1961).
M. Sluyters-Rehbach and J. H. Sluyters, *Electroanal. Chem.* 4, 1 (1969).
D. E. Smith, *Electroanal. Chem.* 1, 1 (1966).

Oscillopolarography

J. Heyrovský and R. Kalvoda, "Oszillographische Polarographie mitt Wechselstrom." Akademie-Verlag, Berlin, 1960.
M. Heyrovský and K. Micka, *Electroanal. Chem.* 2, 193 (1967).
R. Kalvoda, "Technique of Oscillographic Polarography," 2nd ed. Elsevier, Amsterdam, 1965.

Linear Sweep and Cyclic Sweep Voltammetry

R. N. Adams, "Electrochemistry at Solid Electrodes." Dekker, New York, 1969.

P. Delahay, "New Instrumental Methods in Electrochemistry." Wiley (Interscience), New York, 1954.

N. Kemula and Z. Kublick, *Adv. Anal. Chem.* 2, 123–178 (1963).

V. G. Levich, "Physicochemical Hydrodynamics." Prentice-Hall, Englewood Cliffs, New Jersey, 1962.

A. C. Riddiford, *Adv. Electrochem. Electrochem. Eng.* 4, 47 (1964).

Pulse Polarography

G. C. Barker and A. W. Gardner, *Z. Anal. Chem.* 173, 79 (1960).

V. I. Boduy, I. V. Kollova, and U. S. Lyapikov. *Zavod. Lab.* 28, 1042 (1963).

J. G. Osteryoung and R. A. Osteryoung, *Am. Lab.* 4, 8 (1972).

Controlled Potential Electrolysis and Coulometry

A. J. Bard and K. S. V. Santhanam, *Electroanal. Chem.* 4, 215 (1969).

J. J. Lingane, "Electroanalytical Chemistry," 2nd ed. Wiley (Interscience), New York, 1966.

L. Meites, *in* "Physical Methods of Organic Chemistry" (A. Weissberger, ed.), 3rd ed., Part IV, pp. 3281–3333. Wiley (Interscience), New York, 1960.

G. W. C. Milner and G. Philips, "Coulometry in Analytical Chemistry." Pergamon, Oxford, 1967.

G. A. Rechnitz, "Controlled Potential Electrolysis." Pergamon, Oxford, 1963.

Instrumentation

E. J. Bair, "Introduction to Chemical Instrumentation." McGraw-Hill, New York, 1962.

R. W. Landee, D. C. Davies, and A. P. Albrecht, "Electronic Designers' Handbook." McGraw-Hill, New York, 1957.

H. V. Malmstadt and C. G. Enke, "Digital Electronics for Scientists." Benjamin, New York, 1969.

H. V. Malmstadt, C. G. Enke, and E. C. Toren, "Electronics for Scientists." Benjamin, New York, 1962.

L. P. Morgenthaler, "Basic Operational Amplifier Circuits for Analytical Chemical Instrumentation." McKee, Pedersen Instruments, Danville, California, 1967.

C. F. Morrison, "Generalized Instrumentation for Research and Teaching." Washington State University, Seattle, 1965.

D. E. Smith, *Electroanal. Chem.* 1, 1 (1966).

H. A. Strobel, "Chemical Instrumentation." Addison-Wesley, Reading, Massachusetts, 1960.

General Theoretical and Practical Electrochemistry

R. N. Adams, "Electrochemistry at Solid Electrodes." Dekker, New York, 1969.

M. J. Allen, "Organic Electrode Processes." Van Nostrand-Reinhold, Princeton, New Jersey, 1958.

M. M. Baizer, ed., "Organic Electrochemistry; An Introduction and Guide." Dekker, New York, 1973.

A. J. Bard, ed., "Electroanalytical Chemistry." Dekker, New York, 1966 (continuing series of volumes).

H. H. Bauer, "Electrodics." Wiley, New York, 1972.

J. O'M. Bockris, ed., "Modern Aspects of Electrochemistry," vol. 1. Academic Press, New York, 1954 (continuing series).

J. O'M. Bockris and A. K. N. Reddy, "Modern Electrochemistry." Plenum, New York, 1970.

J. A. Butler, ed., "Electrical Phenomena at Interfaces." Macmillan, New York, 1951.

G. Charlot, ed., "Modern Electroanalytical Methods," Proceedings of the International Symposium of Modern Electrochemical Methods of Analysis, Paris, 1957. Am. Elsevier, New York, 1958.

G. Charlot, J. Badoz-Lambling, and B. Tremillon, "Electrochemical Reactions: The Electrochemical Method of Analysis." Am. Elsevier, New York, 1962.

B. E. Conway, "Electrochemical Data." Am. Elsevier, New York, 1951.

B. B. Damaskin, "The Principles of Current Methods for the Study of Electrochemical Reactions." McGraw-Hill, New York, 1967.

B. B. Damaskin, O. A. Petrii, and V. V. Batrakov, "Adsorption of Organic Compounds on Electrodes." Plenum, New York, 1971.

P. Delahay, "New Instrumental Methods in Electrochemistry." Wiley (Interscience), New York, 1954.

P. Delahay, "Double Layer and Electrode Kinetics." Wiley (Interscience), New York, 1965.

P. Delahay and C. W. Tobias, eds., "Advances in Electrochemistry and Electrochemical Engineering." Wiley (Interscience), New York, 1961 (a continuing series).

Faraday Society, "Electrode Processes," Discussions of the Faraday Society, No. 1. Faraday Soc., London, 1947.

I. Fried, "The Chemistry of Electrode Processes." Academic Press, New York, 1973.

A. J. Fry, "Synthetic Organic Electrochemistry." Harper, New York, 1972.

J. B. Headridge, "Electrochemical Techniques for Inorganic Chemists." Academic Press, New York, 1969.

N. S. Hush, ed., "Reactions of Molecules at Electrodes." Wiley (Interscience), New York, 1971.

D. J. G. Ives and G. J. Janz, "Reference Electrodes: Theory and Practice." Academic Press, New York, 1961.

I. M. Kolthoff and P. J. Elving, eds., "Treatise on Analytical Chemistry," Part I, Vol. 4. Wiley (Interscience), New York, 1963.

C. K. Mann and K. K. Barnes, "Electrochemical Reactions in Non-Aqueous Systems." Dekker, New York, 1970.

R. Parsons, "Handbook of Electrochemical Constants." Butterworth, London, 1959.

S. Swann, Jr., in "Technique of Organic Chemistry" (A. Weissberger, ed., 2nd ed., Vol. 2, p. 385. Wiley (Interscience), New York, 1956.

H. R. Thirsk and J. A. Harrison, "A Guide to the Study of Electrode Kinetics." Academic Press, New York, 1972.

E. Yeager and A. J. Salkind, eds., "Techniques of Electrochemistry." Wiley (Interscience), New York, 1972.

3

Purines

I. INTRODUCTION, NOMENCLATURE, AND STRUCTURE

Certain purines occur in every living cell, usually as constituents of large molecules, although in certain instances free purines are found in biological systems.

Purine itself (Fig. 3-1) consists of fused pyrimidine and imidazole rings. It was first named by Emil Fischer,[1] and the most widely employed numbering system, as shown in Fig. 3-1, is that of Fischer.[2] Robins[3] has prepared a very detailed account of the nomenclature of purines which should be consulted by the uninitiated. As pointed out by Robins,[3] a certain amount of confusion sometimes arises in purine nomenclature because of the trivial names assigned to many purine derivatives such as adenine (6-aminopurine), guanine (2-amino-6-oxypurine), and theophylline (1,3-dimethyl-2,6-dioxypurine). Further confusion arises from the fact that many different tautomeric formulas of certain purine derivatives can be written. Thus, occasionally, uric acid is written as 2,6,8-tri-hydroxypurine (**I**) (purine-2,6,8-triol; 2,6,8-purinetriol) or as 2,6,8-trioxypurine

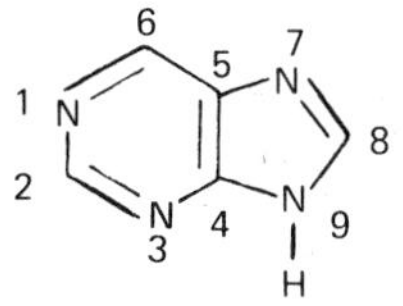

FIG. 3-1. Structure and numbering of purine.

71

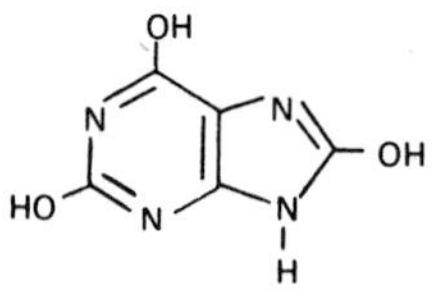

(I)

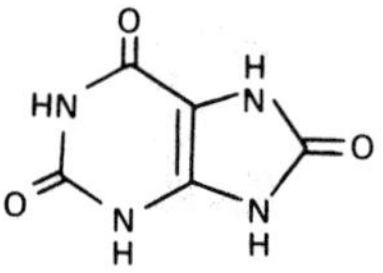

(II)

(II) [purine-2,6,8-trione; purine-2,6,8(1*H*,3*H*,9*H*)-trione]. Such a multiplicity of ways of naming these compounds can lead to misunderstanding. There is reasonably good evidence now that most hydroxypurines in fact exist in the keto form,[4-6] although it is still common practice to refer to such compounds as hydroxypurines and to write the incorrect enol structures when describing reactions of these compounds.

Table 3-1 contains a list of common purines along with their structures, trivial names, and a few of the chemical names of the compounds. From the information in this table it is clear that the chemical name commonly describes

TABLE 3-1   Nomenclature and Structure of Some Common Purines

Trivial name	Structure[a]	Typical chemical name[b]
Purine	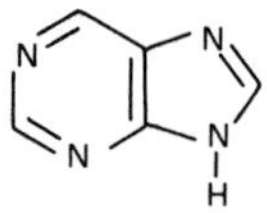	Purine
Hypoxanthine	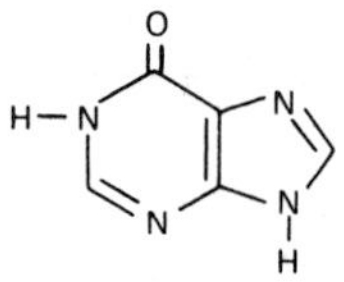	Purine-6-ol; purine-6(1*H*)-one
Xanthine	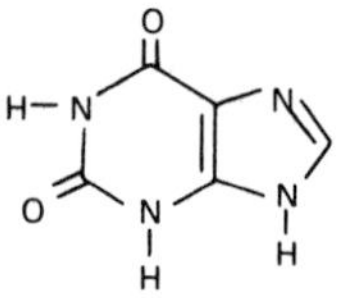	Purine-2,6-diol; 3,8-dihydro-1*H*-purinedione; 8*H*-purine-2,6(1*H*,3*H*)-dione
Uric acid	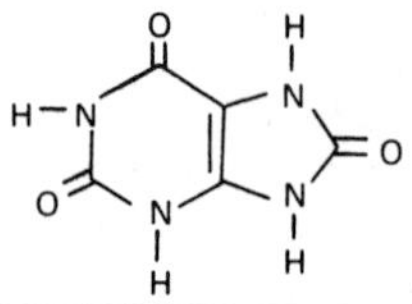	Purine-2,6,8-triol; 8-hydroxy-xanthine; purine-2,6,8 (1*H*,3*H*,9*H*)-trione

TABLE 3-1 *Continued*

Trivial name	Structure[a]	Typical chemical name[b]
Theophylline		1,3-Dimethylxanthine
Theobromine		3,7-Dimethylxanthine
Caffeine		1,3,7-Trimethylxanthine
Adenine		6-Aminopurine
Guanine		2-Amino-6-hydroxypurine; 2-aminohypoxanthine
6-Mercaptopurine		Purine-6-thiol

[a] Probable species existing in aqueous solution; acid–base equilibria are neglected.

[b] As found, for example, in *Chemical Abstracts.*

the structure of a species that does not predominate in solution. In order to simplify nomenclature in this chapter, either trivial names or other nomenclature that more accurately describes the proper structure of the species will be employed. For example, structure **III** will be described as 1,3,7-trimethyluric acid, while **IV** will be described as 8-oxyguanine.

(III)   (IV)

It is more usual to find purines in biological systems as somewhat higher molecular weight derivatives. The purine *nucleosides*, for example, are carbo-hydrate derivatives of purines in which the purine is linked through its N-9 position via a β-*N*-glycosidic bond to either D-ribose or 2-deoxy-D-ribose. As a result of the two types of sugar moity associated with purines in the nucleosides, there are two types of nucleosides: the ribonucleosides and the deoxy-ribonucleosides. Adenine linked to ribose is called adenosine (Fig. 3-2A), while the guanine nucleoside is called guanosine (Fig. 3-2B). It will be noted from Fig. 3-2A that the carbon atoms of the sugar moity are designated by prime numbers in order to distinguish them from the atoms of the purine ring system. The sugar is normally linked through the hydroxyl group originally at C-1′. The analogous adenine and guanine nucleosides formed with deoxyribose are named deoxyadenosine (Fig. 3-2C) and deoxyguanosine (Fig. 3-2D).

Purine nucleotides are the phosphate esters of the nucleosides and are strong acids. Although the esterification occurs primarily at the C-5′ hydroxyl group of the ribose or deoxyribose, in fact a ribonucleoside can alternatively be esterified at hydroxyl groups at the C-2′ and C-3′ positions, while deoxynucleosides can be alternatively esterified at the C-3′ hydroxyl. All of these nucleotides exist in nature. The commonly used nomenclature for nucleotides is derived from the trivial names of the appropriate nucleoside along with the position of the phosphate esterification and the number of attached phosphate groups. The structures of adenosine-5′-monophosphate (AMP) and deoxyadenosine-5′-monophosphate (dAMP) are shown in Fig. 3-3. The purine nucleotides are commonly referred to as adenylic acid, deoxyadenylic acid, guanylic acid, or deoxyguanylic acid.

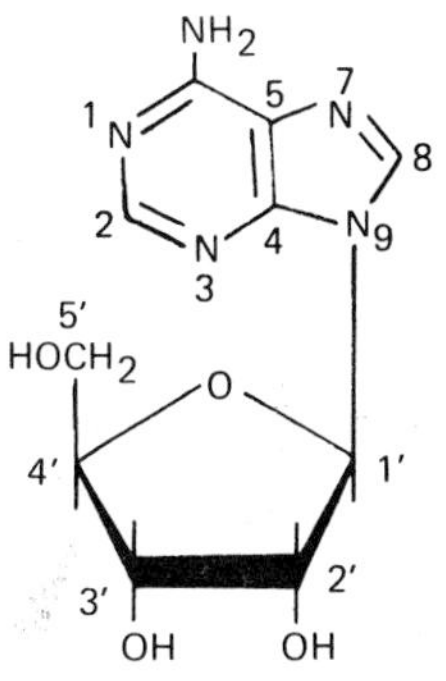

A: Adenosine

(9-β-D-ribofuranosyladenine)

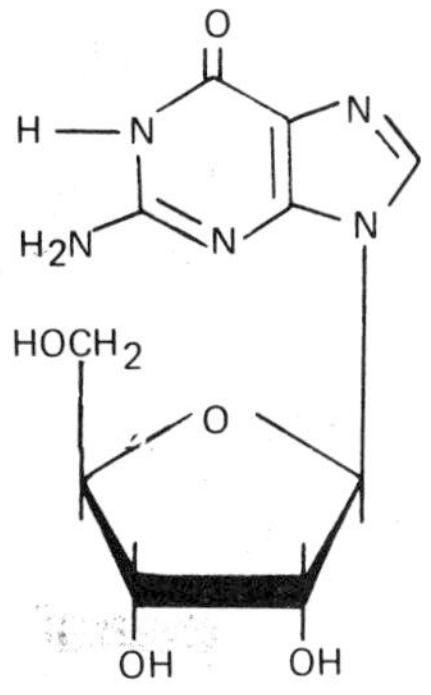

B: Guanosine

(9-β-D-ribofuranosylguanine)

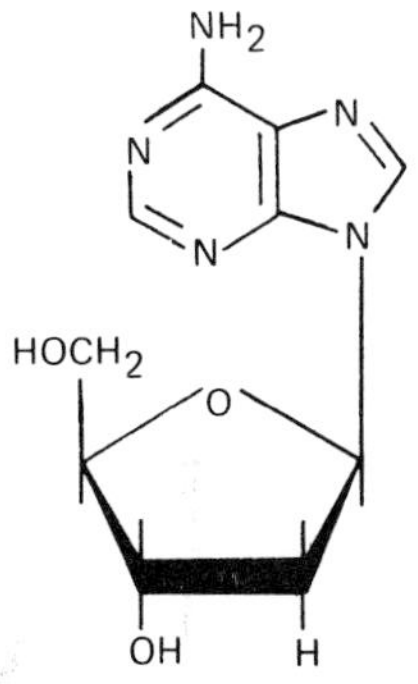

C: Deoxyadenosine

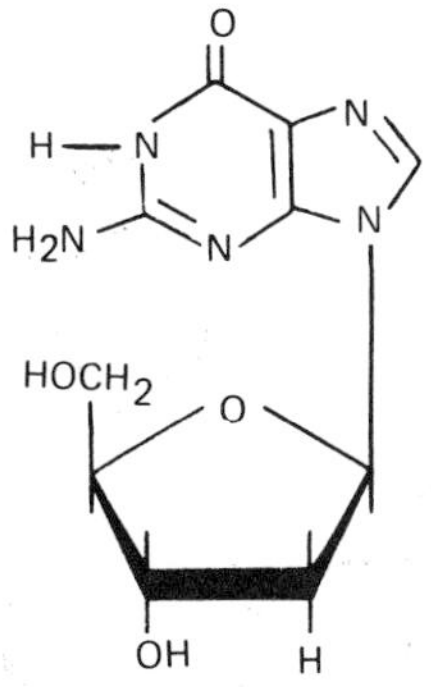

D: Deoxyguanosine

FIG. 3-2. Structure of adenine and guanine nucleosides and deoxynucleosides.

Two derivatives of adenylic acid, adenosine-5'-diphosphate (ADP) (Fig. 3-4) and adenosine-5'-triphosphate (ATP) (Fig. 3-4), are extremely important in intermediary metabolism. Their structures serve to illustrate the general form of the nucleoside polyphosphate esters.

In conjunction with the pyrimidine nucleotides the purine nucleotides can form highly polymerized structures called the nucleic acids, which are discussed in Chapter 5.

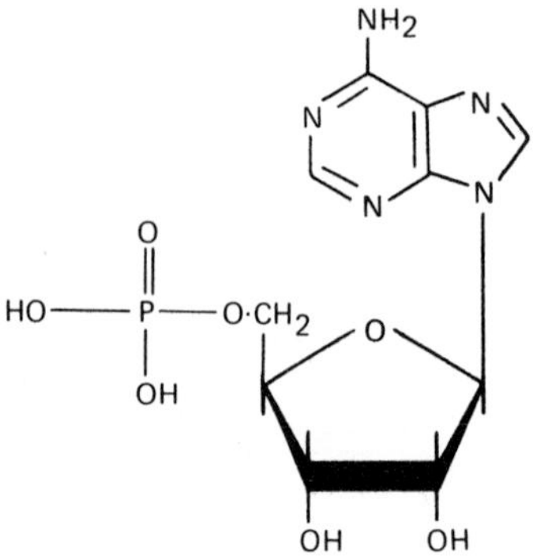

Adenosine − 5′ − monophosphate

(AMP or adenylic acid)

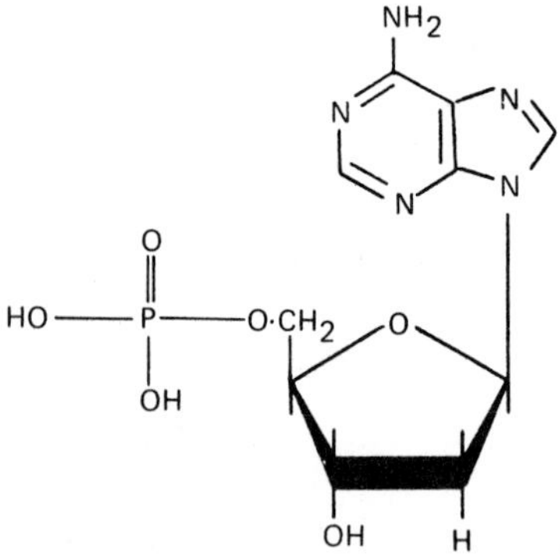

Deoxyadenosine − 5′−monophosphate

(dAMP or deoxyadenylic acid)

FIG. 3-3.   Structure of adenine nucleotides.

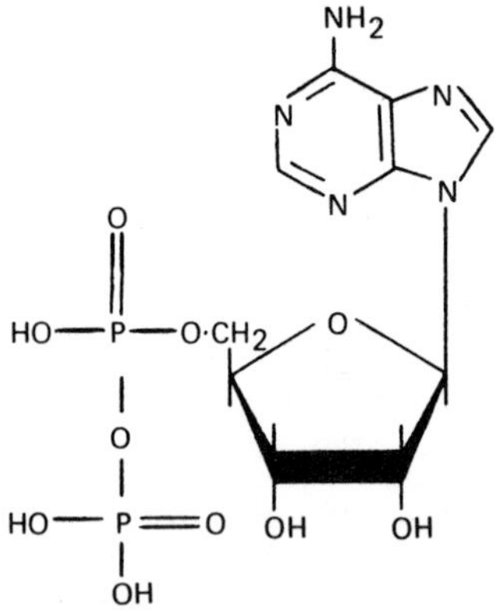

Adenosine −5′− diphosphate

(ADP)

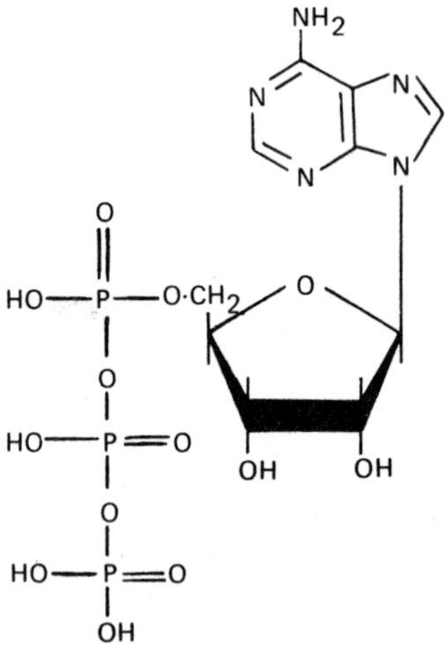

Adenosine − 5′ − triphosphate

(ATP)

FIG. 3-4.   Structure of adenosine polyphosphates.

II. PHYSICAL PROPERTIES OF PURINE DERIVATIVES

In general, purine and its derivatives are a very stable class of compounds. In view of their involvement in biological processes, it is somewhat surprising to note the rather low solubility of many purine derivatives. Information regarding the solubility of purines has been adequately presented elsewhere.[4,7,8-10] Comprehensive dissociation constant data have been prepared by Albert[11] and others[12-22] and these along with the references contained in these reports should be consulted for further information. The ultraviolet absorption spectra of purines have been extensively studied.[4,11,23-28] A rather detailed review of adenosine and adenine nucleotides has been prepared by Phillips.[29] The infrared spectra (IR) of purines have also been studied in some detail.[6,30-35] Of particular interest is the fact that IR evidence in the solid state confirms that hydroxypurines exist predominantly in the keto form.

III. OCCURRENCE AND BIOLOGICAL SIGNIFICANCE OF PURINE DERIVATIVES

Probably the first purine to be discovered was uric acid (2,6,8-trioxypurine); it was discovered in 1776, by Scheele,[36] who found the compound to be a constituent of human urine and of urinary calculi. Uric acid is also found in blood, cerebral spinal fluid, and human and animal milk. Bird excrement, guano, contains a considerable quantity of uric acid. Robins[3] has reviewed these and other natural sources of uric acid and other purines in great detail.

Uric acid and other oxypurines are the principal final products of purine metabolism (catabolism) in man,[37,38] although in many organisms further degradation of the purine molecule occurs. The disease gout is caused primarily by an overproduction of uric acid which, being extremely insoluble, tends to precipitate in joints.

Adenine (6-aminopurine) is one of the two purines most commonly found in ribonucleic acid (RNA) and deoxyribonucleic acid (DNA) and so is intimately involved in the direction of protein synthesis and in the transfer of genetic information. Adenosine triphosphate is one of a group of so-called energy-rich compounds because it exhibits a large decrease in free energy when it undergoes certain hydrolytic reactions. Thus, hydrolysis of ATP to ADP with liberation of a molecule of phosphoric acid (Eq. 1) involves a standard free-energy change, ΔG, of about -8000 cal/mol at pH 7. Under similar physiological conditions the hydrolysis of ADP to AMP is slightly less energetically favorable ($\Delta G = -6500$ cal/mole at pH 7), while hydrolysis of AMP to adenosine and phosphoric acid is considerably less energetically favorable ($\Delta G = -2200$ cal/mole at pH 7).

$$\Delta G = -8000 \text{ cal/mole at pH7} \quad (1)$$

$$(2)$$

Because of the very favorable free-energy change, particularly in the ATP → ADP reaction, ATP is involved in very many biological reactions such as, for example, enzymatic group transfers. Thus, the initial step in the utilization of glucose by many organisms is its phosphorylation by ATP to yield glucose 6-phosphate (Eq. 2). A straightforward and fairly comprehensive description of the biological action of the adenine and other nucleotides has been presented by Conn and Stumpf.[39]

Adenine also occurs as a component of a number of coenzymes. Coenzyme I is nicotinamide adenine dinucleotide (NAD^+) or diphosphopyridine nucleotide (DPN^+). Coenzyme II is nicotinamide adenine dinucleotide phosphate ($NADP^+$) or triphosphopyridine nucleotide (TPN^+). Flavin adenine dinucleotide (FAD) and coenzyme A (pantothenic acid) also contain adenine residues. Coenzymes of these types are often involved in oxidation—reduction processes in biological reactions in conjunction with the appropriate enzyme and therefore have been of particular interest to electrochemists, who have been able to study their electron-transfer properties *in vitro*. The electrochemistry and structures of many of these compounds are discussed in later chapters. Other coenzymes of this type are known; extensive reviews should be consulted for details.[40-44] Adenine is also an important constituent of certain antibiotics such as angustmycin A (V) and C (VI).[3,45]

(V)   (VI)

Guanine is the second purine most commonly found in nucleic acids. The guanine nucleotides, like the adenine nucleotides, are involved in intermediary metabolism, but to a much lesser extent. Guanine is found in a variety of animal excrements and tissues.[3] Some of the biochemical reactions of guanine will be outlined in subsequent discussions.

Xanthine, discovered by Marcet[46] as a constituent of bladder stones, has since been found in tea,[47] cow's milk,[48] human urine,[49,50] and other miscellaneous biological sources.[3] In early studies on the degradation of nucleic acids, xanthine was found to be among the products, along with adenine and guanine. However, it has since been shown that xanthine and hypoxanthine are in fact probably degradation products of the primary nucleic acid purines, adenine and guanine.[3,51] Hypoxanthine (6-oxypurine) is similarly found in a variety of animal tissues, body fluids, and excrement.[3]

N-Methylated xanthines are found extensively in nature. Caffeine (1,3,7-Trimethylxanthine) was discovered by Runge[52] in coffee. Woskresensky[53] found theobromine (3,7-dimethylxanthine) in cocoa beans, and Kossel[54] extracted theophylline (1,3-dimethylxanthine) from tea leaves. 1-Methylxanthine and 7-methylxanthine are common xanthine constituents found in human urine.[55] These, along with 1,7-dimethylxanthine and 3-methylxanthine, are reportedly present in human urine only after intake of coffee and/or large doses of theobromine, theophylline, and caffeine.[56] Several of the methylated xanthines have pharmacological significance. Thus, 1,7-dimethylxanthine possesses marked antithyroid activity.[57] The methylated xanthines are extensively used as diuretics[58]; the relative diuretic activity has been reviewed by Haas.[59]

Caffeine is noted for its stimulation of the central nervous system, as are theobromine and theophylline to a somewhat lesser extent.[60] As a result of both the stimulation of the central nervous system and the stimulation of the heart and vascular system, theophylline is used clinically to increase cardiac output and to alleviate bronchial asthma.[58]

Certain synthetic purines are of considerable interest in cancer chemotherapy. For example, 6-thiopurine (6-mercaptopurine) is one of the most effective drugs available for the treatment of a number of types of leukemia and related neoplastic conditions.[61] A discussion of the therapeutic action of 6-thiopurine has been presented by Murray and co-workers.[62] A number of other thiopurines have been synthesized with a view to developing further antitumor drugs.

A substantial number of other purine derivatives have been employed as drugs, but a detailed discussion of these and their biological effects is beyond the scope of this chapter. A good guide to the natural sources of purines and synthetic purines and their physical and chemical properties is presented by Robins.[3]

It is quite apparent that purines in one way or another are vital components of all living cells, being intimately involved in protein synthesis, storage and transfer of genetic information, and intermediary metabolism, and as constituents of certain coenzymes involved in a variety of oxidation—reduction and electron-transfer processes, as drugs, and as commonly encountered constituents of biological tissues and fluids.

IV. ELECTROCHEMISTRY OF PURINE DERIVATIVES

A. Electrochemical Reduction

1. Historical Background:

In view of the importance of purines in biological processes and the considerable number of biological, biochemical, and clinical studies of these compounds, it is somewhat surprising to find that the first detailed electrochemical investigation of purines themselves was not reported until 1946 and that a really definitive study was not made until the 1960's. The first report of the DC polarography of purines appears to be that of Pech,[63] who found that adenine was reducible at the dropping mercury electrode (DME) in solutions of pH 1.3–2.2. Heath[64] in 1946 also reported that adenine was polarographically reducible and gave rise to a fairly well-formed polarographic wave in acidic solution. By using a plot of E_{DME} versus $\log(i_d - i)/i$, Heath proposed an n value of 1.33, although the ratio of the limiting current to concentration suggested a much larger electron number. Heath[64] also indicated that the reduction of adenine probably involved the N-1=C-6 double bond, although no experimental data to support this view were presented. Subsequent work by Hamer, Waldron, and Woodhouse[65] confirmed the polarographic reducibility of adenine in acidic solution and also indicated that hypoxanthine gave a well-formed polarographic wave, while xanthine and guanine did not. These latter workers found that the solution obtained after prolonged electrolysis of adenine gave positive tests for ammonia (Nessler reagent)[66] and formaldehyde (chromotropic acid reagent)[67] and that the solution contained a diazotizable constituent (Bratton–Marshall reagent).[68] McGinn and Brown[69] also briefly studied the polarography of adenine and some purine *N*-oxides and reported that adenine itself was not reducible above about pH 2–3 (now known to be incorrect), although over the pH range where the wave was observed, the half-wave potentials $(E_{1/2})$ agree fairly well with those of Heath.[64] Palaček[70] also briefly reported some polarographic data on adenine.

2. Purine

In 1962 Smith and Elving[71] reported the first systematic and detailed study of the electrochemical reduction of purine, adenine, and several related compounds. These workers employed polarography at the DME, coulometry and macroscale electrolysis at a large mercury pool electrode, and a spectro-photometric, chemical, and polarographic study of the electrochemical reaction products. Purine itself was reduced in two $2e$ stages i.e., via two separate and distinct diffusion-controlled polarographic waves (Fig. 3-5). The first pH-dependent wave (Table 3-2) involved reduction of the N-1=C-6 double bond of purine (I, Fig. 3-6) in an overall irreversible[72] $2e$ process to give 1,6-dihydropurine (II, Fig. 3-6). The latter compound was found to be rather

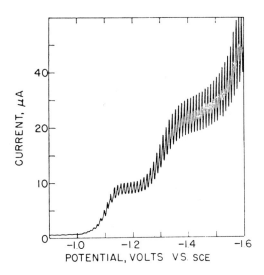

FIG. 3-5. Polarogram of 1.00 m*M* purine in acetate buffer, pH 4.7.[71] [Reprinted with permission from P. J. Elving, *J. Am. Chem. Soc.* **84**, 1412 (1962). Copyright 1962 by the American Chemical Society.]

TABLE 3-2   Half-Wave Potentials for Purine DC Polarographic Reductions at DME in Aqueous Solutions

Name	Wave[a]	pH Range	$E_{1/2}$ (V versus SCE)	Reference
Purine	I	0 to ca. 11[b]	$-0.697 - 0.083$ pH	71, 72
	II	0 to ca. 11	$-0.902 - 0.080$ pH	71, 72
Adenine	I	1–6	$-0.975 - 0.090$ pH	71
Hypoxanthine	I	5.7[c]	-1.61	71
Guanine	NR[d]			
6-Methylpurine	I	1.0–3.9	$-0.820 - 0.079$ pH	73
		3.9–6.0	$-0.745 - 0.091$ pH	73
	II	2.5–6.0	$-0.915 - 0.082$ pH	73
		3.9–5.9	$-0.765 - 0.162$ pH	73
	III	6.0–7.4	$-0.785 - 0.095$ pH	73
		7.4–7.8	$-0.080 - 0.209$ pH	73
6-Methoxypurine	I	2.5–4.2	$-0.825 - 0.105$ pH	73
		4.2–5.5	$-0.535 - 0.174$ pH	73
6-Methylaminopurine	I	1.0–6.5	$-0.995 - 0.081$ pH	73
6-n-Hexylaminopurine	I	1.0–2.5	$-0.995 - 0.076$ pH	73
		2.5–3.7	$-1.105 - 0.047$ pH	73
		3.7–6.5	$-0.995 - 0.076$ pH	73
6-Benzylaminopurine	I	2.0–4.8	$-0.995 - 0.067$ pH	73
		4.8–6.4	$-0.805 - 0.106$ pH	73
6-Phenylaminopurine	I	2.5–4.7	$-0.915 - 0.072$ pH	73
		4.7–7.9	$-0.640 - 0.131$ pH	73
6-Dimethylaminopurine	I	2.0–4.5	$-1.025 - 0.068$ pH	73
		4.5–6.4	$-0.930 - 0.089$ pH	73
2-Oxy-6-aminopurine	I	1.0–4.6	$-0.990 - 0.072$ pH	73
(isoguanine)		4.6–7.2	$-0.820 - 0.109$ pH	73
		7.2–9.6	$-1.210 - 0.055$ pH	73
		8.5–9.1	$-0.705 - 0.104$ pH	73
6-Thiopurine	I	0–5	$-0.79 - 0.116$ pH	84
(6-mercaptopurine)	II	0–2.3	$-1.00 - 0.048$ pH	84
	III	5–8	$-1.29 - 0.027$ pH	84
	IV	9.1[e]	-1.74	84
Purine-6-sulfinic	I	1–9.1	$-0.37 - 0.094$ pH	84
acid	II	8–12.3	$-0.79 - 0.075$ pH	84
	III	3.6–9	$-0.99 - 0.080$ pH	84
Purine-6-sulfonic	I	1–7	$-0.45 - 0.078$ pH	84
acid	II	3.6–12.5	$-0.675 - 0.079$ pH	84
	III	1–9	$-0.98 - 0.064$ pH	84
	IV	9.1[e]	-1.45	84
Purine-2,6-disulfonic	I$_a$[f]	0–4.75	$-0.53 - 0.103$ pH	84
acid	II	0–3.5	$-0.68 - 0.084$ pH	84
	III	0–3.5	$-0.82 - 0.077$ pH	84
	IV$_a$	0–4.75	$-1.01 - 0.062$ pH	84
	I$_b$	6.8–12.75	$-0.89 - 0.058$ pH	84
	IV$_b$	6.8–10.7	$-0.86 - 0.086$ pH	84

TABLE 3-2 *Continued*

Name	Wave[a]	pH Range	$E_{1/2}$ (V versus SCE)	Reference
	V	3.7–4.75	$-0.53 - 0.133$ pH	84
	VI	5.2–6.55	$-0.73 - 0.102$ pH	84
2-Thiopurine	I	1–9	$-0.455 - 0.102$ pH	88
	II	1–4.7	$-0.730 - 0.067$ pH	88
	III	5–9	$-1.05 - 0.060$ pH	88
2,6-Dithiopurine	I	1–5.7	$-0.79 - 0.076$ pH	89
	II	4–8	$-0.97 - 0.102$ pH	89
Adenine 1-*N*-oxide	I	1 to ca. 4	$-0.738 - 0.072$ pH	92
	II	1–5.5	$-0.975 - 0.090$ pH	92
2,6-Diaminopurine	I	1.5[g]	-1.14	64, 69
		0.1 *M* HClO$_4$	-1.40[h]	77
2,6-Diaminopurine 1-*N*-oxide	I	1.5[g]	-1.1	69
6-(*N*-Acetylamino)purine (acetyladenine)	I	0.1 *M* HClO$_4$	-0.85	77
	II		-1.00	77
2,6-Bis(*N*-acetylamino)purine (2,6-diacetyladenine)	I	0.1 *M* HClO$_4$	-1.27[h]	77
	II		-1.50[h]	77

[a] The most positive wave is numbered I; wave II is the next wave to appear at more negative potential; and so on.

[b] Although Smith and Elving[71] reported that purine did not show any polarographic waves above ca. pH 6, subsequent work[72] showed that both waves could be observed up to at least pH 9.

[c] Very indistinct waves were observed at lower pH.

[d] Not polarographically reducible.

[e] Observed at only this pH in ammonia buffer.

[f] Waves I and IV were subdivided into I$_a$, I$_b$ and IV$_a$, IV$_b$, respectively, on the basis of change in the slope of $dE_{1/2}/d$(pH) below and above pH 6 and because the ultimate products of the a and b waves were the same.

[g] An indistinct wave observed at pH 2.3; at higher pH not reducible.

[h] Versus mercury pool anode.

susceptible to air oxidation back to purine. This reaction was thought to proceed by formation of 1-hydroxy-6-hydropurine (II$_a$, Fig. 3-6), which subsequently dehydrates to regenerate the stable purine ring. It was also found that the wave I product in solution, i.e., 1,6-dihydropurine (II, Fig. 3-6), slowly decomposed even in the absence of oxygen to give a yellow-colored, electrochemically inactive species. The nature of this species is not known.

Dryhurst and Elving[72] studied the cyclic voltammetry of purine at the pyrolytic graphite electrode (PGE). At the PGE, purine exhibits only a single voltammetric reduction peak, which corresponds to the first polarographic wave

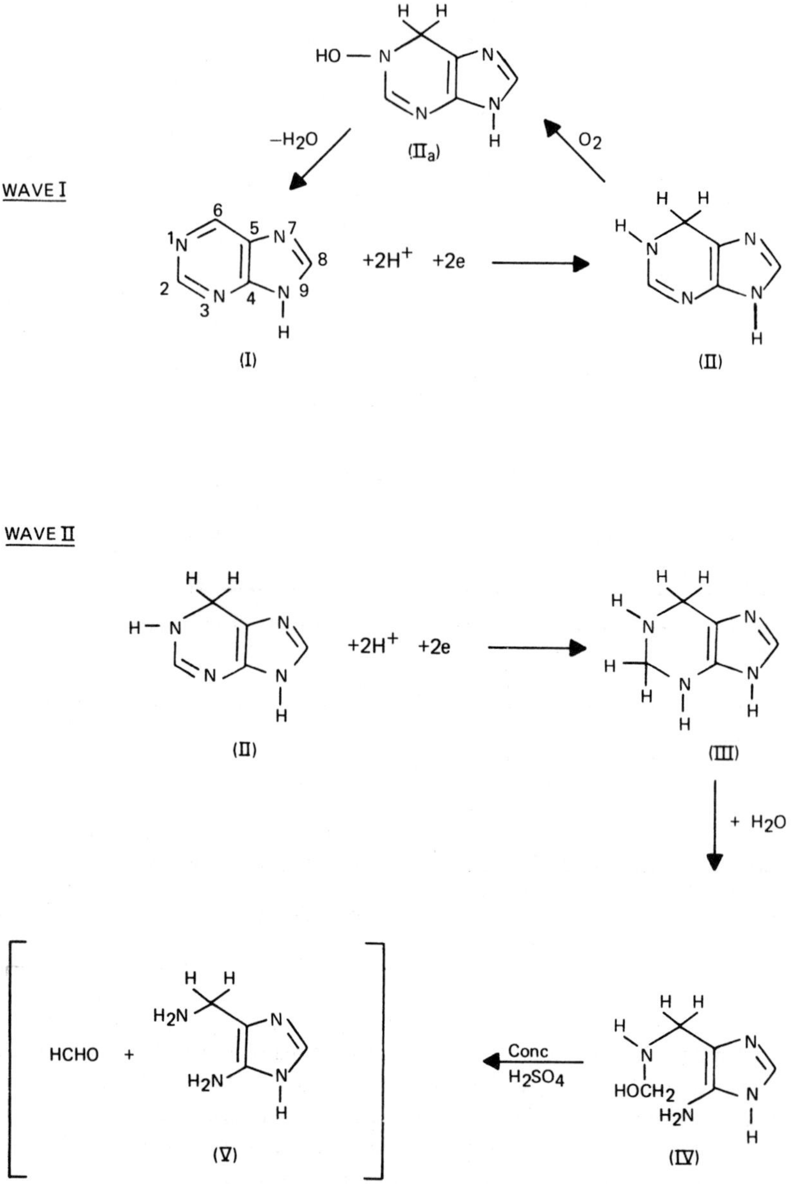

FIG. 3-6.   Reaction scheme for electrochemical reduction of purine.[71] [Reprinted with permission from D. R. Smith and P. J. Elving, *J. Am. Chem. Soc.* **84**, 1412 (1962). Copyright 1962 by the American Chemical Society.]

of purine observed at the DME. The reduction peak observed at the PGE is strongly pH dependent (Table 3-3). Of more interest, however, is the fact that, having once scanned this peak, on sweeping to more positive potential, one observes a well-formed pH-dependent (Table 3-3) oxidation peak. This peak does not appear if the initial voltammetric scan is toward positive potential; i.e., it is necessary to scan to potentials sufficiently negative to carry out the reduction process before the oxidation peak appears. A typical cyclic voltammogram is presented in Fig. 3-7. The pH dependence of these reduction and oxidation peaks observed at the PGE is presented in Table 3-3. The anodic peak was proposed to be due to reoxidation of the product of the cathodic process, i.e., of 1,6-dihydropurine (II, Fig. 3-6) back to purine. This latter study therefore confirmed the nature and ease of oxidation of the product of the purine polarographic wave I. It is noteworthy that above pH 9 no reduction peak could be observed on cyclic voltammetry at the PGE since it is masked by background discharge. However, the oxidation peak could still be observed, provided that the negative-going sweep was continued beyond the background discharge potential.[72]

TABLE 3-3 Linear E_p versus pH Relationships for Purine on Cyclic Voltammetry at PGE and HMDE[a]

Electrode	Scan rate (V sec^{-1})	pH Range	Peak	E_p (V versus SCE)
PGE	0.06	0–12	I[b]	$-0.84 - 0.077$ pH
			I$_a$[c]	$-0.88 - 0.089$ pH
	0.6	0–12	I[b]	$-0.80 - 0.083$ pH
HMDE[d]	0.026	0–9	I[b]	$-0.72 - 0.085$ pH
			II[b]	$-1.02 - 0.075$ pH
	0.26	0–9	I[b]	$-0.74 - 0.086$ pH
			II[b]	$-1.10 - 0.064$ pH

[a] Data from Dryhurst and Elving.[72] Purine concentration, 0.99 mM.
[b] Cathodic (reduction) peak.
[c] Anodic (oxidation) peak.
[d] Hanging mercury drop electrode.

The second pH-dependent polarographic wave of purine (Table 3-2) is a further two-electron, two-proton reduction of 1,6-dihydropurine to 1,2,3,6-tetrahydropurine (III, Fig. 3-6). Since the product of a large-scale controlled potential electrolysis of purine at potentials corresponding to wave II gave a positive test with Bratton—Marshall reagent,[68] i.e., it was diazotizable, and since under the conditions of the chromotropic acid reaction[67] formaldehyde was detected, the primary product of the wave II process, 1,2,3,6-

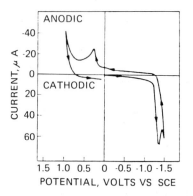

FIG. 3-7.   Cyclic voltammogram of 1 mM purine in pH 7.0 McIlvaine buffer at the PGE.[72] Scan rate, 0.06 V sec^{-1}; scan pattern, 0.0 V → −1.5 V → 1.0 V → 0.0 V. (Reprinted with the permission of Pergamon Press, Inc., New York.)

tetrahydropurine, was thought to hydrolyze to [(5-aminoimidazol-4-ylmethyl)amino] methanol (IV, Fig. 3-6). Under the concentrated sulfuric acid conditions of the chromotropic acid reaction, IV (Fig. 3-6) would be expected to give rise to formaldehyde and 4(or 5)-amino-5(or 4)-(aminomethyl)imidazole (V, Fig. 3-6). Subsequent studies at the stationary hanging mercury drop electrode (HMDE) and the PGE[72] have confirmed the basic mechanism of purine reduction. The two well-formed and very reproducible cathodic peaks produced by purine on cyclic voltammetry at the HMDE gave no evidence for reversibility of the processes. Although linear peak current versus concentration (i_p–C) relationships were observed for both peaks, the i_p/AC ratio [A is electrode area, (millimeters)^2] for peak II tended to decrease slightly with increasing concentration, which suggested the involvement of adsorption in the peak II process. At the PGE the peak current for the single cathodic peak of purine and for the resulting anodic peak observed on cyclic voltammetry (Table 3-3) showed marked decreases in the i_p/AC ratio, and indeed under fairly fast sweep rate conditions the i_p–C plots became almost exponential, with i_p approaching a limiting value at the 1 mM concentration level. Such behavior is indicative of the adsorption of purine or its reduction product, although at high scan rates current limiting effects due to slow chemical processes, e.g., structural rearrangements, or to diffusion-layer limitations could be contributing factors.

Dryhurst and Elving[72] briefly studied the AC polarography of purine and found that two reduction peaks appeared in the potential range from 0.0 to −1.8 V (Fig. 3-8). Although DC polarography indicated that both the first and second reduction steps of purine were overall 2e processes, the AC polarogram showed a much sharper and larger peak for the first peak (equivalent to polarographic wave I) than for the second, more negative peak. This is no doubt a reflection of the relative electron-transfer rates (reversibilities) for the two

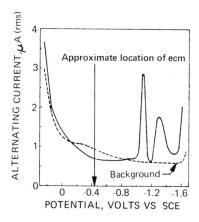

FIG. 3-8. Alternating current polarogram of 2 mM purine in pH 4.6 acetate buffer.[72] Frequency, 50 Hz; amplitude, 4 mV (rms). (Reprinted with the permission of Pergamon Press Inc., New York.)

processes. Both AC polarographic peaks showed a linear pH dependence of E_s (summit potential of the AC peak) between pH 1 and 6 (Table 3-4) and a linear i_p-C relationship at pH 4.6. Purine depresses the AC base current in the region of the electrocapillary maximum (ecm) (i.e., potential of zero charge on the DME), as can be seen in Fig. 3-8. A small but reproducible elevation of the base current occurred at potentials immediately before the appearance of the first AC peak (peak I). Electrocapillary studies also support the fact that in moderately acidic solutions, e.g., pH 4.7 acetate buffer, purine is adsorbed at the electrode in the region of the ecm. Such information supports the view that an uncharged purine species is adsorbed at the electrode. The small base current elevation immediately prior to the appearance of AC peak I may be indicative of desorption of purine or rearrangement on the electrode surface.

Elving and co-workers reexamined the details of the mechanism for the electrochemical reduction of purine. As described earlier, purine is electrochemically reduced at mercury electrodes in two pH-dependent processes, corresponding to the overall irreversible reduction of the 1,6 and 3,2 N=C bonds

TABLE 3-4   Linear E_s versus pH Relationships for Purine on AC Polarography[a]

Concentration (mM)	pH Range	Peak	E_s (V versus SCE)
0.96	1–6	I	$-0.72 - 0.080$ pH
		II	$-1.11 - 0.041$ pH

[a] Data from Dryhurst and Elving.[72]

(see Figs. 3-5 and 3-6) with two electrons and two protons being added to each bond. By use of AC and phase selective AC polarography, phase selective second-harmonic AC polarography, fast sweep cyclic voltammetry, and related electrochemical techniques, Elving *et al.*[72a] concluded that each reduction stage of purine consisted of a very rapid preprotonation of the nitrogen atom at the electroactive site. Thus, in the case of purine reduction wave I a very rapid protonation of N-1 occurs (I → II, Fig. 3-9). Then, two rapid, successive one-electron transfers take place. Addition of the first electron to protonated purine gives a radical (III, Fig. 3-9). Rapid addition of the second electron produces a carbanion (IV, Fig. 3-9). Protonation of the carbanion, giving 1,6-dihydropurine (V, Fig. 3-9), in an irreversible step results in the overall irreversibility of the electrode process. Alternating current polarographic techniques and particularly fast sweep cyclic voltammetry indicate the existence of a very short-lived electrode product that can be reoxidized to protonated purine (II, Fig. 3-9) in a "reversible" process, i.e., species III or more probably IV (Fig. 3-9). The wave II process has been proposed to proceed by an identical mechanism (Fig. 3-9) except that the electrochemical reaction involves the C-2=N-3 double bond. The heterogeneous rate constants ($k_{s,h}$) for the electron-transfer steps for purine waves I and II were calculated to be in excess of 0.1 cm sec^{-1}. The pseudo-first-order homogeneous rate constant, activation energies, and entropies for the follow-up protonation reactions were calculated to be 4.9 x 10^3 sec^{-1}, 1.8 kcal mole^{-1}, and −40 e.u. respectively, for the first reduction process (i.e., IV → V, Fig. 3-9) and 1.5 x 10^4 sec^{-1}, 1.3 kcal mole^{-1}, and −40 e.u., respectively, for the second (i.e., VII → IX, Fig. 3-9).

3. Adenine

Smith and Elving[71] found that adenine gives a single, large, pH-dependent, and largely diffusion-controlled polarographic wave (Table 3-2). In view of the fact that the polarographic reduction of adenine does not occur above pH ca. 6 and the pK_a of adenine is 4.25 (for gain of a proton), it is probably the protonated form of adenine that is electrochemically reducible. Coulometry revealed that a total of six electrons were involved in the complete reduction of adenine, and spectrophotometric and chemical investigation of the product solution revealed that ammonia and the same products obtained from complete reduction of purine were present. On this basis, therefore, the polarography of adenine (I, Fig. 3-10) was proposed to involve a primary potential controlling reduction of the N-1=C-6 double bond in the same way as observed for purine to give 1,6-dihydro-6-aminopurine (II, Fig. 3-10). It was not possible[71] to precisely identify the sequence of steps following the reduction of the N-1=C-6 double bond, but the formation of ammonia, the 6e nature of the wave, and the identical nature of the purine wave II and adenine reduction products indicate

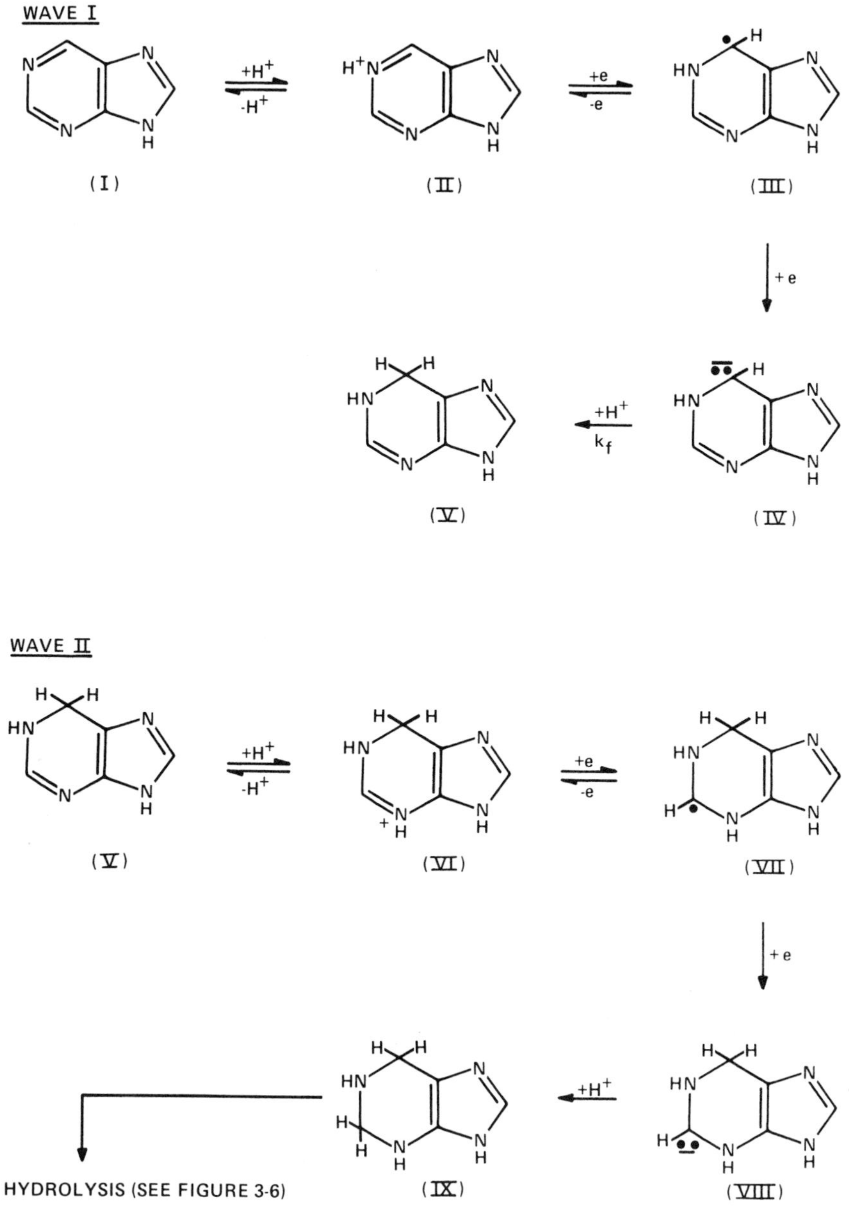

FIG. 3-9. Mechanism for the electrochemical reduction of purine according to Elving *et al.*[72a]

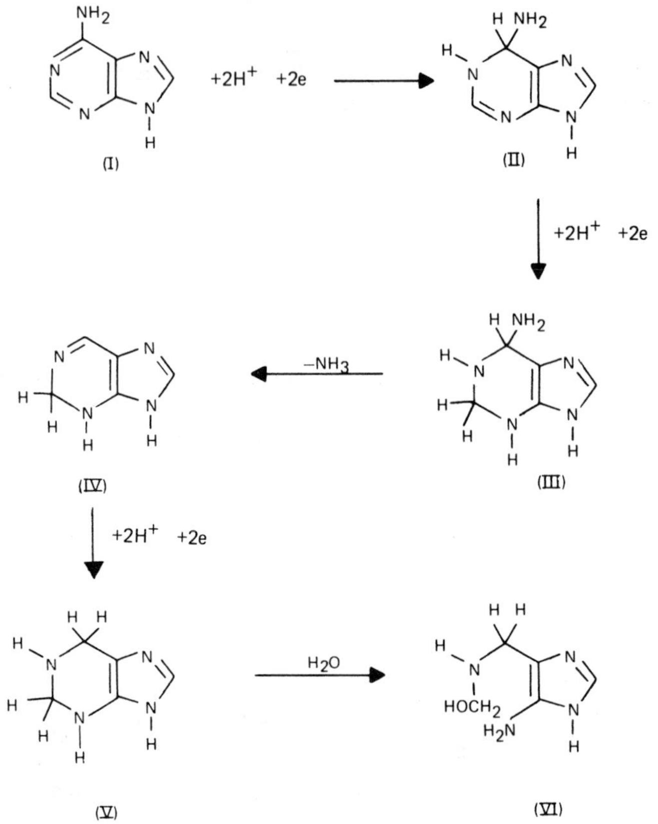

FIG. 3-10.   Reaction scheme for the electrochemical reduction of adenine.[71]

(a) reduction of the C-2=N-3 double bond, (b) deamination of the 6-position, (c) further reduction of the regenerated N-1=C-6 double bond, and (d) hydrolytic cleavage of the 2,3 position. Accordingly, the most likely sequence is that in which 1,6-dihydro-6-aminopurine (II, Fig. 3-10) is rapidly reduced in a $2e-2H^+$ process to give 1,2,3,6-tetrahydro-6-aminopurine (III, Fig. 3-10). The latter species may be sufficiently long-lived compared to normal polarographic drop times that the polarographic reaction ceases at this point. However, under conditions of prolonged electrolysis, e.g., coulometry, compound III (Fig. 3-10) deaminates and regenerates the N-1=C-6 double bond to give 2,3-dihydropurine (IV, Fig. 3-10). This is immediately reduced to 1,2,3,6-tetrahydropurine (V, Fig. 3-10), which, as outlined earlier, hydrolyzes to VI (Fig. 3-10). This process accounts for the overall $6e$ nature of the process and the observed products. The diffusion current constant for adenine in chloride and acetate buffers was found to be 10.2 ± 0.7 for 0.2 mM adenine. However, interpretation of the significance of this value (which corresponds more closely to that expected for a $6e$ process than to the proposed $4e$ process) is difficult because with increasing adenine concentration the diffusion current constant decreases in an approximately linear fashion.[71] This decrease has been interpreted as being due to lowering of the hydrogen overpotential by adenine so that, as the adenine concentration increases, hydrogen ion is more readily reduced, resulting in a lessening of the demarcation between the adenine and discharge waves. The diffusion current constant was also ca. 40% lower in McIlvaine buffers (citrate–phosphate–chloride) than in chloride and acetate buffers, which Smith and Elving[71] interpreted as indicative of a loosely bound complex between adenine and citrate or phosphate having a smaller diffusion coefficient than the noncomplexed adenine.

Dryhurst and Elving[72] studied the cyclic voltammetric behavior of adenine at both the HMDE and PGE. Adenine was not reducible at the PGE. At the HMDE the single, pH-dependent (Table 3-5) cathodic peak of adenine showed no evidence for reversibility of the electrochemical process.

The AC polarography of adenine has been studied.[72] Over the potential range from 0.4 to -1.8 V usually only a single, well-formed, pH-dependent AC peak is observed ($E_s = -1.09 - 0.067$ pH between pH 1 and 6) at about the same potential as that observed for the DC polarographic wave (Table 3-2). However, at concentrations above about 0.5 mM, two or occasionally three additional closely spaced peaks appear at much less negative potentials (Fig. 3-11). The faradaic peak at $E_s = -1.40$ V in pH 4.6 acetate buffer showed an approximately constant i_p/C ratio, but the closely spaced peaks at more positive potential showed no simple concentration dependence. The latter peaks appear primarily to be due to formation of mercury–adenine compounds.[72] Adenine markedly depresses the AC base current in the region of the ecm; a small elevation occurs at more negative potential (Fig. 3-11). This behavior again indicates adsorption

TABLE 3-5 Linear E_p versus pH Relationships for Adenine on Cyclic Voltammetry at HMDE[a]

Adenine concentration (mM)	Scan rate (V sec^{-1})	pH Range[b]	E_p (V versus SCE)
0.94	0.026	0–5	$-1.06 - 0.080$ pH
	0.26	0–5	$-1.06 - 0.089$ pH

[a] Data from Dryhurst and Elving.[72]
[b] E_p not observed above pH 5.5.

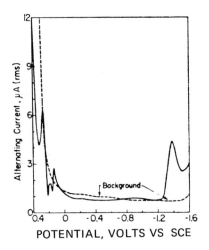

FIG. 3-11.   Alternating current polarogram of 2 mM adenine in pH 4.7 acetate buffer.[72] Frequency, 50 Hz; amplitude, 4 mV (rms). (Reprinted with the permission of Pergamon Press Inc., New York.)

of an uncharged adenine molecule in the region of the ecm, with possibly a desorption or reorientation process occurring prior to the faradaic peak. Adsorption of adenine in the region of the ecm has also been confirmed by electrocapillary studies.[72] In general, the cyclic voltammetric and AC polarographic behavior confirm the overall reaction scheme proposed by Smith and Elving[71] (Fig. 3-10).

Janik and Elving[73] have calculated some of the kinetic parameters for the polarographic reduction of adenine as a function of pH (Table 3-6). The data support the view that the rate-controlling step involves two electrons and between one or two protons.

TABLE 3-6   **Kinetic Parameters for the Polarographic Reduction of Adenine at 25°C at the DME**

pH[a]	αn_a[b]	p[b]
2.0–2.5	1.36 ± 0.09	1.91 ± 0.14
2.5–6.5	1.09 ± 0.07	1.54 ± 0.08
3.9–4.9	0.86 ± 0.08	1.22 ± 0.10

[a] Data from Janik and Elving.[73]
[b] See Chapter 2 for definition of αn_a and p.

4. Hypoxanthine

Hypoxanthine gives a polarographic reduction wave very close to background discharge potentials.[65,71] However, in an acetate buffer, pH 5.7, the wave has a diffusion current constant that is of the approximate magnitude expected for a 2e polarographic process. Although product identification was not attempted,[71] a tentative reaction scheme was proposed (Fig. 3-12) involving a $2e-2H^+$ reduction of the C-2=N-3 double bond of hypoxanthine (I, Fig. 3-12), which, by analogy to the observed behavior of the purine and adenine reduction products, would be expected to hydrolyze to 4(or 5)-amino-5(or 4)-*N*-(hydroxymethyl)-imidazolecarboxamide (III, Fig. 3-12).

5. Guanine

Under normal polarographic conditions guanine is not electrochemically reducible.[64,71] However, under conditions of alternating current oscillographic polarography guanine forms an anodic incision on dE/dt versus $f(E)$ curves provided that the DME is first polarized to cathodic background electrolyte discharge potentials.[74,75] In other words, it appears that an electrochemical reduction of guanine can occur at potentials beyond those at which components of normal aqueous buffer systems are themselves reduced and that at least one of the products of this process can be electrochemically reoxidized at much more positive potential (ca. -0.2 V versus SCE in acetate buffer, pH 3.95), resulting in the observed anodic incision on the oscillopolarogram.

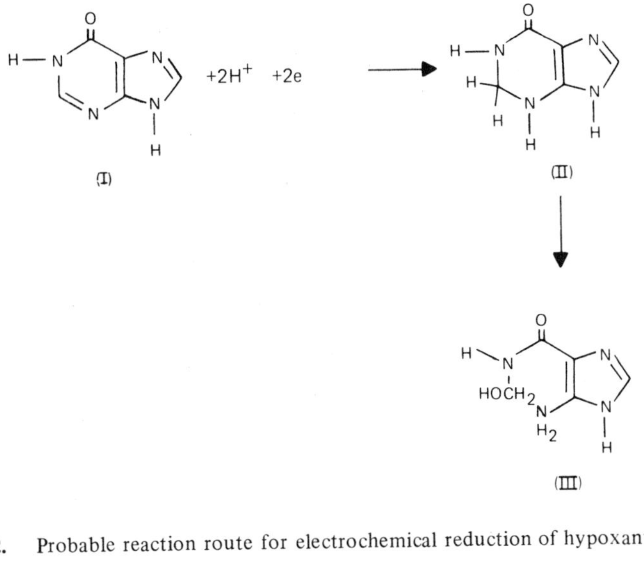

FIG. 3-12.   Probable reaction route for electrochemical reduction of hypoxanthine.[71]

Janik[76] studied this process in some detail using cyclic voltammetry at the HMDE, oscillopolarography, and DC polarography. It was found that an electrooxidizable product of the type produced by guanine was formed in the region of cathodic background discharge potentials only by those purine derivatives that were substituted in the 2 and 6 positions by hydroxy, amino, or substituted amino groups, which were nonreducible under normal polarographic conditions (with the exception of 2,6-diaminopurine) and in which the 7 and 8 positions were not substituted. These findings suggested that the reduction actually occurred at the N-7=C-8 double bond in the imidazole ring moiety of the purine. Substitutions at positions 1,3 and 9 did not interfere with formation of the oxidizable product. The anodic peak did not appear under cyclic voltammetric conditions at slow sweep rates (0.1 Hz; the sweep rate in volts per second was not mentioned) except when the HMDE was maintained for long periods at potentials beyond background discharge, i.e., until an appreciable quantity of product had been generated. Using faster sweep cyclic voltammetry and sweeping to sufficiently negative cathodic potentials, however, the oxidation of the product could be observed. Clearly, then, the oxidizable product of the cathodic reaction is unstable. Although Janik proposed that a chemical reaction of the primary reduction product gives rise to the electrooxidizable material, this conclusion does not seem to be compatible with the observation of the anodic peak under fast cyclic voltammetric and oscillopolarographic conditions. In acetate buffer, pH 3.95, the anodic reactions occurred at ca. -0.2 V versus SCE (presumably).

6. N-Acetylaminopurines

Goetz-Luthy and Lamb[77] observed the polarographic reducibility of adenine but also reported that 2,6-diaminopurine, 6-(N-acetylamino)purine (acetyladenine), and 2,6-bis(N-acetylamino)purine (2,6-diacetyladenine) were reducible Table 3-2).

Acetyladenine gives two polarographic reduction waves, which have been proposed[77] to be due to reduction of the acetyl group. Elving and co-workers[78] have disputed this fact and suggest that the presence of the acetyl group on the 6-amino group promotes a $2e$ reduction of the purine ring itself. This is indicated by the fact that adenine and acetyladenine give approximately equal total currents and that the current for the reduction of 2,6-diacetyladenine [2,6-bis(N-acetylamino)purine] is the same as for acetyladenine. Furthermore, since Goetz-Luthy and Lamb[77] show that 2-N-acetylguanine does not give a polarographic wave, it is unlikely that the acetyl group is involved in the reduction of the acetyladenines.

7. Other 6-Substituted Purines

Purines having substituted amino or methoxy groups at the 6-position exhibit a fairly well-defined DC polarographic reduction wave[73] similar to that observed for adenine which is primarily under diffusion control (Table 3-2). The reducible species are normally the protonated forms of the purines as evidenced by the fact that, at a pH generally 2–3 pH units greater than the pK_a value, the current shows a sigmoidal decrease and exhibits kinetic control. In other words, in these pH regions the current is controlled by the rate of recombination of the uncharged purine with protons to give the electroactive species. The close agreement between the diffusion current constant of these compounds and that of adenine (values of I ranged between 8.3 and 9.8 for adenine) suggests a $4e$ polarographic process for all compounds. 6-Methylpurine shows two nearly equal waves, similar to purine itself, which on the basis of diffusion current constant data both appear to involve two electrons. Accordingly, a reaction scheme has been proposed[73] whereby 6-methylpurine is reduced first at the N-1=C-6 double bond to 1,6-dihydro-6-methylpurine and then at the C-2=N-3 double bond to give 1,2,3,6-tetrahydro-6-methylpurine. In the case of 6-methoxy-, 6-methylamino-, 6-*n*-hexylamino-, 6-benzylamino-, 6-phenylamino-, and 6-dimethylaminopurine, the reaction responsible for the single polarographic wave is proposed to involve a $4e$ reduction of the N-1=C-6 and C-2=N-3 double bonds. The energy or potential controlling step in the reduction of 6-substituted purines is the reduction of the protonated —N-1=C-6— bond.

Janik and Elving[73] calculated αn_a and p values at various pH values for the polarographic reduction of the 6-substituted purines (Table 3-7). Certain regularities are apparent from these data. For example, αn_a decreases with increasing pH, which could indicate a decreasing electron-transfer rate constant (i.e., decreasing reversibility) and/or a decreasing number of electrons involved in the rate-determining step. The latter, however, was thought[73] to be unlikely. The data in Table 3-7 suggest that the rate-controlling step in the polarographic reduction of these purines involves about two electrons and one or two protons.

Janik and Elving[73] also studied the AC polarography of various 6-substituted purines and found that all exhibit a small AC reduction peak with a summit potential (E_s) about 50–90 mV more negative than the corresponding DC polarographic $E_{1/2}$. Such behavior indicates that the oxidized and reduced forms of the depolarizer are not equally strongly adsorbed in the potential region of the half-wave potential.[79] A deep depression of the AC peak was also observed, indicating strong adsorption of the oxidized form of the purine. Data reported for summit potentials in McIlvaine buffer, pH 2.5, are presented in Table 3-8.[73,79a] Actually, the AC base current for 6-substituted purine solutions in the potential range more positive than the faradaic AC reduction peak forms two minima (Fig. 3-13). The more positive minimum occurs close to

TABLE 3-7   Effect of pH on the Rate-Determining Step in the Polarographic Reduction of 6-Substituted Purines[a]

Substituent	Wave	pH[b]	$\alpha n_a{}^c$	p^c
6-Methyl	I	1.0–6.0	1.64	2.33
	II	2.0–6.0	0.99	1.46
		3.9–4.9	1.11	2.01
	III	6.0–7.5	0.80	1.20
6-Methoxy	I	1.0–3.9	1.18	2.10
		3.9–4.5	0.77	1.35
6-Methylamino	I	2.0–3.8	1.52	2.07
		3.8–6.5	1.00	1.36
6-n-Hexylamino	I	1.0–2.5	1.63	2.10
		3.7–6.5	1.10	1.42
6-Benzylamino	I	2.5–4.5	1.54	1.75
		4.5–6.4	1.07	1.66
6-Dimethylamino	I	1.0–6.4	1.25	1.60
2-Oxy-6-amino	I	1.0–4.2	1.05	1.35
(isoguanine)		4.2–7.2	1.11	1.96
		7.2–9.6	1.22	1.13

[a] Data adapted from Janik and Elving.[73]
[b] For $E_{1/2}$ data see Table 3-2.
[c] Standard deviations for αn_a and p were of the order of 5–12%.

TABLE 3-8   Summit Potentials for 6-Substituted Purine AC Polarographic Reductions at DME in Aqueous Solution[a]

Name	Peak[b]	pH	E_s (V versus SCE)
6-Methylpurine	I	2.5	−1.05
	II	2.5	∼−1.17
6-Methoxypurine	I	2.5	−1.15
6-Methylaminopurine	I	2.5	−1.28
	A[c]	2.5	−1.37
6-n-Hexylaminopurine	I	2.5	−1.27
	A[c]	2.5	∼−1.35
6-Benzylaminopurine	I	2.5	−1.25
6-Phenylaminopurine	I	2.5	−1.15
6-Dimethylaminopurine	I	2.5	−1.25
Isoguanine	I	2.5	−1.25
(2-oxy-6-aminopurine)	A[c]	2.5	∼−1.30

[a] Data from Janik and Elving.[73]
[b] Roman numeral indicates successive normal[73] AC polarographic peaks.
[c] Abnormal wave due to solution streaming and apparent reduction of a depolarizer produced by the normal (preceding) peak product in a chemical step.[79a]

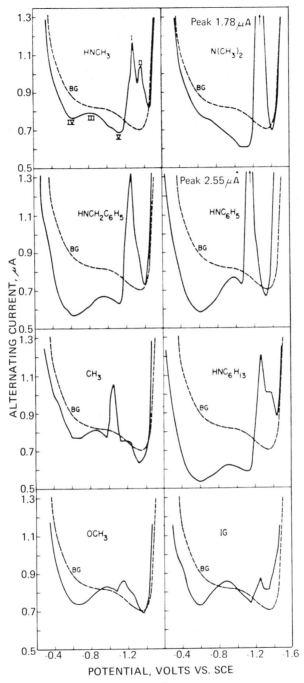

FIG. 3-13. Alternating current polarograms for 0.25 mM 6-substituted purines in pH 2.5 McIlvaine buffer; substituents are indicated. Key: IG, isoguanine; BG, background electrolyte base current.[73] Frequency, 50 Hz; amplitude, 4 mV (rms). (Reprinted with the permission of The Electrochemical Society, Inc.)

the electrocapillary maximum of the DME, i.e., the potential of zero charge, which suggests that an uncharged species is adsorbed. In view of the fact that the pK_a (for a proton gained) for the 6-substituted purines studied falls in the range 2.5–4.5, the protonated species is predominant at pH 2.5. Janik and Elving[73] interpreted this fact as indicating that an uncharged portion of the purine molecule, rather than the nonprotonated purine, is involved in the adsorption at potentials in the region of the electrocapillary maximum. A broad desorption peak and the subsequent adsorption minimum (Fig. 3-13) at more negative potential are ascribed to the adsorbed molecule gradually rearranging its position on the electrode surface so that the positively charged site on the molecule becomes attached to the electrode surface.

Vetterl[80,81] employed an AC bridge technique and a modified tensammetric (i.e., AC polarographic) method to study the effects of purines on the differential capacity of a DME in 1 M sodium chloride solution. All of the purines examined by Vetterl were adsorbed at the DME, as evidenced by the decrease in the differential capacity of the DME in the presence of fairly low concentrations of purines compared to the value in the absence of the purine (Figs. 3-14 and 3-13). However, the purines that are normally found as components of DNA and RNA (e.g., adenine and guanine) gave very sharply defined and pronounced minima or pits on the differential capacity–electrode

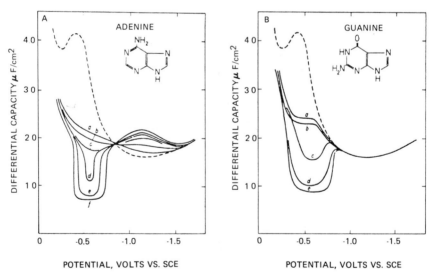

FIG. 3-14.   Differential capacitance curves for adenine (A) and guanine (B) at the DME in 1 M NaCl. (A) Dashed line is background curve. Concentration of adenine: (a) 1.1 mM, (b) 2.0 mM, (c) 4.2 mM, (d) 4.8 mM, (e) 5.5 mM, (f) 6.3 mM. (B) Dashed line is background curve. Concentration of guanine: (a) 0.066 mM, (b) 0.073 mM, (c) 0.086 mM, (d) 0.099 mM, (e) 0.11 mM.[81] (Reprinted with the permission of Academia Publishing House, Prague.)

potential curves above certain concentrations (Fig. 3-14). Vetterl interpreted these minima as being indicative of intermolecular interactions (i.e., association) of the adsorbed purines. The absorbed and associated molecules presumably form a very compact film on the surface of the electrode at potentials corresponding to the base of the minima (i.e., at ca. −0.5 V for adenine and guanine). At potentials corresponding to the steeply rising or falling portions of the minima the film begins to tear or otherwise become disrupted.

Subsequently, Vetterl[82] examined the effect of guanine on polarographic alternating current–time curves in 0.2 M H_2SO_4. Alternating current polarography of guanine in the latter medium indicates that up to a concentration of 1.25 mM guanine is adsorbed but no intermolecular association occurs at the DME (Fig. 3-15). At higher guanine concentrations pronounced, sharply defined minima appear on the AC polarograms (Fig. 3-15) indicative of association of the adsorbed guanine molecules. Below 1.25 mM guanine the alternating current–time curves in the potential region where guanine is adsorbed (−0.4 to −0.8 V versus SCE) followed a relationship expected for normal alternating capacity current [i.e., $I_{AC} = 7.5C(mt)^{2/3}$, where I_{AC} is the alternating current, C is the double-layer capacity, m is the flow rate of mercury through the capillary in milligrams per second, and t is the drop time in seconds]. Above 1.25 mM guanine a maximum appeared on the I_{AC}–time curve, indicating

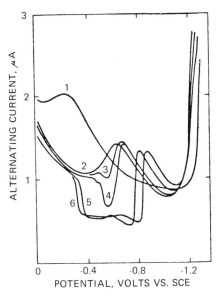

FIG. 3-15. Alternating current polarograms of guanine in 0.2 M H_2SO_4. Guanine concentration: (1) zero, (2) 1.25 mM, (3) 1.7 mM, (4) 1.9 mM, (5) 3.75 mM, (6) 7.5 mM.[82] Frequency, 78 Hz; amplitude, 18 mV. (Reprinted with the permission of Academia Publishing House, Prague.)

association of the adsorbed guanine molecules in the same potential regions. By careful analysis of the I_{AC}–time curves it is possible to characterize the rate of attainment of adsorption equilibrium and the rate of attainment of association equilibrium.[82] The rate of attainment of association equilibrium is at its maximum at −0.55 V versus SCE (i.e., at an uncharged electrode). At other potentials the molecules are repelled from the electrode by ions of the supporting electrolyte (0.2 M H_2SO_4), which are electrostatically attracted to the electrode. Thus, association is hindered and the attainment of association equilibrium requires a longer time.[83] The rate of attainment of association equilibrium is approximately two orders of magnitude slower than the rate of adsorption equilibrium. In other words, association does not occur immediately after adsorption but after an appreciable delay. A second, later maximum was observed by Vetterl[82] on several I_{AC}–time curves which he suggested might be caused by reorientation of the adsorbed/associated guanine molecules or by crystallization within the surface film.

8. Anomalous Wave of 6-Aminopurines

In addition to the normal 4*e* DC polarographic reduction wave, some 6-amino- and 6-alkylaminopurines, including the biologically important adenine nucleotides (see Chapter 5), give a more negative wave of anomalous behavior.[79a] This anomalous wave has some of the characteristics of a maximum of the second kind and is primarily due to an enhanced supply of purine as a result of solution streaming over the electrode surface. It was thought possible[79a] that the anomalous wave might also involve reduction of the N-1=C-6 bond that is regenerated by deamination of the initial reduction product (see Fig. 3-10). Isoguanine (2-oxy-6-aminopurine) also exhibits a second ill-defined DC polarographic wave at more negative potential than the normal wave (Table 3-2),[73] but the second wave differs from that of other 6-aminopurines in that it is primarily due to catalytic hydrogen ion reduction.[79a] Experimental conditions favoring appearance of the anomalous 6-aminopurine wave at normal polarographic operating temperatures (25°C) are high ionic strength of the supporting electrolyte solution and concentrations of electroactive species above 0.1 mM. In order to avoid the complicating presence of the anomalous wave, test solutions should have a low ionic strength (ca. 0.1 M) and be run at temperatures below 5°C.[79a]

9. 6-Thiopurine

Dryhurst[84] has reported the mechanism of electrochemical reduction of 6-thiopurine, which shows a total of four DC polarographic waves between pH 0 and 9.1 (Table 3-2). Below pH 1 and up to a pH a little below 3, 6-thiopurine shows two polarographic waves (wave I and II). Both waves have rather large values for the diffusion current constant, which indicates that for both processes there is a contribution from catalytic hydrogen reduction. These two waves merge at about pH 3 and up to pH 5 give a single wave that, having the same $E_{1/2}$ pH dependence as the first wave observed at lower pH, is still designated wave I. Between pH 5 and 8 a further wave appears (wave III), and a final wave (wave IV) appears in ammonia buffer, pH 9.1. Coulometry at potentials corresponding to wave I revealed that H_2S was immediately evolved but, owing to the catalytic reduction of hydrogen ions, it was not possible to determine electron numbers using this technique. Liberation of H_2S suggests attack at the N-1=C-6 double bond and, since 1,6-dihydropurine is the ultimate product, a probable reaction scheme involves a slow, potential controlling $2e-2H^+$ reduction of the N-1=C-6 bond of 6-thiopurine (I, Fig. 3-16A) to give the 1,6-dihydro derivative (II, Fig. 3-16A). This then rapidly regenerates the N-1=C-6 bond by loss of H_2S to give purine (III, Fig. 3-16A), which at wave I

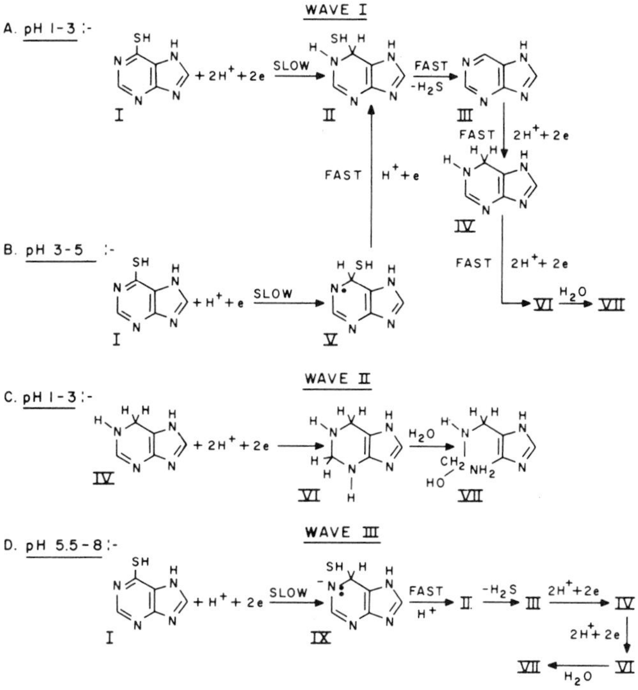

FIG. 3-16. Interpretation of electrochemical and chemical behavior observed for polarographic reduction of 6-thiopurine.[84] (Reprinted with the permission of The Electrochemical Society, Inc.)

potentials is immediately reduced to 1,6-dihydropurine (IV, Fig. 3-16A). Analysis of the slope and pH dependence of the wave confirmed the $2e-2H^+$ nature of the rate-controlling process. The second wave (wave II) observed at low pH is due to a further $2e-2H^+$ reduction of 1,6-dihydropurine to 1,2,3,6-tetrahydropurine (VI, Fig. 3-16C), which no doubt hydrolyzes to the 4-aminoimidazole (VII, Fig. 3-16C), as is observed in the case of complete reduction of purine and adenine.[71] Above pH 3 polarographic waves I and II merge to form a single wave up to pH 5, an overall $6e$ process, to give the same products observed from wave II at lower pH. However, the potential controlling reaction appears to change from a $2e-2H^+$ process at low pH to a $1e-1H^+$ process between pH 3 and 5, so that the primary electrode product is a radical species (V, Fig. 3-16B) that is rapidly reduced and rearranged to the final electrochemical product, 1,2,3,6-tetrahydropurine (VI, Fig. 3-16B).

Above pH 5.5 a single new wave appears (wave III, Table 3-2). The potential controlling step for this reaction appears to involve a $2e-1H^+$ process, and accordingly the reaction can be represented as an initial reduction of 6-thiopurine to an anionic species (IX, Fig. 3-16D) that is rapidly protonated to give again the 1,6-dihydro derivative of 6-thiopurine, which then undergoes the same reactions as outlined for waves I and II. Thus, again, six electrons are finally consumed. The nature of the process observed in ammonia buffer, pH 9.1 (wave IV, Table 3-2), was not investigated in detail,[84] although it is probably specifically associated with the ammonia present in the buffer system.

Other reports on the electrochemical reduction of 6-thiopurine have appeared; Humlová[85] reported the oscillographic polarography of 6-thiopurine in an analytical study of this technique for nucleic acid components. Vaček[86] developed a method for the determination of 6-thiopurine by utilizing the height of the polarographic wave he observed in McIlvaine buffer, pH 7.1.

10. Purine-6-sulfinic Acid

Purine-6-sulfinic acid gives rise to three polarographic waves,[84] all of which show $E_{1/2}$ values that shift linearly more negative with increasing pH (Table 3-2). At low pH additional waves are observed, but these are not due to purine-6-sulfinic acid itself, but rather to its decomposition products, i.e., principally to 6-thiopurine. The first wave (wave I, Table 3-2), due to the monoanionic form of purine-6-sulfinic acid (I, Fig. 3-17A), is a $2e-2H^+$ reduction of the N-1=C-6 bond to give purine-1,6-dihydrosulfinic acid (II, Fig. 3-17A). The magnitude of the polarographic diffusion current constant for wave I between pH 2 and a little below pH 7 indicates that polarographically the reaction ceases at this stage; i.e., the lifetime of the 1,6-dihydrosulfinic acid (II, Fig. 3-17A) is long compared to the polarographic drop time. However, under conditions of prolonged electrolysis, e.g., coulometry, four electrons are

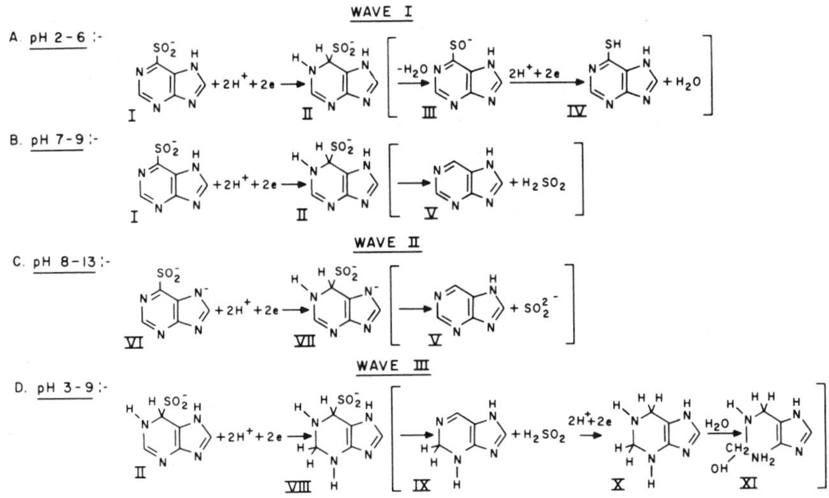

FIG. 3-17. Interpretation of electrochemical and chemical behavior for the three polarographic waves observed for purine-6-sulfinic acid. Compounds and reactions enclosed in brackets are those that occur after the initial polarographic reactions (e.g., under coulometric conditions).[84] (Reprinted with the permission of The Electrochemical Society, Inc.)

transferred and the ultimate product is 6-thiopurine (IV, Fig. 3-17A). Accordingly, the reaction scheme proposed[84] is the relatively slow decomposition of the primary polarographic product (II, Fig. 3-17A) to give purine-6-sulfenic acid (III, Fig. 3-17A), which is further reduced in a $2e-2H^+$ process to 6-thiopurine. At pH 7–9 the polarographic process responsible for wave I remains the same, i.e., a $2e-2H^+$ reduction of monoanionic purine-6-sulfinic acid to the 1,6-dihydrosulfinate, but coulometry reveals that even upon prolonged electrolysis only two electrons are transferred. However, the ultimate product at these pH values is purine, *not* 6-thiopurine. Accordingly, between pH 7 and 9 the decomposition of the 1,6-dihydrosulfinate (II, Fig. 3-17B) changes and sulfoxylic acid, H_2SO_2, and purine are produced.

As the pH increases above 8, wave I decreases in height owing to further ionization of the monoanionic form of purine-6-sulfinic acid to give the dianionic form (VI, Fig. 3-17C). It is the $2e-2H^+$ reduction of this species that is responsible for the second wave that appears at pH 8 and above. The 1,6-dihydrosulfinate that is produced decomposes under conditions of prolonged electrolysis to give purine (V, Fig. 3-17C) and the sulfoxylate dianion.

The third wave observed polarographically (wave III, Table 3-2) between pH 3 and 9 involves two electrons at the DME, although coulometrically four electrons are transferred. Accordingly, the polarographic wave III process was proposed[84] to be a $2e-2H^+$ reduction of the C-2=N-3 bond of the purine 1,6-dihydrosulfinate produced from the wave I (or wave II) polarographic process to give the 1,2,3,6-tetrahydrosulfinate (II → VIII, Fig. 3-17D). Under conditions of prolonged electrolysis this latter species desulfoxylates to give 2,3-dihydropurine (IX, Fig. 3-17) which is immediately further reduced at the available N-1=C-6 bond to give 1,2,3,6-tetrahydropurine (X, Fig. 3-17D). This is hydrolyzed to give the 4-aminoimidazole (XI, Fig. 3-17D) characteristic of purine reduction.[71]

11. Purine-6-sulfonic Acid

Purine-6-sulfonic acid gives four polarographic waves between pH 1 and 13[84] (Table 3-2). Between pH 1 and 7 the polarographic diffusion current constant indicates that the first wave of purine-6-sulfonic acid involves two electrons. However, below pH 3 coulometry reveals that a nonintegral number of electrons are transferred and that two products are formed: purine and 6-thiopurine. Accordingly, the primary polarographic process probably involves a $2e-2H^+$ reduction of the N-1=C-6 bond of the monoanion of purine-6-sulfonic acid (I, Fig. 3-18A) to give the corresponding 1,6-dihydro derivative (II, Fig. 3-18A). The nonintegral number of electrons transferred coulometrically and the fact that two products are observed suggest that this compound decomposes by two routes, one by loss of water to give purine-6-sulfinic acid (II → IV, Fig. 3-18A), which is then reduced by the mechanism outlined previously (Fig. 3-17) to give 6-thiopurine (IV → VII, Fig. 3-18A), and the other by loss of sulfurous acid to give purine (II → III, Fig. 3-18A). Between pH 3 and 7 both the polarographic and coulometric processes involve two electrons and the only product is purine, so that the primary electrochemical product, 1,6-dihydropurine-6-sulfonate (II, Fig. 3-18B), quantitatively loses sulfurous acid to give purine.

Wave II, which appears between pH 3.6 and 12.5 (Table 3-2), is due to reduction of the dianionic form of purine-6-sulfonic acid (VIII, Fig. 3-18C). The polarographic process involves two electrons over the entire range of pH that wave II appears, as did the coulometric reaction between pH 3.6 and below pH 6.9 and between pH greater than 8 and 12.5. The product of controlled potential electrolysis under these conditions is purine. However, coulometry between pH 6.9 and 8 involves the transfer of four electrons. Accordingly, within these pH regions the primary polarographic process is a $2e-2H^+$ reduction of the sulfonic acid to the 1,6-dihydro derivative (VIII → IX, Fig. 3-18C) followed by loss of sulfurous acid to give purine, which is further reduced in a $2e-2H^+$ reaction to 1,6-dihydropurine (X, Fig. 3-18C). The latter product was identified spectrally and by cyclic voltammetry at the PGE.[72] At all other pH values the potential employed in electrolyses at wave II was insufficiently negative to cause reduction of purine.

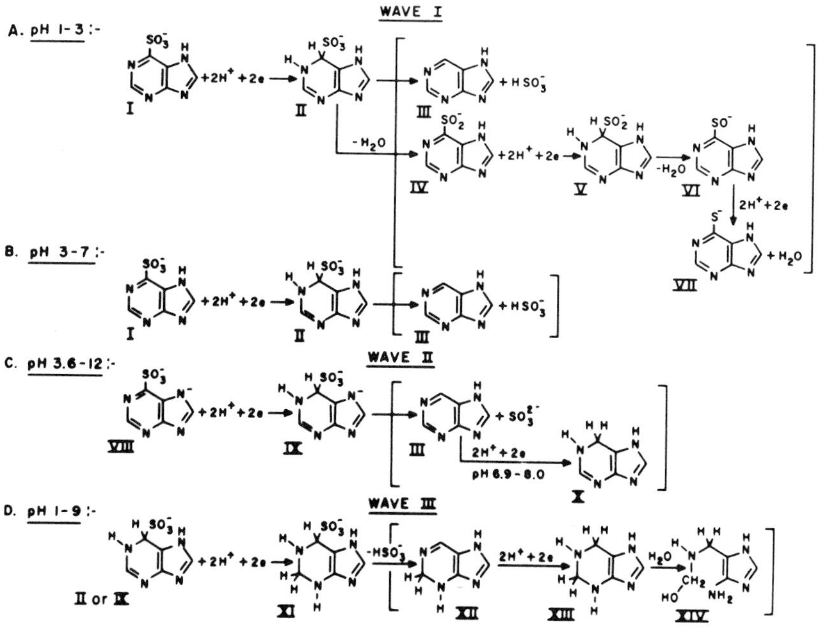

FIG. 3-18. Interpretation of electrochemical and chemical behavior observed for polarographic reduction of purine-6-sulfonic acid. Compounds and reactions enclosed in brackets are those that occur after the initial polarographic reactions (e.g., under coulometric conditions).[84] (Reprinted with the permission of The Electrochemical Society, Inc.)

The third wave observed (wave III, Table 3-2) between pH 1 and 9 under coulometric conditions gives the same product observed upon complete reduction of purine and adenine, although the value of the diffusion current constant indicates that polarographically only two electrons are transferred. The reaction sequence proposed, therefore, for the wave III process is that under polarographic conditions 1,6-dihydropurine-6-sulfonic acid (II or IX, Fig. 3-18D), produced from the wave I or II processes, is reduced at the available C-2=N-3 bond to the corresponding tetrahydro derivative (XI, Fig. 3-18D). Under prolonged electrolysis conditions this slowly decomposes by loss of sulfurous acid to regenerate the N-1=C-6 bond (XII, Fig. 3-18D), which is immediately reduced ($2e-2H^+$) to 1,2,3,6-tetrahydropurine (XIII, Fig. 3-18D). The elucidation of the wave III process by controlled potential electrolysis methodology was complicated by the tendency for tetrahydropurine to facilitate catalytic hydrogen evolution.

The fourth polarographic wave of purine-6-sulfonic acid appears only in

ammonia buffer, pH 9.1, and is presumably specifically associated in some way
with ammonia.[84]

12. Purine-2,6-disulfonic Acid

The electrochemical reduction of purine-2,6-disulfonic acid is extremely
complex.[87] Between pH 0 and 13 purine-2,6-disulfonic acid shows a total of
eight polarographic reduction waves at the DME (Table 3-2, Fig. 3-19). The
nature of these waves and the products were investigated by DC polarography,
cyclic voltammetry at the pyrolytic graphite electrode, coulometry, and
macroscale electrolysis and by spectral and chemical studies. The variations of
the half-wave potentials and diffusion current constants for each of the eight
polarographic waves are shown in Figs. 3-19 and 3-20, respectively.

At low pH (0 to ca. 3.5) purine-2,6-disulfonic acid (I, Fig. 3-21) is reduced
polarographically in three main steps. The first is a 2e process (wave I_a) to

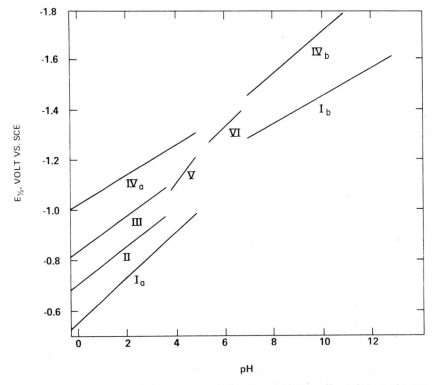

FIG. 3-19.   Linear variation of $E_{1/2}$ (volts versus SCE) with pH for purine-2,6-
disulfonic acid. Roman numerals refer to waves described in text.[87] (Reprinted with the
permission of Elsevier Publishing Company, Amsterdam.)

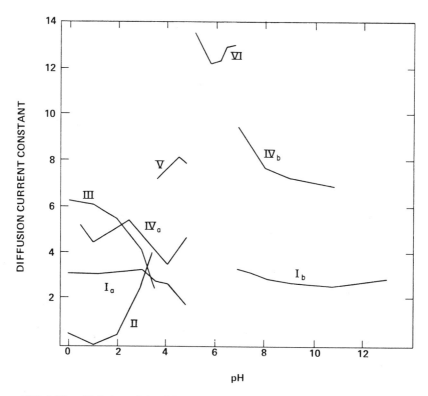

FIG. 3-20. Variation of the diffusion current constant ($I = i_1/Cm^{2/3}t^{1/6}$) with pH for purine-2,6-disulfonic acid. Roman numerals refer to waves described in Table 3-2 and text.[87] (Reprinted with the permission of Elsevier Publishing Company, Amsterdam.)

1,6-dihydropurine-2,6-disulfonic acid (II, Fig. 3-21). The second wave is an overall $4e$ reduction (wave III) of the 1,6-dihydropurine-2,6-disulfonic acid, produced in the wave I_a process, to give 1,6-dihydropurine (XI, Fig. 3-21). This reaction proceeds by an initial $2e-2H^+$ reduction of the C-2=N-3 bond of 1,6-dihydropurine-2,6-disulfonic acid to the corresponding tetrahydro compound (XIII, Fig. 3-21), which eliminates sulfurous acid, forming purine (X, Fig. 3-21). This is immediately further reduced in a $2e-2H^+$ reaction to give 1,6-dihydropurine. The third wave observed at low pH (wave IV_a) is a $2e-2H^+$ reduction of 1,6-dihydropurine (XI, Fig. 3-21) to 1,2,3,6-tetrahydropurine (XIV, Fig. 3-21), which then hydrolyzes to a 4-aminoimidazole (XV, Fig. 3-21). A fourth and very small kinetic wave (wave II) is also observed at low pH. This wave is due to the reduction of purine-2-sulfonic acid (III, Fig. 3-21), which is produced via a slow chemical elimination of sulfurous acid from the wave I_a product, 1,6-dihydropurine-2,6-disulfonic acid (i.e., II → III, Fig. 3-21). Under coulometric conditions with an applied potential corresponding to wave I_a,

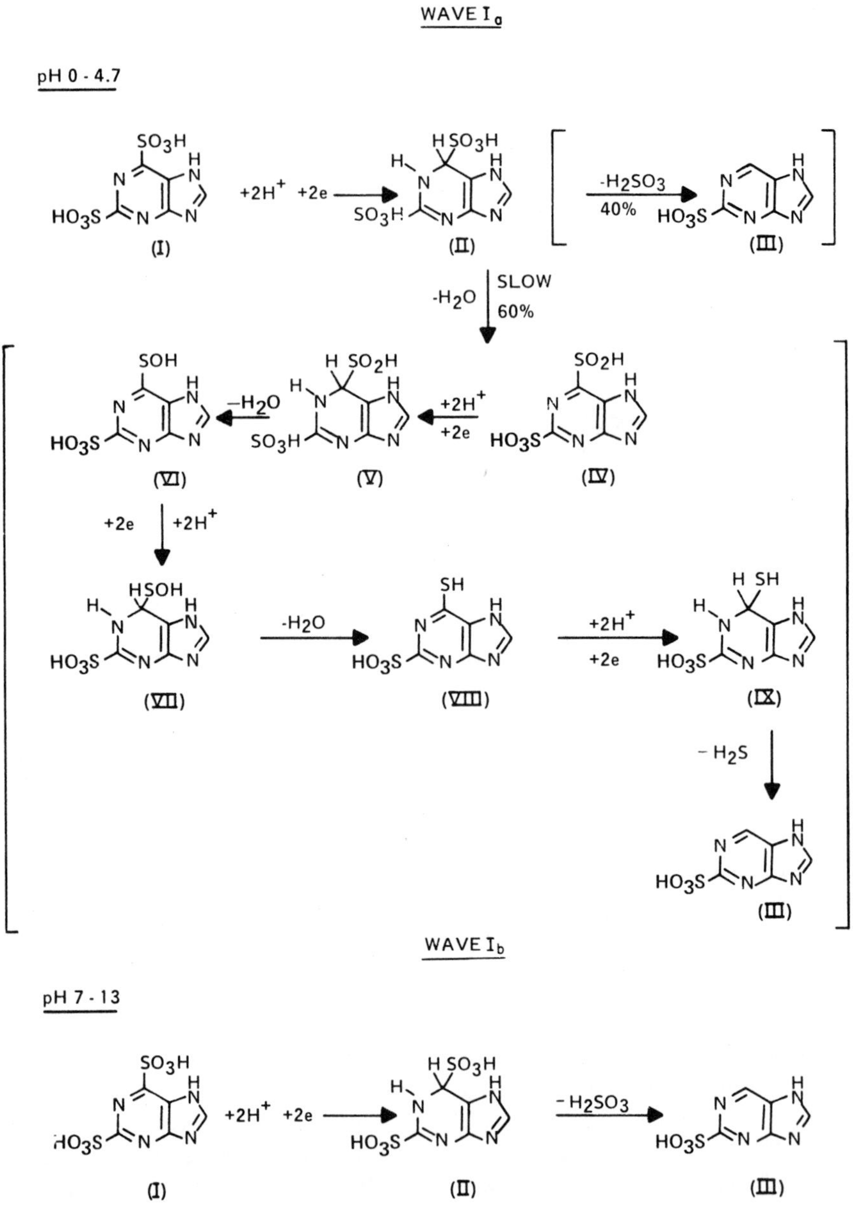

FIG. 3-21. Proposed reaction scheme for the electrochemical reduction of purine-2,6-disulfonic acid. The equations in brackets are those that occur only under conditions of prolonged electrolysis, e.g., coulometric conditions. All compounds are shown without regard to dissociation.[87]

WAVE II

pH 0 - 3.5

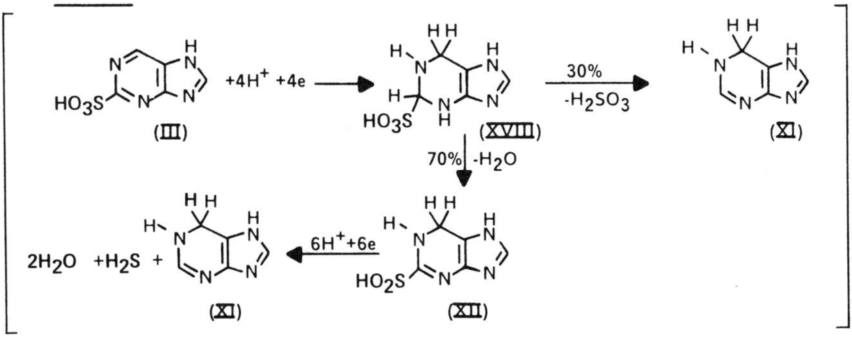

WAVE III

pH 0 - 3

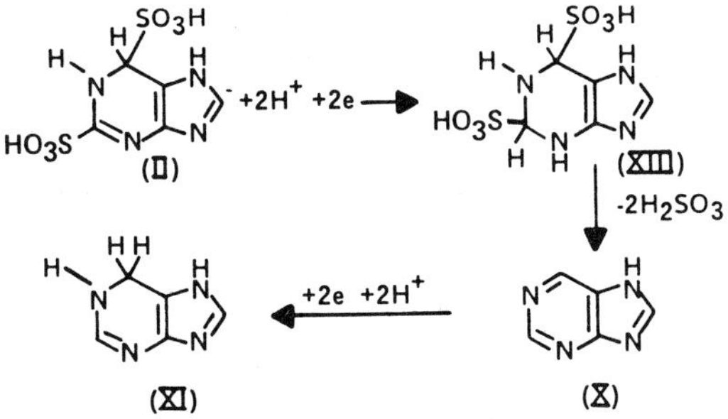

WAVE IV$_a$

pH 0 - 5

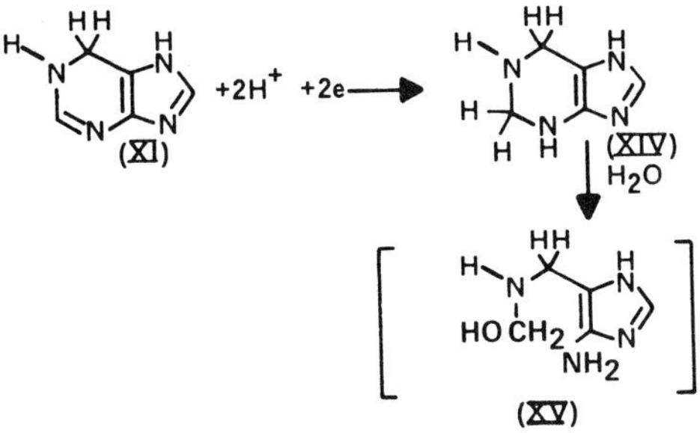

FIG. 3-21. *Continued.*

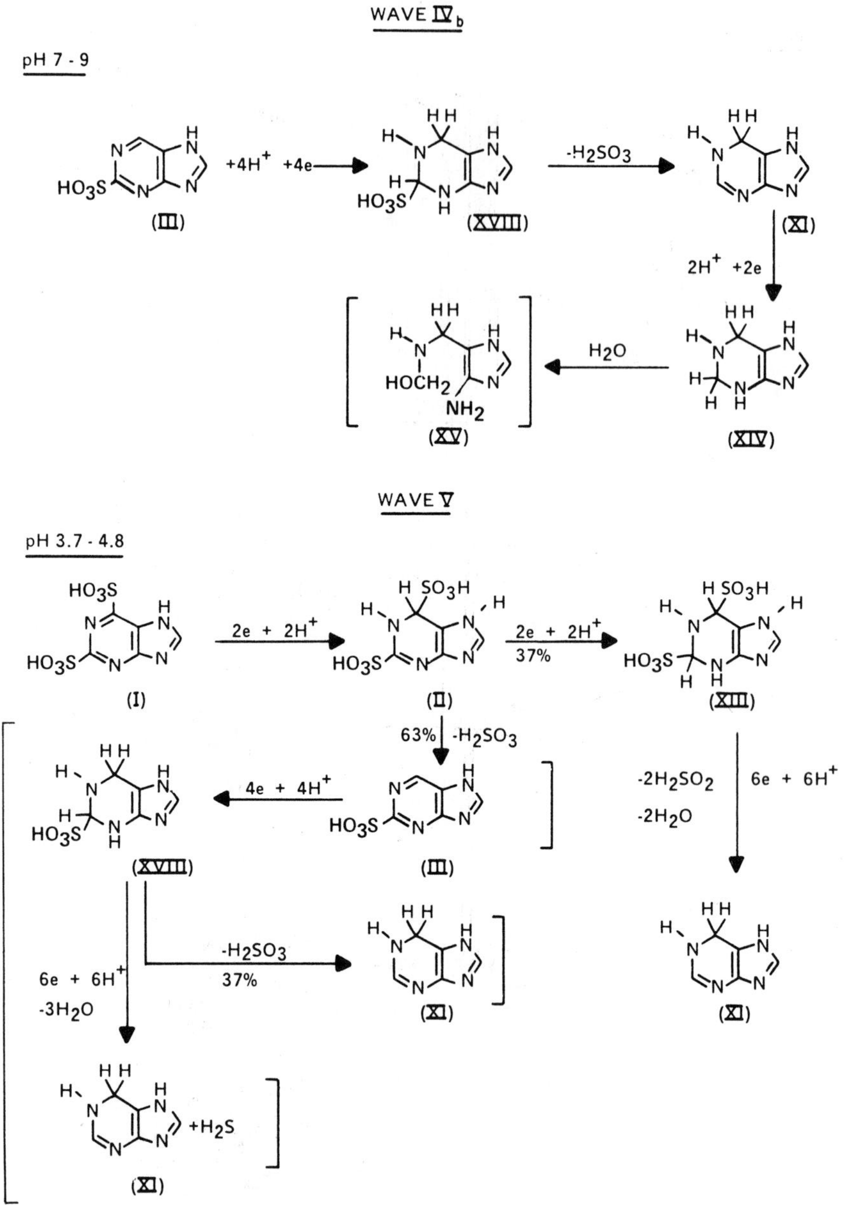

WAVE $\underline{IV}_b$

pH 7 - 9

WAVE $\underline{V}$

pH 3.7 - 4.8

FIG. 3-21   *Continued*

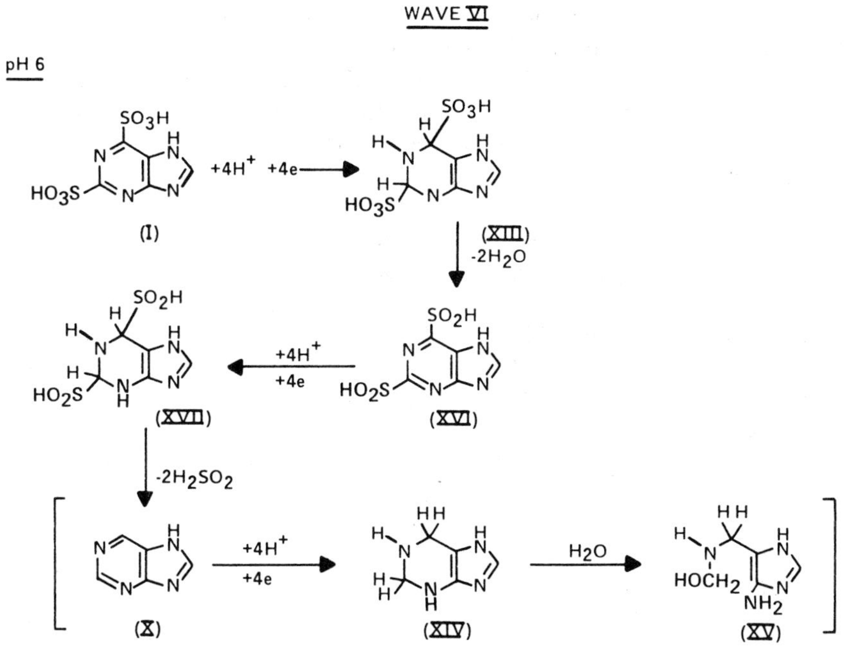

WAVE $\underline{\text{VI}}$

<u>pH 6</u>

FIG. 3-21 *Continued*

purine-2,6-disulfonic acid is reduced in a process involving about five electrons. This rather odd electron number is caused by the primary $2e-2H^+$ reaction product, 1,6-dihydropurine-2,6-disulfonic acid (II, Fig. 3-21), chemically decomposing by two routes. One route is loss of H_2SO_3 to give purine-2-sulfonic acid (III, Fig. 3-21), which is not electroactive at wave I_a potentials. The other route is by loss of water to give purine-2-sulfonic acid—6-sulfinic acid (IV, Fig. 3-21), which is further reduced to the corresponding 1,6-dihydro derivative (V, Fig. 3-21). By a series of similar dehydration and reduction steps (V → IX, Fig. 3-21), 1,6-dihydro-6-thiopurine-2-sulfonic acid is produced (IX, Fig. 3-21). Hydrogen sulfide is readily lost to give purine-2-sulfonic acid. By measuring quantitatively the amount of H_2S produced and the total electrons transferred, the reaction scheme presented for wave I_a in Fig. 3-21 under both polarographic and coulometric conditions was developed.

After complete reduction of purine-2,6-disulfonic acid at wave I_a potentials, the sole purine remaining is purine-2-sulfonic acid. This, as outlined earlier, is the species that gives rise to the small kinetic wave II observed on polarography of the disulfonic acid. The mechanism of the reduction of purine-2-sulfonic acid under conditions of controlled potential electrolysis in acid solution was elucidated (Fig. 3-21). Complete reduction of the 2-sulfonic acid requires seven electrons. This odd electron number is again explained by the decomposition of the initial $4e-4H^+$ product, 1,2,3,6-tetrahydropurine-2-sulfonic acid (XVIII, Fig. 3-21) by two routes. One is by loss of H_2SO_3 to give 1,6-dihydropurine (XI, Fig. 3-21) directly, which is not electroactive at the potentials involved; the other is by sequential dehydrations and reductions and loss of H_2S to give purine, which is then reduced to 1,6-dihydropurine.

Between about pH 3.8 and 4.8 waves II and III merge to give wave V (Fig. 3-19). Coulometric reduction of purine-2,6-disulfonic acid on the crest of wave V requires 11 electrons and the ultimate product is 1,6-dihydropurine. The reactions proposed to account for this electron number and the observed product are shown in Fig. 3-21. At intermediate pH (5.1—6.6) only one large wave (wave VI) is observed. Under polarographic conditions this wave corresponds to the $8e-8H^+$ reduction of purine-2,6-disulfonic acid to 1,2,3,6-tetrahydropurine-2,6-disulfonic acid (XVII, Fig. 3-21). However, under conditions of controlled potential electrolysis the latter compound decomposes to give purine and sulfoxylic acid. Purine (X, Fig. 3-21) is then further reduced in a $4e-4H^+$ process to the tetrahydro derivative (XIV, Fig. 3-21), which hydrolyzes to a 4-aminoimidazole (XV, Fig. 3-21). Overall, then, coulometry at wave VI involves a total of 12 electrons.

Between pH 6.8 and 13 identical two-step reductions occur under both polarographic and coulometric conditions. The first polarographic wave observed (wave I_b) is a $2e$ reduction of purine-2,6-disulfonic acid (I, Fig. 3-21) to purine-2-sulfonic acid with loss of sulfurous acid. The second wave (wave IV_b) is

TABLE 3-9   **Peak Potentials (E_p) for Electrochemical Reduction of Purines at the Stationary Pyrolytic Graphite Electrode**

Compound	Peak	pH Range	E_p (V versus SCE)	Reference
Purine[a]	I	1—12	$-0.84 - 0.077$ pH	72
Adenine	NR[b]			72
2-Thiopurine	I_c	1—9	$-0.525 - 0.090$ pH	88
Purine-2,6-disulfonic acid[c]	$I_a{}^d$	0—6	$-0.68 - 0.072$ pH	
	II	0—4	$-0.77 - 0.096$ pH	87
	III	0—4	$-1.20 - 0.110$ pH	
	VI	5.6—8	$-0.71 - 0.10$ pH	
2,6-Dithiopurine	I	1—7	$-0.96 - 0.068$ pH	89

[a] Scan rate, 60 mV sec^{-1}.

[b] Not reducible.

[c] Scan rate, 200 mV sec^{-1}.

[d] Numbering of peaks corresponds to that for polarography at DME (see also Table 3-3 and McAllister and Dryhurst[87]).

a 6*e* reduction of purine-2-sulfonic acid (III, Fig. 3-21) to 1,2,3,6-tetra-hydropurine (XIV, Fig. 3-21), which hydrolyzes to a 4-aminoimidazole (XV, Fig. 3-21).

Purine-2,6-disulfonic acid is also reduced at the PGE (Table 3-9). It was assumed that the mechanisms for the peaks observed were the same as those for the waves observed at the DME at approximately the same potentials.

13. 2-Thiopurine

2-Thiopurine exhibits three pH-dependent polarographic waves at the DME between pH 1 and 9[88] (Table 3-2, Fig. 3-22). The first wave (wave I) is a 1*e*

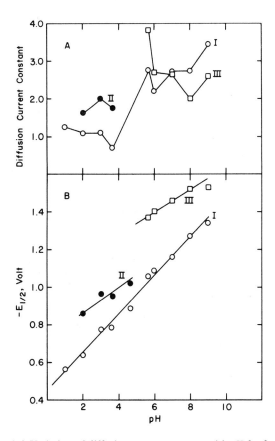

FIG. 3-22. (A) Variation of diffusion current constant with pH for 2-thiopurine waves I, II, and III at the DME. (B) Variation of $E_{1/2}$ (versus SCE) with pH for 2-thiopurine waves I, II, and III at the DME.[88] (Reprinted with the permission of Elsevier Publishing Company, Amsterdam.)

reduction of 2-thiopurine (I, Fig. 3-23A) to a radical species (II, Fig. 3-23A). By analogy with other purine reductions (*vide supra*) it is no doubt the N-1=C-6 double bond of 2-thiopurine that is reduced to the radical. Under conditions of controlled potential electrolysis the radical dimerizes, probably to give 6,6′-bis(1,6-dihydro)-2-thiopurine (III, Fig. 3-23A). This is shown not only by the 1*e* nature of the process, but also by the fact that after exhaustive electrolysis at potentials corresponding to the first wave, the second wave (wave II) below pH 4.7 is also eliminated. Wave II, which is observed only below pH 5, represents further reduction of the radical product of wave I (II, Fig. 3-23B) in a

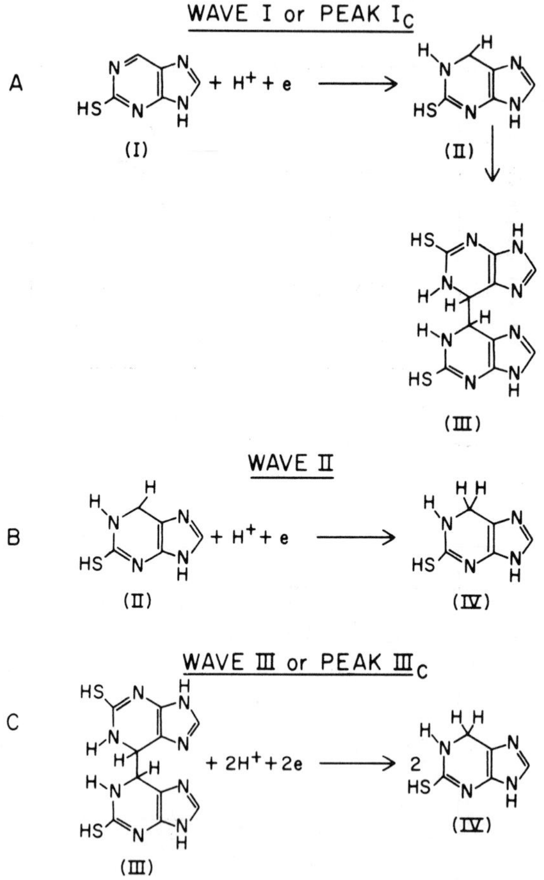

FIG. 3-23. Proposed pathways for the electrochemical reduction of 2-thiopurine at the DME (waves I, II, and III) and PGE (peaks I_c and III_c).[88] (Reprinted with the permission of Elsevier Publishing Company, Amsterdam.)

further $1e-1H^+$ process to give 1,6-dihydro-2-thiopurine (IV, Fig. 3-23B). This second step disappears above pH 5 and is replaced by a more negative wave (wave III). Since the latter wave is also due to a single-electron process, it is probably due to reduction of 6,6'-bis(1,6-dihydro)-2-thiopurine (III, Fig. 3-23C) to 1,6-dihydro-2-thiopurine (IV, Fig. 3-23C).

Under conditions of slow scan voltammetry (scan rate, $3-5$ mV sec^{-1}) 2-thiopurine shows only a single voltammetric reduction peak at the stationary pyrolytic graphite electrode (peak I_c, Table 3-9). The current observed for this peak is of the magnitude expected for a one-electron process, and accordingly the electrode reaction assigned[88] was identical to that of wave I observed at the DME. Under conditions of fast sweep voltammetry, however, up to two further cathodic peaks can be observed, peaks II_c and III_c. These peaks occur close to background discharge potentials. The least negative of these two peaks (peak II_c) is clearly observed only when saturated solutions of 2-thiopurine are employed.[88] The exact nature of the peak II_c process has not been ascertained, but in view of the latter finding it seems likely that it might be some type of adsorption process. The process responsible for peak III_c could not be studied in detail, but it was suspected[88] to be the same as that proposed for polarographic wave III (Fig. 3-23C).

14. 2,6-Dithiopurine

The electrochemical reduction of 2,6-dithiopurine has been investigated polarographically at the DME and voltammetrically at the stationary PGE.[89] At the DME, 2,6-dithiopurine exhibits two reduction waves (Table 3-2, Fig. 3-24B); the first wave (wave I) occurs between pH 1 and 5.65, and the second wave (wave II) occurs between pH 4 and 8. The diffusion current constant for wave I is 28 at pH 1 and decays at pH values above 2, in a more or less regular fashion, to zero at pH 6 (Fig. 3-24A). The drop time dependence of the wave did not support the view that the wave was entirely under diffusion control,[89] and the temperature coefficient was intermediate between that expected for diffusion control (ca. $1.5\%/°C$) and that for a catalytic process (ca. $6-10\%/°C$). Since 2,6-dithiopurine also has a tendency to react quite rapidly with mercury, attempts to determine faradaic n values or products were not possible.[89] The evidence, however, favors a process involving two simultaneous processes: a catalytic hydrogen discharge process and a faradaic process involving reduction of 2,6-dithiopurine. As the pH increases the decrease in the wave height (Fig. 3-24A) was thought to be primarily due to a decrease in the contribution of the catalytic process, which in turn could be at least in part related to the state of ionization of 2,6-dithiopurine.[89] The first pK_a of 2,6-dithiopurine is close to 5[89,90] and, accordingly, the catalytic contribution to wave I could be

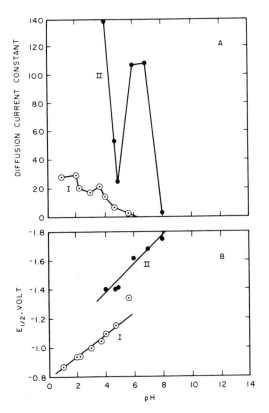

FIG. 3-24. (A) Variation of diffusion current constant for waves I and II of 2,6-dithiopurine with pH at the DME. (B) Variation of $E_{1/2}$ (versus SCE) for waves I and II of 2,6-dithiopurine with pH at the DME.[89] (Reprinted with the permission of The Electrochemical Society, Inc.)

represented by the equation shown in Fig. 3-25A, a process typical of many thiols.[91] Thus, as the solution pH increases, the concentration of neutral 2,6-dithiopurine decreases, resulting in the decreased current for wave I. Above pH 5 most of the 2,6-dithiopurine is in the monoanionic form and, accordingly, the catalytic current would be expected to decay to zero.

The wave II process is polarographically quite typical of a catalytic wave and shows, in particular, an extraordinarily large diffusion current constant (Fig. 3-24A). Since it is not observed below pH 4, it is likely that wave II is associated with the monoanionic form of 2,6-dithiopurine, which reaches significant concentrations at this pH. Accordingly, the probable catalytic reaction could be that shown in Fig. 3-25B. The observed shifts in the magnitude of the diffusion current constant between pH 4 and 8 (Fig. 3-24A) were

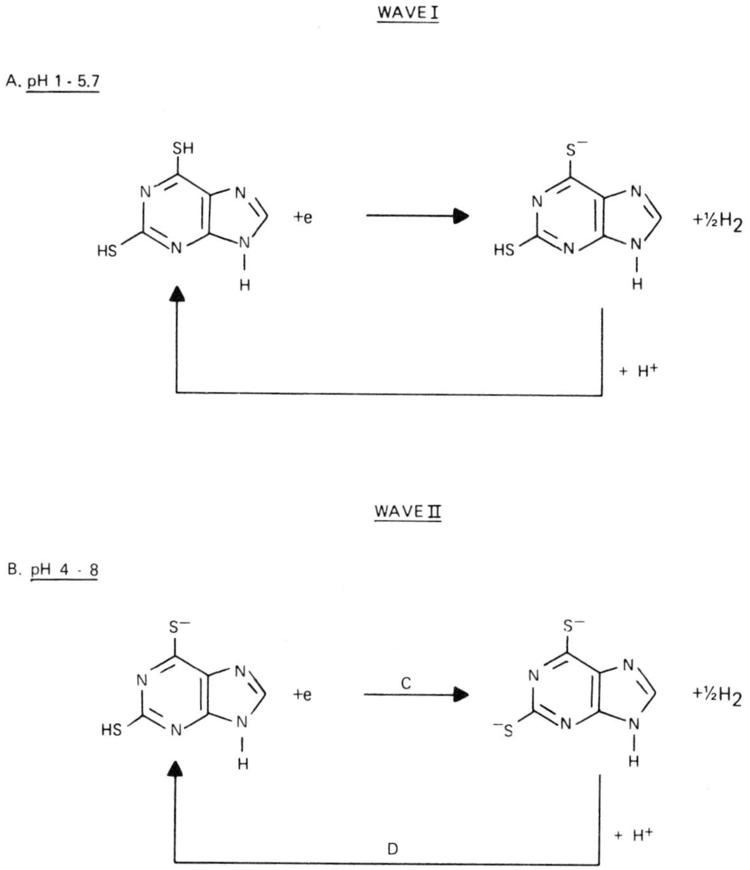

WAVE I

A. pH 1 - 5.7

WAVE II

B. pH 4 - 8

FIG. 3-25. Interpretation of processes responsible for the catalytic contribution to polarographic wave I and catalytic wave II on polarography of 2,6-dithiopurine.[89]

explained by the fact that at pH 4 (assuming pK_{a1} = 5.0) about 10% of the 2,6-dithiopurine exists in the monoanionic form. However, as a result of the high proton concentration and buffer capacity of the supporting electrolyte employed (McIlvaine, ionic strength 0.5 M), a large catalytic current is observed that is probably limited by reaction D (Fig. 3-25), reaction C being a fast electron-transfer process. As the pH increases to pH 4.9, the concentration of the monoanionic 2,6-dithiopurine still accounts for considerably less than 50% of the compound, whereas the proton concentration has decreased an order of magnitude and the buffer capacity has decreased. As a result, the systematic current decrease up to pH 4.9 is observed. At pH 6 in excess of 90% of the

2,6-dithiopurine is in the monoanionic form, which, as a result of increased catalyst concentration, more than compensates for the decrease in proton concentration, and between pH 6 and 7 the catalytic current is again high. At pH 8 the combination of decreased proton concentration and further ionization of the monoanionic 2,6-dithiopurine (pK_{a2} = 10.06[90] or 8.8[89]) results in the drastic decay of the observed catalytic current.

2,6-Dithiopurine is also reducible at the stationary PGE[89] (Table 3-9) but shows only a single, sharp, pH-dependent peak. Again, two simultaneous processes are responsible for this peak. The major process was thought[89] to be catalytic hydrogen discharge catalyzed by adsorbed neutral 2,6-dithiopurine, as evidenced by the rapid decay of the current density for the process with increasing pH and the pK_{a1} of 2,6-dithiopurine. The minor process was thought to be electrochemical reduction of 2,6-dithiopurine to 1,6-dihydro-2-thiopurine and H_2S. The mechanism responsible for this reaction was proposed to be a primary $2e-2H^+$ reduction of the N-1=C-6 bond of 2,6-dithiopurine to give the 1,6-dihydro derivative (I → II, Fig. 3-26), followed by loss of H_2S to regenerate the N-1=C-6 bond to 2-thiopurine (III, Fig. 3-26). 2-Thiopurine is immediately reduced to 1,6-dihydro-2-thiopurine (IV, Fig. 3-26), the latter being identified spectrally and by means of some very complex cyclic voltammetric curves taken at the PGE.

The process responsible for the catalytic hydrogen contribution to the peak at PGE was thought to be identical to that proposed for the first polarographic wave at the DME (Fig. 3-25A).

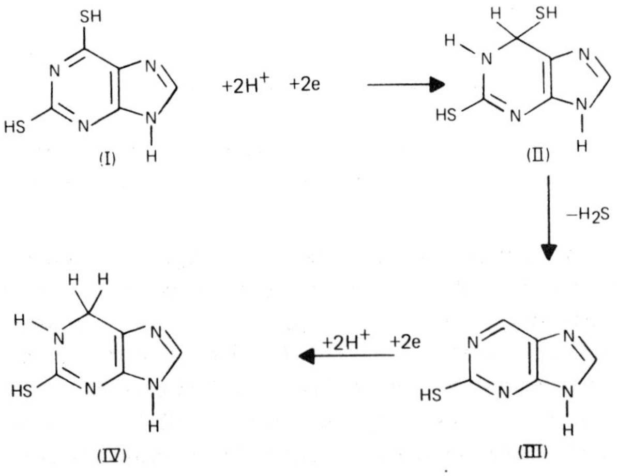

FIG. 3-26.   Mechanism for faradaic contribution to the reduction peak of 2,6-dithiopurine at PGE.[89]

15. Purine *N*-Oxides

McGinn and Brown[69] reported that adenine 1-*N*-oxide gives two polarographic waves at pH 1.5, the first of which was proposed to be due to reduction
of the *N*-oxide moiety, principally on the basis of the fact that adenine could be
identified as the product of electrolysis of the *N*-oxide. The $E_{1/2}$ for the second
wave corresponded to that of adenine itself and therefore was proposed to be
due to reduction of the adenine formed in the first wave. Warner and Elving,[92]
however, considered that the data of McGinn and Brown[69] were not entirely
tenable since equal concentrations of adenine and its *N*-oxide gave equal total
currents. The more detailed and systematic study of Warner and Elving[92]
revealed that adenine 1-*N*-oxide gives three pH-dependent polarographic reduction waves between pH 1.4 and 5.6 (Table 3-2). The first diffusion-controlled
wave is due to a $2e-2H^+$ reduction of the *N*-oxide (I, Fig. 3-27) to give adenine
(II, Fig. 3-27). The second wave is, as proposed previously,[69] a further $6e$
reduction of adenine to 1,2,3,6-tetrahydropurine (III, Fig. 3-27), which presumably is hydrolyzed to a 4-aminoimidazole, the typical final reduction product of
adenine and many other purines. Strangely, the diffusion current constant for
the second adenine *N*-oxide wave was appreciably larger than that for adenine
itself.[71] The authors[92] claim that some catalytic hydrogen ion reduction
involving "one or more" of the adenine species contributes to the observed
current. The third wave is due to a catalytic hydrogen ion reduction, which again
is proposed to involve both the adenine species and their reduction products.
Horn[93] has also briefly reported that adenine *N*-oxide is polarographically
reducible.

2,6-Diaminopurine and its 1-*N*-oxide have been briefly studied by McGinn
and Brown.[69] Both compounds appear to give only a single polarographic wave
at pH 1.5. However, in view of the incompleteness of these authors' data for
adenine 1-*N*-oxide, it may indeed be probable that waves over a much larger pH
range may occur.

16. Summary

Consideration of the electrochemical data and reaction schemes presented in
the previous sections of this chapter reveals that a number of basic conclusions
may be drawn.

1. With the possible exception of guaninelike compounds (see Section
IV-5), the electrochemical reduction of purines involves a primary hydrogenation of the C-1=N-6 double bond regardless of the substituent present in the
C-6 position, *except* when the substituent at the C-6 position is (formally) a
hydroxyl group or when the N-1 position is *N*-oxidized. In the case of C-6
hydroxyl substitution, the overwhelmingly predominant tautomer of the purine

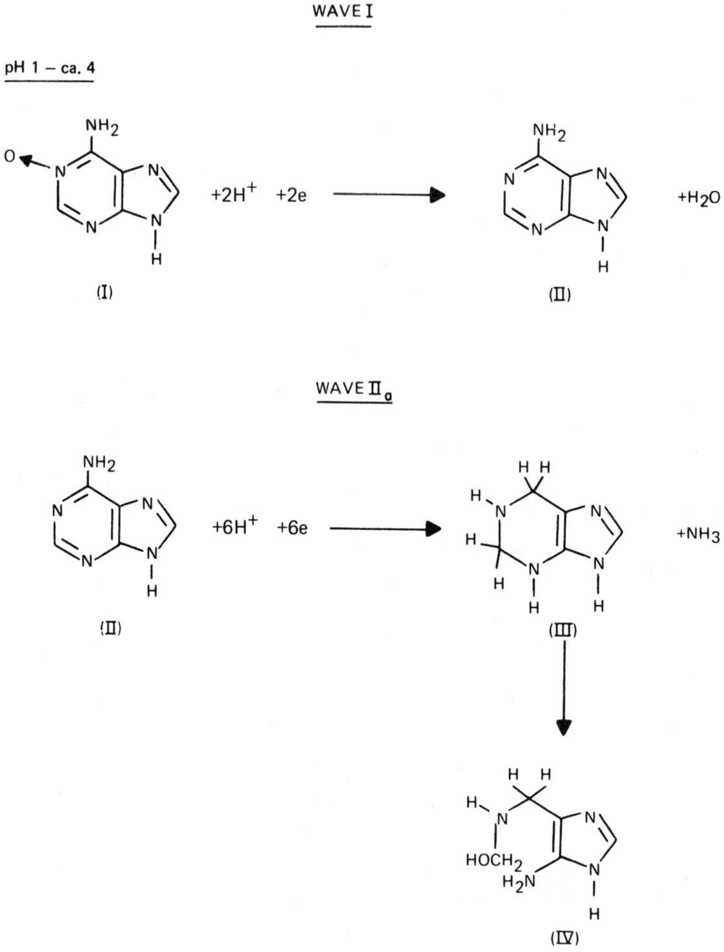

WAVE I

pH 1 − ca. 4

(I)   +2H⁺ +2e ⟶   (II)   +H₂O

WAVE II_a

(II)   +6H⁺ +6e ⟶   (III)   +NH₃

(IV)

FIG. 3-27.   Interpretation of electrochemical reduction of adenine 1-*N*-oxide.[92] (For detailed mechanism of adenine reduction see reference 71 and Fig. 3-10.)

is the keto form. This necessarily removes the N-1=C-6 double bond. Although the electrochemistry of purine *N*-oxides has unfortunately not been extensively studied, it does appear from the information now available that the N-1=C-6 bond is hydrogenated only after the 1-*N*-oxide function has first been reduced. In the case of guaninelike compounds the position of electrochemical reduction has not yet been conclusively ascertained, although in the presence of a 6-hydroxy group it is very unlikely that any reduction at the N-1—C-6 position occurs.

2. In those compounds that have no N-1=C-6 double bond available, either by virtue of the predominance of the keto tautomer of a 6-hydroxyl group (e.g., hypoxanthine) or because the N-1=C-6 bond has been hydrogenated (e.g., 1,6-dihydropurine), the next most reducible position is the C-2=N-3 double bond.

3. Except in the case of guanine there has never been any report of the polarographic or voltammetric reduction of the N-7=C-8 (or N-9=C-8) double bond.

4. The only reported instances of reduction of an exocyclic function on the purine ring are for the 1-*N*-oxide group, which in the case of adenine 1-*N*-oxide is now reasonably well documented,[69,92] for purine-6-sulfenic acid to 6-thiopurine[84] (VI → VII, Fig. 3-18), and for reduction of the acetyl group of acetyladenine.[77] The proposed reduction of the latter group is almost certainly incorrect. Reduction of purine-6-sulfenic acid has not been positively proven, although the extreme reactivity of sulfenates certainly would support such a reaction.

5. Purines having no available N-1=C-6 or C-2=N-3 double bonds are not reducible electrochemically under normal conditions in aqueous solution. Thus, neither 1,2,3,6-tetrahydropurine nor xanthine or any methylated xanthines[94] are polarographically or voltammetrically reducible.

6. The conclusions drawn in points 1−5 confirm the benzenelike inertness of the imidazole ring moiety of the purines.[95,96] Imidazole itself is not reduced polarographically in aqueous solution.[72a,97]

7. There is little doubt that most, if not all, purines are adsorbed at the DME and probably at the PGE. The data at this time do not indicate specifically whether purines are adsorbed at potentials where the electrochemical reduction occurs.

8. The products of many purine reductions seem to be extremely effective in decreasing the activation energy for electrochemical reduction of hydrogen ion.

B. Correlations of Polarographic $E_{1/2}$ with Structural and Electronic Indices

The only extensive work reported on correlations between the polarographic $E_{1/2}$ and theoretically calculated parameters is that of Janik and Elving.[73] They compared polarographic $E_{1/2}$ data for 6-substituted purines with polar substituent constants (e.g., Hammett−Taft plots), with the theoretical energies of the lowest empty molecular orbitals (LEMO energies) and with experimental pK_a values. The $E_{1/2}$ data used by Janik and Elving[73] were taken at pH 2.5 and 4.0 in McIlvaine buffers because the $dE_{1/2}/d(pH)$ slopes were more or less constant for each compound in this region (0.067−0.084 V/pH) and because the waves were not complicated by maxima at these pH values.

A priori, meaningful comparison of experimental $E_{1/2}$ data and calculated parameters requires that the reactions involved be essentially identical. In the case of all 6-substituted purines studied by Janik and Elving,[73] evidence suggests that the primary reaction involves hydrogenation of the N-1=C-6 double bond. In the case of purine and 6-methylpurine, reaction ceases at this stage, whereas in the case of substituted 6-aminopurines further reaction occurs (*vide supra*).

The basis for using $E_{1/2}$ values in linear free-energy relations, e.g., the Hammett—Taft equations, is that the $E_{1/2}$ is a simple function of the logarithm of the heterogeneous rate constant for irreversible electrode processes.[98] It is evident from rate constant and experimental data presented earlier that all purine reductions are indeed, overall, irreversible.

A good linear correlation between $E_{1/2}$ and the total polar substituent constant, σ_p, has been found[73] (Fig. 3-28).[98a] The total polar substituent constant is dependent on the kind and position of the substituent and to some extent the nature of the aromatic-type ring.[98] On the basis that in the 6-substituted alkylaminopurines the N-1=C-6 bond initally reduced polarographically is separated by an amino (—NH—) group from an alkyl or aryl group and that this —NH— group may be considered analogous to the —CH_2— group, in the sense of an additional atom being between the substituent and the reaction site, $E_{1/2}$ can be plotted against the polar substituent constant, $\sigma*$,[98–100] and again a linear relation is obtained (Fig. 3-29).[73] The slopes of the relationships shown in Figs. 3-28 and 3-29 are the so-called ρ values (see reference 98 for discussion). Since both are positive, this may indicate that the mechanism of the potential determining step is a nucleophilic process[99]; i.e., the electron is the nucleophilic agent.[73] Such information is valuable because, it will be recalled, the 6-substituted purines are, for the most part, thought to be reduced only in their protonated form. Accordingly, any primary electrophilic attack such as protonation prior to electron transfer has no significant role in the potential controlling step.[101] Accordingly, it is probably fairly safe to now write a generalized overall potential controlling reaction for the reduction of 6-substituted purines as a $2e–1H^+$ process (Fig. 3-30). It will be recalled that Elving *et al.*[72a] have proposed an even more detailed reaction for the polarographic reduction of purine (see, for example, Fig. 3-9 and associated discussion). It is conceivable that the latter mechanism might be applicable to many purines.

Pullman and Pullman[102] have devoted considerable thought to the electron-acceptor or electron-donor properties of purines and other biologically important molecules as predicted by molecular orbital theory and, to some extent, the relationship of these properties to the biochemistry of the molecules. These workers conclude that purines as a whole should have only very restricted electron-acceptor properties. Using the linear combination of atomic orbitals

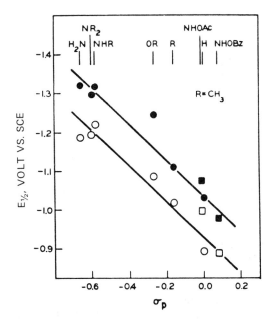

FIG. 3-28. Variation of $E_{1/2}$ of 6-substituted purines with the total polar substituent constant, σ_p (substituent indicated). The $E_{1/2}$ was determined in McIlvaine buffer (open circles, pH 2.5; solid circles, pH 4.0). Purine data from reference 71; 6-acetylamino- and 6-benzoylaminopurine data (open squares, pH 2.5; solid squares, pH 4.0) from Skulacher and Denisovich.[98a] Slope, ρ, is a 0.46 V at pH 2.5 and 4.0.[73] (Reprinted with the permission of The Electrochemical Society, Inc.)

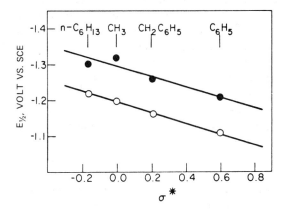

FIG. 3-29. Variation of $E_{1/2}$ of N'-substituted 6-aminopurines with the polar substituent constant, σ^* (substituent indicated). The $E_{1/2}$ was determined in McIlvaine buffer circles, pH 2.5; solid circles, pH 4.0). Slope, ρ is 0.14 V for pH 4.0 and 0.15 V for pH 2.5.[73] (Reprinted with the permission of The Electrochemical Society, Inc.)

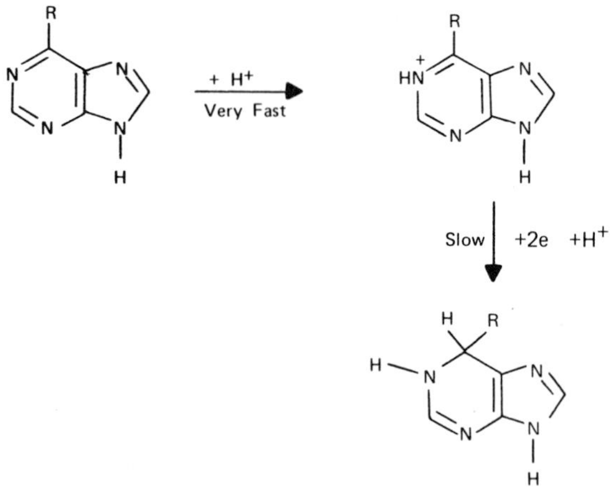

FIG. 3-30.   Probable overall primary reaction scheme for the electrochemical reduction of 6-substituted purines.

approach (LCAO), the energies of the molecular orbitals of the mobile or π electrons can be calculated. Using the approach of the Pullmans[102], the energies are obtained in the form

$$E_i = \alpha + \beta k_i$$

where E_i is the energy, α is a coulomb integral, β is a resonance integral, and k_i is the coefficient or matrix eigenvalue of the resonance integral. Rather than completely evaluating E_i in, for example, units of electron volts, the relative molecular orbital energies are indicated by values of k_i expressed in units of β. Negative values of k_i correspond to empty or antibonding orbitals, and the smallest negative value of k_i for a compound corresponds to the energy of the LEMO. Accordingly, comparison of the smallest negative values of k_i for a series of related molecules should be a theoretical indication of the relative electron-acceptor properties of the species. The polarographic $E_{1/2}$ is also a measure of the electron-acceptor properties of the purines, although the value of $E_{1/2}$ cannot always be expected to be simply related to the LEMO energy because of the fact that the theoretical values are calculated for a neutral gas-phase molecule. The $E_{1/2}$ value for an irreversible process is also affected by the electrode kinetic parameters, solvation effects, adsorption effects, and specific interactions between the purine and the electrode. Some discussion of these effects have appeared in articles by Dryhurst and Elving[72] and Pysh and Yang.[103] Nevertheless, Janik and Elving[73] found that fairly linear relationships were obtained between $E_{1/2}$ for the polarographic reduction of purine and some

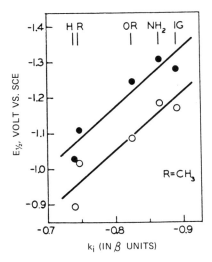

FIG. 3-31. Variation of $E_{1/2}$ of 6-substituted purines (substituent indicated) with LEMO energies (k_i) (IG, isoguanine). The $E_{1/2}$ was determined in McIlvaine buffer (open circles, pH 2.5; solid circles, pH 4.0).[73] (Reprinted with the permission of The Electrochemical Society, Inc.)

of its 6-substituted derivatives and the LEMO energy* (Fig. 3-31). In view of the fairly good correlations it is probably reasonable to predict that the heterogeneous rate constants ($k_{s,h}$, i.e., the rate constant at the formal potential of the couple) or reversibility of the electron-transfer process for all species are close and that major changes in solvation energy, adsorption phenomena, and electrode–purine interactions are relatively constant for the purines examined.

A measure of the double-bond character of a bond that can be calculated theoretically is the *mobile bond order*,[104] or simply the *bond order*.[102] Janik and Elving[73] found that substitution of purines in the 6-position influences the N-1=C-6 bond order and the electronic charge distributions at N-1. Since it has been shown that electrochemical reduction occurs primarily at the N-1=C-6 bond, it is not surprising that $E_{1/2}$ correlates with the bond order for the N-1=C-6 position (Fig. 3-32) and with electronic charge distribution at N-1 (Fig. 3-33), but not with the C-2=N-3 bond order. Rather interestingly, Janik and Elving[73] were able to derive a relationship between the $E_{1/2}$–bond-order data, which at pH 2.5 (McIlvaine buffer) was $E_{1/2} = -3.474 + 3.727$ BO, where BO is the bond order. The standard error of estimate of $E_{1/2}$ for the compounds used to prepare the equation was 0.018 V. It will be interesting to see if this equation is applicable to other purine derivatives.

* Values of k_i from Pullman and Pullman.[102]

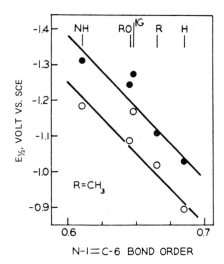

FIG. 3-32. Variation of $E_{1/2}$ of 6-substituted purines (substituent indicated) with calculated N-1=C-6 bond order (IG, isoguanine). The $E_{1/2}$ was determined in McIlvaine buffer (open circles, pH 2.5; solid circles, pH 4.0).[73] (Reprinted with the permission of The Electrochemical Society, Inc.)

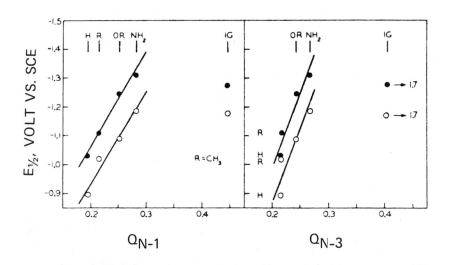

FIG. 3-33. Variation of $E_{1/2}$ of 6-substituted purines (substituent indicated) with net negative charges (Q) on N-1 and N-3 (IG, isoguanine). The $E_{1/2}$ was determined in McIlvaine buffer (open circles, pH 2.5; solid circles, pH 4.0).[73] (Reprinted with the permission of The Electrochemical Society, Inc.)

If indeed the 6-substituted purines are polarographically reducible only in their protonated form, it is somewhat surprising that the $E_{1/2}$–LEMO correlations are so good because, as mentioned earlier, the molecular orbital (MO) calculations are carried out for an uncharged molecule. An even better correlation might be attained if the MO calculations were carried out on the appropriately charged purine species. Thus, Nakajima and Pullman[105] found that the molecular orbital prediction for the species most susceptible to chemical reduction by elemental hydrogen was for the N-1 protonated form of purine. Indeed, it is observed experimentally that, in the presence of hydrogen gas and a palladium–charcoal catalyst at room temperature, only the protonated form of purine is reduced (to 1,6-dihydropurine).[105]

C. Electrochemical Oxidation

In biological situations oxidative processes are among the more important biological reactions associated with the purines. Such reactions are of particular importance in the catabolic degradation of these molecules. A considerable body of information is building up on the routes, ease, and positions of purine biological (i.e., enzymatic) oxidations. In addition, a considerable amount of work on the photochemical, radiochemical, and electrochemical oxidation has been reported. More detailed consideration of theoretical parameters with respect to biological oxidations of purines has also been carried out. Accordingly, in view of the more intensive and detailed studies of purine oxidations and because of their greater biological significance, the behavior of purines under electrochemical and other related oxidative conditions will be considered in some detail.

Until recently little electrochemical work concerned with the oxidation of biologically significant molecules had been carried out. The reasons for this are probably twofold. First, except in the case of the reduced members of reversible redox couples, the electrochemical oxidation of organic compounds, in general, in aqueous media even under controlled potential conditions is not as straightforward or simple a reaction as so frequently is the electrochemical reduction of organic compounds.[78] Second, it is not until comparatively recently that electrodes have become available that are suitable for observing electrochemical electron-transfer reactions at very positive potentials with respect to the saturated calomel electrode. The platinum electrode is not particularly recommended (by the author) for oxidation studies. Its range of positive potentials is too small and the involvement of surface oxide effects at very positive potentials can inhibit or eliminate the voltammetric behavior of the compounds studied. The range of negative potentials available is far too small, so that for purposes of cyclic voltammetry over wide potential ranges the electrode is rather poor. In addition, at sufficiently negative potentials where hydrogen evolution occurs, the platinum electrode absorbs hydrogen with the result that, on polarizing the electrode to positive potentials (e.g., as in cyclic voltammetry), the sorbed hydrogen is oxidized and considerable hydrogen anodic dissolution current is observed. Adams[106] has prepared a review of the properties of a number of solid electrode materials. Probably the best solid electrodes for use at both very positive and negative potentials are those based on carbon. Wax-impregnated spectroscopic carbon electrodes have an excellent potential range[107] but have a rather high and somewhat variable residual current, although procedures for eliminating most of these effects have been proposed. However, the involved nature of the electrode treatment and resurfacing[108] does not recommend this electrode for general use. Carbon paste electrodes have been used by many workers. These were introduced by Adams and co-workers[109,110] and are made by mixing powdered graphite with an inert organic liquid such as Nujol (mineral oil) to a thick paste, which is then packed in a depression in a suitable holder. Carbon paste electrodes have a wide negative and positive potential range.[106] The author has utilized the pyrolytic graphite electrode for many electrochemical studies. This electrode, which was introduced by Beilby and co-workers[111] and Miller and Zittel,[112] also has an excellent potential range,[106,113] has small residual currents, and can be very easily fabricated. Recommended techniques for preparation of pyrolytic graphite electrode surfaces have recently appeared.[114]

1. Uric Acid

Fichter and Kern[115] first reported the electrochemical oxidation of uric acid. The reaction was studied at a lead oxide (PbO_2) electrode but without control of the anode potential. Thus, it is likely that considerable oxygen evolution occurred during the electrolysis so that chemical oxidations may have occurred along with electrochemical processes. However, electrolyzing uric acid in a lithium carbonate solution at 40–60°C under such uncontrolled conditions, Fichter and Kern[115] obtained an approximately 70% yield of allantoin, while in sulfuric acid solution they obtained a 63% yield of urea. In 1962 Smith and Elving[116] reported that uric acid gave a well-formed voltammetric oxidation peak at a wax-impregnated spectroscopic graphite electrode in 2 M H_2SO_4 and in acetate buffers of pH 3.7 and 5.7. Later, Struck and Elving[117] examined the mechanism of the latter process in some detail and found that uric acid gives a single, well-formed faradaic oxidation peak at a stationary spectroscopic graphite electrode. By use of large graphite electrodes in dilute acetic acid solutions, it was revealed coulometrically that a nonintegral number of electrons (ca. 2.2) were transferred upon exhaustive electrolysis of uric acid. Analysis of the product solution revealed that 0.25 mole of CO_2, 0.25 mole of a precursor of allantoin, 0.75 mole of urea, 0.3 mole of parabanic acid, and 0.3 mole of alloxan simultaneously appeared per mole of uric acid oxidized. On the basis of the products observed, a reaction scheme was postulated whereby uric acid (I, Fig. 3-34) was oxidized in a primary $2e$ process to a short-lived dicarbonium

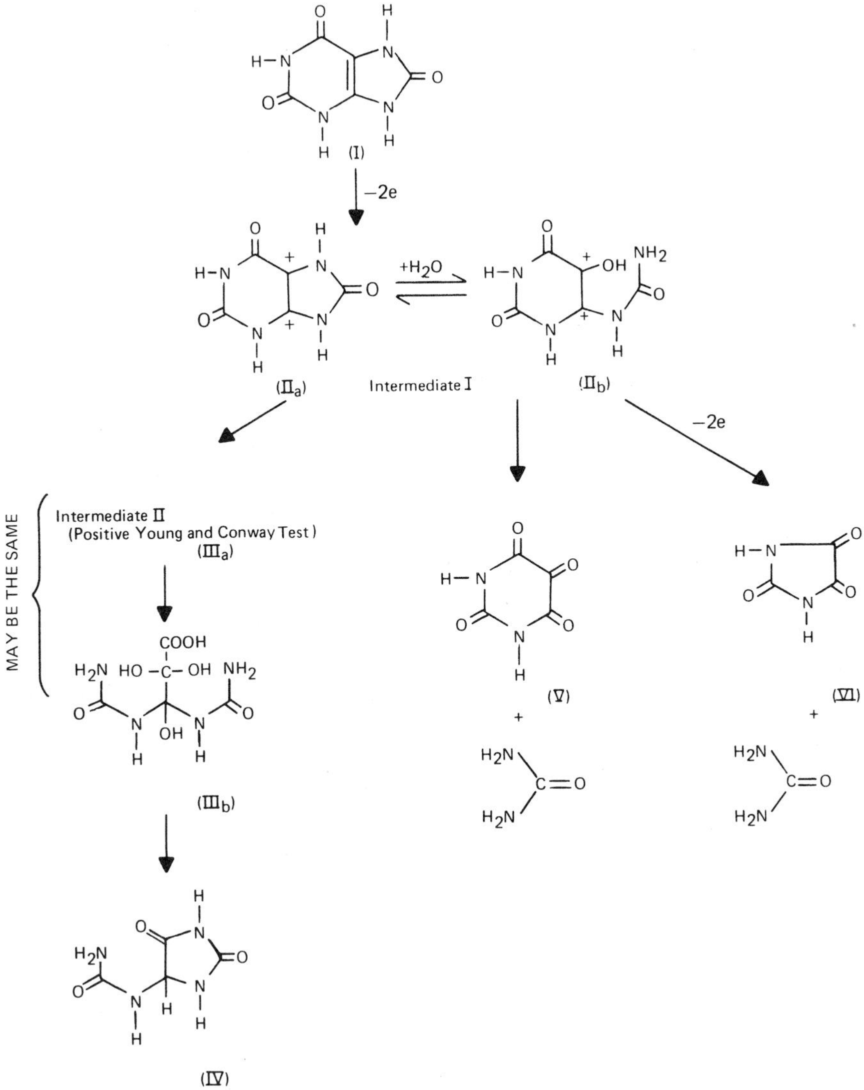

FIG. 3-34. Proposed pathway for electrolytic oxidation of uric acid in acetic acid solution.[117]

ion (II_a, II_b, Fig. 3-34), which undergoes three simultaneous transformations: (a) hydrolysis to an allantoin precursor (III_a, III_b, Fig. 3-34), i.e., a species that, upon subjection to the isolation procedures employed, gave a positive allantoin reaction in the Young and Conway test[118]; (b) hydrolysis to alloxan (V, Fig. 3-34) and urea; and (c) further oxidation and hydrolysis leading to parabanic acid (VI, Fig. 3-34) and urea. The nonintegral overall electron number (2.2) was claimed to be accounted for by the further oxidation of the dicarbonium ion (II_a, II_b, Fig. 3-34) with formation of 0.3 mole of parabanic acid (VI, Fig. 3-34). This, however, cannot be true because such a transformation requires an additional 0.6 electron above the primary $2e$ process. The intermediates III_a and III_b (Fig. 3-34) were not isolated and subjected to structural analysis.

There are several major objections to the scheme proposed by Struck and Elving.[117] First, there is no substantial evidence in favor of the dicarbonium ion intermediate. Indeed it is more plausible, if a positively charged species is produced, to localize any positive charge on surrounding nitrogen atoms, not at the C-4 and C-5 carbon atoms. Second, the peak for oxidation of uric acid is strongly pH dependent[117,119] (Table 3-10), and hence protons must be involved in the electrode reaction, which the mechanism does not indicate. Third, it was claimed that the dicarbonium ion primary electrode product was further oxidized at the potentials where uric acid itself is oxidized to give parabanic acid. However, the probability of a doubly positively charged ion readily losing two further electrons to give parabanic acid is remote. Finally, as mentioned earlier, oxidation of 1 mole of uric acid to 0.3 mole of parabanic acid plus other products requires a total transfer of 2.6 electrons, not 2.2 electrons.

In view of these facts and because all other purine oxidations have been studied at pyrolytic graphite rather than spectroscopic graphite electrodes, a further investigation of the electrooxidation of uric acid at the PGE was carried out by Dryhurst.[119] This study reveals that between pH 1 and 7 uric acid is electrochemically oxidized in a $2e-2H^+$ reaction. Fast sweep cyclic voltammetry indicates that the primary product of the $2e-2H^+$ reaction is very unstable but very easily reducible.[119-121] A fast sweep cyclic voltammogram of uric acid in acetate buffer, pH 4.7, is shown in Fig. 3-35. On the first potential sweep at a clean stationary PGE a single voltammetric oxidation peak is observed (peak I_a, Fig. 3-35). If the positive-going potential sweep is reversed after having scanned anodic peak I_a, then on the negative-going sweep two well-formed cathodic peaks are observed (peaks I_c and II_c, Fig. 3-35). Peak I_c cannot be observed unless the voltage sweep rate exceeds about 0.5 V sec^{-1}; i.e., the species responsible for the peak is unstable. Controlled potential electrolysis of uric acid at various pH values between 1 and 7 reveals that the nature of the products depends on the pH (Table 3-11). Thus, at low pH the only products of oxidation

TABLE 3-10 Linear E_p versus pH Relationships for Oxidation of Some Purines at the Stationary Pyrolytic Graphite Electrode[a]

Compound	Peak	pH Range	E_p (V versus SCE)	Reference
Purine		0–14	NO[b]	116
Theobromine (3,7-dimethyl-xanthine)	I	2.3–5.5	1.67 − 0.064 pH	121, 170
Caffeine (1,3,7-trimethyl-xanthine)	I	2.3–5.5	1.59 − 0.042 pH	121, 170
Adenine	I	3.6–10	1.39 − 0.051 pH[c]	190
	I	0–12	1.338 − 0.063 pH[d]	192
Theophylline (1,3-dimethyl-xanthine)	I	4–9	1.35 − 0.069 pH[e]	121, 172
	II	2.3–8.5	1.45 − 0.056 pH	
1,7-Dimethyl-xanthine	I	0–12.5	1.31 − 0.059 pH	121
Hypoxanthine	I	0–5.7	1.27 − 0.067 pH[f]	116
3-Methylxanthine	I	5.5–12.5	1.20 − 0.056 pH[e]	121
	II	0–11.9	1.27 − 0.050 pH	
7-Methylxanthine	I	7–12.5	1.19 − 0.049 pH[e]	121
	II	0–12.5	1.22 − 0.042 pH	
Guanine	I	0–12.5	1.12 − 0.065 pH	207
Xanthine	I	0–12.5	1.07 − 0.060 pH	121
Isoguanine	I	2 M H_2SO_4	1.05[f,g]	116
1-Methylxanthine	I	0–12.5	1.05 − 0.049 pH	121
6-Thiopurine	I	2–8	0.51 − 0.047 pH[e]	
	II	0–12	0.81 − 0.052 pH	218
	III	2–10	1.88 − 0.136 pH	
2,6-Dithiopurine	I	1–8	0.61 − 0.057 pH[e]	226
	II	4.7–9	1.26 − 0.062 pH	
	III	4.7–12.5	1.86 − 0.100 pH	
Uric acid	I	2.3–5.7	0.59 − 0.073 pH[f,h]	116
	I	0–12	0.76 − 0.069 pH	119
2-Thiopurine	I	0–9	0.36 − 0.049 pH	225
	II	4–13	1.83 − 0.082 pH	

[a] Except where otherwise stated the scan rate was 3.3 mV sec^{-1}.

[b] Not oxidized.

[c] Scan rate, 60 mV sec^{-1}.

[d] Scan rate, 5 mV sec^{-1}.

[e] Adsorption peak.

[f] Equation for the half-peak potential, $E_{p/2}$, at wax-impregnated spectroscopic graphite electrode.

[g] Only one data point available.

[h] Equation based on three data points.

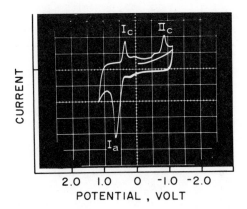

FIG. 3-35.   Cyclic voltammogram of uric acid in pH 4.7 acetate buffer at a clean stationary PGE. Scan pattern, $0.0 \text{ V} \rightarrow -1.10 \text{ V} \rightarrow 1.20 \text{ V} \rightarrow -1.10 \text{ V} \rightarrow 0.0 \text{ V}$; scan rate, 4.6 V sec^{-1}; current sensitivity, 200 μA per division. Current above axis marker is cathodic; that below axis marker is anodic. (Reprinted with permission of Springer-Verlag Publishers, New York.)

of uric acid are alloxan and urea. With increasing pH the yield of alloxan decreases while that of allantoin increases. At pH 7 the major product is allantoin with smaller amounts of urea and alloxanic acid (not reported in Table 3-11). Parabanic acid could be detected only in exceedingly small amounts in the products of electrooxidation of uric acid in 1 M HOAc, pH 2.3.

Summarizing the electrochemical and analytical data on the electrochemical oxidation of uric acid at the PGE reveals that uric acid is oxidized in a pH-dependent process (Table 3-10). The oxidation is almost reversible, as

TABLE 3-11   **Quantitative Analytical Data on Products of Electrochemical Oxidation of Uric Acid at the PGE[a]**

		Product (moles per mole of uric acid electrooxidized[b])			
pH	Buffer system	Alloxan	Urea	Allantoin	Parabanic acid
1.0	Chloride	1.0	1.0	ND[c]	ND
2.3	1 *M* HOAc	0.8	0.8	0.2	ca. 0.03
4.7	Acetate	0.37	0.37	0.63	ND
7.0	McIlvaine	ND	0.18	0.82	ND

[a] Data from Dryhurst[119].

[b] Generally, between 0.15 and 0.30 mmole of uric acid in 150 ml of buffer was electrolyzed.

[c] Not detected.

evidenced by fast sweep cyclic voltammetry (Fig. 3-35). The pH dependence of the peak potential for the almost reversible electrode reaction is in accord with that expected for a reaction involving an equal number of electrons and protons. Coulometry reveals that two electrons and hence two protons are involved in the electrode reaction. Finally, the primary product of the electrode reaction is very readily reducible (peak I_c, Fig. 3-35) but very unstable.

In view of these facts and the nature of the ultimate products a mechanism for the electrooxidation of uric acid has been proposed.[119,122] The primary electrochemical reaction is proposed to be a $2e-2H^+$ oxidation of uric acid (I, Fig. 3-36) to give species II (Fig. 3-36), which will be referred to as a diimine. The diimine formed from uric acid could exist in two tautomeric forms (II_a, II_b, Fig. 3-36). Such a system of conjugated double bonds would be expected to be very electrochemically reducible. Indeed, molecules having somewhat similar diimine structures such as riboflavin,[123-128] quinoxalines,[129] and bisquinoxalines[130] are rather readily reduced electrochemically. Aldimines and ketimines are also very readily electrochemically reduced.[131-133] The expected ease of reduction of structures II_a and/or II_b along with their expected facile hydration across the imine $-N=C-$ double bonds to give a 4,5-diol accounts for part of the observed cyclic voltammetry of uric acid. Thus, provided the sweep rate is fast enough, II_a or II_b formed as the primary products of oxidation of uric acid can be detected as a very reducible species (peak I_c, Fig. 3-35). At slow scan rates II_a or II_b cannot be detected because they are hydrated too rapidly. Cathodic peak II_c (Fig. 3-35) is observed at slower scan rates than is peak I_c. Peak II_c was originally thought to be due to reduction of parabanic acid, a reaction that occurs at similar potentials to peak I_c.[120,121,134] However, at completion of an exhaustive electrolysis of uric acid only a very small quantity of parabanic acid is detected (and that only in 1 M HOAc solution), quite insufficient to account for the relative peak current observed for peak II_c. Accordingly, it has been proposed[119] that the diimine hydrates in two stages, the first fast and the second slower. Addition of the elements of one molecule of water gives rise to the tertiary alcohol III (Fig. 3-36). It may be this compound that gives rise to peak II_c of uric acid (Fig. 3-35), being due to reduction of III possibly to IV (Fig. 3-36). Addition of a second molecule of water to III would give uric acid-4,5-diol (V, Fig. 3-36). (Many papers of Dryhurst and co-workers[120,121,135] proposed that the diol was the primary $2e-2H^+$ electrochemically reducible product of oxidation of uric acid that gives rise to peak I_c on cyclic voltammetry. However, this is highly unlikely since such a diol would not be expected to be readily reducible electrochemically.)

Uric acid-4,5-diol is a typical intermediate of an iminelike hydrolysis and would be expected to readily fragment to the observed products. There is some definite evidence for the formation of a product having the 4,5-diol type of

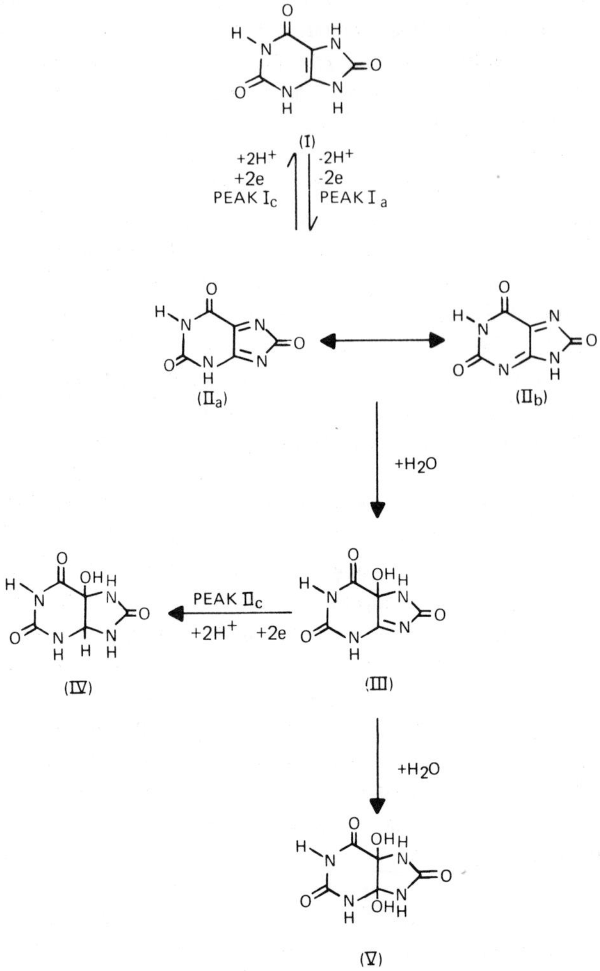

FIG. 3-36. Primary electrochemical oxidation and reactions of cyclic voltammetric peaks observed for uric acid at the PGE.[119]

structure.[135] This evidence is based on a study of the electrochemistry of uric acid in acetate buffer, pH 3.7–4.7, containing methanol. In purely aqueous media allantoin is the major product of the $2e$–$2H^+$ reaction (Table 3-11), and a large, well-formed diimine reduction peak is observed by fast sweep cyclic voltammetry (peak I_c, Fig. 3-35). In the presence of increasing concentrations of methanol a systematic decrease in the height of the cathodic diimine peak is observed, until in 50% methanol solutions the latter peak cannot be observed. In 50% methanol solutions the oxidation still involves two electrons,[119] but the yield of allantoin decreases to less than one-third of its value in completely aqueous solution. The nature of these observations in methanol has been rationalized[135] by assuming that under voltammetric conditions the greater nucleophilicity of methanol results in preferential and very rapid formation of 4,5-dimethoxyuric acid rather than uric acid-4,5-diol. The former compound would not be expected to fragment in the same way as the diol and hence give appreciable yields of allantoin.

A plausible mechanism for formation of allantoin from uric acid-4,5-diol (I, Fig. 3-37) involves cleavage of the C-5–C-6 bond of the diol, giving an imidazole isocyanate (II, Fig. 3-37). A hydrogen shift reaction must occur in compound II to give III, which upon hydrolysis forms allantoin (V, Fig. 3-37) and CO_2. A simple fragmentation of the 4,5-diol to alloxan (VI, Fig. 3-37) and urea (VII, Fig. 3-37) can also be written. Although only very minor amounts of parabanic acid are formed upon electrochemical oxidation of uric acid at the PGE in $1 M$ HOAc, large amounts are formed at spectroscopic graphite electrodes in the same medium.[117] Formation of parabanic acid necessarily involves some secondary electrochemical oxidation. A mechanism has been proposed in Fig. 3-37[119] whereby uric acid-4,5-diol undergoes a ring-opening reaction to give structures VIII_a or VIII_b (Fig. 3-37). In acid solution VIII_a (Fig. 3-37) should readily cleave across the original C-5–C-6 bond to give an isocyanate (IX, Fig. 3-37) and 2-oxy-4,5-dihydroxyimidazole (X, Fig. 3-37). Alternatively, compound VIII_b (Fig. 3-37) could undergo ring closure to XI, which in turn should cleave to X and IX (Fig. 3-37). Simple hydrolysis of the isocyanate (IX, Fig. 3-37) would give urea and CO_2. 2-Oxy-4,5-dihydroxyimidazole (X, Fig. 3-37) is an enediol which, since enediols are normally readily oxidizable to 1,2-diketones even by such weak oxidizing agents as cupric ion or oxygen,[136] should be electrochemically oxidized in a $2e$–$2H^+$ reaction to parabanic acid (XIV, Fig. 3-37). The pronounced difference between the yield of parabanic acid formed on electrooxidation of uric acid at spectroscopic and pyrolytic graphite electrodes can be rationalized only by invoking some type of specific electrode effects related to the nature of the electrode materials.

a. Adsorption of Uric Acid.   Vetterl[81] has demonstrated that uric acid is quite strongly adsorbed at a dropping mercury electrode in $1 M$ NaCl solution. Maximum adsorption in this medium appears to occur around −0.5 V versus SCE. Uric acid is not electrochemically reducible or oxidizable at the DME.

Overproduction of uric acid in man can lead to gout. One of the most effective drugs for treatment of gout is allopurinol (1*H*-pyrazolo[3,4-*d*]-pyrimidin-4-ol). In order to understand the effectiveness of allopurinol in gout

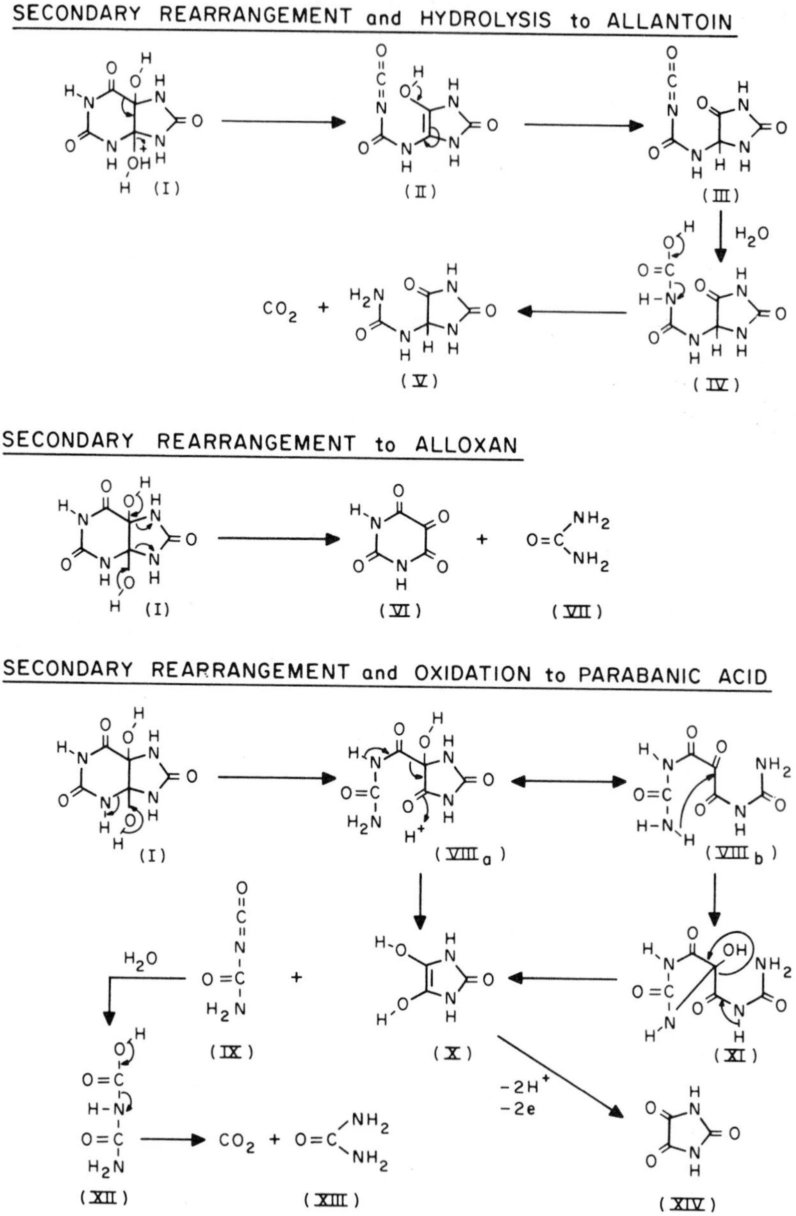
SECONDARY REARRANGEMENT and HYDROLYSIS to ALLANTOIN
(I)
(II)
(III)
H2O
(IV)
CO2 +
(V)
SECONDARY REARRANGEMENT to ALLOXAN
(I)
(VI)
+
(VII)
SECONDARY REARRANGEMENT and OXIDATION to PARABANIC ACID
(I)
(VIIIa)
(VIIIb)
H2O
(IX)
+
(X)
(XI)
-2H+
-2e
(XII)
CO2 +
(XIII)
(XIV)

FIG. 3-37. Mechanisms for decomposition of uric acid-4,5-diol to allantoin (V), alloxan (VI), urea (VII), parabanic acid (XIV), and CO_2.[119]

therapy it is desirable to have a simple and direct method for monitoring the concentration of allopurinol and uric acid. Dryhurst and De[136a] have developed a method for such an analysis based on the voltammetric oxidation peak of uric acid at the PGE and the polarographic reduction wave of allopurinol at the DME.[136a,136b] During the course of this study it became apparent that uric acid is strongly adsorbed at the PGE so that concentration versus anodic peak current curves were not linear, but when the solution is saturated with allopurinol, these plots become linear. By use of scan rate and concentration studies and AC voltammetry at the PGE it was shown that allopurinol is also adsorbed at the PGE and in saturated solutions displaces adsorbed uric acid from the electrode surface.

b. Biochemical Oxidation of Uric Acid.   As mentioned earlier, oxypurines and particularly uric acid are the principal final products of purine metabolism in man,[37,38] although in many other organisms further degradation of the purine molecule occurs. One of the important enzymes involved with uric acid oxidation is uricase, which has been studied extensively *in vitro* although it acts quite well, for example, when injected intravenously into man.[137] Uricase, however, does not occur naturally in man. In much of the early work on the oxidation of uric acid (I, Eq. 3) by uricase, it was proposed that the reaction

$$+ O_2 + H_2O \xrightarrow{\text{Uricase}} \qquad + CO_2 + H_2O_2 \tag{3}$$

(I)   (II)

proceeded quantitatively with formation of allantoin (II, Eq. 3), carbon dioxide, and hydrogen peroxide.[138–142] However, later workers[143,144] showed that under certain conditions the uptake of oxygen was significantly more rapid than the production of carbon dioxide or that the yield of CO_2 was much less than the oxygen consumed. This was interpreted to mean that some moderately stable intermediate was produced in the reaction and, on the basis of pH changes that occurred in the reaction, that the primary product was of greater acid strength than uric acid itself.

By using water and gaseous oxygen labeled with ^{18}O, Bentley and Neuberger[145] were able to show that the oxygen atoms of the hydrogen peroxide formed in the oxidation of uric acid in the presence of uricase were derived from molecular oxygen. It was thus concluded that uricase simply catalyzes the transfer of two electrons from a urate ion to oxygen. In other words, uricase acts effectively as a catalytic surface for transfer of two electrons from uric acid to (presumably dissolved) oxygen. It was also found by labeling the C-6 position of uric acid with ^{14}C that the CO_2 formed in the enzymatic oxidation was derived from that carbon atom. On the basis of their experiments and others,[140] Bentley and Neuberger[145] found that the optimum reactivity of uricase was at about pH 9.25 which, since uric acid is a dibasic acid (pK_{a_1} = 5.4, pK_{a_2} = 10.6[24] or pK_{a_1} = 5.75, pK_{a_2} = 10.3[16]), indicates that uric acid would be primarily in the monoanionic form. Transfer of two electrons from the uric acid monoanion (I, Fig. 3-38) would give a carbonium ion (II, Fig. 3-38). It was proposed that the electron-transfer process was the only step that involved the enzyme and that further changes were purely chemical reactions arising from the reactivity of the intermediates. It was further proposed that in the carbonium ion (II, Fig. 3-38) the carbon atom at position 5 being strongly electrophilic would have a tendency to interact with the N-1 nitrogen atom to give III (Fig. 3-38), which, being unstable, interacts with hydroxyl ion to give IV (Fig. 3-38). This latter species could be transformed in a variety of ways, but simply by the opening of one ring and addition of the elements of water, it could give rise to allantoin (V, Fig. 3-38) and CO_2.

The studies of Agner,[146,147] Paul and Avi-Dor,[148] and Canellakis et al.[149]

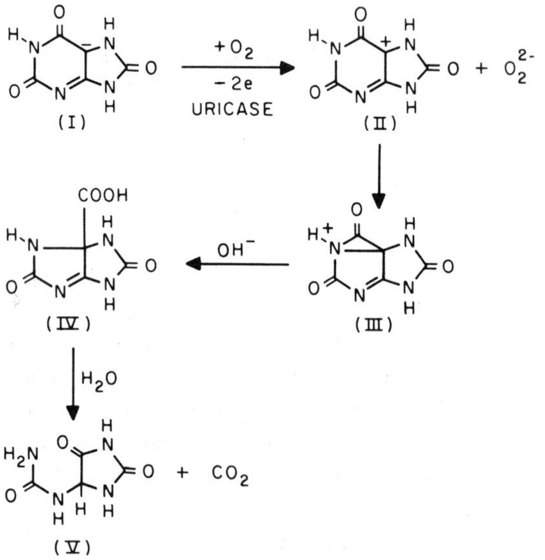

FIG. 3-38.   Mechanism of uricase-catalyzed oxidation of uric acid according to Bentley and Neuberger.[145]

on the mechanistics of the oxidation of uric acid by various peroxidase enzymes indicate that the products obtained depend on the buffer system and pH employed. With horseradish peroxidase[148] it is found that 1-methyluric acid is oxidized, in the presence of hydrogen peroxide, most rapidly when the molecule is uncharged. The primary oxidation product obtained at pH 3–5 can be reversibly transformed into two other spectrophotometrically active forms by changing the pH to >6 or <1. Depending on the pH, the primary oxidation product undergoes a nonenzymatic decomposition to allantoin (pH 3-6) or alloxan (pH ≤ 1). In a general sense the reaction could be represented as in Eq. 4.

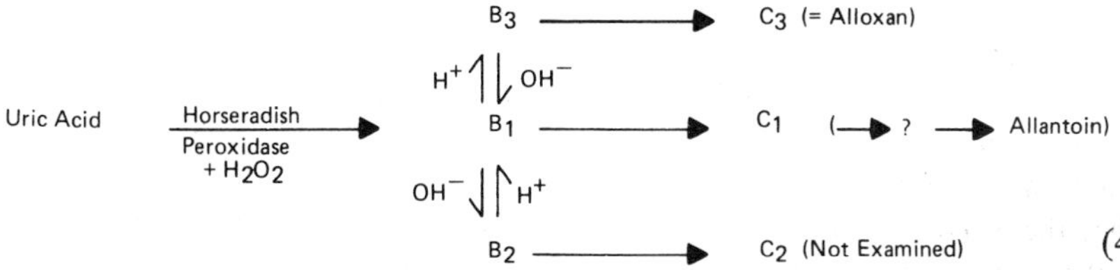

$$
\text{Uric Acid} \xrightarrow[\substack{\text{Horseradish} \\ \text{Peroxidase} \\ + \text{H}_2\text{O}_2}]{} \quad
\begin{array}{l}
B_3 \longrightarrow C_3 \ (= \text{Alloxan}) \\[4pt]
H^+ \Big\uparrow\Big\downarrow OH^- \\[4pt]
B_1 \longrightarrow C_1 \ (\longrightarrow ? \longrightarrow \text{Allantoin}) \\[4pt]
OH^- \Big\downarrow\Big\uparrow H^+ \\[4pt]
B_2 \longrightarrow C_2 \ (\text{Not Examined})
\end{array}
\qquad (4)
$$

Paul and Avi-Dor[148] suggest that the possible structures of B are those shown in Fig. 3-39. Blitz and Max[150] many years earlier had suggested that the 4,5-diol (Fig. 3-39B) was an intermediate in the nitric acid oxidation of uric acid to alloxan or the oxidation with alkaline permanganate to allantoin, although the product supposedly isolated by Blitz and Max has been shown[151] to be some adduct of alloxan and urea rather than a true diol. However, Blitz[152] has reported that the dimethyl ether of B (Fig. 3-39) could be isolated after oxidation of uric acid with chlorine in methanol.

Canellakis *et al.*[149] studied the oxidation of uric acid with the catalase–hydrogen peroxide system and with the lactoperoxidase–H_2O_2, verdoperoxidase–H_2O_2, and horseradish peroxidase–H_2O_2 systems. On the basis of

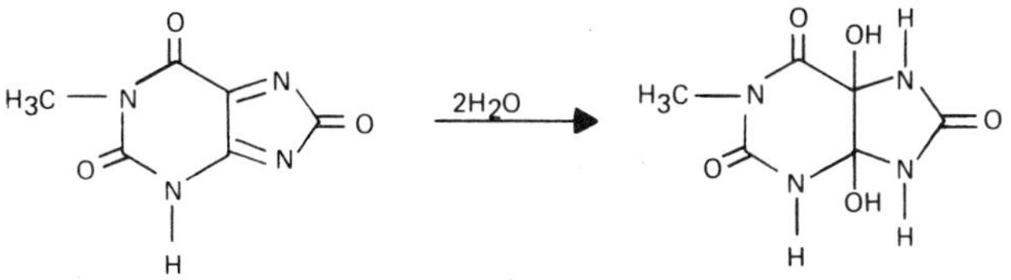

FIG. 3-39. Representation of Paul and Avi-Dor[148] intermediates in the oxidation of 1-methyluric acid in the presence of horseradish peroxidase.

their findings and related studies,[148] they propose that uric acid (I, Fig. 3-40) is oxidized to uric acid-4,5-diol (II, Fig. 3-40). Then, possibly via another intermediate species D, it decomposes to allantoin (VI, Fig. 3-40) at moderate pH. At low pH decomposition to alloxan (V, Fig. 3-40) occurs, possibly via an unspecified intermediate B. At moderately high pH it has been suggested that the ultimate product observed, alloxanic acid (IV, Fig. 3-40), is not produced via the well-known hydrolysis of alloxan, but rather via formation of 5-ureido-2-imidazolidone-4,5-diol-4-carboxylic acid (III, Fig. 3-40).

Howell and Wyngaarden[153] studied the oxidation of uric acid with methemoglobin (a heme protein) and hydrogen peroxide. At pH 5 the rate of oxidation is at its optimum and methemoglobin is as active as horseradish peroxidase. Allantoin is one of the main oxidation products. For the reaction to occur there has to be at least one unsubstituted imidazole nitrogen. Further, it is found that methemoglobin forms a peroxide complex and that, if luminol dye is present during formation of the complex, a light flash is observed in the solution; i.e., hydroxyl radicals are probably formed. It has thus been suggested that the 8-oxo group of uric acid renders the N-9 hydrogen labile and therefore makes this position susceptible to dehydrogenation by the methemoglobin–peroxide complex. It was considered possible that the resultant uric acid radical

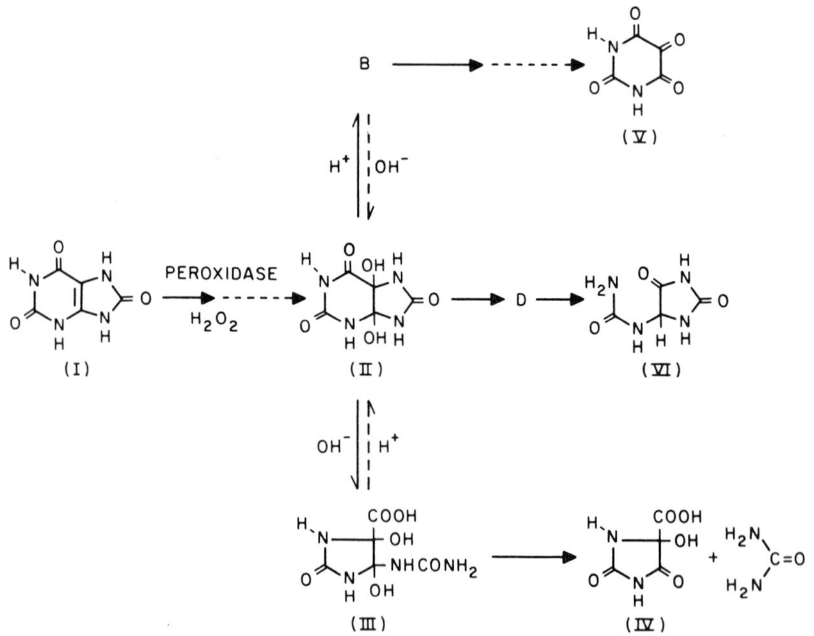

FIG. 3-40.   Proposed generalized mechanism of oxidation of uric acid by peroxidase enzymes according to Canellakis *et al.*[149] and Paul and Avi-Dor.[148]

could exist in several structural forms (II or III, Fig. 3-41), one of which (III) might react in one of two ways: (a) If the structure loses an electron to a hydroxyl radical, a carbonium ion of uric acid (IV, Fig. 3-41) would result and therefore might undergo the Bentley—Neuberger transformation[145] giving allantoin, or (b) if III (Fig. 3-41) adds a hydroxyl radical, the sequence $V \rightarrow X$ (Fig. 3-41) might result; loss of CO_2 from X (Fig. 3-41) would give rise to allantoin. It is particularly interesting to note, however, that hydration of the C-4=N-9 double bond could occur at any stage involving structures III–IX (Fig. 3-41), resulting in certain instances, e.g., III and V (Fig. 3-41), in the

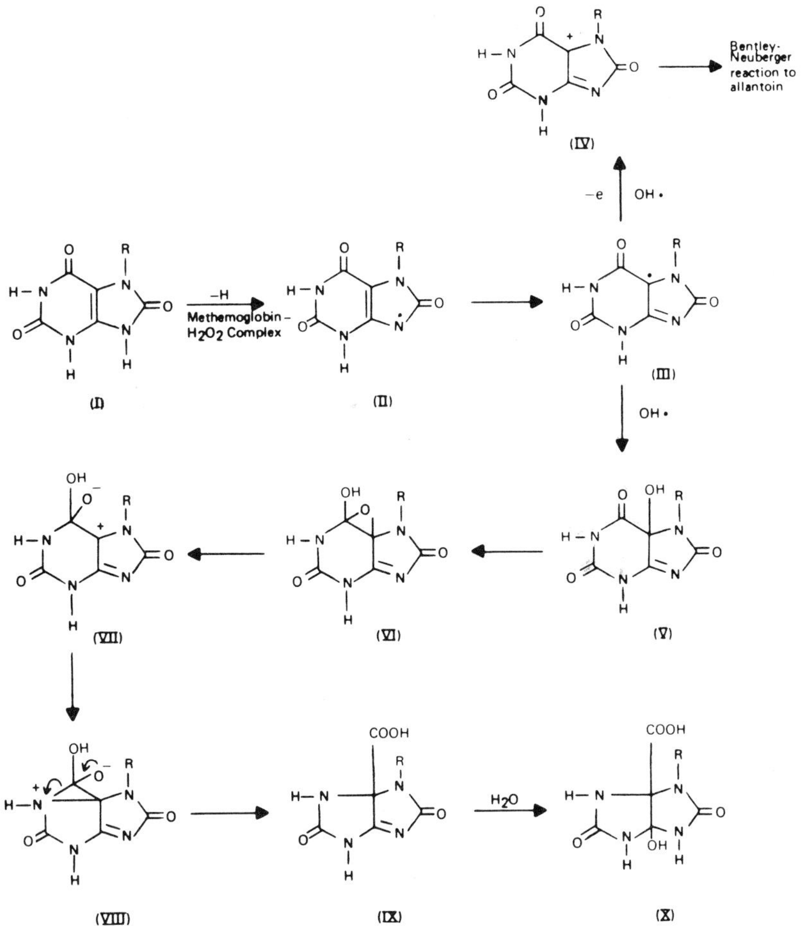

FIG. 3-41. Proposed mechanism of oxidation of uric acid by methemoglobin and H_2O_2 according to Howell and Wyngaarden.[153]

4,5-diol structures outlined previously. The overall electron-transfer sequence probably involves the iron moiety of the heme protein, the sequence being represented in Eq. 5 and effectively representing oxidation of uric acid in a $2e-1H^+$ process.

$$[Fe_p(OH)]^{2+} \xrightarrow{H_2O_2} [Fe_pO]^{2+} + OH\cdot + H_2O$$

$$[Fe_pO]^{2+} \xrightarrow[+e]{+H^+} [Fe_p(OH)]^{2+} \qquad (5)$$

$$OH\cdot \xrightarrow{+e} OH^-$$

Soberon and Cohen[154] have shown that uric acid is oxidized in the presence of myelo-peroxidase and peroxide to alloxan and other products by mechanisms that might be similar to those previously discussed.

c. Photochemistry and Radiation Chemistry.   Because of the increasing awareness of the implications of photochemical processes in mutagenesis and other biological processes, and because of the similarity between the results of electrochemical, biochemical, and photochemical reactions of the purines, a brief outline of some of the photochemistry of the purines is presented here.

Purine bases, in general, are practically unaffected by doses of ultraviolet light that are sufficient to destroy their companions in nucleic acids, namely, the pyrimidines.[155,156] However, certain purines in the presence of a variety of dyes and oxygen, when exposed to visible light of a wavelength absorbed by the dye, are fairly readily oxidized. This effect is often referred to as *photodynamic action*[157] and, although observed for uric acid many years ago,[158] is only recently receiving intensive study.

Simon and Van Vanukis[159] examined the dye-sensitized photooxidation of a number of purines and found that uric acid was the most reactive of all the compounds tested. In general, it appears that the greater the number of oxygen atoms in the purine ring, the more susceptible to photooxidation is the purine. Matsuura and Saito[160,161] irradiated oxygenated alkaline solutions of uric acid with visible light in the presence of the dye rose bengal. A little over 1 mole equivalent of oxygen was consumed, and isolation of the products at pH 2 yielded triuret (V, Fig. 3-42), sodium oxonate (VI, Fig. 3-24), allantoxaidin (VII, Fig. 3-42), and CO_2. Isolation of the products at pH 5 yielded only sodium oxonate and CO_2, leading to the conclusion that allantoxaidin was a secondary product formed during the course of the isolation at low pH. It was concluded that the mechanism involves attack by a reactive form of oxygen, either in its

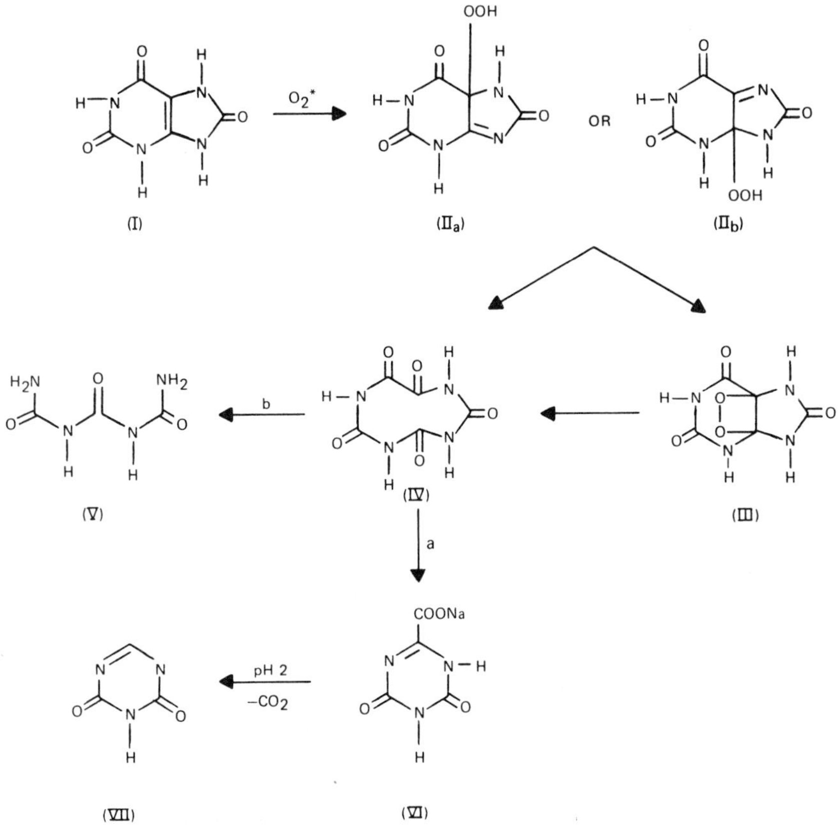

FIG. 3-42. Proposed pathways for the photodynamic oxidation of uric acid in alkaline solution in the presence of rose bengal.[160,161]

singlet or triplet excited state, or in a complex form with the excited sensitizer (dye), on the ground-state uric acid to give a C-4 or C-5 hydroperoxide intermediate (II$_a$ or II$_b$, Fig. 3-42). The latter could cleave concertedly or via a four-membered cyclic peroxide (III, Fig. 3-42) to form a nine-membered ring intermediate (IV, Fig. 3-42) which could be hydrolyzed to sodium oxonate via path a (Fig. 3-42) or to triuret via path b.

However, a further report by Matsuura and Saito[162] reveals that the photosensitized oxidation of 1,3,7,9-tetramethyluric acid (I, Fig. 3-43) in methanol containing rose bengal gives 4,5-dimethoxy-1,3,7,9-tetramethyluric acid (II, Fig. 3-43) as the major product, along with a small amount of allocaffeic acid (III, Fig. 3-43). When the photosensitized oxidation is carried

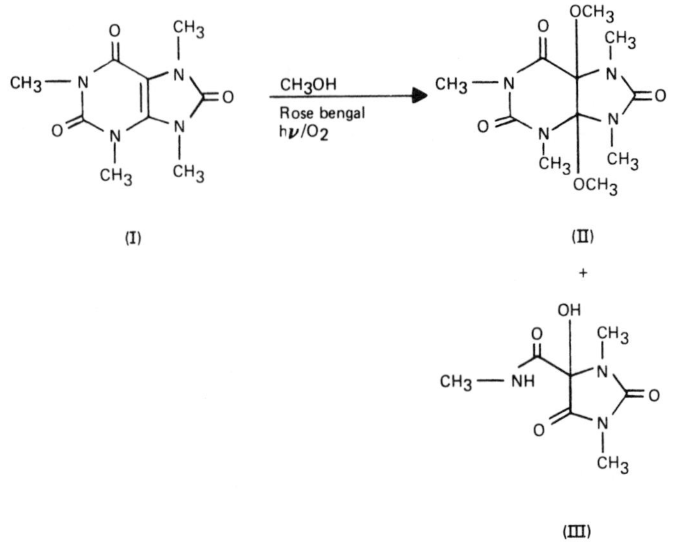

FIG. 3-43.   Photodynamic oxidation of tetramethyluric acid in methanol with a rose bengal sensitizer.[162]

out in chloroform in the presence of a methylene blue photosensitizer, tetramethyluric acid gives a mixture of 1,3,7-trimethylcaffolide (II, Fig. 3-44) and 1,3-dimethylparabanic acid (III, Fig. 3-44).

In the case of photodynamic oxidation of tetramethyluric acid in both methanol and chloroform, a peroxide intermediate structurally similar to II_a or II_b (Fig. 3-42), obtained by attack of excited oxygen on the C-4=C-5 bond, was proposed to be the primary product of the reaction.

There have been other reports of the photosensitized oxidation of uric acids.[163] A particularly interesting one that demonstrates the biological utility of such studies is that of Anmann and Lynch,[164] who observed that the

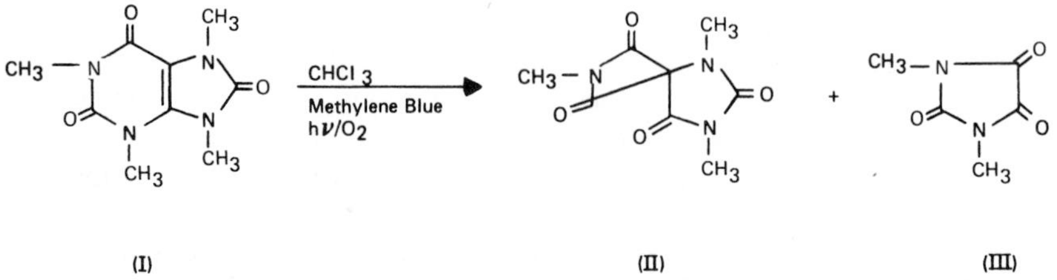

FIG. 3-44.   Photodynamic oxidation of tetramethyluric acid in chloroform.[162]

unicellular alga *Chlorella pyrenoidosa* could utilize uric acid as its sole nitrogen source for growth. In an attempt to elucidate the biochemical pathway for nitrogen incorporation, they found that uric acid could be oxidized *in vitro* in the presence of chlorophyll (which acted as a photosensitizer) and visible light to give allantoin, cyanuric acid, parabanic acid, urea, and other unidentified lower molecular weight products.

There has been relatively little work done on the effects of high-energy ionizing radiation on uric acid solutions.[155,165] However, Holian and Garrison[166] found that, in oxygenated solution, X radiation, which under appropriate conditions generates hydroxyl radicals in aqueous solution, leads to preferential attack at the C-4=C-5 double bond to give an intermediate which possibly has structure **VII**.[167]

(**VII**)

d. Correlations between Electrochemical, Biochemical, and Photochemical Oxidations of Uric Acid. From the considerable body of experimental work reported on uric acid oxidations, it is clear that the preponderance of evidence favors some sort of attack at the C-4=C-5 double bond. Most proposed mechanisms invoke the existence of some type of hydroxy or hydroperoxide unstable primary product or intermediate at some stage in the reaction. For the sake of comparison and discussion the structures of some of these proposed intermediates are presented in Table 3-12.

In the electrochemical and enzymatic oxidations the uric acid-4,5-diol is postulated to be either directly formed or formed by simple addition of the elements of water to some intermediate structure. In the case of photodynamic oxidation the fact that tetramethyluric acid gives the 4,5-dimethoxy derivative is strongly suggestive of a similar intermediate, even though much more complicated multistep mechanisms are proposed. In the case of the oxidation of uric acid under the influence of γ rays, the evidence for the hydroxy–hydroperoxy intermediate is minimal, and detailed analyses of reaction products have not been performed or mass balances attempted. However, in view of the multiplicity of reactive species formed on irradiation of aqueous solutions,[168,169] it is not altogether unlikely that very different processes may occur.

TABLE 3-12   Structures of Proposed Primary Products or Intermediates Formed on Oxidation of Uric Acid

Oxidizing system	Oxidizing agent	Structure	Reference
Enzymatic	Uricase		145
	Peroxidases + H_2O_2		148, 149
		or	
	Heme proteins + H_2O_2		153
Photodynamic	Rose bengal, $h\nu/O_2/H_2O$		160, 161
	Rose bengal, $h\nu/O_2/MeOH$ or $CHCl_3$		162
Ionizing radiation	γ Rays		166, 167

TABLE 3-12 *Continued*

Oxidizing system	Oxidizing agent	Structure	Reference
Electrochemical	Pyrolytic graphite electrode		120, 121, 134
		or	
			119

Comparison of the ultimate products obtained by the various oxidative processes reveals that the products formed upon enzymatic oxidation, particularly with peroxidase enzymes, and electrochemical oxidation at the PGE are essentially identical with respect to the nature, yields, and effects of pH on these. Accordingly, it is pertinent to consider what the electrochemical information reveals about the biological processes that is not available from the enzymatic and other related studies. First, not only can some type of unstable primary product or intermediate be inferred from the nature of the products formed electrochemically, but cyclic voltammetry allows one to physically detect it and confirm that it is extremely unstable and that it is very reducible, i.e., the diimine. The electrochemical data also positively indicate that two electrons and two protons are involved in the primary reaction. In addition, the electrochemical data support the view that a uric acid-4,5-diol is almost certainly formed at some stage of the reaction by hydration of the diimine because of the formation of the 4,5-dimethoxy derivative when methanol is present. Finally, the ultimate products observed electrochemically can all be explained by secondary reactions of the uric acid-4,5-diol using currently acceptable organic mechanisms. The enzymatic reactions, which give essentially the same products, are based on the existence of a reactive unstable intermediate. It seems reasonable, therefore, in view of these findings to propose that the enzymatic and electrochemical mechanisms are essentially identical.

It appears clear that in this, the first instance of a detailed comparison of the electrochemical and enzymatic oxidation of a purine, not only are the processes superficially the same, but electrochemical techniques allow one to sort out a considerable amount of detail on the fine points of the mechanistic processes involved.

2. Xanthines

The earliest report of the electrochemical oxidation of a xanthine is that of Fichter and Kern.[115] They found that theobromine (3,7-dimethylxanthine) and caffeine (1,3,7-trimethylxanthine) were oxidized at a PbO_2 anode in $4 N$ H_2SO_4. The products identified from theobromine were methyl alloxan, 3,7-dimethyluric acid, methylparabanic acid, ammonia, methylamine, and CO_2. The products of oxidation of caffeine were dimethyl alloxan and apocaffeine (**VIII**). When oxidized in a cell without a diaphragm separating the anode and cathode compartments, caffeine gave 1,1′,3,3′-tetramethyl alloxantin (**IX**).

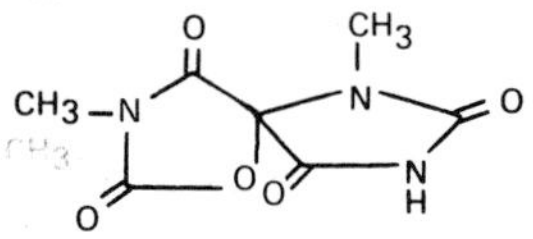

(**VIII**)

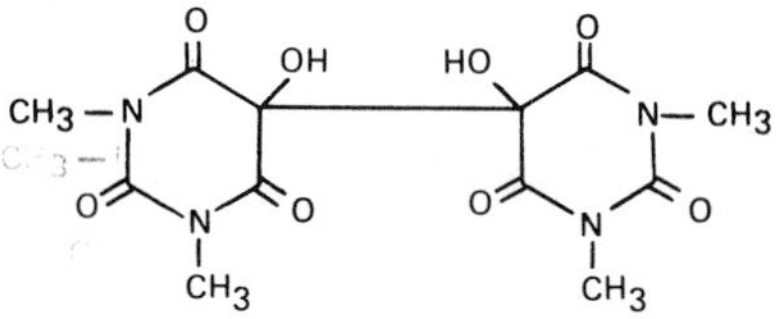

(**IX**)

Smith and Elving[116] reported that xanthine itself is oxidized at a wax-impregnated spectroscopic graphite electrode in $2 M$ H_2SO_4 between pH 3.7 and 4.7, but no mechanistic details were given.

Xanthine is electrochemically oxidized at the PGE by way of a single, pH-dependent voltammetric peak (Table 3-10). Coulometry of xanthine in solutions of pH 1–7 reveals that complete oxidation involves four electrons.[119,122] The electrode reaction appears to proceed via two $2e-2H^+$ oxidations. The first and potential controlling reaction is a $2e-2H^+$ oxidation of the N-7=C-8 (or C-8=N-9) bond of xanthine (I, Fig. 3-45) to give uric acid (II, Fig. 3-45). Since uric acid is more readily oxidized than is xanthine (Table 3-10) the former compound is immediately oxidized further in a $2e-2H^+$ process to uric acid diimine (III$_a$, III$_b$, Fig. 3-45), which then undergoes the same secondary reactions as described previously (Fig. 3-37), so that the same products and about the same yields are observed from xanthine as are observed from uric acid.[119] The involvement of uric acid and its diimine in the electrochemical oxidation of xanthine is clearly established from fast sweep cyclic voltammetry.[119,121] A cyclic voltammogram for xanthine at a clean PGE is shown in Fig. 3.46. The first sweep toward positive potentials shows only a single peak (peak I_a) corresponding to the $4e-4H^+$ oxidation of xanthine to the uric acid diimine (I–III_a/III_b, Fig. 3-45). Provided that the sweep rate is fast enough the diimine can be detected as a reduction peak (peak I_c, Fig. 3-46).

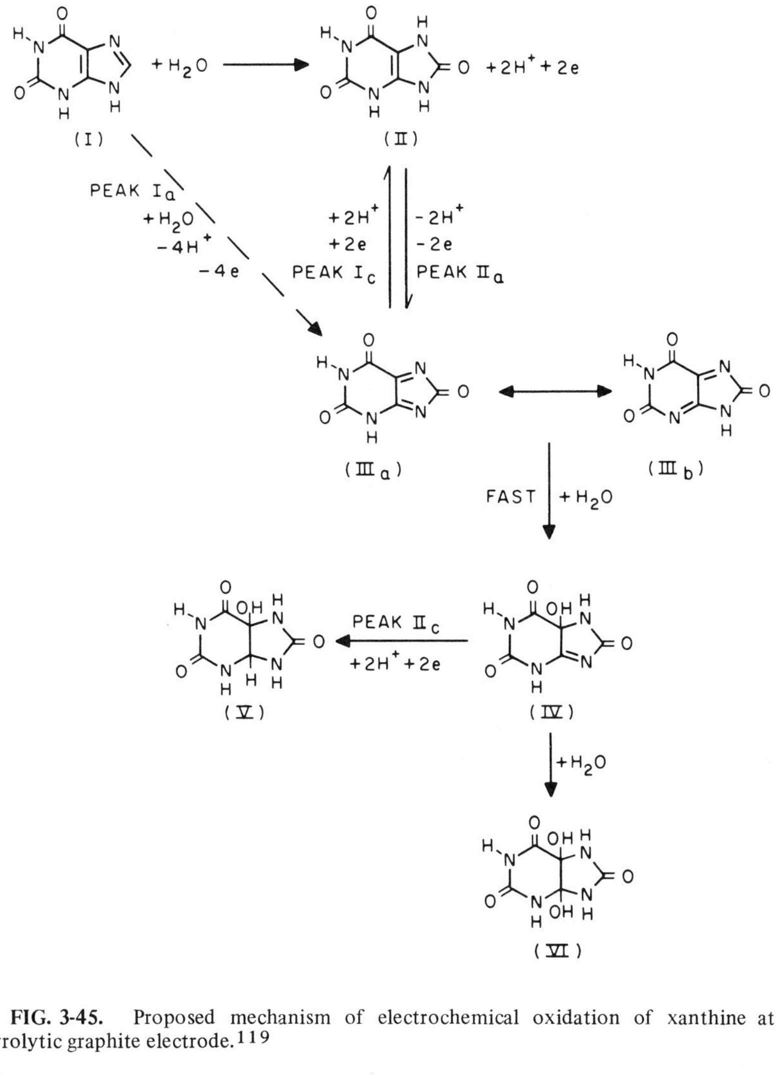

FIG. 3-45. Proposed mechanism of electrochemical oxidation of xanthine at the pyrolytic graphite electrode.[119]

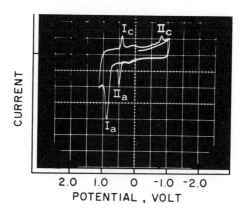

FIG. 3-46. Cyclic voltammogram of xanthine in pH 4.7 acetate buffer at a clean PGE. Scan pattern, $0.0 \text{ V} \rightarrow -1.10 \text{ V} \rightarrow 1.20 \text{ V} \rightarrow -1.10 \text{ V} \rightarrow 0.6 \text{ V}$; scan rate, 4.6 V sec[-1]; current sensitivity, 200 μA per division. Current above axis marker is cathodic; that below axis marker is anodic.[121] (Reprinted with the permission of Springer-Verlag Publishers, New York.)

Peak I_c corresponds to reduction of the diimine to uric acid ($III_a/III_b \rightarrow II$, Fig. 3-45). On the second positive-going sweep the uric acid formed in the latter reaction is reoxidized to the diimine and gives rise to peak II_a (Fig. 3-46). The more negative, second cathodic peak observed in Fig. 3-46 (peak II_c) has been proposed[119] to be due to reduction of the partially hydrated diimine (IV, Fig. 3-45) to compound V (Fig. 3-45). The slight disagreement between the peak potentials for peaks due to uric acid and its diimine in Figs. 3-35 and 3-46 is due to the shift of these peaks as a function of concentration.[119]

Studies of the linear sweep and cyclic voltammetric behavior of N-methylated xanthines[121,170] reveal that they undergo electrochemical oxidation over a fairly wide pH range at the PGE (Table 3-10). All but three of the xanthines studies show just a single voltammetric oxidation peak, although it is probable that the additional peaks observed at most negative potentials (peak I, Table 3-10) for some xanthines are due to adsorption processes.[121] Fast sweep cyclic voltammetry of many methylated xanthines reveals that methylation of the xanthine at N-7 causes a very pronounced decrease, perhaps even elimination, of the cathodic peak corresponding to reduction of the diimine primary oxidation product to the appropriate uric acid (i.e., equivalent to peak I_c of Figs. 3-35 and 3-46). In the case of 3,7-dimethylxanthine (theobromine) and 1,3,7-trimethylxanthine (caffeine) there was no evidence at all from fast sweep cyclic voltammetry that peaks equivalent to peaks I_c and II_a of xanthine (see Fig. 3-46) were present.

In order to demonstrate that the effect of N-3 and N-7 dimethylation of xanthines was principally to alter the stability or lifetime of a diimine type of intermediate such that it could not be detected cyclic voltammetrically, rather than to completely alter the overall mechanism, Hansen and Dryhurst[170] examined the nature and amounts of the products of electrochemical oxidation of theobromine and caffeine. Both of these compounds are oxidized by way of a single voltammetric oxidation peak at the PGE in a process that involves, overall, four electrons. In view of the pH dependence of the peak observed for both compounds (Table 3-10) and the nature of the products formed (Figs. 3-48—3-50) a mechanism can be proposed in which, as with xanthine, the initial $2e-2H^+$ reaction involves oxidation of the N-9=C-8 double bond of the methylated xanthine (I, Fig. 3-47) to give the corresponding methylated uric acid (II, Fig. 3-47). The follow-up electrochemical reactions can be described as removal of a proton and two electrons from the C-4=C-5 double bond of the uric acid to give not a diimine but rather a diiminium ion (II → III, Fig. 3-47). This iminium ion would be expected to be extraordinarily susceptible to hydration[171] so that the corresponding methylated uric acid-4,5-diol (IV, Fig. 3-47) would be formed very rapidly. It is logical, therefore, that the failure to detect an unstable, reducible product by fast sweep voltammetry of theobromine and caffeine is not due to the fact that a reducible species is not produced, but rather to the fact that the reducible iminium ion (III, Fig. 3-47) would be extremely rapidly hydrated to give the electrochemically inactive substituted uric acid-4,5-diol (IV, Fig. 3-47).

The observed electron number and amount of substituted parabanic acids formed upon oxidation of theobromine and caffeine indicate that the primary

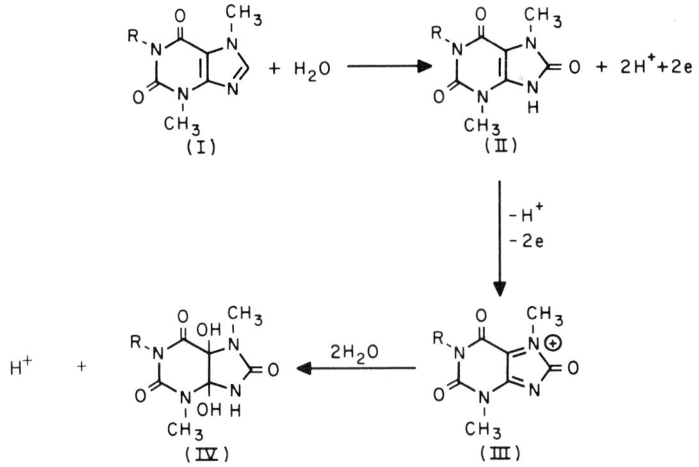

FIG. 3-47. Mechanism of primary electrochemical oxidation of theobromine (I, R = H) and caffeine (I, R = CH$_3$) at the PGE.[170]

route of decomposition of a 3,7-dimethyluric acid-4,5-diol is to allantoin and alloxan derivatives. Nevertheless, small amounts of parabanic acid are produced. In the case of theobromine the appropriate uric acid-4,5-diol (IV, Fig. 3-48) could undergo ring opening across the N-3—C-4 bond to give V_a (Fig. 3-48), which upon protonation and fragmentation would give 2-oxy-3-methyl-4,5-dihydroxyimidazole (VII, Fig. 3-48) and the isocyanate VIII (Fig. 3-48). Electrochemical oxidation of the former would lead to methylparabanic acid (IX Fig. 3-48) derived from the imidazole moiety of the original compound, while hydrolysis of VIII (Fig. 3-48) would yield CO_2 and N-methylurea. The same products could be obtained via intermediates V_b and VI (Fig. 3-48), although in this case the N-methylparabanic acid that results would originate from the pyrimidine ring moiety of the original compound. In the case of caffeine only dimethylparabanic acid is produced.[170] A mechanism analogous to IV → V_a (Fig. 3-48) is not possible with caffeine, which is methylated at N-1. The only route to parabanic acid is first by opening of the N-3—C-4 bond of X (Fig. 3-48) to give XI (Fig. 3-48) followed by ring closure to XII (Fig. 3-48). This, upon protonation and fragmentation, would give 2-oxy-1,3-dimethyl-4,5-dihydroxyimidazole (XIII, Fig. 3-48), which should be readily oxidized to dimethylparabanic acid (XV, Fig. 3-48). The isocyanate XIV (Fig. 3-48) would hydrolyze to CO_2 and methylurea.

Formation of methylated allantoins from the uric acid-4,5-diols would likely proceed by different mechanisms. Protonation of the diol IV, (Fig. 3-49) derived from theobromine would lead to ring opening at the C-6—C-5 position, giving an imidazole isocyanate (XVI, Fig. 3-49). This could readily form XVII (Fig. 3-49) which after hydrolysis and loss of CO_2 would give dimethyl allantoin (XVIII, Fig. 3-49). On the other hand, the uric acid diol derived from caffeine (X, Fig. 3-49) cannot fragment by this mechanism. Accordingly, either or both of the processes could occur via the form of the diol hydrated at the C-6 carbonyl group (XIX, Fig. 3-49), which could readily lose CO_2 to give XX (Fig. 3-49) followed by rearrangement to trimethyl allantoin (XXI, Fig. 3-49).

Secondary rearrangement of the uric acid-4,5-diol derived from theobromine or caffeine to give a methylated alloxan is shown in IV → XXII and X → XXIII (Fig. 3-50), respectively.

Hansen and Dryhurst[172] found that theophylline (1,3-dimethylxanthine) is electrochemically oxidized by way of a single pH-dependent (Table 3-10) voltammetric peak at the PGE in a process that in 1 M HOAc involves about three electrons per molecule of theophylline oxidized. The products and their quantitative yields are shown in Fig. 3.51. These products differ from those obtained from other xanthines in that a theophylline dimer, 8-(1,3-dimethylxanthyl)-1,3-dimethylxanthine (V, Fig. 3-51) is formed. All of the remaining products are similar to those obtained from other xanthines with the obvious methylation differences. In view of the dimer formation it is obvious

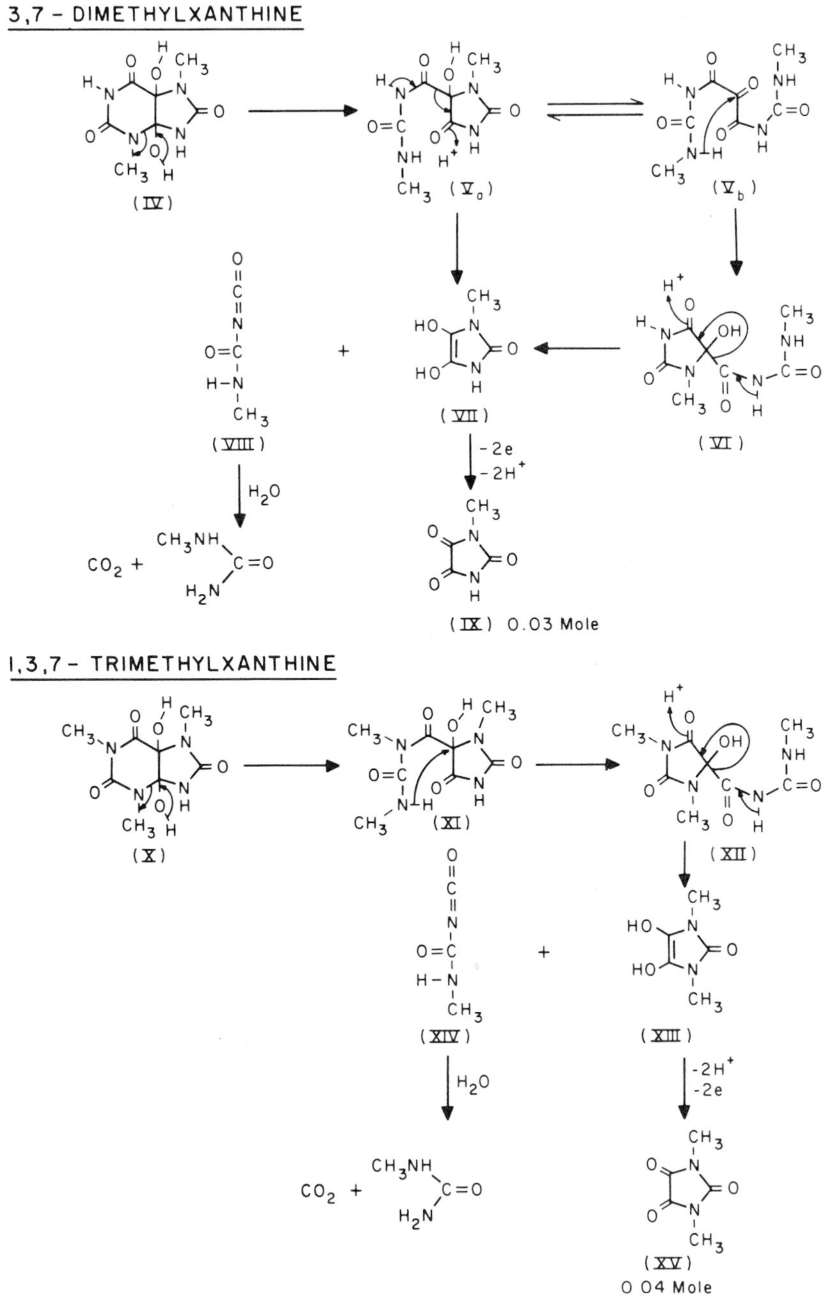

3,7 - DIMETHYLXANTHINE

1,3,7 - TRIMETHYLXANTHINE

FIG. 3-48. Mechanism of formation of parabanic acids from the methylated uric acid-4,5-diols derived from theobromine and caffeine. Molar amounts of products are those formed in $1 M$ HOAc.[170] (Reprinted with the permission of Springer-Verlag Publishers, New York.)

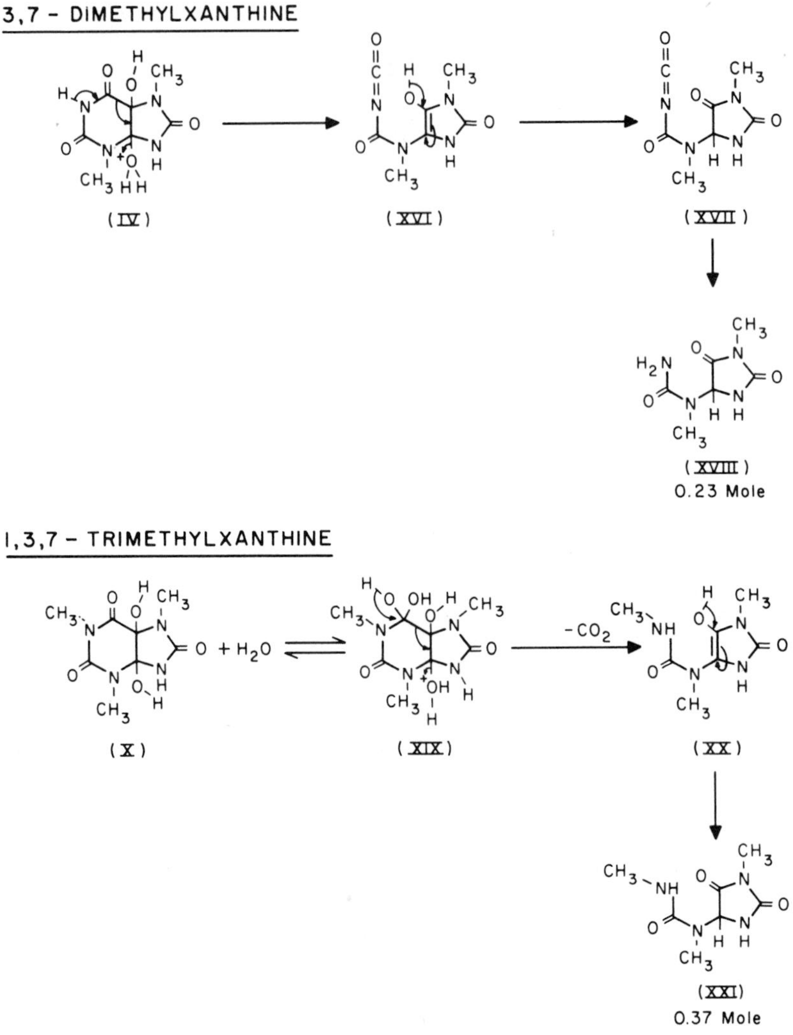

3,7 – DIMETHYLXANTHINE

(IV)   (XVI)   (XVII)

(XVIII)
0.23 Mole

1,3,7 – TRIMETHYLXANTHINE

(X)   (XIX)   (XX)

(XXI)
0.37 Mole

FIG. 3-49. Mechanism of formation of allantoins from the methylated uric acid-4,5-diol derived from theobromine and caffeine. Molar amounts of products are those formed in 1 *M* HOAc.[170] (Reprinted with the permission of Springer-Verlag Publishers, New York.)

that the first two electrons must be removed from theophylline (I, Fig. 3-51) in a stepwise manner, resulting first in the formation of a free radical (II, Fig. 3-51). About 40% of the free radical dimerizes to give V, Fig. 3-51, while the remainder is further oxidized to 1,3-dimethyluric acid (III, Fig. 3-51). This is then further oxidized to the diimine (IV, Fig. 3-51). The diimine is susceptible

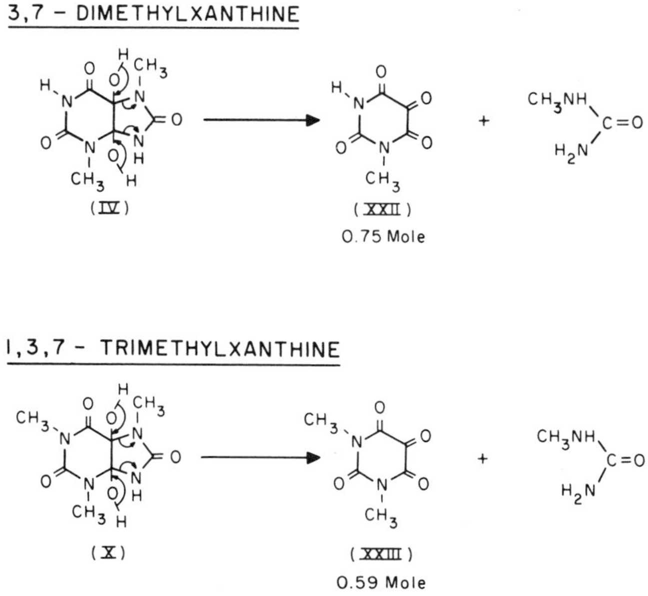

3,7 – DIMETHYLXANTHINE

(IV)

(XXII)

0.75 Mole

1,3,7 – TRIMETHYLXANTHINE

(X)

(XXIII)

0.59 Mole

FIG. 3-50. Mechanism of formation of alloxans from the methylated uric acid-4,5-diols derived from theobromine and caffeine. Molar amounts of products are those formed in 1 M HOAc.[170] (Reprinted with the permission of Springer-Verlag Publishers, New York.)

to hydrolysis, but it is sufficiently stable to be detected by fast sweep cyclic voltammetry as a cathodic peak corresponding to reaction IV → III (Fig. 3-51). The 1,3-dimethyluric acid (III, Fig. 3-51) so formed can be detected by cyclic voltammetry since it is more readily oxidizable than theophylline.[121,172] Hydration of IV (Fig. 3-51) results in formation of 1,3-dimethyluric acid-4,5-diol (VI, Fig. 3-51), which can undergo a ring opening at the N-3—C-4 bond to give the substituted imidazole (VII$_a$, Fig. 3-51) or its hydrated derivative (VIII$_b$, Fig. 3-51). Protonation and fragmentation of the latter would yield dimethyl-urea and CO_2 and the 4,5-dihydroxyimidazole (VIII, Fig. 3-51), which should be readily electrochemically oxidized to parabanic acid (IX, Fig. 3-51). Again, unsubstituted parabanic acid can result only from the original imidazole moiety of the purine ring. The hydrated form of 1,3-dimethyluric acid (VI$_a$, Fig. 3-51) upon protonation can give rise to X (Fig. 3-51), which should readily form dimethyl allantoin (XI, Fig. 3-51). Simple cleavage of VI (Fig. 3-51) yields dimethyl alloxan (XII, Fig. 3-51) and urea.

a. Biochemical Studies. Most studies concerned with the biochemical oxidation of oxypurines other than uric acid have centered around the enzyme

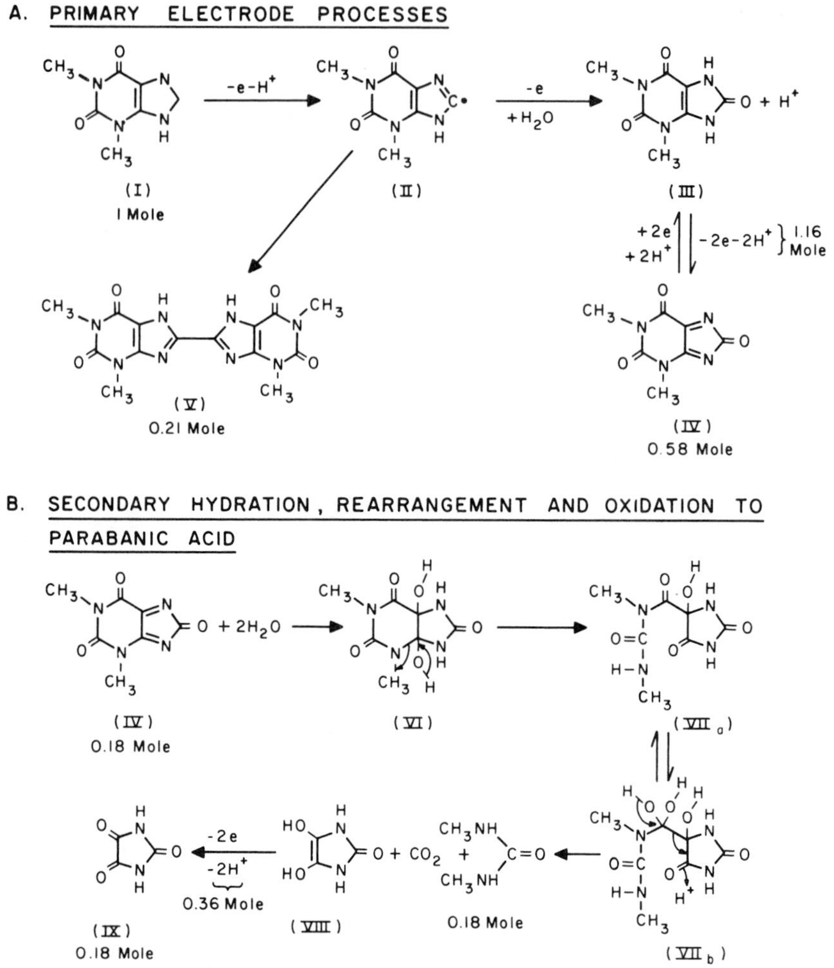

A. PRIMARY   ELECTRODE   PROCESSES

B. SECONDARY   HYDRATION ,   REARRANGEMENT   AND   OXIDATION   TO
PARABANIC   ACID

FIG. 3-51.   Products and mechanism of the electrochemical oxidation of theophylline at the PGE. Molar amounts of products are those formed in 1 *M* HOAc.[172] (Reprinted with the permission of Springer-Verlag Publishers, New York.)

xanthine oxidase. This enzyme has been known for a considerable time[173] and contains a flavin adenine dinucleotide prosthetic group (i.e., a cofactor or coenzyme firmly bound to the enzyme protein without which xanthine oxidase cannot function).[174] Molybdenum is apparently present as a functionally important component of the enzyme.[175,176] Xanthine oxidase is a rather nonspecific enzyme.[37] Bergmann and Dikstein[177] examined the main

C. SECONDARY HYDRATION AND FRAGMENTATION TO DIMETHYL ALLANTOIN

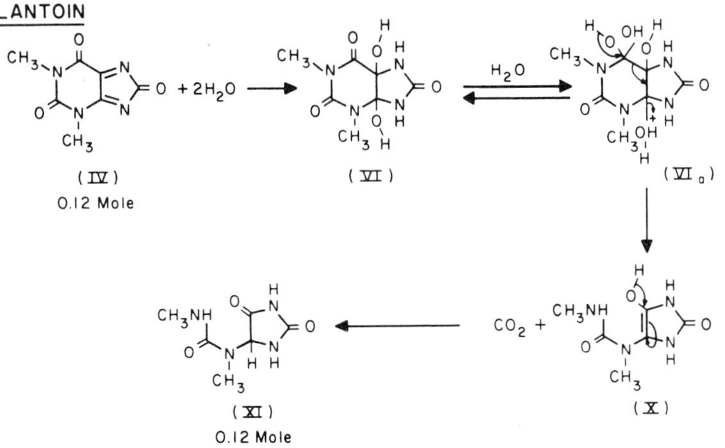

D. SECONDARY HYDRATION AND FRAGMENTATION TO DIMETHYL ALLOXAN

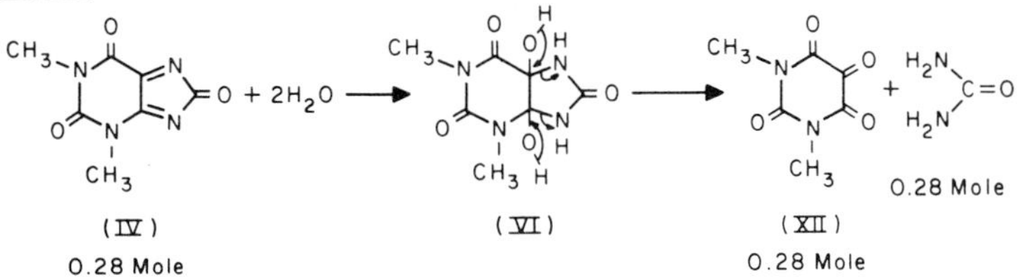

FIG. 3-51. *Continued.*

oxidative pathways in the xanthine oxidase oxidation of purines and concluded that the rate and direction of the oxidative attack depends on the position of the oxygen introduced previously. With purine itself the initial attack is at C-6 to give hypoxanthine. Hypoxanthine is then oxidized at the C-2 position to give xanthine. However, positions C-2 and C-8 form a closely related pair so that oxidation at one of these carbons always leads to attack at the other. By studying a large number of purines, Bergmann and Dikstein[177] found that, on the basis of their response to xanthine oxidase, there were three distinct groups of purines:

1. Those that reacted with xanthine oxidase at a rate of oxidation comparable to that of xanthine, e.g., 1-methylxanthine, 6,8-dioxypurine, hypoxanthine, and purine

2. Those that were attacked at one-tenth to one-ten thousandth the rate of xanthine, e.g., 2- and 8-oxypurine and 2,8-dioxypurine

3. Those that were not attacked at a measurable rate, e.g., 1- and 7-methylhypoxanthine, 3-, 7-, and 9-methylxanthine, and caffeine (1,3,7-trimethylxanthine)

The mechanism for conversion of purine to uric acid (Fig. 3-52) was formulated as a series of hydration, dehydrogenation, and hydrogen shift reactions. Thus, purine (I, Fig. 3-52) could be hydrated at the C-6 and N-3 positions to give II (Fig. 3-52), which by dehydrogenation across the $-N$-3$-C$-4$=C$-5$-N$-7$- grouping leads to III (Fig. 3-52). Then follows a hydrogen shift to give hypoxanthine (IV or V, Fig. 3-52). Evidence for this type of mechanism was that xanthine (IX, Fig. 3-52) could be oxidized if methylated at N-1 but not if methylated at N-3, N-7, or N-9. Some later studies by Bergmann and co-workers examined the action of mammalian xanthine oxidase on 8-azapurines[178] and pteridines.[179] They concluded that the mechanism proposed earlier for purines[177] (Fig. 3-52), which involved loss of two hydrogen atoms from the HN-3$-C$-4$=C$-5$-N$-7H grouping (to the N-1 and N-10 positions of the flavin moiety of xanthine oxidase) was not completely correct. Accordingly, a further series of methyl derivatives of purines was examined,[180] and indeed a compound such as 7-methyl-8-oxypurine was found to be oxidized at C-2 in spite of the blockage of the N-7 position by a methyl group. The modified mechanism that resulted from this latter study was that the enzyme binds a specific tautomeric form of the substrate, regardless of whether or not that form represents the major structure present in solution. By analogy with the oxidation of aldehydes with xanthine oxidase, it was assumed that purines undergo hydration at one $-HC$=N$-$ grouping either prior to or simultaneously with a dehydrogenation step. Accordingly, the process would involve either pathway a or b of Fig. 3-53. Route a would give a lactim form of the oxidized purine, while b would give the corresponding lactam. It was proposed that, in multistage oxidations such as the conversion of purine to uric acid, the enzyme–substrate complex dissociates after each individual step to recombine with the newly oxidized purine in a different fashion. This was evident because intermediates in multistage oxidations accumulated in solution and could be detected spectrophotometrically or chromatographically.

Trotter and co-workers[181] found that, if luminol dye was present during the oxidation of hypoxanthine by oxygen in the presence of calf liver xanthine oxidase, light was emitted from the solution. The intensity of the light was found to be directly related to the reaction velocity of the enzyme-catalyzed reaction. The results were tentatively interpreted as indicating that hypoxanthine is oxidized in more than one step, with the production of hydroxyl or oxygen radicals.

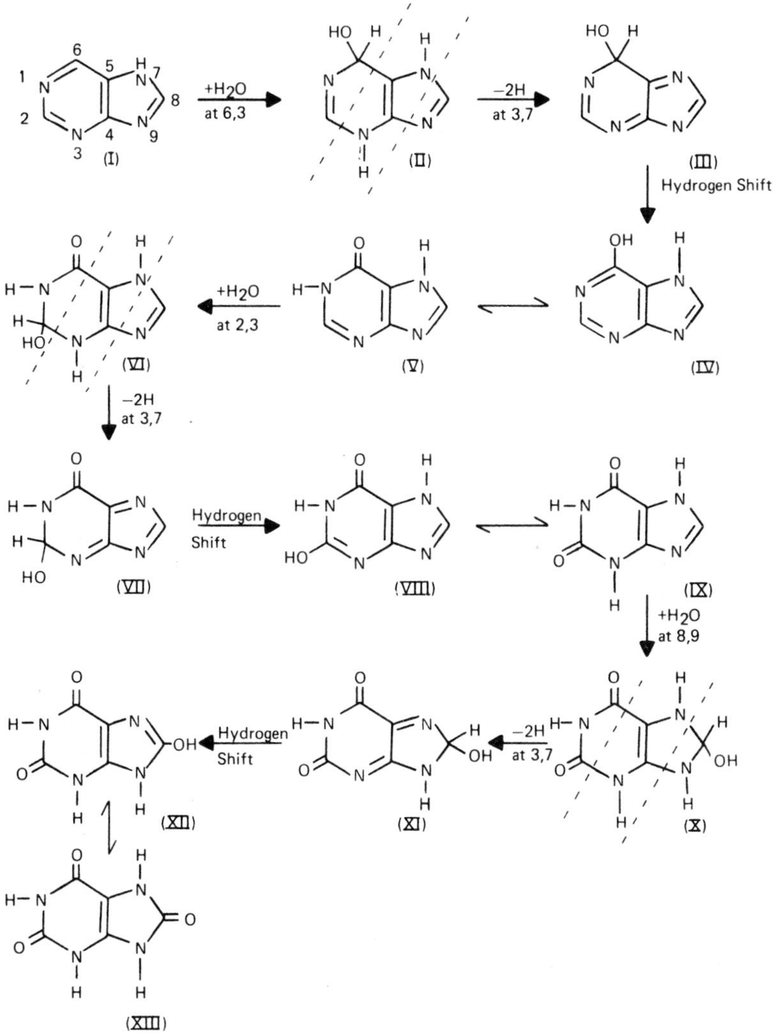

FIG. 3-52. First proposed mechanisms for the oxidation of purines in the presence of xanthine oxidase.[177]

A considerable amount of work has been concerned with the biochemistry of the methylated xanthines caffeine, theophylline, and theobromine. Many of the early studies suggested that all three compounds were metabolized in man to uric acid.[182,183] However, Buchanan et al.[184] later reported that uric acid is not excreted as a result of metabolism of caffeine and theophylline, but rather

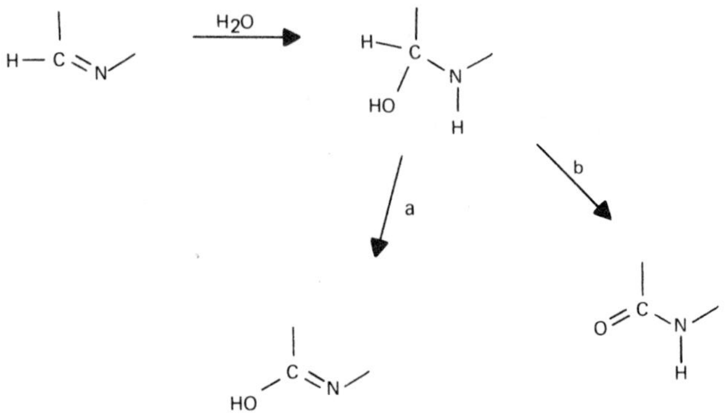

FIG. 3-53.   Generalized route for oxidation of purines by xanthine oxidase.[180]

1-methyl-, 3-methyl-, and 1,3-dimethyluric acids are produced. Brodie *et al.*[185] showed that the primary metabolic oxidation product of theophylline (1,3-dimethylxanthine) is 1,3-dimethyluric acid. Weinfeld and Christman[186] succeeded in isolating 1-methyluric acid from human urine after ingestion of caffeine, and both 1-methyl- and 1,3-dimethyluric acid after ingestion of theophylline. A report by Cornish and Christman[56] details the probable metabolic pathways for caffeine, theophylline, and theobromine. In summary, 62% of theobromine, 77% of theophylline, and 66% of caffeine appear to be excreted in the form of methylxanthines and methyluric acids within 48 hr. There is a considerable amount of demethylation in man, the order of demethylation being N-3, N-7, N-1, although total demethylation does not appear to occur.

    b. Photochemistry and Radiation Chemistry.   Matsuura and Saito[160,161] studied the photodynamic oxidation of xanthine in alkaline solution in the presence of rose bengal under bubbling oxygen and visible light. A little over 1 mole equivalent of oxygen was consumed and the major isolated product was allantoin, along with CO_2 and a small amount of triuret. The formation of allantoin was rationalized by proposing that xanthine (I, Fig. 3-54) formed a cyclic peroxide (II, Fig. 3-54) by reaction with an excited form of oxygen. Rearrangement of the cyclic peroxide was postulated to give rise to an alloxanimide derivative (III, Fig. 3-54), which would presumably undergo a benzylic acid type of rearrangement (III–IV, Fig. 3-54) followed by decarboxylation to allantoin (V, Fig. 3-54). It had been previously reported that theophylline (I', Fig. 3-54) gave dimethyl allantoin (V', Fig. 3-54) under similar conditions,[187] so that a generalized mechanism can be proposed (Fig. 3-54). This mechanism is supported by Friedman's[188] observation that a prerequisite

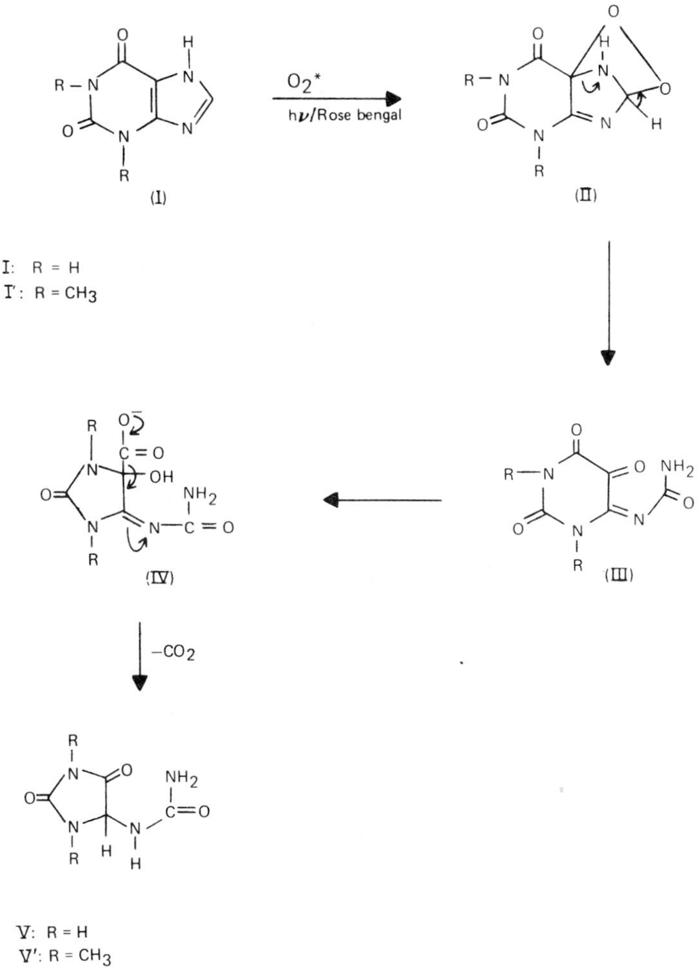

FIG. 3-54. Proposed mechanism of photodynamic oxidation of xanthine and theophylline in alkaline solution.[160,161,187] (I : R = H, I' : R = CH$_3$; V : R = H, V' : R = CH$_3$).

for the photosusceptibility of a purine to photodynamic action is a chemical structure that can be oxidized at C-8. In this respect, Elad *et al.*[189] studied the γ-ray- and UV-induced reactions of caffeine in alcohols and found that a reaction could be induced directly, or sensitized radiolytically or photochemically with a ketone, which resulted in substitution at the C-8 position. For example, with 2-propanol both the γ-ray-induced and photochemical reactions lead to **X**. Zenda *et al.*,[163] however, also systematically studied the photo-dynamic oxidation of purines in the presence of methylene blue at 0°C and concluded that, in order for the purine to be easily photodecomposed, it was

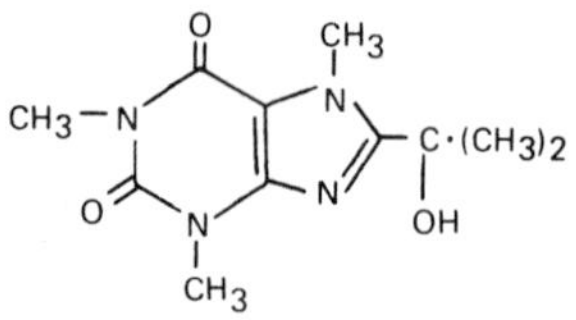

(**X**)

necessary to have the lactim structure with respect to N-1 and N-3 and that an imidazole ring was essential for photodegradation. Substituents at the C-8 and N-9 positions apparently did not affect the photochemical reaction. The partial structure **XI** is that associated with the greatest ease of photodegradation, X and Y being without influence, and R and R' being O or NH.

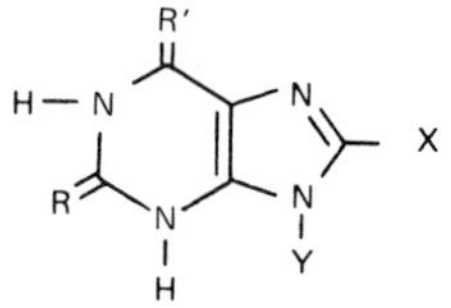

(**XI**)

Holian and Garrison[166] claim that the radiolytic oxidation (γ rays) of xanthine and hypoxanthine involves preferential attack at the C-4=C-5 double bond to give a labile hydroxyhydroperoxide intermediate, which undergoes hydrolytic degradation to yield alloxan, ammonia, formic acid, and hydrogen peroxide from xanthine, and mesoxalic acid, oxalic acid, ammonia, urea, formic acid, and hydrogen peroxide from hypoxanthine.

3. Adenine

Adenine is electrochemically oxidized by way of a single, well-defined peak in aqueous solution at the pyrolytic graphite electrode. Dryhurst and Elving[190] studied the mechanism of this reaction and found that the electrochemical oxidation of adenine involves transfer of close to six electrons per molecule oxidized. Analysis of the product solution after an electrolysis in 1 *M* HOAc revealed the presence of parabanic acid, oxaluric acid, urea, ammonia, allantoin, and 4-aminopurpuric acid. Initially,[190] a mechanism was proposed that involved two sequential $2e-2H^+$ oxidations to give first 2-oxy- and then 2,8-

dioxyadenine (or vice versa). Then, since 2,8-dioxyadenine is more easily oxidized than adenine,[191] a further 2e oxidation at the C-4=C-5 double bond was thought to occur to give a dicarbonium ion intermediate similar to that originally proposed for uric acid.[117] Subsequent studies[120] using fast sweep cyclic voltammetry revealed that a reducible short-lived electrode product could be detected at about the same potential as was observed upon cyclic voltammetry of xanthine and uric acid (*vide supra*). The pH dependence of the cathodic peak of the reducible product, and of the anodic peak for reoxidation of the product of the latter cathodic process (which formed an almost reversible couple), suggested that an equal number of electrons and protons were involved in the redox processes. It was initially proposed that the reducible species was 6-amino-2,8-dioxypurine-4,5-diol; however, in view of the more explicit mechanisms now proposed for uric acid and xanthines, the reducible species is probably a diimine. Accordingly, the probable mechanism[122] for oxidation of adenine (I, Fig. 3-55) is, as originally proposed, three primary $2e-2H^+$ oxidations to give 2-oxyadenine (II, Fig. 3-55), 2,8-dioxyadenine (III, Fig. 3-55), and the diimine (IV, Fig. 3-55). It is probably the diimine that is detected by fast sweep cyclic voltammetry as a reduction peak to give 2,8-dioxyadenine (III, Fig. 3-55), which is then detected as an anodic peak (at potentials more negative than the peak of adenine) due to its oxidation back to the diimine. Secondary hydrolysis of the diimine could give the 4,5-diol (V, Fig. 3-55), which upon hydrolysis and fragmentation gives the 2-oxy-4,5-dihydroxyimidazole (VI → VII, Fig. 3-55). This would be expected to be readily oxidized to parabanic acid (X, Fig. 3-55), part of which was hydrolyzed to oxaluric acid (X_a, Fig. 3-55). The remaining fragment (VIII, Fig. 3-55) could be readily hydrolyzed to ammonia, urea, and CO_2. Similarly, fragmentation of (V, Fig. 3-55) should lead to allantoin (XI → XIV, Fig. 3-55). Under the experimental conditions employed by Dryhurst and Elving,[190] the counterelectrode was unavoidably present in the same solution as the working electrode so that a portion of the diimine (IV, Fig. 3-55) could undergo reduction at the counterelectrode, resulting in formation of 4-aminopurpuric acid. It is possible that the nonappearance of alloxan in the oxidation of adenine might be related to the secondary reduction of the diimine (IV, Fig. 3-55) to 4-aminopurpuric acid.

a. Adsorption of Adenine at the PGE. In the course of developing an analytical method for determination of adenine and its nucleoside adenosine in mixtures, Dryhurst[192] found that both compounds are strongly adsorbed at the PGE. In the presence of adenosine, adsorbed adenine is displaced from the electrode surface by adsorbed adenosine; i.e., a competitive adsorption occurs. The peak potential for oxidation of adenine occurs at more negative potentials than that of adenosine[192] (i.e., for adenine $E_p = 1.338 - 0.063$ pH and for adenosine $E_p = 1.778 - 0.087$ pH at 5 mV sec^{-1} scan rate). One of the effects,

PRIMARY ELECTROCHEMICAL OXIDATION

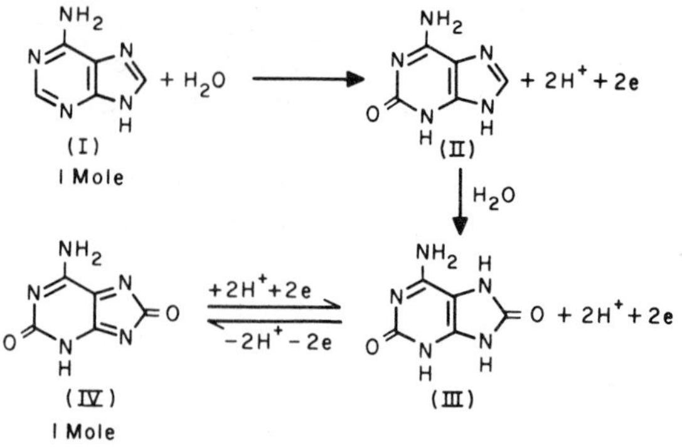

(I)
1 Mole
+ H₂O ⟶ (II) + 2H⁺ + 2e

(II) + H₂O ⟶ (III) + 2H⁺ + 2e

(IV) ⇌ (III) (+2H⁺+2e / −2H⁺−2e)

(IV)
1 Mole

SECONDARY HYDRATION, FRAGMENTATION AND OXIDATION TO PARABANIC ACID

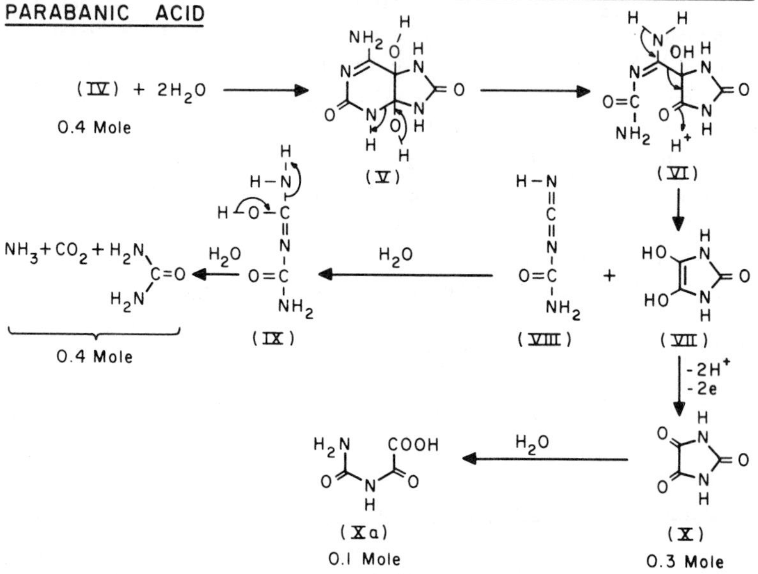

(IV) + 2H₂O ⟶ (V) ⟶ (VI)
0.4 Mole

(VI) ⟶ (VIII) + (VII)

NH₃ + CO₂ + H₂N–C(=O)–NH₂ ⟵ (IX) ⟵ (VIII) + (VII)
0.4 Mole

(VII) −2H⁺ −2e ⟶ (X)

(X) + H₂O ⟶ (Xa)
0.3 Mole → (Xa) 0.1 Mole

(IX)
(VIII)
(VII)

(Xa)
0.1 Mole

(X)
0.3 Mole

SECONDARY HYDRATION AND FRAGMENTATION TO ALLANTOIN

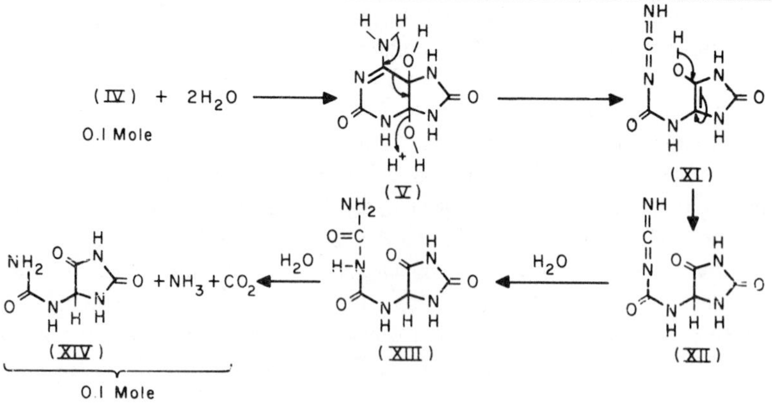

(IV) + 2H₂O ⟶ (V) ⟶ (XI)
0.1 Mole

(XI) ⟶ (XII)

(XII) + H₂O ⟶ (XIII)

(XIII) + H₂O ⟶ (XIV) + NH₃ + CO₂
0.1 Mole

(XIV)

(XIII)

(XII)

therefore, of the displacement of adsorbed adenine by adenosine is that the peak current for the anodic peak of adenine decreases as the concentration of adenosine increases. It was found that, when the concentration of adenosine exceeds 6 mM, adenine is essentially completely desorbed from the electrode surface and the peak current for adenine is that expected for a diffusion-controlled electrode reaction.

b. Biochemical Studies.   The oxidation of adenine in animal tissues is now known to be the result of the action of the enzyme xanthine oxidase.[193-195] The product of the oxidation is 2,8-dioxyadenine.[195,196] Wyngaarden and Dunn[197] studied the oxidation of adenine and 2-oxy- and 8-oxyadenine and concluded that the 8-oxy derivative was the principal intermediate in the oxidation by xanthine oxidase. Bergmann and co-workers[198] studied adenine and some related derivatives and concluded that adeninelike derivatives are oxidized in the presence of xanthine oxidase only if specific structural requirements are fulfilled: (a) at least one hydrogen atom must be present in the 6-amino group; and (b) the imidazole ring must contain a free —NH— group. Under these conditions attack was first at C-8 and then at C-2. Other details of the biochemistry of adenine are more extensively covered in the reviews of Lister,[199] Robins,[3] and Balis.[37]

c. Photochemistry and Radiation Chemistry.   The relative rates of photo-oxidation of a number of purine derivatives, by use of visible light and thiopyronine as the photosensitizer, have been reported by Wacker *et al.*[201] Adenine is attacked only very slowly (see discussion on p. 177). Uehara *et al.*[201] irradiated adenine with visible light in the presence of riboflavin for a long period of time, and a low yield of hypoxanthine was produced. However, Dellweg and Opree[202] have reported that adenine irradiated with visible light in the presence of thiopyronine is quite stable.

Radiolysis of adenine in water gives 8-oxyadenine and 4,6-diamino-5-formamidopyrimidine as the major oxidation products.[203,204] Scholes and co-workers[205] found that adenine is more resistant to oxidation by X rays in pH 5.1 aqueous solution in the presence of oxygen than are xanthine and uric acid. Holian and Garrison,[166] however, found that X radiation of oxygen-saturated solutions of adenine did not result in appreciable damage to the adenine molecule. Weiss[206] has reviewed the effects of ionizing radiation on adenine in oxygenated solutions and has concluded that if hydroxyl radicals are the active

---

FIG. 3-55.   Probable mechanism for the electrochemical oxidation of adenine at the PGE. Molar amounts of products are those formed in 1 M HOAc.[122,190] (Reprinted with the permission of Springer-Verlag Publishers, New York.)

oxidant, and if they attack the C-4=C-5 bond, then an organic peroxy radical should be produced according to a reaction of the type shown in Eq. 6.

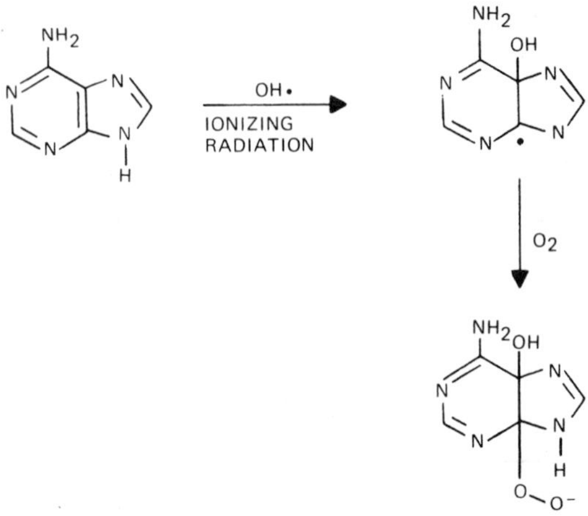

$$(6)$$

Further reaction of such a peroxy radical should result in considerable degradation of the adenine molecule. However, no such intermediate has ever been detected, and it is concluded that either it does not exist or it is extremely unstable.

4. Guanine

Guanine (2-amino-6-oxypurine) shows a single, well-defined voltammetric oxidation peak at the PGE between pH 0 and 12.5 (Table 3-10).[207] Examination of the process by linear and cyclic sweep voltammetry and by controlled potential electrolysis reveals that guanine (I, Fig. 3-56) is oxidized by an initial $2e-2H^+$ potential controlling attack at the N-7=C-8 bond to give 8-oxyguanine (II, Fig. 3-56), which is immediately oxidized in a further $2e-2H^+$ process to the diimine (III, Fig. 3-56). Hydration of III should give the proposed[207] 4,5-diol (IV, Fig. 3-56), which could rearrange to give 2-oxy-4,5-dihydroxyimidazole (VI, Fig. 3-56). This is further oxidized to parabanic acid (VIII, Fig. 3-56). The remaining fragments of IV give the isocyanate (VII, Fig. 3-56), which hydrolyzes to guanidine (IX, Fig. 3-56) and CO_2. Fragmentation of IV results in formation of oxalylguanidine (XII, Fig. 3-56) and CO_2. Close to 4.7 electrons are transferred during oxidation of guanine in 1 M HOAc, which accounts for the four electrons involved in the primary electron-transfer process (Fig. 3-56) plus

FIG. 3-56. Proposed mechanism for the electrochemical oxidation of guanine at the PGE. Molar amounts of products are those formed in 1 M HOAc.[122,207] (Reprinted with the permission of Springer-Verlag Publishers, New York.)

PRIMARY ELECTROCHEMICAL OXIDATION

SECONDARY HYDRATION, FRAGMENTATION AND OXIDATION TO PARABANIC ACID

SECONDARY HYDROLYSIS TO OXALYL GUANIDINE

the extra electrons required to oxidize 2-oxy-4,5-dihydroxyimidazole (VI, Fig. 3-56) to parabanic acid (VIII, Fig. 3-56). The diimine (III, Fig. 3-56) could be readily detected by cyclic voltammetry.[207]

a. Adsorption of Guanine at the PGE. Dryhurst[208] has shown that guanine and guanosine are both electrochemically oxidized at the PGE in aqueous solution but at different potentials, guanosine being more difficult to oxidize than guanine. By use of DC voltammetric concentration and scan rate studies and AC voltammetry it was shown that both guanine and guanosine are adsorbed at the PGE. In the presence of guanosine, adsorbed guanine is displaced from the electrode surface, resulting in a decrease in the guanine voltammetric peak current. The extent of this decrease depends on the relative concentrations of guanine and guanosine. Complete replacement of adsorbed guanine by guanosine appears to occur when greater than a fivefold amount of guanosine is present. The oxidation of guanine then becomes diffusion-controlled.

b. Biochemical Studies. Surprisingly, there appears to have been very little work done on the mode of biological or biochemical oxidation of guanine. Generally, those studies that have been carried out are characterized in terms of the major or most easily detected products. In many animals guanine appears to be converted to allantoin,[209] although in xanthinuric man guanine appears to be primarily oxidized to xanthine.[210] Since 8-oxyguanine is known to exist in nature,[211,212] Wyngaarden[213] examined the possibility of xanthine oxidase being the agent responsible for such oxidation of guanine. Although guanine was oxidized in the presence of large amounts of xanthine oxidase, the product was not 8-oxyguanine but uric acid. It was therefore concluded that the xanthine oxidase was contaminated with guanase, which was responsible for the observed reaction. In rat liver, guanine is converted to hypoxanthine.[214]

c. Photochemistry and Radiation Chemistry. Guanine is extremely susceptible to photodynamic oxidation.[200] Simon and Van Vanukis[159] concluded that under photodynamic conditions disubstituted purines in general react rapidly and that guanine 9-riboside (guanosine) reacts the most rapidly. Sussenbach and Berends[215,216] studied the photodynamic degradation of guanine in the presence of a lumichrome sensitizer and visible light. Using ^{14}C-labeled guanine, it was found that C-2 was recovered primarily as guanidine and C-6 as CO_2. Parabanic acid was also formed along with some unidentified products. Urea has also been found as a photooxidation product.[217]

FIG. 3-57.   Reaction scheme for the electrochemical oxidation of 6-thiopurine at the PGE.[218] (Reprinted with the permission of Springer-Verlag Publishers, New York.)

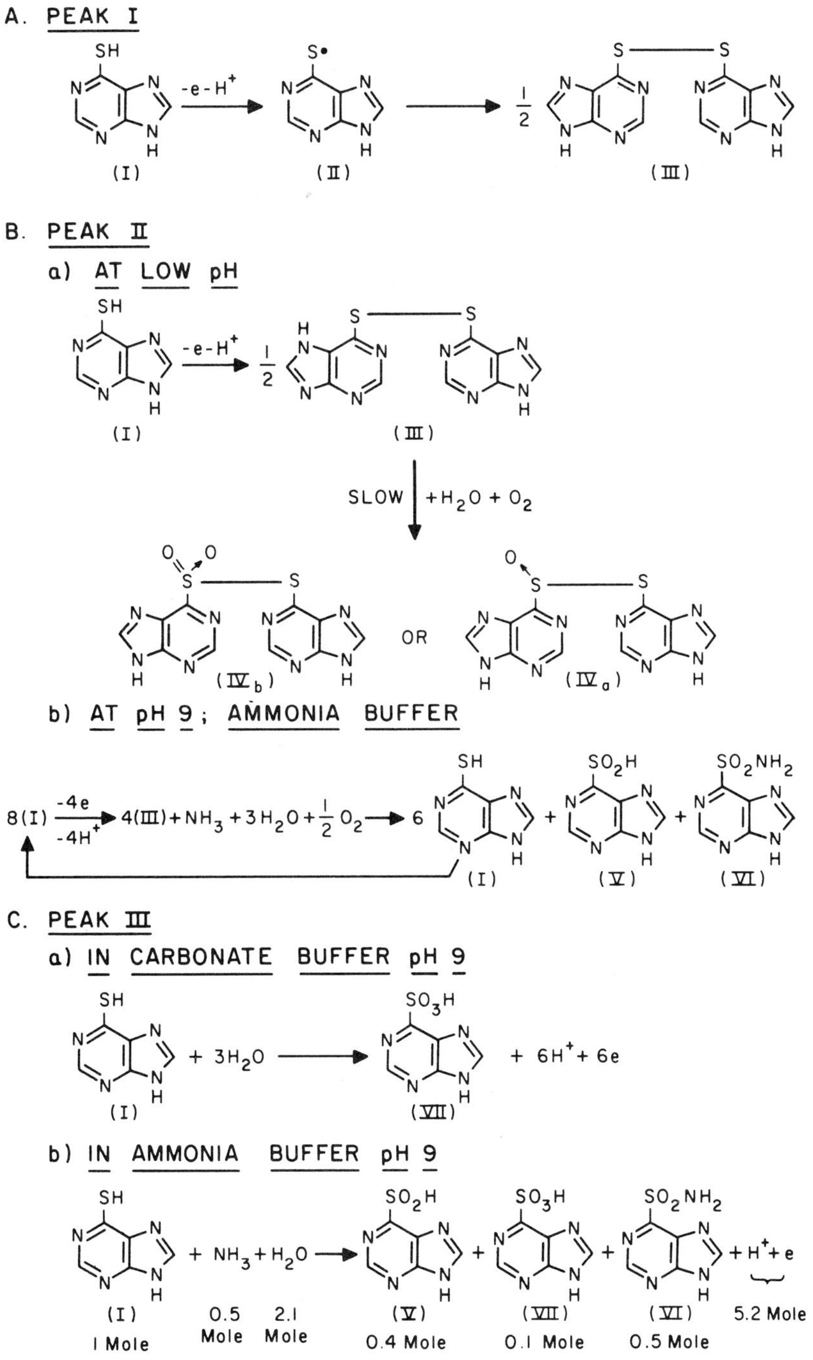

A. PEAK I

SH

$-e-H^+$

S•

$\frac{1}{2}$

S——S

(I)   (II)   (III)

B. PEAK II

a) AT LOW pH

SH

$-e-H^+$

$\frac{1}{2}$

S——S

(I)   (III)

SLOW $\Big\downarrow$ $+H_2O + O_2$

$\overset{O}{\underset{O}{\overset{\|}{S}}}$——S   OR $\overset{O}{\overset{\uparrow}{S}}$——S

(IV_b)   (IV_a)

b) AT pH 9 ; AMMONIA BUFFER

$$8(I) \xrightarrow[-4H^+]{-4e} 4(III) + NH_3 + 3H_2O + \frac{1}{2}O_2 \longrightarrow 6$$

SH SO_2H SO_2NH_2

(I)   (V)   (VI)

C. PEAK III

a) IN CARBONATE BUFFER pH 9

SH

$+ 3H_2O \longrightarrow$

SO_3H

$+ 6H^+ + 6e$

(I)   (VII)

b) IN AMMONIA BUFFER pH 9

SH

$+ NH_3 + H_2O \longrightarrow$

SO_2H SO_3H SO_2NH_2

$+ H^+ + e$

(I)   0.5   2.1   (V)   (VII)   (VI)   5.2 Mole

1 Mole   Mole   Mole   0.4 Mole   0.1 Mole   0.5 Mole

5. 6-Thiopurine

At the stationary PGE, 6-thiopurine gives rise to three voltammetric oxidation peaks[218] (Table 3-10). The first (least positive) pH-dependent peak is an adsorption peak due to a 1*e* oxidation of 6-thiopurine (I, Fig. 3-57) to an adsorbed layer of product, bis(6-purinyl) disulfide (III, Fig. 3-57A), presumably via the free-radical species (II, Fig. 3-57A). The second pH-dependent peak is a 1*e* oxidation of 6-thiopurine to the dissolved form of III (Fig. 3-57B). However, at low pH a further slow chemical oxidation of bis(6-purinyl) disulfide occurs, probably to give either a sulfone (IV$_a$, Fig. 3-57B) or a sulfoxide (IV$_b$, Fig. 3.57B). At higher pH, e.g., at pH 9 in an ammonia buffer, the bis(6-purinyl) disulfide decomposes quite rapidly under conditions of prolonged electrolysis to give 75% of the original 6-thiopurine and 25% of a mixture of purine-6-sulfonamide (VI, Fig. 3-57B) and purine-6-sulfinic acid (V, Fig. 3-57B), so that a cyclic process occurs, and for complete oxidation of 6-thiopurine close to four electrons are transferred. The third pH-dependent peak is observed only at high pH. In a non-ammonia-containing buffer such as carbonate, pH 9, a straight-forward 6*e*−6H$^+$ oxidation to purine-6-sulfonic acid (VII, Fig. 3-57C) occurs, while in ammonia buffer at the same pH a mixture of purine-6-sulfinic acid, purine-6-sulfonic acid, and purine-6-sulfonamide is produced (Fig. 3-57C), with an overall number of ca. 5.2 electrons being transferred in the process.

a. Biochemical Studies.   6-Thiopurine is one of the most effective drugs available for treating a number of types of leukemia and related neoplastic conditions.[61] The therapeutic action of 6-thiopurine has been discussed elsewhere.[62] The mechanism of metabolic breakdown of 6-thiopurine, even in terms of the complete picture of the metabolites formed in man and other systems, is apparently not available. However, various studies in man have revealed that 6-thiopurine (I, Fig. 3-58) is at least partially oxidized to 6-thiouric acid (II, Fig. 3-58), although inorganic sulfate and other unidentified products are obtained.[219,220] Xanthine oxidase catalyzes formation of 6-thiouric acid as a major metabolite from 6-thiopurine in bacteria,[221] mice,[222] and man.[220]

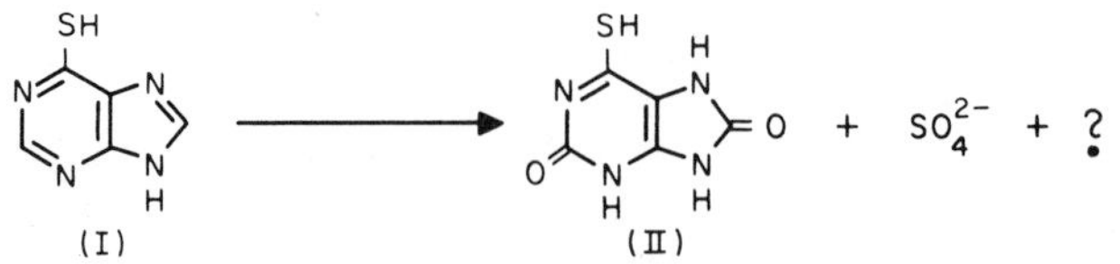

FIG. 3-58.   Partial description of the metabolism of 6-thiopurine in mammals.[219,220] (Reprinted with the permission of Springer-Verlag Publishers, New York.)

Bergmann and Ungar[223] have shown that 6-thiopurine is attacked in the presence of xanthine oxidase first at C-8 and then at C-2. However, in the purine-oxidizing system of *Pseudomonas aeruginosa* 6-thiopurine is attacked first at C-2 and then at C-8, but further oxidation to unidentified products occurs.[224]

b. Photochemistry and Radiation Chemistry. Friedman,[188] using methylene blue as a photosensitizer and irradiating 6-thiopurine with visible light in oxygenated solutions, found that photodynamic oxidation occurred, possibly at the C-8 position.

6. 2-Thiopurine

2-Thiopurine gives rise to two pH-dependent voltammetric peaks at the stationary PGE[225] (Table 3-10). The first (most negative) peak is a $1e-1H^+$ oxidation of 2-thiopurine (I, Fig. 3-59) to a free-radical species (II, Fig. 3-59), which dimerizes to give bis(2-purinyl) disulfide (III, Fig. 3-59). The second peak is a $6e-6H^+$ oxidation of 2-thiopurine to purine-2-sulfonic acid (IV, Fig. 3-59).

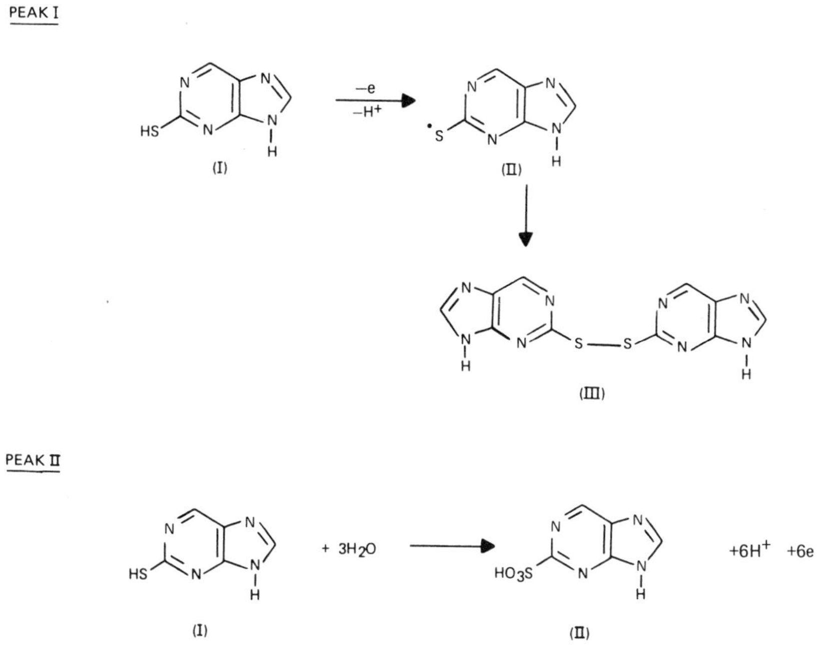

PEAK I

PEAK II

FIG. 3-59. Proposed mechanism for the electrochemical oxidation of 2-thiopurine at the pyrolytic graphite electrode.[225]

7. 2,6-Dithiopurine

2,6-Dithiopurine is oxidized by way of three voltammetric oxidation peaks at the stationary PGE^{226} (Table 3-10). The first peak appears to be due to oxidation of weakly adsorbed 2,6-dithiopurine (I, Fig. 3-60) in a $1e-1H^+$ process to give bis(6-purinyl)disulfide 2,2'-dithiol (III, Fig. 3-60), probably via a radical species (II, Fig. 3-60). The bis(6-purinyl)disulfide 2,2'-dithiol (III, Fig. 3-60) appears to be very strongly adsorbed at the electrode surface. The second peak was observed only indistinctly and was thought to be due to a further $1e-1H^+$ oxidation of III (Fig. 3-60) to bis(6-purinyl)disulfide 2,2'-disulfide (IV, Fig. 3-60), which is also strongly adsorbed. The third peak is due overall to a $12e$ oxidation of 2,6-dithiopurine to purine-2,6-disulfonic acid (V, Fig. 3-60).

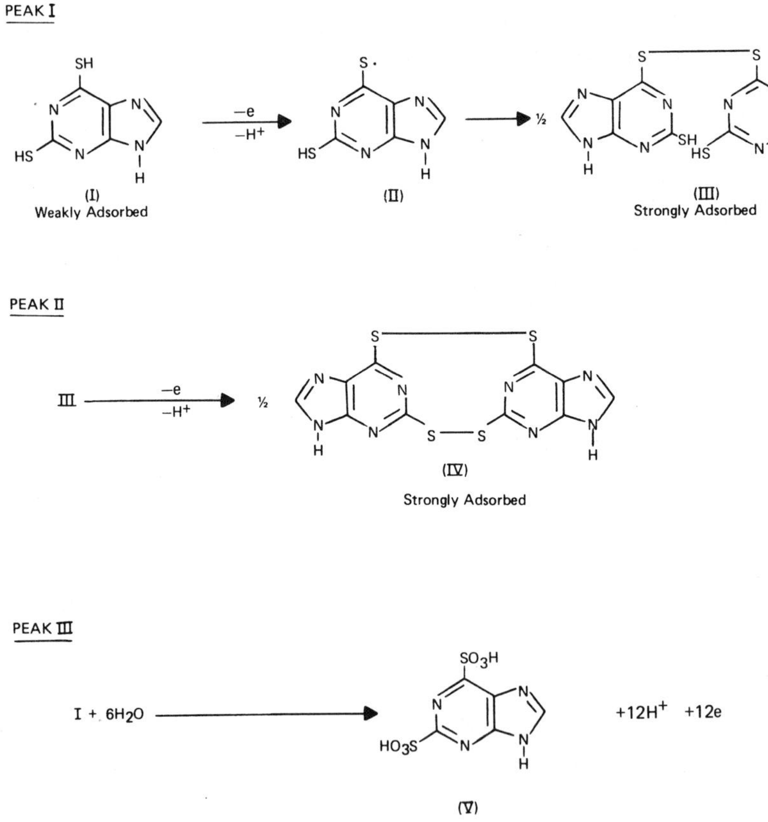

PEAK I

(I)
Weakly Adsorbed

(II)

(III)
Strongly Adsorbed

PEAK II

(IV)

Strongly Adsorbed

PEAK III

(V)

FIG. 3-60. Proposed mechanism of electrochemical oxidation of 2,6-dithiopurine at the pyrolytic graphite electrode.[226]

Biochemical Studies. Bergmann and co-workers[224] have reported that in the purine-oxidizing system *Pseudomonas aeruginosa*, which has a high xanthine oxidase activity, 2,6-dithiopurine is very slowly oxidized at C-8 to give 2,6-dithio-8-oxypurine.

D. Summary of Electrochemical Data

According to information available on their electrochemical behavior, it seems that purines can be divided into two main groups: (a) those without an exocyclic sulfur atom, and (b) those with an exocyclic sulfur atom. When a sulfur atom is substituted into the purine ring, it appears that all of the electro-oxidation behavior involves conversion of the sulfur atom to higher oxidation states, the purine ring itself being unattacked. This statement, however, may not be valid in the case of compounds such as 6-thiouric acid and 6-thioxanthine in which oxidations of both the thiol group and the purine ring might occur.

1. Non-sulfur-containing Purines

A number of generalizations may be made regarding the position and ease of oxidation of purines:

a. Purine is not electrochemically oxidizable. Based on the available evidence a prerequisite for electrochemical oxidizability of a purine is that it must be at least monosubstituted with an amino or oxy group in the pyrimidine ring moiety of the purine molecule.

b. The more highly oxygenated the purine ring, the more readily oxidized is the purine. Thus, uric acid is more readily oxidized than xanthine, which in turn is more readily oxidized than hypoxanthine.

c. In all cases purine oxidation appears to proceed initially by oxidation of any unoxidized and unsubstituted $-N=C-$ bonds present in the purine ring. If such a bond is present, its oxidation appears to be the potential controlling reaction. The ultimate or final site of electrochemical attack is the C-4=C-5 bond.

d. The product of the primary electrochemical oxidation of purines appears to be some type of diimine. This diimine is unstable but can be occasionally detected by fast sweep cyclic voltammetry at the PGE as a well-formed cathodic peak that results from reduction of the diimine and regeneration of the C-4=C-5 double bond.

e. Substitution of a purine at both the N-3 and N-7 (or probably N-9) positions, e.g., by methylation, appears to result in formation of a diiminium ion which, as might be expected, is far more reactive than the diimine, so that it cannot be detected under normal cyclic voltammetric conditions.

    f. Hydrolysis of the diimine or diiminium ion appears to lead to a 4,5-diol, which can undergo a variety of secondary reactions to give the final products detected in solution.

    g. Many purines are adsorbed at pyrolytic graphite electrodes.

2. Thiopurines

    Neglecting secondary complicating chemical and electrochemical reactions, thiopurines appear to be oxidized first to disulfides and then to the corresponding sulfonic acids. As might be expected with thiols, adsorption plays a major role in the overall electrochemical behavior.

E. CORRELATION OF ELECTROCHEMICAL REACTIONS WITH BIOLOGICAL AND RELATED REACTIONS AND WITH MOLECULAR ORBITAL PARAMETERS

    The parameter that has been calculated theoretically and that should be related to the relative ease of electron removal, e.g., electrooxidation, is the energy of the highest occupied molecular orbital (HOMO energy). As outlined in Section IV-B, by use of the linear combination of atomic orbitals approach, the energies of the mobile or π electrons are obtained in the form

$$E_i = \alpha + k_i\beta$$

Positive values of k_i correspond to occupied or bonding orbitals; the smallest positive value of k_i corresponds to the HOMO energy. It is the comparison of the latter value for a purine series that gives a theoretical prediction of the relative ease of removal of an electron from the compounds, i.e., of their relative electron-donor properties.

    The available data for the HOMO energies of a number of purines discussed earlier are presented in Table 3-13. Although this is a somewhat incomplete list of compounds in terms of the present discussion, there are several trends that can be observed. First, the HOMO energies indicate that purine should be the poorest electron donor, while uric acid should be the best donor. In other words, if all other things are equal, the molecular ionization potentials should decrease on passing down Table 3-13 from purine to uric acid and, accordingly, so also should the relative ease of chemical, biological, and electrochemical oxidation. The equations for the first voltammetric oxidation peak potentials for purines are also presented in Table 3-13. Comparison of the data in this table reveals that the general agreement between the predicted and experimental electron-donor properties is very poor indeed, although the most difficultly oxidized and most easily oxidized purines (purine and uric acid, respectively) are correctly

TABLE 3-13 Energies of Highest Occupied Molecular Orbitals for Some Purines[a]

Compound[b]	pK_a[c] Proton lost	pK_a[c] Proton gained	HOMO energy, k_i (β units)[d]	Peak potential, E_p, for first anodic peak[e] (V versus SCE)
Purine	8.93	2.39	0.69	NO[f]
8-Oxypurine	8.24	2.58	0.49	NA[g]
Adenine	9.8	4.22	0.49	1.338 − 0.063 pH
2-Oxypurine	8.4	1.69	0.45	NA
Isoguanine	8.99	4.51	0.43	1.05 (2 M H$_2$SO$_4$)
Hypoxanthine	8.94	1.98	0.40	1.27 − 0.067 pH
Xanthine	7.44		0.40	1.07 − 0.060 pH
1-Methylxanthine	7.7		0.40	1.05 − 0.049 pH
9-Methylxanthine	6.3		0.39	NA
3-Methylxanthine	8.5		0.35	1.27 − 0.050 pH
2,8-Dioxypurine	7.45		0.32	NA
Guanine	9.2	3.3	0.31	1.12 − 0.065 pH
6,8-Dioxypurine	7.65		0.22	NA
6-Thiopurine	7.77	<2.5	0.20	0.81 − 0.052 pH
Uric acid	5.4		0.17	0.76 − 0.069 pH

[a] Data on HOMO energies from Pullman and Pullman[102]; calculations based on the LCAO approximation.

[b] Where appropriate, data are for the 9-H derivative.

[c] Data taken from Albert and Brown[4] and Cavalieri and co-workers.[20]

[d] For method of calculation and definition of units see text and Pullman and Pullman.[102]

[e] The peak potential equations quoted are taken from Table 3-10 and reflect the first "normal" wave, i.e., not an adsorption prewave.

[f] Not oxidizable.

[g] Data not available.

predicted. In general, the theoretical HOMO energies also correctly predict that increasing oxygenation of a purine ring makes oxidation easier. Thus, 2,8-dioxypurine has a much lower HOMO energy than 2-oxypurine or 8-oxypurine. However, hypoxanthine (6-oxypurine) and xanthine (2,6-dioxypurine) are predicted to have equal electron-donor properties, but electrochemically they have quite different oxidation peak potentials.

A priori, it might be expected that the relative rates of enzymatic oxidation of purines are related to the electron-donor properties of the purine and, therefore, to HOMO energies. Some typical data for the xanthine-oxidase-catalyzed oxidation of purines are presented in Tables 3-14 and 3-15. It appears from these data that xanthine is oxidized most rapidly and 2,8-dioxypurine the least rapidly (Table 3-14), which is quite opposite to the order

TABLE 3-14   Pathways and Relative Rates of Oxidation of Purines with Mammalian Xanthine Oxidase[a]

Compound	Oxidation at position	Relative initial rate
Xanthine	8	100
6,8-Dioxypurine	2	100
Hypoxanthine	2	70
1-Methylxanthine	8	45
Purine	6	20
2-Oxypurine	8	16
8-Oxypurine	2	1.5
2,8-Dioxypurine	6	0.2

[a] Taken from Bergmann and co-workers.[177,180]

TABLE 3-15   Rates of Xanthine-Oxidase-Catalyzed Oxidation of Purines in Air[a]

Compound	Initial oxidation rate (μmoles/hr)
Xanthine	11.0
Hypoxanthine	8.9
Adenine	0.068
Isoguanine	0.067

[a] Taken from data of Wyngaarden.[213]

one might expect for these compounds based on the HOMO energies. Indeed, most of the remaining rates of enzymatic oxidation are not at all in accord with those HOMO energies.

There also seems to be little correlation between the susceptibility of purines to photodynamic oxidation (Tables 3-16 and 3-17) and the order one might expect based on HOMO energies (Table 3-13).

It has been suggested by Pullman[102,227] and others[228] that a better index of the reactivity of purines toward various enzymes is the nucleophilic localization energies of carbon atoms associated with various —N=CH— bonds. In essence, if more than one —N=CH— bond is available in a purine ring, the carbon atom preferentially attacked by a nucleophilic agent, such as OH^-, would be that having the smallest nucleophilic localization energy. There are some interesting correlations between the latter energies of the constituent

TABLE 3-16 Photodynamic Decomposition of Various Purines After Irradiation with Visible Light in the Presence of Thiopyronin[a]

Compound	Relative rates of decomposition
Guanine	80
Isoguanine	44
Xanthine	30
Adenine	4
Hypoxanthine	3

[a] Taken from data of Wacker *et al.*[200]

TABLE 3-17 Susceptibility of Purines to Photodynamic Action[a]

Substrate	Time required for 50% destruction (min) or % destroyed after 120 min exposure to methylene blue and light in O_2-saturated solution		
	pH 6.8	pH 8.8	pH 10.5
Uric acid	1.2	0.4	0.3
Theophylline	101	2.3	1.3
Xanthine	37	3.9	2.5
Guanine	19%	10	3.1
Theobromine	8%	63	10
Hypoxanthine		12%	77
Caffeine			17%
Purine		<5%	<5%

[a] Taken from data of Simon and Van Vanukis.[159]

carbon atoms of purines and the susceptibility of those atoms to enzymatic attack (i.e., oxidation) by xanthine oxidase.[102] The positions of initial attack by xanthine oxidase for a number of purines are shown in Table 3-14, and indeed in the case of purine, adenine, and hypoxanthine the carbon atom having the lowest calculated nucleophilic localization energy is the one that is oxidized,[102] although other predictions are not completely correct. In the case of electrochemical oxidation of purines such as adenine and hypoxanthine, which have two —N=CH— bonds available, it has not been possible to exactly define the position of initial attack, although it does occur preferentially at either C-2 or C-8 and there is no reason to doubt that the reaction does not parallel the enzyme process. The fact that the nucleophilic localization energies correlate at all is rather curious and it might indeed shed some further light on the nature of the primary electrode process. Thus, attack on a —N=CH— bond by hydroxyl ion or water behaving as a nucleophile might be the controlling factor that decides the site of oxidation of the purine. Indeed, in the case of the enzymatic oxidation of purines the addition of the elements of water across an —N=CH— bond as proposed by Bergmann and co-workers[180] could also be regarded as a type of nucleophilic attack.

At this stage it is worthwhile summarizing what electrochemical studies in particular reveal about the probable reaction routes for the biological oxidative transformations of the purines.

In the case of uric acid there is an impressive degree of parallelism between the enzymatic and electrochemical products and, as outlined earlier, the electrochemical information allows the existence of unstable products to be determined and a much more detailed mechanistic route to be written. In the case of less oxygenated purines, as far as can be judged, the initial positions of electrochemical and enzyme-catalyzed (xanthine oxidase) attack appear to be the same. However, on reaching the point where the purine ring is oxidized at all available carbon atoms with the exception of those at the 4 and 5 positions, xanthine oxidase itself ceases to promote any further oxidation.

In the case of purines in a lower state of oxidation than uric acid, the electrochemical oxidation appears to parallel quite closely the behavior of a dual enzyme system, e.g., xanthine oxidase—peroxidase, which is a system found in many organisms. Although biochemical studies of the complete oxidation of, for example, xanthine or methylated xanthines have apparently not been conducted, it appears likely that the detailed mechanisms proposed for the electrochemical oxidations should greatly facilitate the interpretation of such studies. It is particularly significant in this respect that the electrochemical data rather clearly define the effect of substitution (i.e., methylation) on the oxidation potentials of xanthines and, more interestingly, on the apparent reactivity of the primary electrode product. In the case of xanthines or, no doubt, other purines substituted at N-3 and N-7, the diimine primary product cannot be formed; rather, a diiminium ion is evolved, which results in the increased reactivity of the primary electrode product.

In the case of thiopurines it is significant that the electrochemical processes do not appear to agree at all with the known biochemical oxidations. However, it must be pointed out that in the case of 6-thiopurine, for example, a complete picture of the metabolites is not available. The electrochemical data indicate that 6-thiopurine, 2-thiopurine, and 2,6-dithiopurine are very easily oxidized to the corresponding disulfide, and the disulfides, particularly in the case of bis(6-purinyl) disulfide, are readily hydrolyzed to sulfinic or possibly sulfonic acids. In view of the well-known sulfide—disulfide transformation in biological situations (e.g., L-cysteine to L-cystine), it is not unlikely that part of the metabolic degradation pathway for thiopurines might proceed via reactions of the disulfide moiety. Because of this possibility, electrochemical methods have been developed to analyze mixtures of 6-thiopurine, bis(6-purinyl)disulfide, purine-6-sulfinic acid, and purine-6-sulfonic acid.[229]

The only disappointing area of correlation occurs between the observed electrochemical ease of oxidation (and the rates of enzymatic and photodynamic oxidation) of the purines and the predicted orders based on theoretical molecular orbital calculations. In the case of the correlations involving electrochemical potentials and HOMO energies, some thought has been given to the problem.[103] For the energy of the HOMO to be simply related to the oxidation potential of a molecule, the electrode process should *a priori* be reversible, i.e., $(E_{1/2})_{Ox} = E_0'$, where E_0' is the formal potential of the couple involved (due allowance being made for the ratio of the square roots of the diffusion coefficients of the oxidized and reduced species), since then $E_{1/2}$ is simply related to the standard free-energy change of the process, ΔG^0. (For a discussion of the relationship between a polarographic type $E_{1/2}$ and stationary electrode peak potentials, E_p, or half-peak potentials, $E_{p/2}$, see Nicholson and Shain[230] or Adams.[106]) Any deviation of $(E_{1/2})_{Ox}$ from E_0' can be expressed as

$$E_{1/2} = E_0' + \eta \tag{7}$$

where η is the overpotential, i.e., the extra energy or electrical potential required to cause the electrode reaction to proceed at a reasonable rate and, accordingly, is a measure of the departure from electrochemical reversibility. In other words, provided that the electrode reaction is reversible, i.e., the electron-transfer process is rapid at the formal potential of the couple involved, the calculated HOMO energy should be directly proportional to $(E_{1/2})_{Ox}$. If, however, the electrode reaction is not reversible, i.e., the electron transfer is slow at the formal potential of the couple, then a knowledge of the electrode rate constant and associated parameters (or η) is required in order to explain apparent discrepancies in the $E_{1/2}$–HOMO correlations.

There are other factors, however, that must be considered. For example, account should be taken of the contribution of the solvation energy to $(E_{1/2})_{Ox}$. By solving the appropriate thermodynamic equations, one can obtain a relationship between $(E_{1/2})_{Ox}$ and the ionization potential of the type[72,103] shown in Eq. 8, where IP is the ionization potential of the compound, SE_{Red}

$$(E_{1/2})_{Ox} = IP + (SE_{Red} - SE_{Ox}) - \frac{T \Delta S^\circ}{nF} - \frac{RT}{nF} \ln \frac{f_{Ox}}{f_{Red}} - \frac{RT}{nF} \ln \frac{D_{Red}^{1/2}}{D_{Ox}^{1/2}} \tag{8}$$

and SE_{Ox} are the solvation energies of the reduced and oxidized species, respectively, f_{Ox} and f_{Red} are the activity coefficients, and D_{Ox} and D_{Red} are the diffusion coefficients. Other symbols have their usual thermodynamic significance. For most purposes $(E_{1/2})_{Ox}$, IP, SE_{Red}, and SE_{Ox} would be expressed in electron volts. Combining this equation with that describing the energy of a molecular orbital, $E_i = \alpha + \beta k_i$, and neglecting the small entropy term, one can derive the expression

$$(E_{1/2})_{Ox} = \beta k_i + (SE_{Red} - SE_{Ox}) + \text{const} \qquad (9)$$

Accordingly, for a reversible electrode process, linear plots of $(E_{1/2})_{Ox}$ versus HOMO energy should be expected only when the solvation energy terms for the series of electroactive species are either constant or vary in a regular fashion. That this is so for purine systems is rather unlikely because of the variation in the nature, number, and strength of the hydrogen-bonding donor and acceptor groups in the different compounds and in their oxidation products. The fact that in some instances an uncharged primary electrode product is obtained and in other instances a charged product is obtained could also have profound effects on the relative solvation energies of the reduced and oxidized species. Thus, even in terms of solvation energy, one can expect some deviation from linear $(E_{1/2})_{Ox}$–HOMO relationships. Coupled with conceivable large differences in the overpotential or electrode kinetics for purine derivatives, it is not altogether unexpected that poor correlations exist. It is also known that many purines are oxidized in processes where either the reactant purine or its oxidation product is at least partially adsorbed at the electrode.[136a,192,208] Adsorption of a depolarizer can result in appreciable shifts of potential for electrochemical oxidation or reduction.[231] Thus, the kinetics of an electrode reaction, solvation, and adsorption effects can also affect the value of $(E_{1/2})_{Ox}$.

It is also clear from data quoted in this chapter that in many instances the electrochemically oxidized or reduced purine is not a neutral species but occasionally a protonated (i.e., positively charged) species. In fact, it is often found that only the protonated form of a purine is electroactive (e.g., polarographic reduction of adenine and substituted 6-aminopurines[73]). It may prove necessary to consider the effect of the charged state of a molecule on the theoretically calculated parameters before more satisfactory correlations are obtained. With regard to the correlations between the theoretically predicted sites of oxidation and those observed, it would also appear to be important to perform the theoretical calculations on the appropriate ionic form of the purine.

REFERENCES

1.   E. Fischer, *Ber. Dtsch. Chem. Ges.* 17, 329 (1884).
2.   E. Fischer, *Ber. Dtsch. Chem. Ges.* 30, 558 (1897).
3.   R. K. Robins, *Heterocycl. Compd.,* 8, 162 (1967).
4.   A. Albert and D. J. Brown, *J. Chem. Soc.* p. 2060 (1954).

5. C. H. Willits, J. C. Decius, K. L. Dille, and B. E. Christiensen, *J. Am. Chem. Soc.* 77, 2569 (1955).
6. C. L. Angell, *J. Chem. Soc.* p. 504 (1961).
7. "Properties of Nucleic Acid Derivatives," 5th rev. ed. Calbiochem, Los Angeles, California, 1964.
8. F. Bergmann and S. Dikstein, *J. Am. Chem. Soc.* 77, 691 (1955).
9. J. M. Gulland, E. Holiday, and T. Macrae, *J. Chem. Soc.* p. 39 (1934).
10. L. F. Cavalieri, J. J. Fox, A. Stone, and N. Chang, *J. Am. Chem. Soc.*, 76, 1119 (1954).
11. A. Albert, *Phys. Methods Heterocycl. Chem.* 1, 1 (1963).
12. A. Giner-Sorolla and A. Bendich, *J. Am. Chem. Soc.* 80, 3932 (1958).
13. A. Albert and E. Serjeant, *Biochem. J.* 76, 621 (1960).
14. P. Levene and H. Simms, *J. Biol. Chem.*, 65, 519 (1925).
15. P. Levene and H. Simms, *J. Biol. Chem.* 65, 527 (1925).
16. P. Levene, H. Simms, and L. Bass, *J. Biol. Chem.* 70, 243 (1926).
17. R. A. Alberty, *J. Biol. Chem.* 193, 425 (1951).
18. A. Albert, *Biochem. J.* 54, 646 (1953).
19. H. F. W. Taylor, *J. Chem. Soc.* p. 765 (1948).
20. A. G. Ogston, *J. Chem. Soc.* p. 1376 (1935).
21. A. Turner and A. Osol, *J. Am. Pharm. Assoc., Sci. Ed.* 38, 158 (1949).
22. W. Pfleiderer, *Justus Liebigs Ann. Chem.* 647, 167 (1961).
23. S. F. Mason, *J. Chem. Soc.* p. 2071 (1954).
24. D. J. Brown and S. F. Mason, *J. Chem. Soc.* p. 682 (1957).
25. H. C. Mautner and G. Bergson, *Acta Chem. Scand.* 17, 1694 (1963).
26. L. B. Clark and I. Tonoco, *J. Am. Chem. Soc.* 87, 11 (1965).
27. G. B. Brown and V. S. Welkiky, *J. Biol. Chem.* 204, 1019 (1953).
28. E. Johnson, *Biochem. J.* 51, 133 (1952).
29. R. Phillips, *Chem. Rev.* 66, 502 (1966).
30. W. C. Coburn, M. C. Thorpe, J. A. Montgomery, and K. Hewson, *J. Org. Chem.* 30, 1110 (1965).
31. E. R. Blout and M. Fields, *J. Biol. Chem.* 178, 335 (1949).
32. E. R. Blout and M. Fields, *J. Am. Chem. Soc.* 72, 479 (1950).
33. C. H. Willits, J. C. Decius, K. L. Dille, and B. E. Christensen, *J. Am. Chem. Soc.* 77, 2569 (1955).
34. M. Tsuboi, Y. Kyogoku, and T. Shimanouchi, *Biochim. Biophys. Acta* 55, 1 (1962).
35. J. R. Lacher, J. L. Bitner, D. J. Emery, M. E. Seffl, and J. D. Park, *J. Phys. Chem.* 59, 615 (1955).
36. Scheele, *Opuscula* 2, 73 (1776).
37. M. E. Balis, *Adv. Clin. Chem.* 10, 157 (1967).
38. W. T. Caraway, *Stand. Methods Clin. Chem.* 4, 239 (1963).
39. E. E. Conn and P. K. Stumpf, "Outlines of Biochemistry," 2nd ed. Wiley, New York, 1967.
40. B. Lythgoe, *Annu. Rep. Prog. Chem.* 42, 175 (1945).
41. D. M. Needham, *Adv. Enzymol.* 13, 151 (1952).
42. J. Baddiley, *Adv. Enzymol.* 16, 1 (1955).
43. W. S. McNutt, *Fortschr. Chem. Org. Naturst.* 9, 401 (1952).
44. J. F. Henderson and G. A. LePage, *Chem. Rev.* 58, 645 (1958).
45. A. H. Yuntsen, *J. Antibiot., Ser. A* 9, 196 (1956).
46. Marcet, "An Essay on the Chemical History and Medical Treatment of Calculous Disorders," London, 1817, pp. 95–107 (cited in Robins[3]).

47.   F. B. Brown, J. C. Cain, D. E. Gant, L. F. J. Parker, and E. L. Smith, *Biochem. J.* 59, 82 (1955).

48.   R. Burian and H. Schur, *Hoppe-Seyler's Z. Physiol. Chem.* 23, 55 (1897).

49.   A. Strecker, *Ann. Chem. Pharm.* 108, 151 (1858).

50.   G. Salmon, *Z. Physiol. Chem.* 11, 410 (1887).

51.   P. A. Levene and L. W. Bass, "Nucleic Acids," ACS Monogr. No. 56. Chem. Catalog Co. (Tudor), New York, 1931.

52.   Runge, "Neuste phytochemische Entdeckungen," Vol. I, p. 144. Berlin, 1820 (cited in Robins[3]).

53.   A. Woskresensky, *Ann Chem. Pharm.* 41, 125 (1942).

54.   A. Kossel, *Ber. Dtsch. Chem. Ges.* 21, 2164 (1888).

55.   E. H. Rodd, "Chemistry of Carbon Compounds," Vol. IVc, pp. 1663–1671. Am. Elsevier, New York, 1960.

56.   H. H. Cornish and A. A. Christman, *J. Biol. Chem.* 228, 315 (1957).

57.   F. G. Mann and J. W. Porter, *J. Chem. Soc.* p. 751 (1945).

58.   V. Papesch and E. F. Schroeder, *Med. Chem.* (*N.Y.*) 3, 175–237 (1956).

59.   T. A. Haas, *Pharmazie* 3, 97 (1948) (see Papesch and Schroeder,[58] p. 214).

60.   L. S. Goodman and A. Gilman, eds., "The Pharmacological Basis of Therapeutics," 3rd ed., p. 355. Macmillan, New York, 1967.

61.   D. A. Karnofsky, *in* "Drugs of Choice, 1968–1969" (W. Model, ed.), p. 540. Mosby, St. Louis, Missouri, 1967.

62.   A. W. Murray, D. C. Elliot, and M. R. Atkinson, *Prog. Nucleic Acid Res. Mol. Biol.* 10, 103 (1970), and references contained therein.

63.   J. Pech, *Collect. Czech. Chem. Commun.* 6, 126 (1934).

64.   J. C. Heath, *Nature* (*London*) 158, 23 (1946).

65.   D. Hamer, D. M. Waldron, and D. L. Woodhouse, *Arch. Biochem. Biophys.* 47, 272 (1953).

66.   M. J. Taras, *in* "Colorimetric Determination of Non-Metals" (D. F. Boltz, ed.), p. 75. Wiley (Interscience), New York, 1958.

67.   G. Dryhurst, "Periodate Oxidation of Diol and Other Functional Groups. Analytical and Structural Applications," p. 141. Pergamon, Oxford, 1970.

68.   A. C. Bratton and E. K. Marshall, *J. Biol. Chem.* 128, 537 (1939).

69.   F. A. McGinn and G. B. Brown, *J. Am. Chem. Soc.* 82, 3193 (1960).

70.   E. Palaček, *Naturwissenschaften* 45, 186 (1958).

71.   D. L. Smith and P. J. Elving, *J. Am. Chem. Soc.* 84, 1412 (1962).

72.   G. Dryhurst and P. J. Elving, *Talanta* 16, 855 (1969).

72a. P. J. Elving, S. J. Pace, and J. E. O'Reilly, *J. Am. Chem. Soc.* 95, 647 (1973).

73.   B. Janik and P. J. Elving, *J. Electrochem. Soc.* 116, 1087 (1969).

74.   E. Paleček and B. Janik, *Chem. Zvesti* 16, 406 (1962).

75.   B. Janik and E. Paleček, *Abh. Dtsch. Akad. Wiss. Berlin* 4, 513 (1966).

76.   B. Janik, *Z. Naturforsch., Teil B* 24, 539 (1969).

77.   N. Goetz-Luthy and B. Lamb, *J. Pharm. Pharmacol.* 8, 410 (1956).

78.   P. J. Elving, W. A. Struck, and D. L. Smith, *Mises Point Chim. Anal. Org., Pharm. Bromatol.* 14, 141 (1965).

79.   B. Breyer, T. Biegler, and H. H. Bauer, *in* "Modern Aspects of Polarography" (T. Kambara, ed.), pp. 50–57. Plenum, New York, 1966.

79a. B. Janik and P. J. Elving, *J. Electrochem. Soc.* 117, 457 (1970).

80.   V. Vetterl, *Abh. Dtsch. Akad. Wiss. Berlin, Kl. Med.* 4, 493 (1966).

81.   V. Vetterl, *Collect. Czech. Chem. Commun.* 31, 2105 (1966).

82.   V. Vetterl, *Collect. Czech. Chem. Commun.* 34, 673 (1969).

83. W. Lorenz, *Z. Elektrochem.* **62**, 192 (1958).
84. G. Dryhurst, *J. Electrochem. Soc.* **116**, 1357 (1969).
85. A. Humlová, *Collect. Czech. Chem. Commun.* **29**, 182 (1964).
86. V. Vaček, *Cesk. Farm.* **9**, 126 (1960).
87. D. L. McAllister and G. Dryhurst, *J. Electroanal. Chem.* **32**, 387 (1971).
88. G. Dryhurst, *J. Electroanal. Chem.* **28**, 33 (1970).
89. G. Dryhurst, *J. Electrochem. Soc.* **117**, 1118 (1970).
90. A. Viout and P. Rumpf, *Bull. Soc. Chim. Fr.* p. 1123 (1959).
91. S. G. Maĭranovskiĭ, "Catalytic and Kinetic Waves in Polarography," Chapter IX, p. 269. Plenum, New York, 1968.
92. C. R. Warner and P. J. Elving, *Collect. Czech. Chem. Commun.* **30**, 4210 (1965).
93. G. Horn, *Monatsber. Dtsch. Akad. Wiss. Berlin* **3**, 380 (1961).
94. B. H. Hansen and G. Dryhurst, unpublished data.
95. K. Hofmann, "Imidazole and Derivatives," Part I. Wiley (Interscience), New York, 1953.
96. E. S. Schipper and A. R. Day, *Heterocycl. Compd.* **5**, 194 (1957).
97. L. F. Cavalieri and B. A. Lowy, *Arch. Biochem. Biophys.* **35**, 83 (1952).
98. P. Zuman, "Substituent Effects in Organic Polarography." Plenum, New York, 1967.
98a. V. P. Skulacher and L. D. Denisovich, *Biokhimiya* **31**, 132 (1966).
99. P. Zuman, *Ric. Sci., Suppl. Contrib. Teor. Sper. Polarogr.* **5**, 229 (1960).
100. P. Zuman, *in* "Modern Aspects in Polarography" (T. Kambara, ed.), pp. 102–116. Plenum, New York, 1966.
101. P. Zuman, *Collect. Czech. Chem. Commun.* **25**, 3225 (1960).
102. B. Pullman and A. Pullman, "Quantum Biochemistry." Wiley (Interscience), New York, 1963.
103. E. S. Pysh and N. C. Yang, *J. Am. Chem. Soc.* **85**, 2124 (1963).
104. C. A. Coulson, *Proc. R. Soc. London, Ser. A* **169**, 413 (1939).
105. T. Nakajima and B. Pullman, *J. Am. Chem. Soc.* **81**, 3876 (1959).
106. R. N. Adams, "Electrochemistry at Solid Electrodes," Chapter 2. Dekker, New York, 1969.
107. J. B. Morris and J. M. Schempf, *Anal. Chem.* **31**, 286 (1959).
108. P. J. Elving and D. L. Smith, *Anal. Chem.* **32**, 1849 (1960).
109. R. N. Adams, *Anal. Chem.* **30**, 1576 (1958).
110. C. Olson and R. N. Adams, *Anal. Chim. Acta* **22**, 582 (1960).
111. A. L. Beilby, W. Brooks, and G. L. Lawrence, *Anal. Chem.* **36**, 22 (1964).
112. F. J. Miller and H. E. Zittel, *Anal. Chem.* **35**, 1866 (1963).
113. L. Chuang, I. Fried, and P. J. Elving, *Anal. Chem.* **36**, 2426 (1964).
114. R. E. Panzer and D. J. Elving, *J. Electrochem. Soc.* **119**, 864 (1972).
115. F. Fichter and W. Kern, *Helv. Chim. Acta* **9**, 429 (1926).
116. D. L. Smith and P. J. Elving, *Anal. Chem.* **34**, 930 (1962).
117. W. A. Struck and P. J. Elving, *Biochemistry* **4**, 1343 (1965).
118. E. G. Young and C. F. Conway, *J. Biol. Chem.* **142**, 839 (1942).
119. G. Dryhurst, *J. Electrochem. Soc.* **119**, 1659 (1972).
120. G. Dryhurst, *J. Electrochem. Soc.* **116**, 1411 (1969).
121. B. H. Hansen and G. Dryhurst, *J. Electroanal. Chem.* **30**, 417 (1971).
122. G. Dryhurst, *Top. Curr. Chem.* **34**, 47 (1972).
123. R. Brdička and E. Knoblock, *Z. Elektrochem.* **47**, 721 (1941).
124. R. Brdička, *Collect. Czech. Chem. Commun.* **12**, 522 (1947).
125. Y. Asahi, *J. Pharm. Soc. Jpn.* **76**, 378 (1956).
126. J. R. Merkel and W. J. Hickerson, *Biochim. Biophys. Acta* **14**, 303 (1954).

127. R. C. Kay and H. I. Stonehill, *J. Chem. Soc.* p. 3244 (1952).
128. J. J. Lingane and O. L. Davis, *J. Biol. Chem.* **137**, 567 (1941).
129. J. Pinson and J. Armand, *Collect. Czech. Chem. Commun.* **36**, 585 (1971).
130. S. Kwee and H. Lund, *Acta Chem. Scand.* **25**, 1813 (1971).
131. P. Zuman, *Chem. Listy* **46**, 688 (1952).
132. P. L. Pickard and S. H. Jenkins, *J. Am. Chem. Soc.* **75**, 5899 (1953).
133. L. Holleck and B. Kastening, *Z. Elektrochem.* **60**, 127 (1956).
134. G. Dryhurst and G. F. Pace, *J. Electrochem. Soc.* **117**, 1259 (1970).
135. G. Dryhurst, *J. Electrochem. Soc.* **118**, 699 (1971).
136. L. F. Fieser and M. Fieser, "Reagents for Organic Synthesis," pp. 159–160. Wiley, New York, 1967.
136a. G. Dryhurst and P. K. De, *Anal. Chim. Acta* **58**, 183 (1972).
136b. P. K. De and G. Dryhurst, *J. Electrochem. Soc.* **119**, 837 (1972).
137. M. London and P. B. Hudson, *Science* **125**, 937 (1957).
138. W. Wiechowski, *Beitr. Chem. Physiol. Pathol.* **9**, 295 (1907).
139. F. Batelli and L. Stern, *Biochem. Z.* **19**, 219 (1909).
140. D. Keilin and E. F. Hartree, *Proc. R. Soc. London, Ser. B* **119**, 114 (1936).
141. C. G. Holmberg, *Biochem. J.* **33**, 1901 (1939).
142. J. N. Davidson, *Biochem. J.* **36**, 252 (1942).
143. F. Felix, F. Scheel, and W. Schuler, *Hoppe-Seyler's Z. Physiol. Chem.* **180**, 90 (1929).
144. W. Schuler, *Hoppe-Seyler's Z. Physiol. Chem.* **208**, 237 (1932).
145. R. Bentley and A. Neuberger, *Biochem. J.* **52**, 694 (1952).
146. K. Agner, *Acta Physiol. Scand., Suppl.* **8**, 5 (1941).
147. K. Agner, *Acta Chem. Scand.* **12**, 89 (1958).
148. K. G. Paul and Y. Avi-Dor, *Acta Chem. Scand.* **8**, 637 (1954).
149. E. S. Canellakis, A. L. Tuttle, and P. P. Cohen, *J. Biol. Chem.* **213**, 397 (1955).
150. H. Blitz and F. Max, *Ber. Dtsch. Chem. Ges.* **54**, 2451 (1912).
151. R. M. Archibald, *J. Biol. Chem.* **158**, 347 (1945).
152. H. Blitz and M. Heyn, *Ber. Dtsch. Chem. Ges.* **47**, 459 (1914).
153. R. R. Howell and J. B. Wyngaarden, *J. Biol. Chem.* **235**, 3544 (1960).
154. G. Soberon and P. P. Cohen, *Arch. Biochem. Biophys.* **103**, 331 (1963).
155. J. G. Burr, *Adv. Photochem.* **6**, 193 (1968).
156. A. D. McLaren and D. Shugar, "Photochemistry of Proteins and Nucleic Acids." Pergamon, Oxford, 1964.
157. J. D. Spikes and B. W. Glad, *Photochem. Photobiol.* **3**, 471 (1964).
158. C. W. Carter, *Biochem. J.* **22**, 575 (1928).
159. M. I. Simon and H. Van Vanukis, *Arch. Biochem. Biophys.* **105**, 197 (1964).
160. T. Matsuura and I. Saito, *Chem. Commun.* p. 693 (1967).
161. T. Matsuura and I. Saito, *Tetrahedron* **24**, 6609 (1968).
162. T. Matsuura and I. Saito, *Tetrahedron* **25**, 549 (1969).
163. K. Zenda, M. Saneyoshi, and G. Chihara, *Chem. Pharm. Bull.* **13**, 1108 (1965).
164. E. C. B. Anmann and V. H. Lynch, *Biochim. Biophys. Acta* **120**, 181 (1966).
165. J. J. Weiss, *Prog. Nucleic Acid Res. Mol. Biol.* **3**, 103 (1964).
166. J. Holian and W. M. Garrison, *J. Phys. Chem.* **71**, 462 (1967).
167. J. Holian and W. M. Garrison, *Chem. Commun.* p. 676 (1967).
168. H. A. Schwarz, *Annu. Rev. Phys. Chem.* **16**, 347 (1965).
169. A. O. Allen, "Radiation Chemistry of Water and Aqueous Solutions." Van Nostrand-Reinhold, Princeton, New Jersey, 1961.
170. B. H. Hansen and G. Dryhurst, *J. Electroanal. Chem.* **30**, 407 (1971).
171. J. March, "Advanced Organic Chemistry: Reactions, Mechanisms and Structure," pp. 656–660. McGraw-Hill, New York, 1968.

172. B. H. Hansen and G. Dryhurst, *J. Electroanal. Chem.* **32**, 405 (1971).
173. J. Horbaczewski, *Monatsh. Chem.* **12**, 221 (1882).
174. H. S. Corran, J. G. Dewan, A. H. Gordon, and D. E. Green, *Biochem. J.* **33**, 1694 (1939).
175. D. E. Green and H. Beinert, *Biochim. Biophys. Acta* **11**, 599 (1953).
176. J. R. Trotter, W. T. Burnett, R. A. Monroe, I. B. Whitney, and C. L. Comer, *Science* **118**, 555 (1953).
177. F. Bergmann and S. Dikstein, *J. Biol. Chem.* **223**, 765 (1956).
178. F. Bergmann, G. Levin, and H. Kwietny, *Arch. Biochem. Biophys.* **80**, 318 (1959).
179. F. Bergmann and H. Kwietny, *Biochim. Biophys. Acta* **33**, 280 (1959).
180. F. Bergmann, H. Kwietny, G. Levin, and D. J. Brown, *J. Am. Chem. Soc.* **82**, 598 (1960).
181. J. R. Trotter, E. C. de Dugros, and C. Riveiro, *J. Biol. Chem.* **235**, 1839 (1960).
182. L. B. Mendell and E. L. Wardell, *J. Am. Med. Assoc.* **68**, 1805 (1917).
183. V. C. Meyers and E. L. Wardell, *J. Biol. Chem.* **77**, 697 (1928).
184. O. H. Buchanan, W. D. Block, and A. A. Christman, *J. Biol. Chem.* **157**, 181 (1945).
185. B. B. Brodie, J. Axelrod, and J. Reichenthal, *J. Biol. Chem.* **194**, 215 (1952).
186. H. Weinfeld and A. A. Christman, *J. Biol. Chem.* **200**, 345 (1953).
187. M. I. Simon and H. Van Vanukis, *Arch. Biochem. Biophys.* **105**, 197 (1964).
188. P. Friedman, *Biochim. Biophys. Acta* **166**, 1 (1968).
189. D. Elad, I. Rosenthal, and H. Steinmaus, *Chem. Commun.* p. 305 (1969).
190. G. Dryhurst and P. J. Elving, *J. Electrochem. Soc.* **115**, 1014 (1968).
191. G. Dryhurst, unpublished data.
192. G. Dryhurst, *Talanta* **19**, 769 (1972).
193. H. M. Kalckar, N. O. Kjeldgaard, and H. Klenow, *Biochim. Biophys. Acta* **5**, 525 (1950).
194. H. Klenow, *Biochem. J.* **50**, 404 (1952).
195. A. Bendich, G. B. Brown, F. S. Philips, and J. B. Thiersch, *J. Biol. Chem.* **183**, 267 (1950).
196. A. Nicolaier, *Z. Klin. Med.* **45**, 359 (1902).
197. J. B. Wyngaarden and J. T. Dunn, *Arch. Biochem. Biophys.* **70**, 150 (1957).
198. F. Bergmann, H. Kwietny, G. Levin, and H. Engelberg, *Biochim. Biophys. Acta* **37**, 433 (1960).
199. J. H. Lister, *Adv. Heterocycl. Chem.* **6**, 1 (1966).
200. A. Wacker, H. Dellweg, L. Tragar, A. Kornhauser, E. Lodeman, G. Turck, R. Selzer, P. Chandra, and N. Ishimoto, *Photochem. Photobiol.* **3**, 369 (1964).
201. K. Uehara, T. Mizoguchi, and S. Hosomi, *J. Biochem. (Tokyo)* **59**, 550 (1966).
202. K. Dellweg and W. Opree, *Biophysik* **3**, 241 (1966).
203. C. A. Ponnamperuma, R. M. Lemmon, and M. Calvin, *Radiat. Res.* **18**, 540 (1963).
204. E. R. Lochmann, D. Weinblum, and A. Wacker, *Biophysik* **1**, 396 (1964).
205. G. Scholes, J. F. Ward, and J. Weiss, *J. Mol. Biol.* **2**, 379 (1960).
206. J. J. Weiss, *Prog. Nucleic Acid Res. Mol. Biol.* **3**, 120 (1964).
207. G. Dryhurst and G. F. Pace, *J. Electrochem. Soc.* **117**, 1259 (1970).
208. G. Dryhurst, *Anal. Chim. Acta* **57**, 137 (1971).

4

Pyrimidines

I. INTRODUCTION, NOMENCLATURE, AND STRUCTURE

Pyrimidines in various guises appear in all living cells and are vitally involved in many biological processes, although like purines (which themselves can be considered as pyrimidine derivatives) they are rarely found in the free state. Rather, they are found as constituents of much larger molecules.

Pyrimidines have a six-membered unsaturated ring containing two heterocyclic nitrogen atoms separated by a single carbon atom. For this reason pyrimidines are occasionally known as *m*-diazines. Structure **I** illustrates the

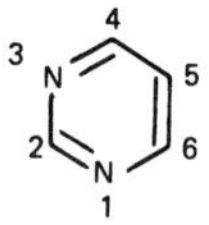

(I)

pyrimidine ring structure and the numbering system that will be employed here. Several numbering systems have been used, but for the sake of clarity the structure and numbering system shown here will be employed throughout this and other relevant chapters. Many pyrimidines, especially those of biological interest, are known by trivial names. A number of these compounds, along with their structures and chemical and trivial names, are shown in Table 4-1. It can be seen from these data that in general the oxygen-containing pyrimidines exist

TABLE 4-1   Nomenclature and Structure of Some Common Pyrimidines

Trivial name	Structure	Typical chemical name[a]
Pyrimidine		Pyrimidine; 1,3-diazine
Uracil		2,4-Dioxypyrimidine; 2,4-pyrimidinediol; 2,4(1*H*,3*H*)-pyrimidine-dione
Cytosine		2-Oxy-4-aminopyrimidine; 4-amino-2-pyrimidinol; 4-amino-2(1*H*)-pyrimidinone
Thymine		2,4-Dioxy-5-methyl-pyrimidine; 5-methyluracil
Alloxan (mesoxalylurea)		2,4,5,6-Tetraoxypyrimidine; 2,4,5,6(1*H*,3*H*)-pyrimidinetetrone
Barbituric acid		Malonylurea; 6-hydroxyuracil; 2,4,6-Pyrimidinetriol; 2,4,6(1*H*,3*H*,5*H*)-pyrimidinetrione

[a] As found, for example, in *Chemical Abstracts*.

predominantly in the keto form. The spectra,[1-3] ionization constants,[1] synthesis,[4,5] theoretical aspects,[4,5] and general chemistry of pyrimidines[6-8] have been reviewed exhaustively.

II. OCCURRENCE AND BIOLOGICAL SIGNIFICANCE OF PYRIMIDINES

A. Pyrimidine Nucleosides

The pyrimidine nucleosides are naturally occurring carbohydrate derivatives in which the pyrimidines are linked to a sugar by a β-*N*-glycosyl bond to either D-ribose (nucleosides) or 2-deoxy-D-ribose (deoxynucleosides). The N-1 nitrogen of the pyrimidine is attached to the sugar and the sugar is attached by the C-1' carbon atom. The structures of some common pyrimidine nucleosides are shown in Table 4-2.

TABLE 4-2   Names and Structures of Some Common Pyrimidine Nucleosides

Name	Structure	Parent pyrimidine
Cytidine		Cytosine
Deoxycytidine		Cytosine

TABLE 4-2 *Continued*

Name	Structure	Parent pyrimidine
Uridine		Uracil
Deoxyuridine		Uracil
Thymine riboside		Thymine
Thymidine		Thymine

B. Pyrimidine Nucleotides

The pyrimidine nucleotides are phosphate esters of the nucleosides. As described earlier for purine nucleotides, the ribose portion of a riboside can be esterified at the 2'-, 3'-, or 5'-hydroxyl, and the deoxyriboside can be esterified at the 3' and 5' positions. Table 4-3 lists the names and structures of some typical pyrimidine nucleotides.

In conjunction with the purine nucleotides the pyrimidine nucleotides can form the highly polymerized nucleic acids.

C. Other Pyrimidines

Because of their involvement in nucleic acids, cytosine, uracil, and thymine have been studied most extensively. However, a number of other pyrimidine derivatives play a vital role in many biological processes. Another pyrimidine derivative that occurs in nature is orotic acid (**II**), an intermediate in the

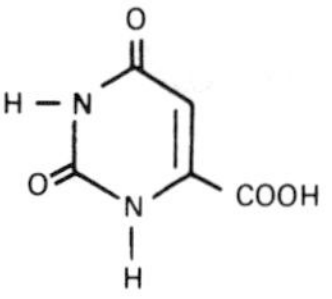

(II)

biosynthesis of uridine 5'-monophosphate (UMP), which is also the starting point for biosynthesis of the cytidine and thymidine nucleotides.[4,5,9]

Vitamin B_1 (**III**) (thiamine, aneurin), the antineuritic or antiberiberi

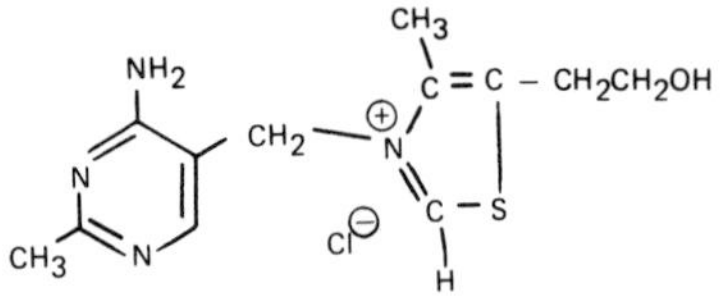

(III)

vitamin,[10] is a derivative of 2-methyl-4-aminopyrimidine. In most biological situations vitamin B_1 occurs as its coenzyme, thiamine pyrophosphate (**IV**).[13,14] It is likely that this esterified form of the vitamin is the species that functions in biological systems.

TABLE 4-3   Names and Structures of Some Common Pyrimidine Nucleosides

Name as an acid	Name as a phosphate	Structure	Parent pyrimidine
3'-Uridylic acid	Uridine 3'-mono-phosphate[a]		Uracil
3'-Cytidylic acid	Cytidine 3'-mono-phosphate[a]		Cytosine
Thymidylic acid	Thymidine 5'-mono-phosphate[b]		Thymidine

[a] Each of the 3'-monophosphates can also occur as the 5'-monophosphate, diphosphate, and triphosphate.

[b] The 5'-monophosphates also occur as the diphosphate and triphosphate.

(IV)

Barbituric acid[15-17] (Table 4-1) has a well-known hypnotic effect. This effect is even more pronounced with many of the alkyl derivatives of barbituric acid and thiobarbituric acid,[5] which therefore find extensive use as anesthetic agents. The fully oxygenated pyrimidine alloxan (Table 4-1) has been the subject of intensive study because, when administered to experimental animals, it causes necrosis of the pancreatic islets and consequent diabetes.[18-20] A number of pyrimidine derivatives have found extensive application as antibiotics.[21] In addition, many quite simple pyrimidine derivatives are antimetabolites and have therefore been of considerable clinical interest in cancer chemotherapy.[22] The metabolism of pyrimidines has been reviewed by Smellie.[23]

III. ELECTROCHEMISTRY OF PYRIMIDINES

A. Electrochemical Reduction

Heath[24] was probably the first to examine the polarographic reducibility of a pyrimidine derivative. He reported that cytidine, cytidylic acid, and uracil were not reducible at the DME in 0.1 M HClO$_4$. Although it was not specifically examined, he proposed that cytosine would not be reducible. It will be shown later in this chapter that Heath's conclusions about cytosine, its nucleosides, and nucleotides were completely wrong. Cavalieri and Lowy[25] subsequently used a polarographic method to study the tautomerism of a large number of hydroxy- and amino-substituted pyrimidines. Some typical data are presented in Table 4-4. By plotting the values of E_{DME} versus $\log i/(i_d - i)$ at pH 2.3, a straight line was obtained for all the compounds in Table 4-4, and the reciprocal of the slopes indicated that the processes responsible for the first polarographic waves were one-electron, reversible reactions. However, the $E_{1/2}$ versus pH plots of these workers suggest that the processes are more probably irreversible.

Pyrimidine itself gave only a single wave at pH 1.2 and 6.8. At the latter pH the limiting current was nearly twice that at pH 1.2. At intervening pH values (3.6, 4.5) two waves were observed which were of about equal height to the single wave observed at pH 1.2. In the case where two waves were observed, the one occurring at more negative potential was almost pH independent, i.e., protons

TABLE 4-4   Polarographic Data for Reduction of Some Pyrimidines[a]

Compound	pH 1.2[b] $E_{1/2}$[e]	pH 1.2[b] I[f]	pH 2.3[c] $E_{1/2}$	pH 2.3[c] I	pH 3.6[c] $E_{1/2}$	pH 3.6[c] I	pH 4.5[c] $E_{1/2}$	pH 4.5[c] I	pH 5.6[d] $E_{1/2}$	pH 5.6[d] I	pH 6.8[d] $E_{1/2}$	pH 6.8[d] I
Pyrimidine[g]	−0.68	1.7	−0.79	1.5	−0.92	1.7	−0.99	1.2	−1.23	3.9	−1.30	3.5
					−1.24	1.8	−1.23	2.4				
2-Aminopyrimidine[h]	−0.82	1.8	−0.87	1.8	−0.98	1.7	−1.08	1.7	−1.19	1.6	−1.4	3.5
					−1.40	1.7	−1.42	1.7	−1.43	1.9		
4-Aminopyrimidine	−1.13	5.7	−1.23	7.5	−1.32	5.0	−1.38	4.3	−1.41	4.3	−1.48	2.8
	−1.23	3.0										
4-Oxypyrimidine	−1.16	9.5	−1.25	8.7	ID[i]		−1.45	5.7	−1.55	5.5	−1.60	3.8
2-Amino-4-methoxy-pyrimidine	−1.21	4.3	−1.30	4.5	ID		−1.52	4.5	−1.54	4.5	−1.61	4.6
2,4-Diamino-pyrimidine	−1.19	4.8	−1.28	4.2	ID		ID		−1.58	6.7	−1.66	6.8
4,6-Diamino-pyrimidine	−1.16	2.7	−1.38	40.2	−1.36	19.8	−1.39	19.7	−1.44	19.3	−1.52	15.4
4,5,6-Triamino-pyrimidine	−1.17	7.4	ID		−1.37	16.8	−1.42	18.1	−1.47	18.3	−1.57	14.3
6-Amino-4-oxypyrimidine	−1.18	3.5	−1.28	3.3	−1.42	3.54	ID		−1.57	8.3	−1.62	5.3
1,4-O-Dimethyl-thymine	−1.18	6.0	−1.24	5.4	−1.38	4.0	−1.53	3.4	−1.60	2.4	−1.65	1.2

[a] Data from Cavalieri and Lowy.[25]
[b] Background 0.09 M hydrochloric acid–potassium chloride.
[c] Background, 0.09 M citrate buffer.
[d] Background, 0.09 M phosphate buffer.
[e] Potentials versus SCE.
[f] $I = i_l/Cm^{2/3} t^{1/6}$.
[g] Ill-defined waves at pH 9.2.
[h] Two waves observed at pH 9.2: $E_{1/2}$ −1.54 V and −1.91 V, with I = 0.9 and 2.4, respectively.
[i] Ill-defined waves.

were probably not involved in the potential controlling reaction, while the one at more positive potential was strongly pH dependent. By appropriate substitution into the Ilkovič equation, the n value at pH 1.2 was calculated to be 1.1 and, at pH 3.6 and 4.5, n was 2.2 for the combined heights of both waves. At pH 5.6 and 6.8 the n value for the single wave was 2.4 and 2.2, respectively. The possibility of two mechanisms for the polarographic reduction was suggested, one that could involve a 1,4-hydrogenation at the N-3 and C-6 positions or a 1,2-hydrogenation at the N-3=C-4 double bond. That the reduction involved hydrogenation of the N-3=C-4 bond was indicated by the fact that isocytosine (**V**) was not reducible at the DME and possessed no N-3=C-4 double bond, while O-methylisocytosine (**VI**) did possess such a bond and was reducible. Similar behavior was noted for 1-methyl-*O*-methylthymine (**VII**), thymine (**VIII**), and 1-methylthymine (**IX**). The latter two were not reducible, while the former, which possesses the N-3=C-4 bond, was reducible.

(**V**)   (**VI**)

(**VII**)   (**VIII**)   (**IX**)

The work of Cavalieri and Lowy suffers from several weaknesses. First, the $\log i/(i_d - i)$ versus E_{DME} plot method is not a reliable index of reversibility, and indeed the reciprocal slopes obtained by these workers varied by as much as 15 mV from the theoretical 59 mV expected for a 1e reversible reaction at 25°C. Second, the very sketchy mechanistic details proposed for pyrimidine did not appear to be applicable to the majority of other polarographically reducible pyrimidine derivatives since the diffusion current constants were generally very large and indicative of either multielectron processes or of other complicating factors.

Cavalieri and Lowy[25] also reported a number of pyrimidines that were apparently not polarographically reducible under the conditions that they employed (Table 4-5). These data have considerable potential analytical utility for the determination of mixtures of pyrimidines.

TABLE 4-5   Pyrimidines   Exhibiting   No   Polarographic
Reduction   Waves   at   the   Dropping   Mercury
Electrode [a,b]

Uracil	2,4,6-Triaminopyrimidine
1,3-Dimethyluracil	4,6-Diamino-2-oxypyrimidine
Thymine	2,6-Diamino-4-oxypyrimidine
O,O-Dimethylthymine	2,4,5,6-Tetraaminopyrimidine
1-Methylthymine	2,5,6-Triamino-4-oxypyrimidine
Isocytosine	4,5,6-Triamino-2-oxypyrimidine
Barbituric acid	5,6-Diamino-2,4-dioxypyrimidine
2,4,6-Trimethoxypyrimidine	5,6-Diamino-4-oxypyrimidine

[a] Data from Cavalieri and Lowy.[25]

[b] In buffers at pH 2.3, 4.5, and 6.8. Note that the numbering system
here is different from that in the original reference.

Hamer, Waldron, and Woodhouse[26] also studied the polarographic and
chemical reductions of some pyrimidines and found that 2-aminopyrimidine was
readily polarographically reduced and that cytosine and 4-aminopyrimidine gave
waves very close to background discharge potentials. Cytosine gave a wave at pH
2.2–5.0 and also a fairly well-defined wave at pH 9. At pH values greater than 9
and also at pH 7 there was no indication of the presence of a reduction wave for
cytosine. Although specific experimental details were not given with respect to
the mechanism and the reduction products, cytosine and 4-aminopyrimidine
apparently gave evidence indicating that cleavage and fragmentation of the
pyrimidine ring occurred.

Sugino and Shirai[27] demonstrated that 2-amino-6-chloropyrimidine and its
4-methyl derivative, when suspended in dilute aqueous ethanolic alkali, are
reduced   at   a   lead   cathode   to   2-aminopyrimidine   and   2-amino-4-
methylpyrimidine, respectively, along with some more highly reduced product.
A subsequent study by Sugino and co-workers[28] showed that 2-amino-6-
chloropyrimidine and its 4-methyl derivative give two polarographic reduction
waves in the pH region 7.4–8.9. Under similar conditions 2-aminopyrimidine
and 2-amino-4-methylpyrimidine give only a single wave. Some typical half-wave
potential data are presented in Table 4-6. It was concluded that the first wave of
the chloro compounds was caused by a reductive dehalogenation process. The
$E_{1/2}$ of the second wave of the chloro compounds coincided fairly closely with
that of the single wave of the parent compounds (Table 4-6) and hence was
ascribed to reduction of the pyrimidine nucleus in both cases. The product of
macroscale reduction of 2-aminopyrimidine in ammoniacal medium was thought
to be a dihydro-2-aminopyrimidine. This product had previously been obtained
upon   catalytic   hydrogenation   of   2-aminopyrimidine   by   Smith   and
Christensen.[29]

TABLE 4-6   Half-Wave Potentials for the Polarographic Reduction of Some Pyrimidines[a]

Compound	Background pH	$E_{1/2}$ (V versus SCE)	
		Wave I	Wave II
2-Amino-6-chloropyrimidine	7.4	−1.35	−1.48
	8.9	−1.40	−1.57
2-Amino-6-chloro-4-methylpyrimidine	7.4	−1.36	−1.56
	8.9	−1.46	−1.63
2-Aminopyrimidine	7.4	−1.45	
	8.9	−1.52	
2-Amino-4-methylpyrimidine	7.4	−1.53	
	8.9	−1.62	

[a] Data from Sugino, Shirai, and Odo.[28]

1. Pyrimidine

In order to clarify and more definitively interpret the electrochemical reduction of pyrimidines, Smith and Elving[30] investigated the processes by use of polarography over a wide pH range, coulometry, and macroscale electrolysis. The reduction products were examined chemically, polarographically, and spectrophotometrically. Data that will allow an understanding of the following discussion are presented in Table 4-7 and Fig. 4-1.

Pyrimidine shows a total of five polarographic waves between pH 0 and 13.[30] In very acidic solution only a single wave is observed (wave I). At about pH 3 a second wave (wave II) appears which is of about the same height as wave I. Since wave I is pH dependent, while wave II is essentially pH independent, the two waves merge near pH 5 to give wave III, which is also pH dependent. The current for wave III is about equal to the sum of those of waves I and II. At about pH 7 wave IV appears. It is approximately the same height as wave III and is almost pH independent. Accordingly, wave IV merges with wave III at about pH 9 to form pH-dependent wave V, the height of which is about twice that of wave III. Coulometry at controlled potential on each wave revealed[30] that the *n* values for waves I–V were 1, 1, 2, 2, and 4, respectively. There was no evidence for reversibility, in aqueous solution, of any of the waves, which are all essentially under diffusion control.

After electrolysis of pyrimidine at a potential corresponding to the crest of wave I, wave II also disappeared, indicating that the wave I product that gives rise to wave II is very reactive. This fact strongly supports the 1*e* nature of wave I being a free-radical mechanism. Since the pK_a of pyrimidine is 1.30,[31] a

TABLE 4-7 Linear $E_{1/2}$ versus pH Relationships for Pyrimidines and Derivatives[a]

Compound	Wave	Approximate pH range	$E_{1/2}$ (V versus SCE)
Pyrimidine	I	0.5–5	$-0.576 - 0.105$ pH
	II	3–5	$-1.142 - 0.011$ pH
	III	5–8	$-0.680 - 0.089$ pH
	IV	7–8	$-1.600 - 0.005$ pH
	V	9–13	$-0.805 - 0.079$ pH
2-Aminopyrimidine	I	2–3	$-0.685 - 0.049$ pH
		4–7	$-0.425 - 0.121$ pH
	II	4–7	$-1.360 - 0.004$ pH
	III	7–9	$-0.680 - 0.090$ pH
2-Amino-4-methylpyrimidine	I	2–4	$-0.770 - 0.063$ pH
		4–7	$-0.550 - 0.113$ pH
	II	5–7	$-1.424 - 0.008$ pH
	III	7–9	$-0.745 - 0.094$ pH
2-Hydroxypyrimidine		2–9	$-0.530 - 0.078$ pH
4-Amino-2-hydroxypyrimidine (cytosine)		4–6	$-1.125 - 0.073$ pH
4-Amino-2,6-dimethylpyrimidine		2–8	$-1.130 - 0.073$ pH

[a] Data from Smith and Elving.[30]

neutral species must undergo reduction over most of the pH range studied. Although the initial report by Smith and Elving[30] indicated that the wave I process was a simple $1e-1H^+$ reduction of pyrimidine at the N-3=C-4 bond to a free radical, later work[32] revealed that the mechanism probably involved an initial, very rapid preprotonation of pyrimidine (I, Fig. 4-2) to give I_a (Fig. 4-2). The latter protonated species is then reduced in a $1e$ reaction to give a free radical (II, Fig. 4-2), which dimerizes to give 4,4'-bis(3,4-dihydro)pyrimidine (III, Fig. 4-2). The equilibrium constant for dimerization of the pyrimidine free radical has been estimated to be greater than 10^5 liters mole^{-1}.[32] Wave II, which is also a $1e$ process, disappeared upon electrolysis at potentials corresponding to wave I; therefore, wave II is due to further reduction of the primary wave I product. Since wave II is essentially pH independent, the reduction of II (Fig. 4-2) presumably forms the carbanion (IV, Fig. 4-2), which then abstracts protons from water to give 3,4-dihydropyrimidine (V, Fig. 4-2). The rate constant for protonation of the pyrimidine carbanion (IV, Fig. 4-2) has been estimated to be approximately 5×10^4 sec^{-1}.[32] That III (Fig. 4-2) is produced from wave I, and V (Fig. 4-2) from wave II, was evidenced by the very similar ultraviolet spectra of the two compounds, which is expected on the basis of the conjugation present in both compounds. Wave III, which is simply the

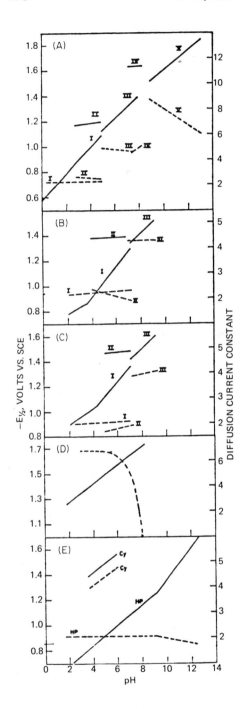

FIG. 4-1. Variation of diffusion current constants (dashed lines) and half-wave potentials (solid lines) with pH: (A) pyrimidine, (B) 2-amino-pyrimidine, (C) 2-amino-4-methyl-pyrimidine, (D) 4-amino-2,6-di-methylpyrimidine, (E) cytosine (Cy) and 2-hydroxypyrimidine (HP). [Reprinted with permission from D. L. Smith and P. J. Elving, *J. Am. Chem. Soc.* **84**, 2741 (1962). Copyright 1962 by the American Chemical Society.]

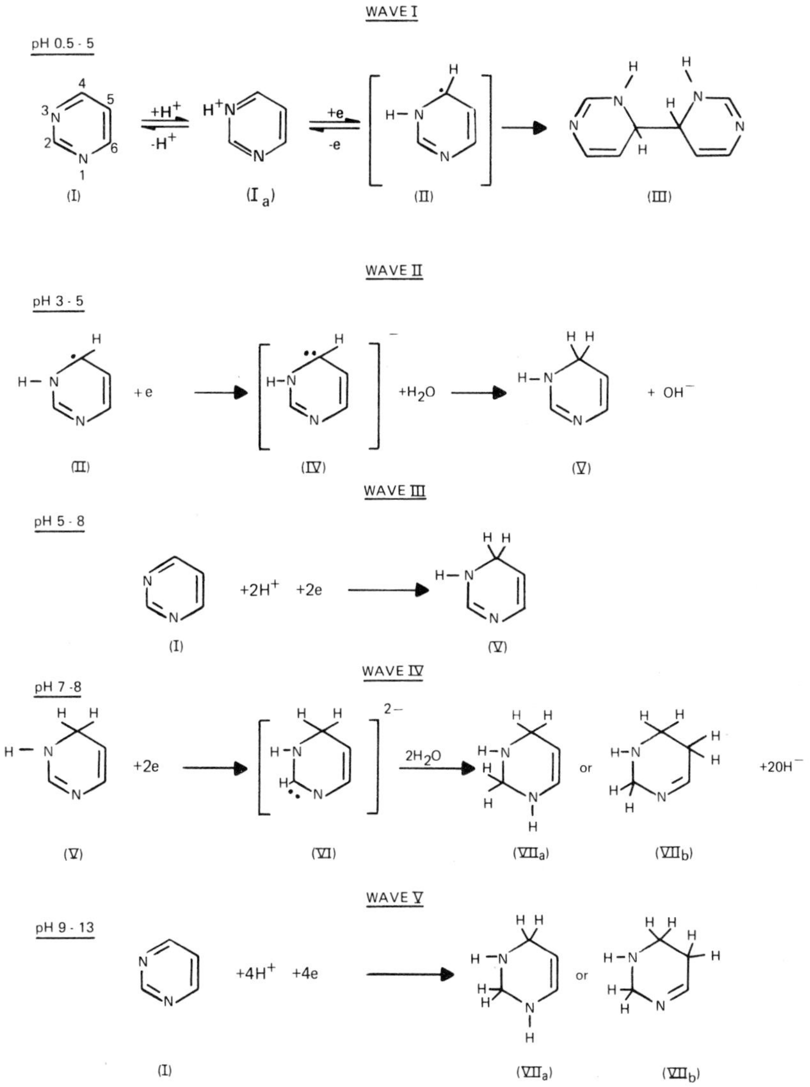

FIG. 4-2. Interpretation of the electrochemical reduction of pyrimidine.[30]

sum of the wave I and II processes, is overall a $2e-2H^+$ reduction of the $-N-3=C-4-$ bond of pyrimidine to give 3,4-dihydropyrimidine (V, Fig. 4-2). Since wave IV occurs at more negative potential than wave III, yet is observed even after wave III has been eliminated by controlled potential electrolysis, it was thought[30] to be due to further reduction of the reasonably stable product of wave III, 3,4-dihydropyrimidine (V, Fig. 4-2). The process is essentially pH independent and involves two electrons, so presumably V (Fig. 4-2) is reduced to give a dianion (VI, Fig. 4-2), which abstracts protons from water to give 1,2,3,4-tetrahydropyrimidine (VII$_a$, Fig. 4-2) or 2,3,4,5-tetrahydropyrimidine (VII$_b$, Fig. 4-2). The location of the isolated double bond in the wave IV product is uncertain, and in view of the fact that the product was unstable under the alkaline conditions of its formation, detailed examination of this product was not possible. The wave V process is simply an overall $4e-4H^+$ reduction of pyrimidine to VII$_a$ or VII$_b$ (Fig. 4-2).

Thevenot and co-workers[33,34] repeated much of the work of Smith and Elving[30] on pyrimidine. Their data agree essentially with those of the latter workers. However, at pH 3–5 waves I and II appear (see Figs. 4-1A and 4-2) in the absence of complications only at pyrimidine concentrations smaller than 1 mM. At pyrimidine concentrations greater than 1 mM wave II splits into two waves, the limiting current of the first (least negative) being concentration independent, which indicates that adsorption processes were involved. Dryhurst and Elving,[35] using cyclic voltammetry at sweep rates up to 600 mV sec^{-1}, found that the reduction of pyrimidine at the hanging mercury drop electrode and pyrolytic graphite electrode in aqueous solution shows no evidence for reversibility. Some typical results from these studies are presented in Table 4-8. The results of the latter workers confirmed the basic mechanisms for the reduction of pyrimidine.[30,32]

2. 2-Aminopyrimidine and 2-Amino-4-methylpyrimidine

2-Aminopyrimidine shows three polarographic reduction waves at the DME[30] (Fig. 4-1, Table 4-7), which correspond in behavior to those observed for pyrimidine in acidic and moderately alkaline solutions. Wave I has been proposed to be due to a 1e reduction of 2-aminopyrimidine (I, Fig. 4-3) in a pH-dependent process to a free radical (II, Fig. 4-3). In fact, on the basis of more recent studies[32] it may be more likely that the wave I process involves an initial, rapid preprotonation of 2-aminopyrimidine followed by a 1e reduction to II (Fig. 4-3) (see pyrimidine wave I mechanism, Fig. 4-2). Dimerization of the free radical II (Fig. 4-3) was proposed to give 4,4′-bis(3,4-dihydro)-2,2′-diaminopyrimidine (III, Fig. 4-3). This dimer, however, was never isolated and subjected to structural analysis. The pH-independent wave II process is due to a further 1e reduction of the free radical II (Fig. 4-3) to a carbanion (IV, Fig. 4-3), which upon protonation gives 3,4-dihydro-2-aminopyrimidine. The wave III process involves, overall, two electrons and two protons to give V (Fig. 4-3); i.e., it is due to the combined process responsible for waves I and II at lower pH. The

TABLE 4-8   Linear E_p versus pH Relationships for Pyrimidine and Cytosine on Cyclic Voltammetry[a]

Electrode	Scan rate (V sec^{-1})	pH Range	Wave	E_p (V versus SCE)
Pyrimidine (0.62 mM)				
HMDE[b]	0.026	0–5.5	I	$-0.52 - 0.125$ pH
		4.5–5.5	II	$-1.11 - 0.036$ pH
		6–9	III	$-0.73 - 0.082$ pH
		8	IV	-1.66
		8.5–13	V	$-0.76 - 0.094$ pH
	0.26	0–5.5	I	$-0.57 - 0.113$ pH
		4.5–5.5	II	$-1.22 - 0.031$ pH
		6–9	III	$-0.74 - 0.087$ pH
		8	IV	-1.71
		9–13	V	$-0.61 - 0.126$ pH
PGE[c]	0.06	0–4	I	$-0.61 - 0.126$ pH
		4.5–6	II	$-0.77 - 0.086$ pH
		7–8	III	$-1.23 - 0.036$ pH
	0.6	0–3	I	$-0.67 - 0.123$ pH
		3.5–6	II	$-0.65 - 0.122$ pH
Cytosine (0.53 mM)				
HMDE	0.026	3.7–7	I	$-1.14 - 0.084$ pH
	0.26	3.7–7	I	$-1.17 - 0.082$ pH

[a] Data taken from Dryhurst and Elving.[35]

[b] Hanging mercury drop electrode.

[c] Pyrolytic graphite electrode.

$E_{1/2}$–pH relationship for wave I consists of two linear segments (Fig. 4-1, Table 4-7); the inflection at pH 3.6 corresponds closely to the pK_a of 2-aminopyrimidine, which is 3.54.[31] 2-Amino-4-methylpyrimidine shows an identical set of polarographic processes (Fig. 4-1C, Table 4-7, Fig. 4-3).

3. 2-Oxypyrimidine (Pyrimidone-2)

2-Oxypyrimidine exhibits a single, linearly pH-dependent polarographic reduction wave (Table 4-7).[30,36] Smith and Elving[30] suggested that electrochemical reduction of 2-oxypyrimidine is a $1e-1H^+$ process giving a free radical that dimerizes. However, the dimer was not isolated nor was its structure determined. Janik and Paleček[36] subsequently proposed that 2-oxypyrimidine is reduced in a $2e$ reaction with formation of 3,4-dihydro-2-oxypyrimidine. The product, identified as a dimer by Smith and Elving[30] and as 3,4-dihydro-2-oxypyrimidine by Janik and Paleček,[36] was claimed to be unstable. However,

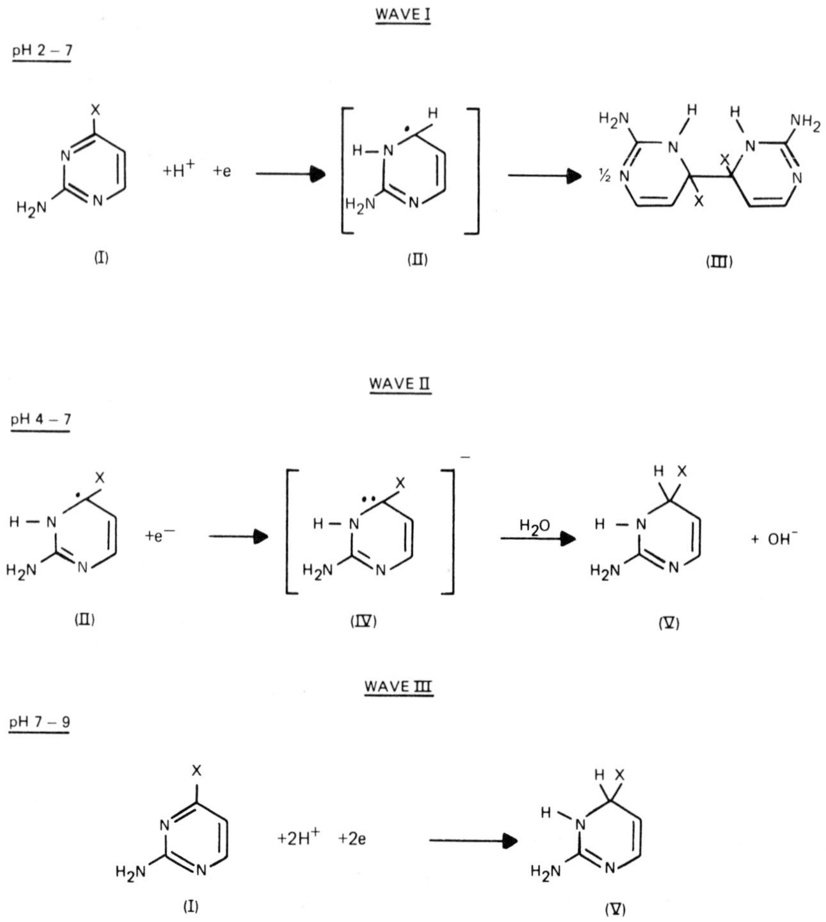

WAVE I

pH 2 – 7

(I)   (II)   (III)

WAVE II

pH 4 – 7

(II)   (IV)   (V)

WAVE III

pH 7 – 9

(I)   (V)

FIG. 4-3.   Interpretation of the electrochemical reduction of 2-aminopyrimidines (X = H or CH$_3$).[30]

Czochralska and Shugar[37] have shown more recently by polarography, coulometry, and macroscale electrolysis followed by product isolation and identification that 2-oxypyrimidine is reduced in a $1e-1H^+$ process. The product isolated and identified by elemental analysis, ultraviolet (UV), nuclear magnetic resonance (NMR), and mass spectroscopy was 6,6'-bis(3,6-dihydropyrimidone-2). Accordingly, a mechanism has been proposed whereby 2-oxypyrimidine (I, Fig. 4-4) is reduced in a $1e-1H^+$ process to give a free radical (II, Fig. 4-4), which then dimerizes to 6,6'-bis(3,6-dihydropyrimidone-2) (III, Fig. 4-4). In 0.1 *M* tetramethylammonium bromide, 2-oxypyrimidine gives a

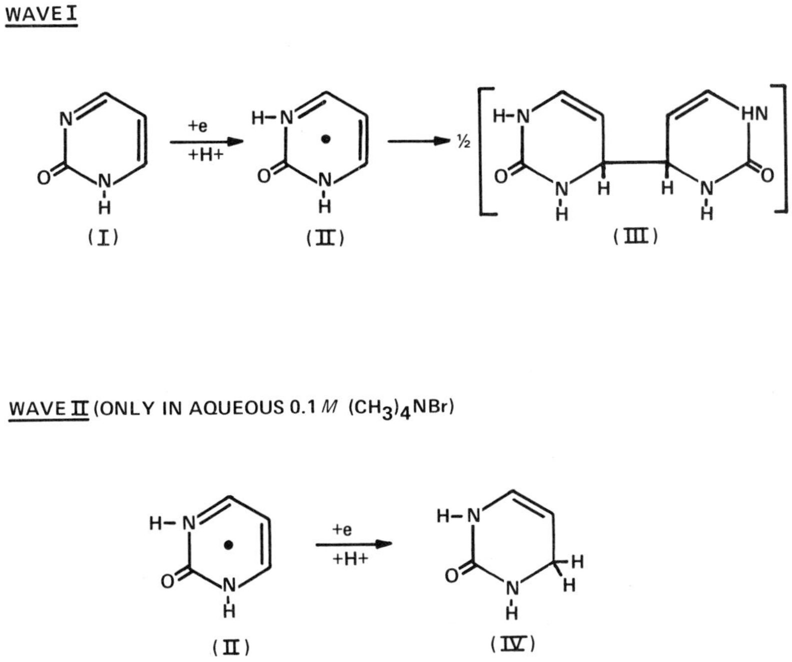

WAVE I

WAVE II (ONLY IN AQUEOUS 0.1 *M* (CH$_3$)$_4$NBr)

FIG. 4-4. Interpretation of the electrochemical reduction of 2-oxypyrimidine.[37]

second wave (wave II, $E_{1/2}$ = −1.53 V versus SCE) equal in height to wave I ($E_{1/2}$ = −0.75 V).[37] Controlled potential electrolysis at potentials corresponding to wave II in the latter supporting electrolyte followed by product separation and identification revealed that 3,6-dihydro-2-oxypyrimidine (IV, Fig. 4-4) and 6,6′-bis(3,6-dihydropyrimidone-2) were formed. Accordingly, it was proposed[37] that wave II is due to the 1e−1H$^+$ reduction of the 2-oxypyrimidine free radical (II, Fig. 4-4) to 3,6-dihydro-2-oxypyrimidine (IV, Fig. 4-4).

4. Cytosine (2-Oxy-4-aminopyrimidine)

Cytosine, a compound of considerable biological interest, exhibits a single, pH-dependent polarographic reduction wave close to background discharge between about pH 3.7 and 5.7 (Fig. 4-1E, Table 4-7). Coulometry by Smith and Elving[30] indicated that three electrons were transferred. The product of the reduction was not isolated, but it had a UV spectrum that was very similar to that observed for the wave I product of 2-oxypyrimidine (λ_{max} = 246 nm at pH 4.7). Ammonia was also liberated during the electrolysis. Subsequent studies by Czochralska and Shugar[37] revealed that the products formed on controlled potential electrolysis of cytosine were 6,6'-bis(3,6-dihydropyrimidone-2) and some 3,6-dihydropyrimidone-2. The final details of the mechanism were worked out by Elving and co-workers,[38] who employed a variety of sophisticated electrochemical techniques. These workers concluded that the basic reaction pattern involved an initial, rapid protonation of cytosine (I, Fig. 4-5) at the N-3 position[39] to form the electroactive species (II, Fig. 4-5). A 2e reduction of the N-3=C-4 double bond then occurs to form a carbanion (III, Fig. 4-5). Protonation of the latter (rate constant, 5×10^4 sec^{-1}), followed by deamination, regenerates the N-3=C-4 bond (rate constant, 10 sec^{-1}), giving 2-oxypyrimidine (V, Fig. 4-5). Protonation and further 1e reduction of 2-oxypyrimidine (i.e., 2-oxypyrimidine is reducible at potentials where cytosine is reduced) gives a free radical (VII, Fig. 4-5), which then dimerizes to 6,6'-bis(3,6-dihydropyrimidone-2) (VIII, Fig. 4-5). It is probable that some further 1e−1H$^+$ reduction of the free radical (VII, Fig. 4-5) occurs since 3,6-dihydropyrimidone-2 (IX, Fig. 4-5) is detected as a reaction product.[37] Janik and Paleček[36] have claimed that the electrochemical reduction of cytosine is a 4e process based on comparison of polarographic wave heights for cytosine and compounds of known faradaic n values. Since the latter workers neither employed coulometry nor properly isolated and identified the reaction products, it is reasonable to assume that their claim for a 4e process is incorrect.

Dryhurst and Elving[35] have reported some voltammetric data on cytosine at the hanging mercury drop electrode and the pyrolytic graphite electrode. Some typical results are presented in Table 4-8. These workers also found, by use of alternating current polarography and electrocapillary studies, that cytosine is adsorbed at the DME at potentials in the vicinity of the electrocapillary maximum (see Section III,A,7).

5. 4-Amino-2,6-dimethylpyrimidine

Smith and Elving[30] found that the single, pH-dependent polarographic reduction wave (Table 4-7, Fig. 4-1D) of 4-amino-2,6-dimethylpyrimidine (I, Fig. 4-6) was due, overall, to a 4e process proceeding by an initial 2e−2H$^+$ reduction of the N-3=C-4 bond to give the 3,4-dihydro compound (II, Fig. 4-6), which, similar to cytosine, deaminated to give 2,6-dimethylpyrimidine (III, Fig. 4-6). This species was then proposed to be immediately reduced at the regenerated N-3=C-4 bond to 3,4-dihydro-2,6-dimethylpyrimidine (IV, Fig. 4-6).

6. Pyrimidine in Nonaqueous Solution

Since, in acidic solution, the first wave of pyrimidine is a one-electron process, a number of workers have studied the electrochemical reduction of pyrimidine in nonaqueous solution in an attempt to detect the presence of

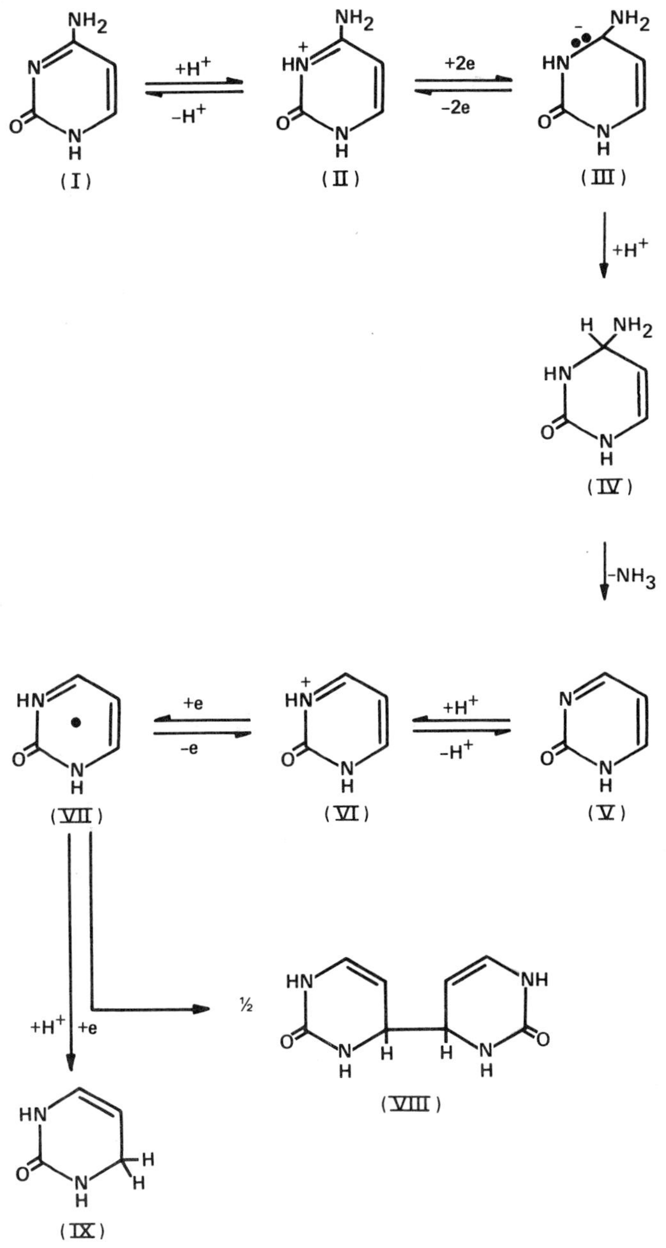

FIG. 4-5. Interpretation of the electrochemical reduction of cytosine according to Elving and co-workers.[38]

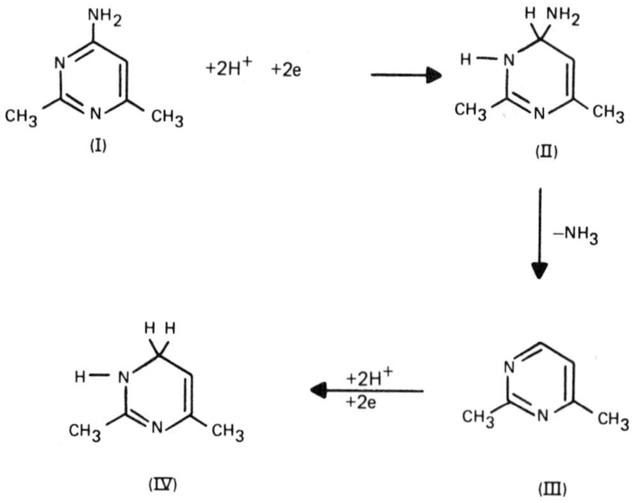

FIG. 4-6.   Interpretation of the electrochemical reduction of 4-amino-2,6-dimethyl-pyrimidine.[30]

radicals, generally by use of electron spin resonance (ESR) spectroscopy. Henning,[40] for example, electrochemically reduced pyrimidine, and many other azine compounds, at a tungsten cathode in acetonitrile containing tetra-butylammonium iodide. These electrolyses were carried out without control of the working electrode potential. Radical anions of pyrimidine could not be detected using ESR spectroscopy. Stone and Maki[41] were similarly unable to detect an ESR response on electrochemical reduction of pyrimidine in dimethyl sulfoxide at a mercury pool electrode. Carrington,[42,43] however, was able to observe an ESR signal upon reduction of pyrimidine with an alkali metal in dimethoxyethane or tetrahydrofuran *in vacuo*. By comparison of the ESR spectrum of reduced pyrimidine with that of similar heterocyclic compounds (e.g., pyridine) it was concluded that the spectrum was due to a bipyrimidyl anion radical. Van der Meer et al.[44-46] have also claimed that, upon prolonged electrolysis of pyrimidine in dimethylformamide (DMF), an ESR spectrum is obtained that is characteristic of the anion radical of 4,4'-bipyrimidine. In other words, the presence of a radical of pyrimidine could not be detected; only that of a product formed on reaction of the initially formed pyrimidine radical was observed. This, of course, implies that the initial pyrimidine radical formed is exceedingly reactive and hence short lived. Fast sweep cyclic voltammetry of pyrimidine at a hanging mercury drop electrode in DMF[46] also reveals that after scanning the initial 1e reduction peak an almost reversible oxidation peak is observed on the reverse sweep. Sweep rates of 18 V sec^{-1} were required to

TABLE 4-9   Polarographic Data for the Electrochemical Reduction of Pyrimidine at the DME in Dimethylformamide[a,b]

Pyrimidine concentration (mM)	I^c	$E_{1/2}$ (V versus Hg pool[d])	$E_{1/2}$ (V versus Ag/AgCl in DMF)	H_2O in DMF (%)
2.82	2.47	−1.78		0.08
1.41	2.45	−1.77		0.09
1.41	2.42	−1.75		0.32
1.48	2.48	−1.70		1.35
1.57	2.64		−2.32	
1.57	2.64		−2.25	

[a] Data from van der Meer and Feil.[44,45]

[b] Supporting electrolyte: 0.1 M tetra-n-ethylammonium perchlorate.

[c] $I = i_1/Cm^{2/3}t^{1/6}$.

[d] Mercury pool electrode in 0.1 M tetra-n-butylammonium iodide.

observe the oxidation peak clearly, which also supports the view that the initial radical anion of pyrimidine is very unstable. Quantum mechanical calculations[46] predict that the proposed radical anion of pyrimidine should have a very high protonation rate constant. Typical half-wave potentials for pyrimidine in DMF are shown in Table 4-9.

O'Reilly and Elving[47] examined the electrochemical reduction of pyrimidine in acetonitrile solution at mercury electrodes in considerable detail. It was found that pyrimidine gave a single, one-electron, diffusion-controlled, reversible DC polarographic reduction wave in acetonitrile plus 0.1 M tetra-n-ethylammonium perchlorate (TEAP) at the DME (Table 4-10).

Using cyclic voltammetry, it proved possible to detect the primary 1e reduction product of pyrimidine, i.e., a pyrimidine radical anion. Thus, at scan rates of less than ca. 3 V sec^{-1} the single reduction peak of pyrimidine at a hanging mercury drop electrode showed little evidence for a corresponding

TABLE 4-10   Polarographic Data for the Electrochemical Reduction of Pyrimidine at the DME in Acetonitrile plus 0.1 M TEAP[a]

$E_{1/2}$ (V versus Ag/AgNO$_3$, 0.01 M)	I
−2.628	3.39

[a] Data from O'Reilly and Elving.[47]

reversible anodic peak. However, at scan rates above 3 V sec^{-1} a distinct reversible anodic peak could be observed, although even at a scan rate of 60 V sec^{-1} the height of the anodic peak was still considerably less than that of the $1e$ cathodic peak. The conclusion was, therefore, that the product of the cathodic peak of pyrimidine (I, Fig. 4-7), a radical anion (II, Fig. 4-7), is extremely unstable. This finding correlated nicely with the inability of earlier workers to detect such a radical anion by ESR spectroscopy. In view of the difference between the heights of the cathodic and anodic peaks observed cyclic voltammetrically, it was concluded that the pyrimidine radical anion (II, Fig. 4-7) was very rapidly deactivated via two competitive pathways. The first of these was a fast dimerization of the radical anion to give a pyrimidine anionic

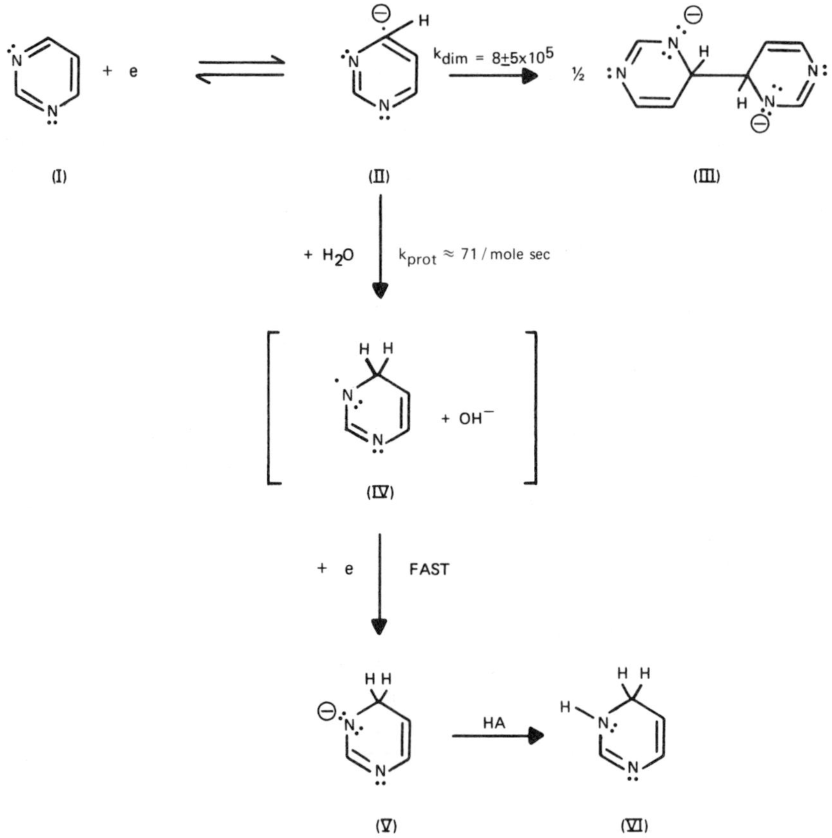

FIG. 4-7.   Interpretation of the electrochemical and chemical behavior observed for pyrimidine in acetonitrile media; HA refers to any proton source.[47]

dimer (III, Fig. 4-7). This was considered to be the 4,4′ species although it was not confirmed by any chemical evidence. The rate constant for this dimerization reaction (k_{dim}) was determined by a cyclic voltammetric method[48] to be $8 \pm 5 \times 10^5$ liters/mole sec. The second deactivation route of the pyrimidine radical anion was considered to occur by proton abstraction from residual water present in the acetonitrile solvent to give a free radical (IV, Fig. 4-7). A further cyclic voltammetric method[49] was employed to calculate the protonation rate constant $(k_{prot} \approx 7$ liters/mole sec). Addition of water to an acetonitrile solution of pyrimidine led to an increase in the polarographic limiting current up to that expected for a $2e$ reduction and also to rather large positive shifts of $E_{1/2}$. Accordingly, it was assumed that an ECE* mechanism was taking place; i.e., the primary electrode product (II, Fig. 4-7) abstracts a proton from the solvent to give a free radical (IV, Fig. 4-7), which is then very rapidly reduced in a $1e$ process to V (Fig. 4-7). Further protonation of the latter product leads to 3,4-dihydropyrimidine (VI, Fig. 4-7). This mechanism corresponds closely to that observed over certain pH regions in aqueous solutions (*vide supra*). In view of the fact that the mechanism shown in Fig. 4-7 does not involve formation of a radical anion of 4,4′-bipyrimidine, for which there is now substantial evidence from the ESR spectra,[46] it is reasonable to expect that some further refinements of the mechanism of pyrimidine electroreduction in nonaqueous solution will appear.

O'Reilly and Elving[47] also found that addition of acid to pyrimidine solutions in acetonitrile resulted in the appearance of a new wave at less negative potential. This wave was attributed to the one-electron reduction of an *N*-protonated pyrimidine species (III, Fig. 4-8) to produce a free radical (IV, Fig. 4-8), which dimerizes very rapidly, probably to the 4,4′ compound (V, Fig. 4-8). This wave corresponds to the first one-electron wave of pyrimidine in aqueous media.[30] The electrochemical reduction of pyrimidine is somewhat unique because it provides the only aromatic system thus far systematically examined that shows initial one-electron reductions in both aqueous (proton-rich) and nonaqueous (proton-poor) media.

7. Adsorption of Pyrimidines

Vetterl[50-52] has investigated the adsorption of a number of pyrimidine (and purine) bases at the DME in 1 M NaCl solutions using both differential capacity and AC polarographic methods. The usual pyrimidine bases found in DNA and

---

* An ECE (electrochemical–chemical–electrochemical) mechanism for a single polarographic wave is one in which an initial electron-transfer reaction produces a species that rapidly undergoes a chemical reaction to form a product that readily undergoes a second electrochemical reaction at the same (or similar) potential to the initial electron-transfer reaction.

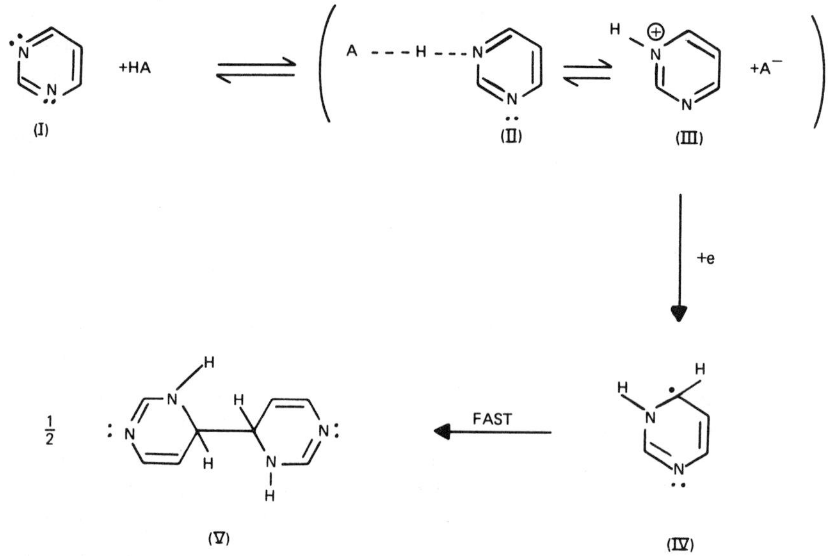

FIG. 4-8. Interpretation of the electrochemical and chemical behavior observed for pyrimidine in acetonitrile media containing an acid, HA, as strong as phenol.[47]

RNA (thymine, uracil, and cytosine) were adsorbed at low concentrations at the electrode surface, as evidenced by depressions in the differential capacity–electrode potential curves (Fig. 4.9). These depressions caused by adsorption of the pyrimidines are centered around −0.5 V versus SCE, which is close to the electrocapillary maximum potential (i.e., potential of zero charge on the electrode). Above certain concentrations, very pronounced pits or wells are formed in the curves (Fig. 4-9). These pits were thought to be the result of strong intermolecular interactions between the adsorbed molecules, i.e., association. It is clear from Fig. 4-9 that the potential region over which association occurs is characteristic of the pyrimidine. Similar adsorption–association phenomena have been noted for the two normal purine bases found in nucleic acids, i.e., guanine and adenine (see Chapter 3). Anomalous DNA components such as 5-methylcytosine and 5-hydroxymethylcytosine did not show any evidence for intermolecular interactions at the concentrations where the normal components showed such behavior. However, at concentrations above 10^{-2} M, intermolecular interactions were in fact observed for the anomalous compounds. In the case of those bases that are adsorbed and also associate on the DME surface, the rate of association was much smaller than the rate of adsorption.[53,54]

Elving and co-workers[38,54a] have confirmed by use of AC polarography and electrocapillary studies that cytosine is adsorbed at the DME between pH 4 and 5.

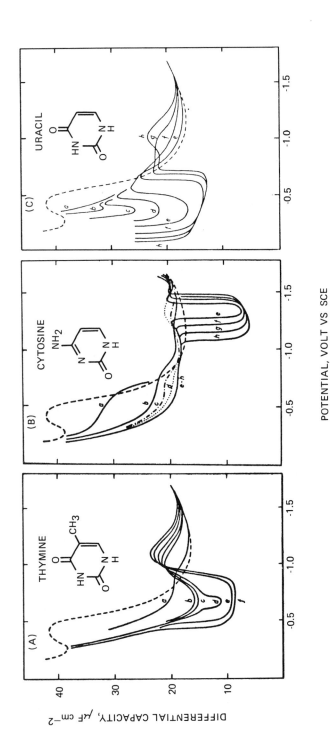

POTENTIAL, VOLT VS SCE

FIG. 4-9. Differential capacity versus electrode potential curves for (A) thymine, (B) cytosine, and (C) uracil at the DME in 1 M NaCl solution.[51] Dashed line in each figure represents trace in the absence of a pyrimidine. (A) Concentrations of thymine: (a) 2.0 mM, (b) 7.1 mM, (c) 9.5 mM, (d) 13 mM, (e) 17 mM, (f) 22 mM. (B) Concentrations of cytosine: (a) 1.8 mM, (b) 9.0 mM, (c) 13 mM, (d) 26 mM, (e) 39 mM, (f) 45 mM, (g) 51 mM, (h) 58 mM. (C) Concentrations of uracil: (a) 4.6 mM, (b) 4.9 mM, (c) 5.3 mM, (d) 6.2 mM, (e) 8.3 mM, (f) 17 mM, (g) and (h) > 20 mM. (Reprinted with the permission of Academia Publishing House, Prague.)

In view of the fact that these workers also observed depression of the differential capacity below that of background at the potential of the electrocapillary maximum (ca. −0.5 V versus SCE), it has been concluded[54a] that adsorption of an unchanged portion of the cytosine molecule occurs.

8. Alloxan—Alloxantin—Dialuric Acid System

Ono, Takagi, and Wasa[55] studied the polarography of alloxan and found that alloxan was reduced in a kinetically controlled process. In a buffer at pH 3.5, $E_{1/2}$ was −0.05 V versus NCE (1 N calomel electrode). The single wave that they observed was considered reversible since upon controlled potential electrolysis the reduction product, which they proposed to be dialuric acid, could be polarographically oxidized with an $E_{1/2}$ the same as that for the reduction of alloxan. The subsequent literature is very confusing when one attempts to unravel the details of the electrochemical and chemical reduction of alloxan. One would expect, *a priori*, that alloxan (I_a, I_b, Fig. 4-10) would be first reduced to alloxantin (II, Fig. 4-10) and then to dialuric acid (III, Fig. 4-10). Indeed, alloxan can be chemically reduced, with hydrogen sulfide, to alloxantin[56] and thence with sodium amalgam[57] to dialuric acid. Tipson[58] and Cretcher[59] have also shown that alloxantin can be prepared by reducing quite concentrated solutions of alloxan with H_2S, the latter being limited to the stoichiometric amount. Alloxantin is rather insoluble in aqueous solutions and

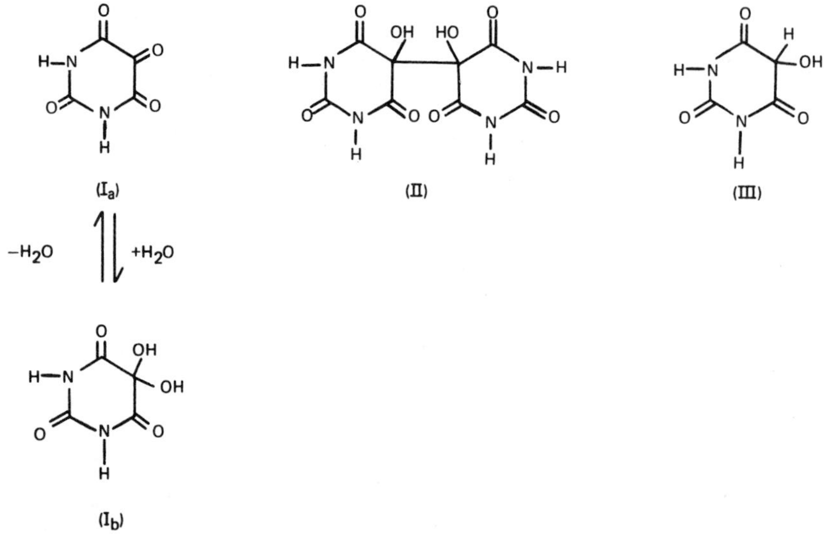

FIG. 4-10.   Structures of alloxan (I_a, I_b), alloxantin (II), and dialuric acid (III).

precipitates under the latter preparative conditions. Hill and Michaelis[60] have reported, on the basis of a potentiometric study that alloxantin is almost completely dissociated in aqueous solution to equimolar alloxan and dialuric acid. Biilmann and Bentzon[61] also concluded that an aqueous solution of alloxantin is ca. 80% dissociated. However, Richardson and Cannan[62] have proposed that the existence of alloxantin is favored in aqueous solution. Thus, questions are raised as to whether the synthesis of alloxantin by reduction of alloxan is successful, simply as a result of the insolubility of alloxantin, and whether in dilute solutions alloxantin is completely dissociated into equimolar alloxan and dialuric acid as might be deduced from the studies of Hill and Michaelis[60] and Biilmann and Bentzon.[61]

Struck and Elving[63] found that alloxan exhibits two polarographic reduction waves at the DME (Fig. 4-11, Table 4-11). The half-wave potential for the first wave (wave I), which was very small and under kinetic control between pH 3.6 and 5.6, followed the equation $E_{1/2} = 0.060 - 0.031$ pH. The upper limit of the pH range was determined by the base-catalyzed rearrangement of alloxan (I, Fig. 4-12) to nonelectroactive alloxanic acid (II, Fig. 4-12), which becomes very rapid above pH 5.6.[64,65] The limiting current for the wave I process showed all

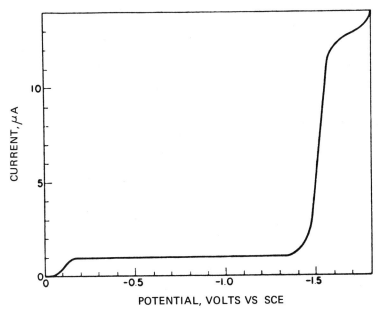

FIG. 4-11. Polarogram of alloxan (2 mM) at pH 5.6. Curve was obtained by combining the average current traces of separate polarograms of wave I and wave II.[63] [Reprinted with the permission from W. A. Struck and P. J. Elving, *J. Am. Chem. Soc.* **86**, 1229 (1964). Copyright 1964 by the American Chemical Society.]

TABLE 4-11   Polarographic Behavior of Alloxan and Its *N*-Methyl Derivatives

Compound	Wave	pH[a]	$E_{1/2}$ (V versus SCE)	i^b
Alloxan[c]	I	3.6 M	−0.053	0.21
		4.0 A	−0.059	0.32
		4.6 A	−0.076	0.21
		4.8 A	−0.090	0.31
		5.6 A	−0.113	0.28
		5.6 M	−0.107	0.25
	II	4.6 M	−1.502	3.63
		4.8 A	−1.515	3.78
		5.6 A	−1.527	d
		5.6 M	−1.512	d
		6.6 M	−1.525	d
Methyl alloxan[e]	I_a	0.0 (2 *M* H_2SO_4)	0.11	0.20
		2.0 M	0.03	0.14
		2.3 (1 *M* HOAc)	0.02	0.22
		2.9 M	0.02	0.16
		3.7 A	−0.02	0.18
		3.9 M	−0.02	0.21
		4.6 A	−0.06	0.31
		4.9 M	−0.06	0.26
		5.8 A	−0.09	0.31
		5.9 M	−0.11	0.36
		7.1 M	−0.12	d
	I_b	0.0 (2 *M* H_2SO_4)	−0.30	0.12
		2.0 M	−0.27	0.08
		2.9 M	−0.25	0.06
	II	4.7 A	−1.47	3.76
		4.9 M	−1.46	2.79
		5.8 A	−1.46	1.91
		5.9 M	−1.46	1.66
Dimethyl alloxan[e]	I_a	0.0 (2 *M* H_2SO_4)	0.11	0.18
		2.2 M	0.03	0.19
		2.3 (1 *M* HOAc)	0.04	0.25
		2.9 M	0.00	
		3.4 A	−0.03	0.36
		3.7 M	−0.06	0.33
		4.2 A	−0.05	0.29
		4.8 M	−0.09	0.41
		5.3 A	−0.09	0.37
		5.8 M	−0.11	0.35
		6.9 M	−0.14	d
		8.0 M	−0.18	d
	I_b	0.0 (2 *M* H_2SO_4)	−0.41	0.19
		2.2 M	−0.29	0.14
		0.5 C	−0.34	0.19
		1.0 C	−0.33	0.16

TABLE 4-11 *Continued*

Compound	Wave	pH[a]	$E_{1/2}$ (V versus SCE)	I[b]
		1.9 C	−0.33	0.18
		3.0 C	−0.36	0.09
	II	4.8 M	−1.40	2.79
		5.3 A	−1.42	2.19
		5.8 M	−1.42	1.59

[a] Buffer system: A, acetate; C, chloride; M, McIlvaine. All buffers 0.5 M in ionic strength except 2 M H_2SO_4 and 1 M HOAc.

[b] $I = i_1/Cm^{2/3}t^{1/6}$.

[c] Taken from Struck and Elving.[63]

[d] Rearrangement to nonelectroactive species too rapid for current measurement.

[e] Taken from Hansen and Dryhurst.[68]

the characteristics of a kinetically controlled process. Coulometry revealed that reduction of alloxan at potentials corresponding to wave I involved transfer of two electrons. The product was dialuric acid, as evidenced by infrared spectrophotometry of the solid product obtained. Dialuric acid also gave an anodic wave having the same $E_{1/2}$ as the first cathodic wave of alloxan. This product, coupled with the fact that the pH dependence of the alloxan wave I process was 0.031 V/pH unit, led to the conclusion that the latter process was due to a $2e-1H^+$ reduction of alloxan (II, Fig. 4-13) to the anion of dialuric acid (III, Fig. 4-13) ($pK_a = 2.8$[64]). The process responsible for the kinetic nature of the wave was thought to be dehydration of the hydrated carbonyl at C-5 of hydrated alloxan (I, Fig. 4-13).

The second wave of alloxan (wave II) could be observed clearly only between about pH 3.6 and 5.6. At higher pH values the rapid rearrangement of alloxan to alloxanic acid occurred. Below about pH 3.6 the second wave was obscured by background discharge. The limiting current for wave II was about 10 times larger than that for wave I and was under overall diffusion control. The value of the diffusion current constant (Table 4-11) was about 3.6, which corresponded to that expected for a $2e$ process. The variation of $E_{1/2}$ for wave II with pH was

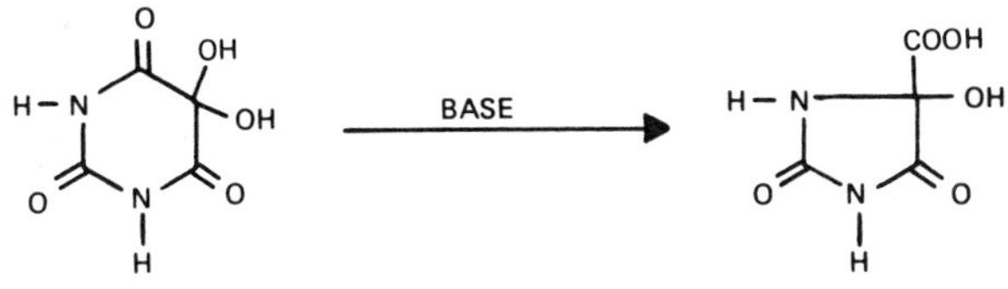

FIG. 4-12. Base-catalyzed rearrangement of alloxan (I) to alloxanic acid (II).

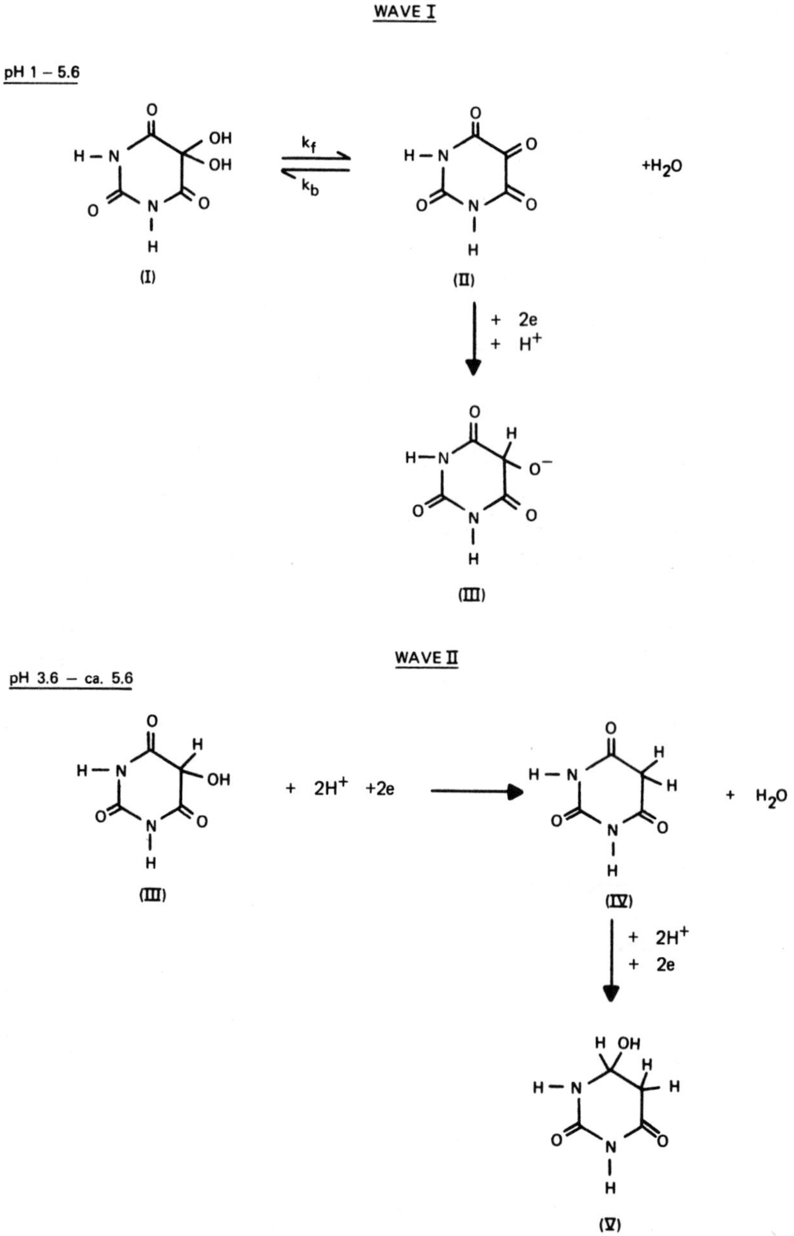

FIG. 4-13.   Interpretation of the mechanism of polarographic reduction of alloxan in aqueous solution according to Struck and Elving.[63]

described by the equation $E_{1/2} = -1.44 - 0.015$ pH. Coulometry was not carried out at wave II potentials, nor was the corresponding product isolated. Nevertheless, an electrode reaction was proposed whereby the dialuric acid (III, Fig. 4-13) formed in the kinetically controlled wave I process was reduced in the diffusion-controlled wave II process to barbituric acid (IV, Fig. 4-13) or further to dihydrouracil (V, Fig. 4-13). There are, however, several serious objections to these reaction schemes. First, if the formation of dialuric acid is limited by the kinetically controlled wave I process, then it seems logical that the wave II process should also be under kinetic control; i.e., it cannot be possible to reduce 10 times as much dialuric acid as is produced from the kinetically controlled wave I process. Second, dialuric acid is not electrochemically reducible. Finally, the wave II process appeared to be almost pH independent, yet two electrons and two protons, at least, are proposed to be involved in the wave II step.

Struck and Elving[63] also investigated the polarography of alloxantin, which gave rise to a combined cathodic–anodic wave whose cathodic portion was very small compared to the anodic portion. The half-wave potential for the combined cathodic–anodic wave of alloxantin was very close to that of the cathodic wave I of alloxan and that of the anodic wave of dialuric acid (Fig. 4-14). Some typical data on the anodic wave of alloxantin are presented in Table 4-12.

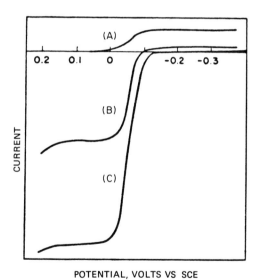

POTENTIAL, VOLTS VS SCE

FIG. 4-14.   Polarograms at equivalent concentrations in pH 4.0 acetate buffer: (A) 0.5 mM alloxan ($E_{1/2} = -0.055$ V), (B) 0.25 mM alloxantin ($E_{1/2} = -0.058$ V); (C) 0.5 mM dialuric acid ($E_{1/2} = -0.056$ V). Figure was constructed from the average current traces of separate polarograms.[63] [Reprinted with the permission from W. A. Struck and P. J. Elving, *J. Am. Chem. Soc.* **86**, 1229 (1964). Copyright 1964 by the American Chemical Society.]

TABLE 4-12   Polarographic Behavior of Alloxantin[a]

pH	Concentration (mM)	$E_{1/2}$ (V versus SCE)	I[b]
		Anodic wave	
3.5	0.30	−0.038	2.46
4.0	1.00	−0.058	2.48[c]
4.0	0.46	−0.052	2.55
5.6	0.30	−0.106	2.29

[a] Data from Struck and Elving [Reprinted with permission from W. A. Struck and P. J. Elving, *J. Am. Chem. Soc.* **86**, 1229 (1964). Copyright (1964) by the American Chemical Society]; all buffers were acetate.

[b] $I = i_1/Cm^{2/3}t^{1/6}$.

[c] Maximum present.

The pH dependence of the combined alloxantin waves was given by the equation $E_{1/2} = 0.078 - 0.031$ pH. Thus, the slope of the $E_{1/2}$ versus pH relationship was exactly the same as that for wave I of alloxan, but the intercept was slightly different. However, the authors[63] indicate that the alloxan data were taken in acetate and McIlvaine buffers, whereas the alloxantin data were obtained only in acetate buffers, which might explain the slightly different intercept obtained. The small cathodic portion of the alloxantin wave was thought to be under kinetic control. The slope of 0.031 V/pH unit was interpreted as indicating that a $2e-1H^+$ reduction of alloxantin to dialuric acid was involved in the kinetically controlled cathodic wave process. This was confirmed by the formation of dialuric acid on controlled potential electrolysis at potentials sufficiently cathodic of the anodic wave. A second wave for alloxantin was observed at much more negative potentials, with an $E_{1/2}$ equivalent to the second wave (wave II) of alloxan but very much smaller than for an equivalent concentration of alloxan. On the basis of this finding Struck and Elving[63] concluded that alloxantin is largely undissociated in aqueous solution. That is, if indeed it were dissociated, a large anodic wave due to oxidation of dialuric acid to alloxan would be observed, along with the small kinetically controlled wave I due to reduction of alloxan to dialuric acid, as well as a large reduction wave due to reduction of dialuric acid to barbituric acid, corresponding to alloxan wave II. In fact, it was found that addition of large amounts of alloxan to alloxantin solutions increased the height of the anodic wave. If the alloxantin solution was indeed an equilibrium mixture of alloxantin, dialuric acid, and alloxan, then addition of alloxan should have decreased the height of the anodic wave if it was due to dialuric acid.

Thus, in summary, alloxantin (I, Fig. 4-15) was proposed to be almost completely undissociated in aqueous solution. It showed a combined anodic—cathodic wave at the same $E_{1/2}$ for the wave I of alloxan, the anodic portion

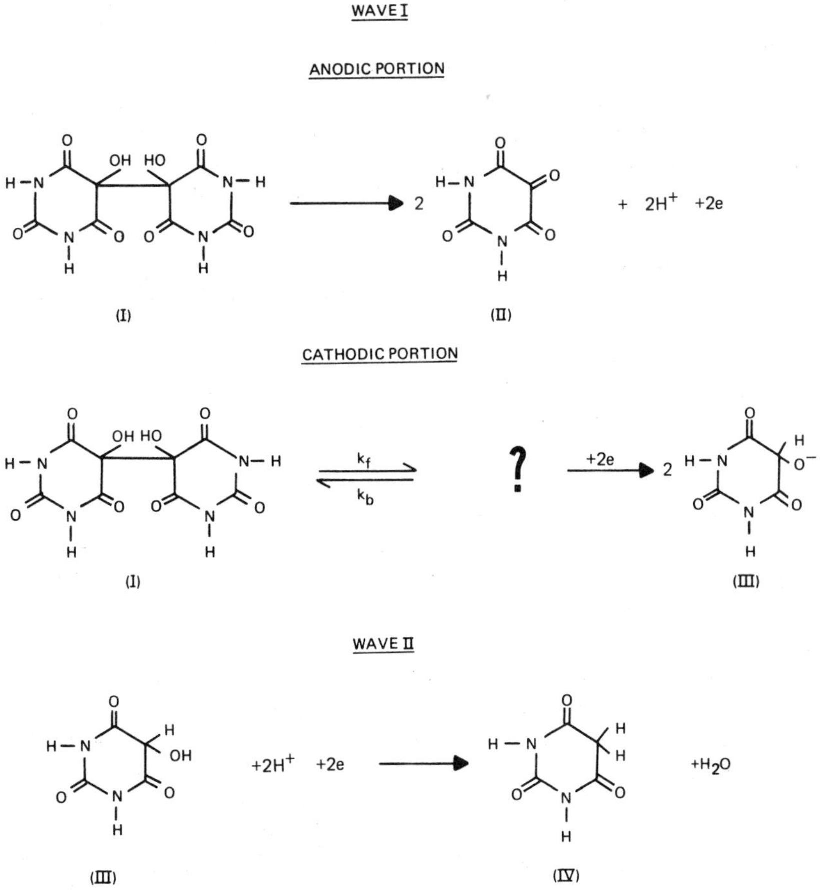

<u>WAVE I</u>

<u>ANODIC PORTION</u>

(I)   (II)

<u>CATHODIC PORTION</u>

(I)   (III)

<u>WAVE II</u>

(III)   (IV)

FIG. 4-15. Interpretation of the polarographic behavior of alloxantin in aqueous solution. Adapted from the data of Struck and Elving. [Reprinted with permission from W. A. Struck and P. J. Elving, *J. Am. Chem. Soc.* **86**, 1229 (1964). Copyright (1964) by the American Chemical Society.]

being due to oxidation of alloxantin, presumably to alloxan (II, Fig. 4-15). The cathodic portion was under kinetic control and was presumably due to reduction of alloxantin to dialuric acid (III, Fig. 4-15). The wave at potentials corresponding to alloxan wave II was presumably due to further reduction of dialuric acid, formed in the kinetically controlled reduction of alloxantin, to barbituric acid (IV, Fig. 4-15).

It is again clear that there are several questionable points in this reaction scheme. First, there are no reasonable structures or mechanisms to explain the fact that the reduction of alloxantin is under kinetic control. Second, if reduction of alloxantin produces dialuric acid, it is not clear, on the basis of earlier arguments, why the reduction wave of dialuric acid (i.e., wave II of alloxan and alloxantin according to Struck and Elving[63]) should be so small. Comparison of the data of Struck and Elving[63] reveals that a 0.25 mM solution of alloxantin gives almost exactly the same anodic current as an equal concentration of dialuric acid and almost the same cathodic current for an equal concentration of alloxan; i.e., the same amount of kinetically produced dialuric acid should be formed on reduction of alloxantin and alloxan, and hence the subsequent diffusion-controlled reduction of the kinetically formed dialuric acid to barbituric acid should be the same.

In subsequent studies by Hansen and Dryhurst on the electrochemical oxidation of theophylline[66] and caffeine,[67] it was found that dimethyl alloxan was produced and that methyl alloxan was an electrooxidation product of theobromine.[67] In order to elucidate the electrochemistry of alloxan, alloxantin, and their N-methylated derivatives, Hansen and Dryhurst reinvestigated the systems.[68]

Aqueous solutions of dimethyl alloxan exhibited three cathodic waves at the DME, waves I_a, I_b, and II (Table 4-11). Wave I_a was a kinetically controlled wave with a half-wave potential described by the equation $E_{1/2} = 0.10 - 0.036$ pH. This equation compares favorably with the equation for alloxan wave I $(E_{1/2} = 0.060 - 0.031$ pH$)$[63] and is in accord with a reversible process involving two electrons and one proton. Wave I_b, which was also under kinetic control, was essentially pH independent and appeared along with wave I_a between pH 0.5 and 3.0 (Table 4-11). with the half-wave potential depending more on the nature of the supporting electrolyte than on pH. The limiting current for wave I_b decreased significantly as the pH approached 3, indicating that the species being reduced exists only at low pH. Coulometry at potentials where both waves I_a and I_b occurred indicated that, overall, two electrons were transferred (Table 4-13). The third wave of dimethyl alloxan (wave II, Table 4-11) was observed only over the limited pH range 4.8–5.8 since it was obscured by background discharge at lower pH and because at higher pH the decomposition of dimethyl alloxan to dimethylalloxanic acid (see Fig. 4-12) was very rapid. The limiting current for wave II was under diffusion control, and the half-wave potential was essentially pH independent. Coulometry at wave II potentials revealed that two electrons were transferred (Table 4-13).

Significantly, it was found that the products of the wave I_a, I_b and II processes were identical, as evidenced by the spectra[69] and the large anodic polarographic waves of the products.[63,68] The product was dimethyldialuric acid.

Aqueous solutions of methyl alloxan also exhibited three cathodic waves at the DME (Table 4-11). Wave I_a appeared between pH 0 and 7.1, wave I_b from pH 0 to 3, and wave II from pH 4.7 to 5.9. Above pH 5 solutions of methyl alloxan were unstable owing to rearrangement to methylalloxanic acid. Waves I_a

TABLE 4-13   Coulometric Determination of the Number of Electrons n Involved in Reduction of Alloxan and Its N-Methyl Derivatives[a]

Compound	Supporting electrolyte	pH	Wave	Controlled potential (V versus SCE)	Amount electrolyzed (mmoles)	n
Alloxan[b]	Acetate	4.0	I	—	1.0	1.9
	McIlvaine	4.8	II[c]	−1.50	0.068	2.2
				−1.50	0.074	2.4
Methyl alloxan[c]	1 M HOAc	2.3	I	−0.40	0.095	1.9
	McIlvaine	4.8	II	−1.50	0.083	1.6
Dimethyl alloxan[c]	1 M HOAc	2.3	I	−0.40	0.150	2.1
				−0.40	0.150	1.8
				−0.40	0.150	1.8
	McIlvaine	4.9	II	−1.40	0.088	1.8
				−1.40	0.074	1.9
				−1.40	0.074	1.9

[a] Determined at a large mercury pool electrode.
[b] Data from Struck and Elving.[63]
[c] Data from Hansen and Dryhurst.[68]

and I_b were both kinetic waves. Wave I_a was pH dependent according to the equation $E_{1/2} = 0.10 - 0.033$ pH. Wave I_b was observed only up to pH 3, above which it disappeared; it was essentially pH independent. Wave II appeared to be diffusion controlled and the $E_{1/2}$ was pH independent. Coulometry at potentials corresponding to wave I_a plus I_b, and to wave II, showed that in both cases two electrons were transferred (Table 4-13).

Hansen and Dryhurst[68] also found that alloxantin and its N-methyl derivatives in aqueous solution do indeed show a large anodic wave and a small kinetically controlled cathodic wave at potentials close to those of alloxan wave I, but they also show a large wave at much more negative potential, the height of which is about 10 times that of the small kinetic cathodic wave. Spectral studies also confirmed that in aqueous solutions alloxantin does dissociate to alloxan and dialuric acid ($\lambda_{max} = 271$ nm at pH 2.3), the latter being the only absorbing species.[69] Hansen and Dryhurst[68] also confirmed that wave I of alloxan was a $2e$ reduction to dialuric acid. They also found that the wave II process was a $2e$ reduction giving the same product. It was concluded, therefore, that the electrochemical reduction of alloxan, methyl and dimethyl alloxan involved the same number of electrons in the wave I and II processes. The products from both waves were identical, and the current and potential controlling factors were essentially the same for all three compounds.

Thus, the kinetic step in the wave I_a process was dehydration of the carbonyl group at C-5.[70] The $E_{1/2}$–pH relationships for all compounds (shift of ca. 30 mV per pH unit) were in accord with a $2e$–$1H^+$ reversible process. A $2e$ reduction was confirmed by coulometry, and the product was the respective dialuric acid. Thus, in support of the Struck and Elving[63] scheme, wave I_a did indeed appear to be a $2e$–$1H^+$ reduction of dehydrated alloxan (II, Fig. 4-16) to the appropriate dialuric acid anion (III, Fig. 4-16).

Wave I_b was observed most distinctly for dimethyl alloxan; it was considerably smaller for methyl alloxan and, except at very high alloxan concentrations and very low pH, it could not normally be observed for alloxan itself. For methyl and dimethyl alloxan, wave I_b was under kinetic control; it

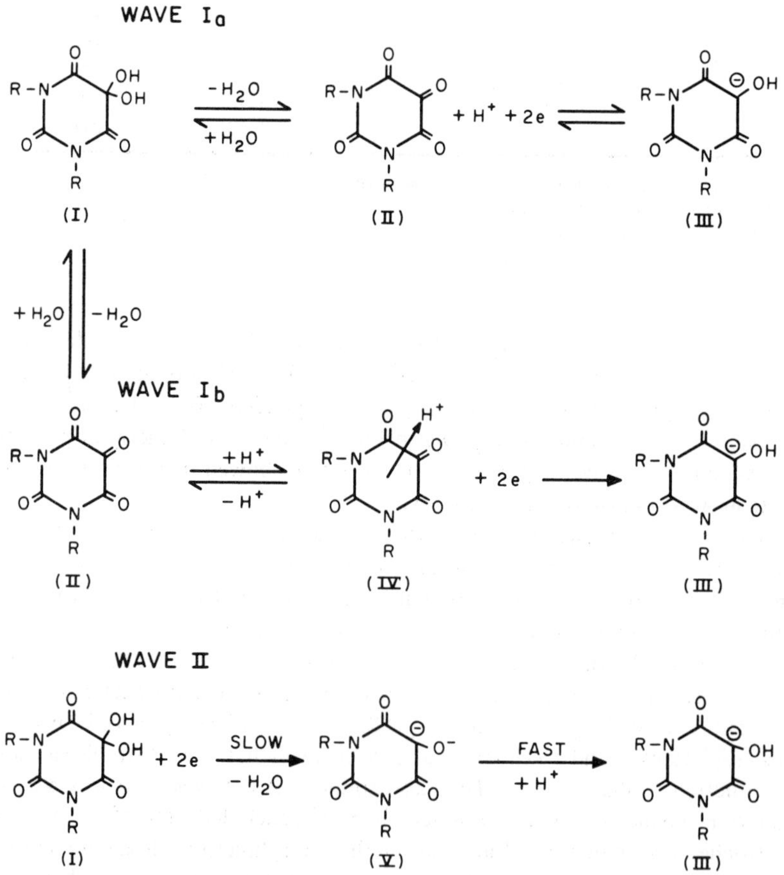

FIG. 4-16.   Proposed reaction pathways for the polarographic reduction of alloxans according to Hansen and Dryhurst.[68]

occurred at more negative potential than wave I_a and was essentially pH independent. The height of wave I_b also decreased with increasing pH and disappeared above pH 3. This behavior was rationalized[68] as follows. At very low pH the alloxans exist at least partially as a cationic species,[64,69] and reduction of this cationic species results in wave I_b. Since both waves I_a and I_b below pH 3 occur to the positive side of the electrocapillary maximum (i.e., the electrode is positively charged), approach of a cationic species toward the electrode would be energetically more difficult than approach of a neutral species, resulting in wave I_b occurring at more negative potentials than wave I_a.[71] The reactions responsible for wave I_b are presented in Fig. 4-16. Again, the dehydration of hydrated alloxan is the current controlling step to give an alloxan cation (IV, Fig. 4-16), which undergoes a $2e$ reduction to give the anion of dialuric acid (III, Fig. 4-16). The fact that the current for wave I_b of methyl alloxan was much smaller than for dimethyl alloxan was accounted for by the inductive effect of the electron-releasing $-CH_3$ groups, which behave in a similar but less pronounced fashion to a negative charge on nitrogen.[72] In other words, a methylated amide carbonyl group is expected to be a more favored proton acceptor than a nonmethylated amide carbonyl. Thus, dimethyl alloxan would be expected to be a better proton acceptor than methyl alloxan. This was reflected in the magnitude of the wave I_b limiting currents or diffusion current constants (Table 4-11) for these two compounds. In the case of unsubstituted alloxan the proton-acceptor properties are the least of all three compounds, such that even in $1\ M\ H_2SO_4$ solution a 10 mM solution of alloxan was necessary before wave I_b could be observed.

The magnitude of the polarographic diffusion current constants (Table 4-11) and coulometry (Table 4-13) indicated that wave II involved transfer of two electrons for all alloxan species. Polarography and spectra indicated that the products from both the wave I and wave II processes were identical, i.e., the appropriate dialuric acid. In view of the virtual pH independence of the wave II processes, the mechanism was ascribed to a direct reduction of the predominant alloxan species in solution, i.e., hydrated alloxan (I, Fig. 4-16), in an irreversible potential controlling $2e$ process to the dianion of dialuric acid (V, Fig. 4-16), which may be protonated to give the monoanion (III, Fig. 4-16).

The data of Hansen and Dryhurst[68] combined with the earlier Struck and Elving[63] reports appear to clarify much of the aqueous electrochemistry of alloxan. The former workers appear to prove that alloxantin species do not enter effectively into the electrochemical processes simply because they are extensively dissociated in aqueous solution. The involvement of alloxantin as a transient, unstable intermediate could not be excluded, however, on the basis of the aqueous electrochemistry of alloxan.

This type of information is pertinent from a biological viewpoint because, although aqueous alloxan administered to animals causes experimental diabetes,[73-76] there are many contradicting reports as to the diabetogenic action of alloxantin and dialuric acid. Thus, some authors[77-81] state that alloxantin solutions are diabetogenic, while others claim that they are not.[73-76,82] Similarly, dialuric acid is,[77,81,83,84] and is not,[73,85] diabetogenic. It would appear from the electrochemical studies that *in dilute aqueous media* alloxantin should be completely dissociated to alloxan and dialuric acid and therefore should be diabetogenic. It has been confirmed by Struck and Elving[63] and Hansen and Dryhurst[68] that dialuric acid is enormously sensitive to air oxidation in aqueous solution, giving alloxantin and alloxan, and accordingly one would again suspect that under physiological conditions dialuric acid would be readily converted to the diabetogen alloxan.

9. Sulfur-Containing Pyrimidines

Barker *et al.*[86] synthesized a number of 5-pyrimidinyl sulfides and found that the best way to detect the presence of disulfides in their product was by polarography. The disulfides all gave characteristic reduction waves, presumably due to reduction to the corresponding thiol. Some typical half-wave potential data are presented in Table 4-14. Subsequently, Luthy and Lamb[87] outlined detailed analytical methods for the determination of 4-pyrimidinyl disulfides. Bardes and co-workers[88] found that 5-thiouracil showed an anodic polarographic wave at $E_{1/2} = -0.5$ V versus SCE, while the corresponding disulfide gave a cathodic wave having the same $E_{1/2}$; the background electrolyte was 0.1 M NH$_4$Cl plus 0.1 M NH$_4$OH.

TABLE 4-14   Half-Wave Potentials for Some 5-Pyrimidinyl Disulfides[a]

Name	Solvent	$E_{1/2}$ (V versus SCE)
Di(2,4,6-trihydroxy-pyrimidin-5-yl)-	0.05 N KOH	−0.5
Di(2,4,6-trihydroxy-6-methylpyrimidin-5-yl)-	0.05 N KOH 0.1 N NaOH	−0.35
Di(6-amino-2,4-dihydroxypyrimidin-5-yl)-	0.1 N NaOH	−0.25
Di(2,4-dihydroxy-5-yl)-	0.1 N NaOH	−0.3
Di(2,4-dihydroxy-6-methylpyrimidin-5-yl)-[b]	0.1 N KOH	−0.1,[c] −0.25,[d] −0.58[e]

[a] Data from Barker *et al.*[86]
[b] Data from Luthy and Lamb.[87]
[c] Anodic wave.
[d] Cathodic (adsorption) wave.
[e] Cathodic diffusion-controlled wave.

Okazaki[89−92] has reported that a large number of pyrimidinylsulfanilamide drugs show polarographic reduction waves at the DME. In view of the similarity between the polarographic behavior of the pyrimidinylsulfanilamides and 2-aminopyrimidine, it was concluded that the pyrimidine nucleus was the electroactive moiety of the drugs.

Thiobarbiturates are widely used as intravenous anesthetics, particularly sodium 5-ethyl-5-(1-methylbutyl) thiobarbiturate. Many thiobarbiturates, like other sulfur-containing compounds,[93] exhibit anodic polarographic waves at the DME corresponding to formation of slightly soluble mercury compounds.[94] Indeed, anodic waves of various thiobarbiturates in 0.1 M sodium hydroxide supporting electrolyte have been used for analysis of pharmaceutical preparations.[95,96] Formation of anodic waves of sulfur-containing compounds is of limited interest from a mechanistic viewpoint. However, Zuman and coworkers[97] have studied the polarography of several of 2-thiobarbiturates which are found to give a number of polarographic reduction waves. Typical half-wave potential data for some 5,5-dialkyl-substituted 2-thiobarbiturates are presented in Table 4-15. Detailed study was confined to 5-ethyl-5-(1-methylpropyl)-2-thiobarbituric acid, although most other compounds (I–VII, Table 4-15) were thought to behave similarly. Basically, the polarographic reduction takes place in two steps, i_1 and i_2 (Fig. 4-17). The first reduction step in acidic media $i_1^{H_2A}$ is a 4e process. At pH greater than about 3 the height of wave $i_1^{H_2A}$ decreased in the form of a dissociation wave until at about pH 5 it reached a value corresponding to a 2e process, this wave being designated i_1^{HA}. The height of this 2e wave remains constant up to pH 9, above which it begins to decrease again in

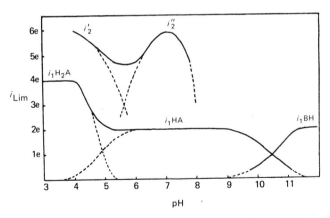

FIG. 4-17. Dependence of the limiting current of 0.5 mM 5-ethyl-5-(1-methylpropyl)-2-thiobarbituric acid on pH. Solid lines, experimental values; dashed lines, theoretical polarographic dissociation curves.[97] (Reprinted with the permission of Elsevier Publishing Company, Amsterdam.)

TABLE 4-15 Half-Wave Potentials of Some 5,5-Dialkyl 2-Thiobarbiturates and Related Species at the DME[a]

$E_{1/2}$ (V versus SCE[b]) in

Compound number	Substituents R₁	R₂	R₃	0.1 M H₂SO₄ C[c]	A[d]	Acetate, pH 4.6 C	A	Phosphate, pH 6.7 C	A	Borate, pH 9.3 C	A	0.1 M NaOH C	A
I	Et	(1-Me)Pr	H	—	—	-1.35	+0.15 / +0.22	-1.37	+0.07	-1.48	-0.06	-1.72	-0.25
II	Et	(1-Me)Bu	H	—	—	e	e	-1.38	+0.09	-1.48	-0.06	-1.72	-0.24
III	Allyl	Isobu	H	—	—	-1.39	e	-1.40	+0.06	-1.52	-0.13	-1.71	-0.26
IV	Allyl	Cyclo-hexen-2-yl	H	-0.97[f]	+0.05 / +0.18	-1.31	+0.12	-1.34	+0.05 / +0.22	-1.46	-0.13 / 0.00 / +0.08	-1.70	-0.35 / -0.24
V	Pr	Pr	Pr	-0.98[f,g]	—	e	e	-1.36[g]	e	-1.42[g]	e	-1.66	-0.31

VI	Et	(1-Me)Pr	Me	—	—	-1.36	-0.02	-1.36	-1.15[h]	-1.43	-0.30[h]	-1.66	-0.31
VII	Allyl	(1-Me)Pr	Me	—	—	-1.35	-0.01[h]	-1.36	-0.13[h]	-1.43	-0.32[h]	-1.65	-0.30
VIII	H	H	H	-1.03[f]	—[i]	—	—	—	+0.01[h]	—	+0.09[h]	—	-0.22
IX	(1-Me)Bu	H	H	-0.9[f]	+0.34[j]	—	+0.17[j]	—	+0.07[h]	—	-0.09[h]	—	-0.27

X

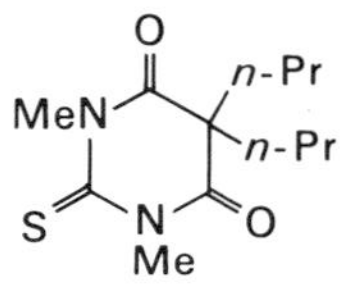

MeN, S, n-Pr, n-Pr, N, Me, O

—	—	-1.27	+0.05	-1.37	+0.05	-1.37	+0.05 / -0.19	—	-0.24

XI

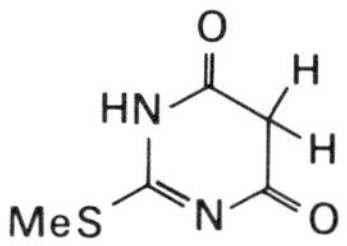

HN, H, H, O, MeS, N, O

-0.98	—	-1.19	+0.24	-1.37 / -1.65	+0.15	—	-0.05	—	—

a Data from Smyth et al.[97]
b For 10^{-4} M solutions.
c Cathodic wave.
d Anodic wave.
e Precipitation observed for all concentrations investigated.

f Massive catalytic wave.
g Despite precipitation a cathodic wave was observed.
h Number of anodic waves observed varies with concentration.
i Complexity of waves.
j $E_{1/2}$ varies with concentration.

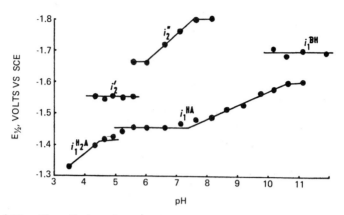

FIG. 4-18. The pH dependence of half-wave potentials of the cathodic waves of 0.5 mM 5-ethyl-5-(1-methylpropyl)-2-thiobarbituric acid.[97] (Reprinted with the permission of Elsevier Publishing Company, Amsterdam.)

the shape of a dissociation curve. As wave i_1^{HA} decreases in height, a more negative wave, i_1^{BH}, appears; the total wave height remains constant. Between about pH 4 and 8 another wave, i_2 (see Fig. 4-17), appears at more negative potentials than waves i_1^{H2A} and i_1^{HA}. In the lower pH region wave i_2' appears and decreases in height with increasing pH, being replaced by wave i_2''. The latter two waves involve transfer of six electrons. The $E_{1/2}$–pH relationships for waves i_1^{H2A}, i_1^{HA}, i_1^{BH}, i_2', and i_2'' are shown in Fig. 4-18. The various breaks in the $E_{1/2}$–pH plots correspond to various acid–base equilibrium processes (*vide infra*). Waves i_1^{H2A}, i_1^{HA}, and i_1^{BH} in the pH range where they reached the value corresponding to four- or two-electron transfer, respectively, were diffusion controlled. Zuman and co-workers[97] thoroughly investigated the spectral properties of the appropriately substituted 2-thiobarbituric acids and combining the spectral, polarographic, and related electrochemical data concluded that the polarographic reduction waves could be interpreted by means of the following scheme:

$$H_2A \underset{k_{-1}}{\overset{k_1}{\rightleftharpoons}} HA + H^+ \qquad\qquad pK_1 \qquad\qquad (1)$$

$$HA \underset{k_{-2}}{\overset{k_2}{\rightleftharpoons}} BH + H^+ \qquad\qquad pK_2 \qquad\qquad (2)$$

$$H_2A + 4e \xrightarrow{E_1^{H_2A}} P_1 \qquad\qquad i_1^{H_2A} \qquad\qquad (3)$$

$$HA + 2e \xrightarrow{E_1^{HA}} H_2P_2 \qquad i_1^{HA} \qquad (4)$$

$$H_2P_2 \underset{k_{-3}}{\overset{k_3}{\rightleftharpoons}} HP_2 + H^+ \qquad pK_A \qquad (5)$$

$$HP_2 \underset{k_{-4}}{\overset{k_4}{\rightleftharpoons}} P_2 + H^+ \qquad pK_B \qquad (6)$$

$$H_2P_2 + 6e \xrightarrow{E_2'} X_1 \qquad i_2' \qquad (7)$$

$$HP_2 + 6e \xrightarrow{E_2''} X_2 \qquad i_2'' \qquad (8)$$

$$BH + 2e \xrightarrow{E_1^{BH}} P_3 \qquad i_1^{BH} \qquad (9)$$

For some compounds, i_2'' decreased and reached a limiting value corresponding to a $2e$ process. Accordingly, it was thought that a further reaction (Eq. 10) could be involved. Approximate values of dissociation constants in the above scheme were $pK_1 < 2$ (for compounds I–IV, Table 4-15) and $pK_1 < 0$ (for V–VII, Table 4-15); $pK_2 = 7–8$; $pK_A < 4$; and $pK_B = 5–7$.

$$P_2 + 2e \xrightarrow{E_2''} X_3 \qquad i_2'' \qquad (10)$$

Summarizing, at $pH < 3$ establishment of equilibrium (Eq. 1) is rapid and species H_2A is reduced in a $4e$ step (Eq. 3). (Based on the studies of Zuman *et al.*[97] the probable structures of the electroactive species H_2A, HA, etc., were deduced and are shown in Table 4-16.) At pH 5.5–9.5 the establishment of the second dissociation step (Eq. 2) is rapid and species HA is reduced in a $2e$ step (Eq. 4). The product of this reduction step, H_2P_2, is reduced at lower pH values when the protonation rate constant k_{-3} is large, in a $6e$ step, like the conjugate base HP_2 at pH 6.5–8 (Eqs. 7 and 8). Above pH 7 the protonation rate constant k_{-4} is no longer large enough to transform all P_2 into HP_2, and wave i_2'' decreases. When P_2 is reducible, this decrease is not complete but stops at a $2e$ process (Eq. 10). In the intermediate pH range 3–5.5, processes 3 and 4 compete and the wave height (for the first wave) is limited by the rate of protonation with constant k_{-1}. In the transitional range between pH 9.5 and 12, processes 4 and 9 compete. Products were not isolated in this study; hence, to some degree the assignment of reduction sites and the identity of products are in some doubt.

Unsubstituted 2-thiobarbituric acid (VIII, Table 4-15) and 5-monoalkyl-2-thiobarbituric acid derivatives (IX, Table 15) did not give any reduction waves in the available potential range. As noted, all compounds gave anodic waves due to

TABLE 4-16   Predominant Forms in Various pH Ranges and the Most Probable Electroactive Forms of 2-Thiobarbiturates[a]

Form	Approx. pH range	Predominant structure	Wave	Approx. pH range	Most probable reducible Structure	Bonds
H₂A	<1	*(structure)*	$_iH_2A$	<4	*(structure)*	$C=\overset{\oplus}{N}H + C=S$
HA	2–7	*(structure)*	$_iHA$	6–9	*(structure)*	$2C=\overset{\oplus}{N}H$
BH	9–11	*(structure)*	$_iBH$	11–13	*(structure)*	$C=\overset{\oplus}{N}H$
B	>13	*(structure)*			*(structure)*	$C=\overset{\oplus}{N}H$

[a] From Smyth et al.[97]

mercury salt formation. The original paper or Table 4-15 should be consulted for details.

(**X**)

2-Methylthiobarbituric acid (**X**) (4,6-dihydroxy-2-methylmercaptopyrimi-dine) exhibits two cathodic polarographic waves at the DME.[98] Between pH 2 and 5.3 a single $2e$ wave is observed (wave I). At higher pH the height of this wave decreases in the shape of a dissociation curve. At pH 5.2 a second, more negative wave is observed (wave II) which corresponds to a $6e$ reduction. The height of wave II also decreases with increasing pH in the form of a dissociation curve. The half-wave potentials for wave I are pH dependent, the $E_{1/2}$–pH plot showing three linear segments related to acid–base dissociation effects (Fig. 4-19). The half-wave potentials for wave II were pH independent below pH 6.6 and shifted by about 50 mV/pH unit at higher pH values. No reduction waves were observed above pH 7.9. The reduction scheme proposed for 2-methylthio-barbituric acid is as follows:

$$HA \underset{k_{-1}}{\overset{k_1}{\rightleftharpoons}} C + H^+ \qquad pK_1 \qquad (11)$$

$$HA + 2e \xrightarrow{E_1} P_1 \qquad \text{wave I} \qquad (12)$$

$$P_1 + 6e \xrightarrow{E_2} P_2 \qquad \text{wave II} \qquad (13)$$

Below pH 5.2, HA, a monoprotonated form of 2-methylthiobarbituric acid, is reduced to unknown products in a $2e$ process. Above pH 5.2 the rate of protonation with rate constant k_{-1} is no longer fast enough to transform all the thiobarbiturate from form C to HA (Eq. 11). Wave I hence decreases as its height becomes limited by the rate of the chemical reaction with rate constant k_{-1}. Apart from the fact that only the reduced form of HA (i.e., P_1) could give wave II and that it was a $6e$ reaction, the products or other details of the electrode mechanisms are unknown. Spectroscopic studies[98] have suggested that the keto form of 2-methylthiobarbituric acid (**X**) is the electroactive species. 1-Ethyl-2-allylthio-5-(2-methylbutyl)barbituric acid (**XI**) is not electrochemically re-ducible, and it has been assumed that the reason for this is that **XI** exists predominantly in the enol form. Similarly it was deduced that, in 2-thiobar-

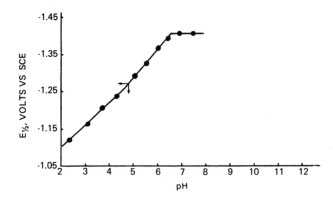

FIG. 4-19.   The $E_{1/2}$ versus pH dependence for the cathodic wave I of 2-methyl-thiobarbituric acid (0.1 m*M*).[98] (Reprinted with the permission of Elsevier Publishing Company, Amsterdam.)

bituric acid and its 5-monoalkyl derivatives (**XII**), the failure to observe polarographic reduction waves is due to a predominance of an enol form (**XIII**) from which the formation of the electroactive keto form is too slow.

(**XI**)

(**XII**)   (**XIII**)

Subsequent studies of Zuman and co-workers[99] were concerned with the polarographic reduction of 5-diethylamino-5-ethyl-2-thiobarbituric acid and 5-isobutyl-5-allyl-1-methyl-6-imino-2-thiobarbituric acid. Polarographic reduction of the diethylamino derivative took place via four cathodic waves. The variation of limiting current and $E_{1/2}$ for each of the four waves with pH is shown in Fig. 4-20. The first and least negative wave, designated i_A in Fig. 4-20,

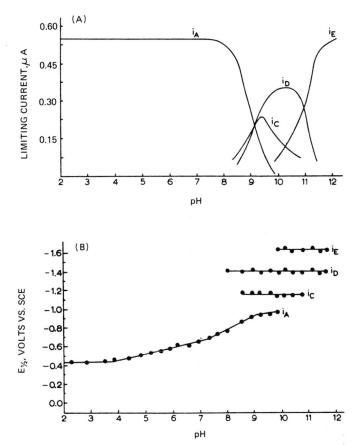

FIG. 4-20. The pH dependence of (A) cathodic limiting currents and (B) half-wave potentials for 1.0 mM 5-diethylamino-5-ethyl-2-thiobarbituric acid.[99] (Reprinted with the permission of Elsevier Publishing Company, Amsterdam.)

was due to a $2e$ process that decreased in height above pH 8 in the form of a dissociation curve. Simultaneously, wave i_C increased, reached a maximum at pH 9.3, and again decreased. The latter decrease was accompanied by an increase of wave i_D, which reached maximum height corresponding to more than a $1e$ reduction at pH 10.2 and then decreased. Finally, an increase of wave i_E was observed in the shape of a dissociation curve. This wave reached a height corresponding to a $2e$ reduction at pH > 11. A small adsorption wave (wave i_B) accompanied wave i_A at pH 4–5, $E_{1/2} = -1.07$ V versus SCE.

The change in polarographic wave heights and half-wave potentials has been interpreted as shown in Eqs. 14–21.[99] Spectroscopic studies were used to deduce the nature of the various electroactive species shown in Eqs. 14–21.

$$\text{(A)} \; \xrightleftharpoons[k_{-1}]{k_1} \; \text{(C)} \qquad +\text{H}^+ \; ; \; \text{p}K_1 \qquad (14)$$

$$\text{A} + 2e \; \xrightarrow{E_A} \; P_1 \qquad i_A \qquad (15)$$

$$\text{C} \; \xrightleftharpoons[k_{-2}]{k_2} \; \text{(D)} \qquad +\text{H}^+ \; ; \; \text{p}K_2 \qquad (16)$$

$$\text{C} + 2e \; \xrightarrow{E_C} \; P_2 \qquad i_C \qquad (17)$$

$$\text{D} \; \xrightleftharpoons[k_{-3}]{k_3} \; \text{(E)} \qquad +\text{H}^+ \; ; \; \text{p}K_3 \qquad (18)$$

$$\text{D} + 2e \; \xrightarrow{E_D} \; P_3 \qquad i_D \qquad (19)$$

$$\text{E} + 2e \; \xrightarrow{E_E} \; P_4 \qquad i_E \qquad (20)$$

$$\text{E} \; \rightleftharpoons \; \qquad +\text{H}^+ \; ; \; \text{p}K_4 \qquad (21)$$

Thus, below pH 8 the triprotonated form A is reduced in a $2e$ (wave i_A) reaction (Eq. 15). At pH >8 the rate constant k_{-1} decreases and so does i_A. Correspondingly, the height of wave i_C, due to reduction of the diprotonated species, increases. However, before the height of wave i_C can reach a maximum value expected for a $2e$ reduction, the rate of protonation with constant k_{-2} decreases so rapidly with increasing pH that reaction 16 cannot transform all species D into C. This decrease of wave i_C is accompanied by an increase in wave i_D, which at pH 10.2 reaches a value corresponding to a $2e$ reduction. Above this pH a decrease in the rate of protonation of species E with rate constant k_{-3} is responsible for the decrease of i_D. Simultaneously, wave i_E increases, corresponding to a $2e$ process. Values of the various pK values are presented in the original paper.

5-Isobutyl-5-allyl-1-methyl-6-imino-2-thiobarbituric acid is reduced in a $2e$ wave (wave i_P) up to pH 8 (Fig. 4-21).[99] The decrease in the height of wave i_P at pH < 5 is due to hydrolysis. At pH > 8 wave i_P decreases and reaches a height corresponding to one-half of the maximum value at pH 9.3. The decrease of wave i_P is accompanied by an increase of wave i_R, which at pH 10.5 reaches a maximum value corresponding to a $4e$ reduction process. At higher pH values wave i_R decreases, reaching one-half of the maximum value at pH 12. The $E_{1/2}$–pH plot (Fig. 4-21B) shows several linear sections, the intersections of which correspond to various dissociation constants. The scheme for polaro-

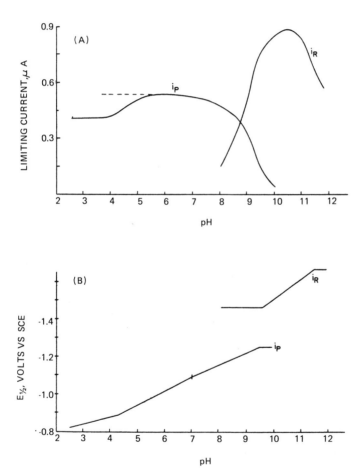

FIG. 4-21. The pH dependence of (A) cathodic limiting currents and (B) half-wave potentials for 0.1 mM 5-isobutyl-5-allyl-1-methyl-6-imino-2-thiobarbituric acid.[99] (Reprinted with the permission of Elsevier Publishing Company, Amsterdam.)

graphic reduction of the imine differs from that of other thiobarbiturates discussed earlier. This difference is apparently due to the presence of the C=N grouping, which in the protonated state is susceptible to reduction[100−102] at much more positive potential than the unprotonated form. The reaction scheme is represented in Eqs. 22–27, Thus, reduction of the diprotonated form, P, and

$$\text{(22)}$$

$$PH + 2e \longrightarrow X_1 \qquad \text{(23)}$$

$$P + 2e \longrightarrow X_2 \qquad \text{(24)}$$

$$\text{(25)}$$

$$R + 4e \longrightarrow X_3 \qquad i_R \qquad \text{(26)}$$

$$\text{(27)}$$

the triprotonated form, PH, occurs in a 2e process (wave i_P, Fig. 4-21). It was proposed that the potentials where these two species are reduced are so close that separation of two waves does not occur.[99] The 4e wave (wave i_R) was proposed to involve reduction of the protonated ketimino group. This reaction, of course, involves only two electrons; the site of the second 2e process is unknown. Both 5-diethylamino-5-ethyl-2-thiobarbituric acid and 5-isobutyl-5-allyl-1-methyl-6-imino-2-thiobarbituric acid gave rise to anodic waves due to mercury salt formation.[99]

10. *N*-Benzoyl Derivatives of Barbituric Acids

Kahl and Pasek[103] have reported that various *N*-benzoyl-5,5-dialkylbarbituric acids are reduced at the DME in aqueous solution between about pH 5 and 7. All of the *N*-benzoyl barbiturates examined gave a single pH-dependent wave (Fig. 4-22). Controlled potential electrolysis of each compound gave the appropriately substituted dialkylbarbituric acid and benzyl alcohol, and four electrons were consumed. Without substantive supporting evidence it was concluded that the electrode reaction proceeded in two sequential stages. The first involved a $2e-2H^+$ reduction of the *N*-benzoyl grouping of the *N*-benzoylbarbituric acid (I, Fig. 4-23) to give the appropriate barbituric acid (II, Fig. 4-23) and benzaldehyde (III, Fig. 4-23). The latter compound was then thought to be immediately further reduced to benzyl alcohol (IV, Fig. 4-23).

Polarography of 1-*p*-nitrobenzoyl-5,5-diethylbarbituric acid[103a] (I, Fig. 4-24) gives rise to a total of three polarographic reduction waves. Typical $E_{1/2}$ values are shown in Table 4-17. The first wave (wave I) is observed between pH 3.6 and 8.3 and is apparently due, overall, to a $4e-4H^+$ reduction of the nitro group of I (Fig. 4-24) to the corresponding hydroxylamine (II, Fig. 4-24). At pH > 5.3 the latter hydroxylamine group is further reduced, in the wave II process, in a $2e-2H^+$ step to the corresponding amine (III, Fig. 4-24). The third wave, also observed only at pH ≥ 5.3, is due to the further $4e-4H^+$ reduction of 1-*p*-aminobenzoyl-5,5-diethylbarbituric acid (III, Fig. 4-24) formed in the wave II process to 5,5-diethylbarbituric acid (IV, Fig. 4-24) and *p*-aminobenzyl

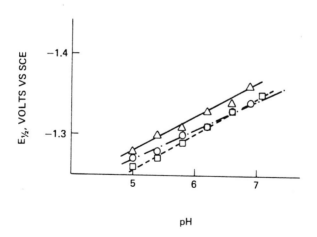

FIG. 4-22. The pH dependence of half-wave potentials for the cathodic polarographic wave of 1-benzoyl-5,5-diethylbarbituric acid (solid line), 1-benzoyl-5-ethyl-5-phenylbarbituric acid (dashed line), and 1-benzoyl-3-methyl-5-ethyl-5-phenylbarbituric acid (dot-dashed line).[103]

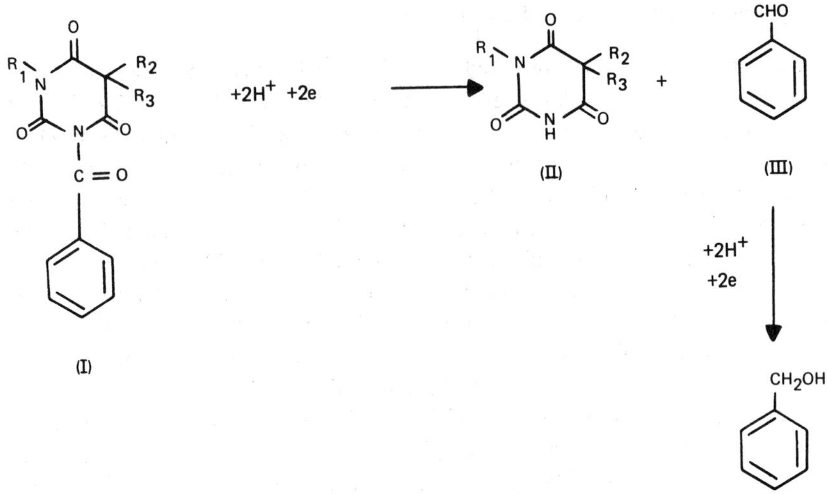

FIG. 4-23. Proposed reaction scheme for the electrochemical reduction of N-benzoylbarbituric acids.[103] R_1 = CH_3 or $-H$; R_2 = C_2H_5 or $-C_6H_5$; R_3 = $-C_2H_5$.

alcohol (V, Fig. 4-24). At pH values less than 5.3 the wave II and wave III processes combine to give a single $6e-6H^+$ wave (referred to as wave II, pH < 5, in Fig. 4-24). In this process the hydroxylamine derivative (II, Fig. 4-24) is reduced directly to 5,5-diethylbarbituric acid (IV, Fig. 4-24) and p-aminobenzyl alcohol (V, Fig. 4-24).

11. 5-Halogenouracils

Certain 5-halogeno derivatives of uracil are important drugs. For example, 5-fluorouracil is an antimetabolite and hence has been used in cancer chemotherapy. Wrona and Czochralska[104] examined the electrochemical reduction of various 5-halogenouracils and found that all showed only a single, irreversible, diffusion-controlled cathodic polarographic wave. Typical half-wave potential data are presented in Table 4-18. With the exception of 5-fluorouracil all 5-halogenouracils (I, Fig. 4-25A) were reduced in a $2e$ process to give uracil (II, Fig. 4-25A) and the appropriate halide ion. In the case of the 5-fluoro derivative (III, Fig. 4-25B), a $4e$ process was observed which resulted in formation of dihydrouracil (IV, Fig. 4-25B) and fluoride ion. In the pH range below pK_a (i.e., where a neutral molecule is reduced; see Table 4-18) the half-wave potential is pH independent only for iodo derivatives. By reference to data in Table 4-18 it is clear that the ease of reduction of the carbon–halogen bond increases in the order F $<$ Cl $<$ Br $<$ I. Substitution of 5-halogenouracils at

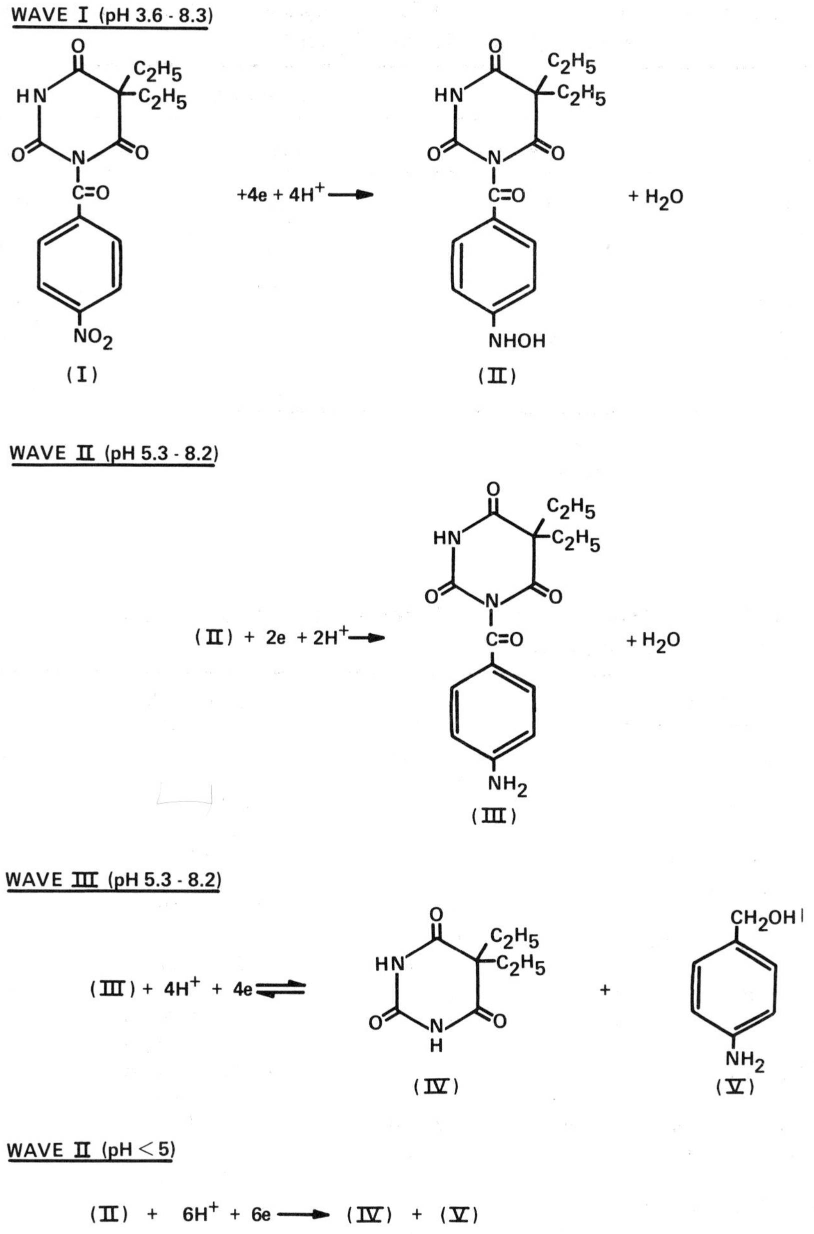

WAVE I (pH 3.6 - 8.3)

$+4e + 4H^+ \longrightarrow$

(I)

$+ H_2O$

(II)

WAVE II (pH 5.3 - 8.2)

$(II) + 2e + 2H^+ \longrightarrow$

$+ H_2O$

(III)

WAVE III (pH 5.3 - 8.2)

$(III) + 4H^+ + 4e \rightleftharpoons$

$+$

(IV)

(V)

WAVE II (pH < 5)

$(II) + 6H^+ + 6e \longrightarrow (IV) + (V)$

FIG. 4-24. Proposed reaction scheme for the electrochemical reduction of 1-*p*-nitrobenzoyl-5,5-diethylbarbituric acid.[103a]

TABLE 4-17   Half-Wave Potentials for Polarographic Reduction of 1-p-Nitro-benzoyl-5,5-diethylbarbituric Acid^a

	$E_{1/2}$ (V versus SCE)		
pH^b	Wave I	Wave II	Wave III
3.6	−0.25	−1.32	−1.32
4.9	−0.30	−1.36	−1.36
5.3	−0.33	−1.28	−1.44
6.0	−0.37	−1.29	−1.49
6.5	−0.39	−1.30	−1.52
6.9	−0.43	−1.31	−1.57
7.6	−0.47	−1.34	−1.63
8.3	−0.54	−1.38	−1.68

^{a} From Kahl and Pasek.[103a]
^{b} In Britton–Robinson buffers.

TABLE 4-18   Half-Wave Potentials for Polarographic Reduction of 5-Halo-genouracils^a

Compound	pH Range	$E_{1/2}$ (V versus SCE)	p$K_a{}^b$
5-Iodouracil	1.0–8.4	−1.02	8.25
	8.4–13	−1.02 − 0.10 pH	
5-Iodouridine	1.0–8.9	−0.92	8.50
	8.9–13	−0.92 − 0.10 pH	
5-Iodouridylic acid	7.4	−0.96	
5-Bromouracil	6.1–8.6	−1.62 − 0.02 pH	8.05
	8.6–11	−1.67 − 0.08 pH	
1-Methyl-5-bromouracil	6.2–8.4	−1.60 − 0.02 pH	8.30
	8.4–11	−1.65 − 0.07 pH	
1,3-Dimethyl-5-bromouracil	6.1–13	−1.60	
5-Bromodeoxyuridine	6.2–8.3	−1.48 − 0.04 pH	8.1
	8.3–11	−1.56 − 0.08 pH	
5-Chlorouracil	6.0–7.9	−1.75 − 0.01 pH	7.95
	7.9–10	−1.77 − 0.05 pH	
1,3-Dimethyl-5-chlorouracil	7.0–10	−1.80	
5-Chlorouridine	6.0–8.6	−1.65 − 0.02 pH	8.50
	8.6–11	−1.70 − 0.07 pH	
5-Chlorouridylic acid	7.4	−1.75	
5-Fluorouracil	7.4	−1.85	7.8

^{a} Data from Wrona and Czochralska.[104]
^{b} Values for pK_a from Berens and Shugar.[105]

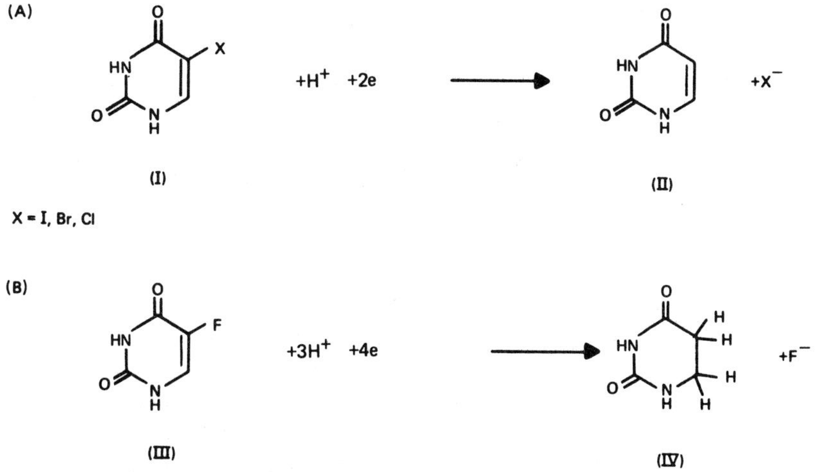

(A)

X = I, Br, Cl

(B)

FIG. 4-25. Reaction pathways for the electrochemical reduction of 5-halogeno-uracils.[104]

N-1 with ribose (i.e., uridine derivatives) only slightly affects the half-wave potential. The corresponding nucleotides are, however, reduced somewhat more readily.

12. Other Pyrimidines

Budnikov[106] has reported the polarographic behavior of some pyrimidones and thiopyrimidones having structures **XIV–XX**. In acidic and neutral buffer

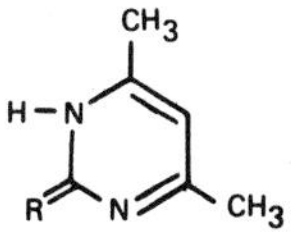

(**XIV**) R = O
(**XV**) R = S

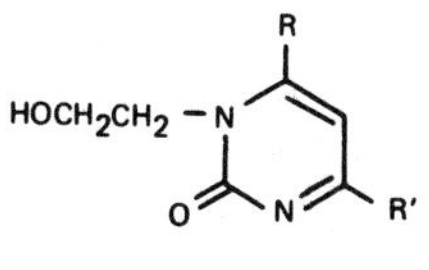

(**XVI**) R = R' = H
(**XVII**) R = R' = CH₃

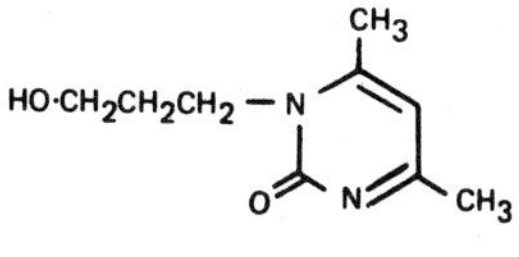

(**XVIII**)

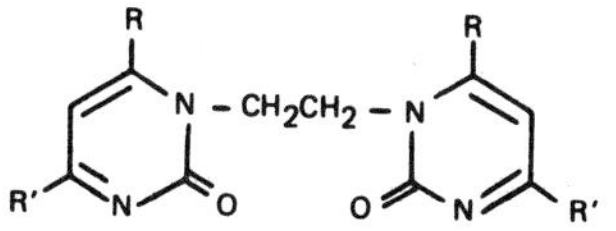

(**XIX**) R = R' = H
(**XX**) R = R' = CH₃

solutions the pyrimidones (**XIV**, **XVI**, and **XVII**) gave one diffusion-controlled wave, corresponding to a 1e process to give a free radical which rapidly dimerized. In weakly alkaline and alkaline solutions the height of the reduction wave decreased; the wave finally disappeared with increasing pH, and a new wave at more negative potential appeared of about the same height. The remaining pyrimidones (**XV**, **XVIII**, **XIX**, and **XX**) showed similar behavior, but the polarograms exhibited a number of adsorption and other anomalous waves.

Jain and Kapoor[107] studied the kinetic parameters and to some extent the mechanism of the polarographic reduction of 5-nitrobarbituric acid.

Although barbituric acid and 5,5-dialkyl barbiturates do not normally give rise to faradaic polarographic waves in aqueous solution, Kissinger and Reilley[108] found that 5,5-diethylbarbituric acid gave a single, irreversible, 1e cathodic peak in acetonitrile solution (containing 0.2 M tetrabutylammonium perchlorate). The peak was observed at platinum, gold, and mercury electrodes (E_p = -1.25, -1.70, and -2.40 V versus SCE, respectively). Specular reflectance techniques revealed that the keto form of the un-ionized barbiturate (I, Eq. 28) undergoes a 1e reduction to give the barbital anion (II, Eq. 28).

$$ \text{(I)} + e^- \longrightarrow \text{(II)} + H\cdot \qquad (28) $$

13. Vitamin B$_1$ and Related Compounds

The polarography of vitamin B$_1$, thiamine, has been extensively reviewed through about 1953 by Březina and Zuman.[93] In summary, thiamine has been shown to give rise to three types of polarographic waves: (a) a catalytic hydrogen discharge wave in buffered and unbuffered solutions, (b) an anodic wave at pH $>$ 9, corresponding to the formation of a mercury salt with the thiol form of thiamine, and (c) a reduction wave in alkaline solution.

Thiamine can also give a catalytic wave in ammoniacal cobalt solutions that is similar to the waves obtained with proteins under similar conditions. The latter process is of no interest in the present discussion; the original literature should be consulted for further information.[93,109-116]

The catalytic wave of thiamine has been observed by many workers.[111,112,117-128] It is a quite typical catalytic wave, its height and half-wave potential being dependent on the pH and buffer capacity of the supporting electrolyte system. In a buffer of pH 7.2 the half-wave potential of the catalytic wave occurs at -1.30 V versus NCE.[109,117]

In alkaline solution above about pH 9, thiamine (I, Eq. 29) is converted to its so-called —SH or thiol form (II, Eq. 29) by rupture of the thiazole ring. Polarography of this —SH form has been shown to give rise to a pronounced

(29)

(I)   (II)

anodic wave ($E_{1/2}$ = —0.42 V versus NCE at pH 11.0,[111,120] and $E_{1/2}$ = —0.5 V versus NCE in 0.1 M LiOH[112]). Pleticha[129] has reported that an adsorption prewave, $E_{1/2}$ = —0.20 V (pH not specified), occurs before the anodic wave. The anodic wave is formed due to oxidation of mercury and formation of a mercurous thiamine complex or salt.[93,130]

If air or oxygen is allowed into a solution of thiamine above pH 9, the thiol form of thiamine is oxidized to the disulfide form (XXI).[111,112,125] The

(XXI)

disulfide form of thiamine does not give the catalytic wave of thiamine.[112,131] The disulfide linkage in the disulfide form of thiamine can be reduced electrochemically at the DME (Table 4-19), presumably regenerating the thiol form.[112]

Although many papers have stated that thiamine itself gives a cathodic polarographic wave, there is very little information regarding the half-wave potential or mechanism associated with the process. Asahi[130] has stated that thiamine shows a cathodic wave at $E_{1/2}$ = —1.1 V at pH 1.0 (reference electrode not stated in abstract but probably NCE). The wave was thought to be caused by the electrochemical reduction of both the thiazole and pyrimidine rings, since on controlled potential electrolysis $H_2\dot{S}$, NH_4Cl, and $CH_3COCH(SH)CH_2CH_2OH$ were liberated. The process was thought to consume about 10 electrons. The wave height decreased with increasing pH above pH 5, and hence the wave was ascribed to reduction of the cationic form of thiamine. Other waves observed at $E_{1/2}$ = —0.14 and —1.0 V were proposed to be due to adsorption and desorption, respectively, of thiamine.

Tikhomirova and Belen'kaya[132] claim that thiamine shows three polarographic waves in KCl solutions, $E_{1/2}$ = —1.20, —1.43, and —1.60 V (reference electrode not known). The processes responsible for these waves are not known.

TABLE 4-19 Half-Wave Potentials for the Polarographic Reduction of the Disulfide Form of Thiamine[a]

pH	$E_{1/2}$ (V versus NCE)
1.08	−0.185
7.08	−0.525
12.11	−0.55
13.0	−0.44

[a] Data from Tachi and Koide.[112]

Palomino and co-workers[133−135] have utilized the cathodic and anodic polarographic waves formed by thiamine and the thiol form of thiamine, respectively, to investigate the kinetics and thermodynamics of the transformation of thiamine to its thiol form in alkaline solution. A number of compounds closely related to thiamine have also been examined. The electrode mechanisms, in as much as they have been elucidated, appear to primarily involve the redox behavior of the sulfur moiety of the compounds. Březina and Zuman[93] have considered the polarographic behavior of several thiamine model compounds in some detail. Some typical half-wave potential data are presented in Table 4-20.[109,111,112,117,120−122,130,132,136−139]

Okamoto[139] examined the electrochemical reduction of thiamine hydrochloride using AC polarography. Between pH 3.0 and 4.75 a single, pH-dependent peak was observed (Table 4-21). At higher pH a new peak appeared at more positive potential. The height of both peaks decreased with increasing pH, and both peaks completely disappeared by pH 10. The behavior of the peak currents for both processes with respect to temperature, mercury column height, and concentration suggested that both waves were essentially under diffusion control, with a small contribution from adsorption−desorption effects. The pK_a values of thiamine hydrochloride were determined to be 4.7−4.8 and 9.3−9.37 at 25°C. Accordingly, the first peak (most positive) was considered to be due to electrochemical reduction of the N=CH double bond of the thiazole moiety of the thiamine chloride molecule (**XXII**). The more negative peak was thought to be a similar reduction but of the more acidic thiamine hydrochloride (**XXIII**). The product was thought to be an appropriate dihydrothiamine for both peaks.

(**XXII**)   (**XXIII**)

TABLE 4-20 **Half-Wave Potentials for the Polarography of Thiamine and Related Compounds**

Compound	Supporting electrolyte	pH	$E_{1/2}$ (V versus NCE^a) for wave		Reference
			Anodic	Cathodic	
Thiamine		1.0		$-0.14^b, -1.0^b,$ -1.14	130
		7.2		-1.30^c	109, 117
		11.0	-0.42		111, 120
	0.1 M LiOH		-0.50		112
	KCl	?		$-1.20, -1.43,$ $1.60^{d,e}$	132
	0.1 M KCl	6.5		-1.25^e	121
	0.1 M KCl	7.5–8		-1.35^e	121
Thiamine monophosphate	25% KCl			$-1.10^e, -1.36^{c,e},$ $-1.48^{b,e}$	136
Thiamine propyl disulfide	Acetate	4		-0.70^e	137
Thiamine tetrahydrofurfuryl disulfide	Phosphate–citrate	6		-0.495^f	138
2-Hydroxyethyl thiamine disulfide		11.1		$-0.62, -1.16$	138
Thiamine disulfide		6		-0.5	138
Thiamine allyl disulfide		6		-0.5	138
Thiamine propyl disulfide		6		-0.50	138
4-Hydroxythiamine			-0.41^g		122
Benzothiazole methiodide			-0.39^g		122

a Except where otherwise stated.
b Adsorption wave.
c Catalytic hydrogen wave.
d Processes responsible for waves not known.
e Reference electrode not known.
f At 15°C.
g Wave pH independent over the range where it is observed.

In a subsequent report Okamoto[140] outlined the AC polarography of thiamine monophosphate. This compound gave a well-defined peak at potentials close to that of the second (more negative) peak of thiamine hydrochloride between pH 1 and 10. A very ill-defined shoulder between pH 1.75 and 3.0 might have corresponded to the first peak of thiamine hydrochloride (Table 4-22). As the pH increased above 1.75, the peak current decreased continuously and the peak disappeared above ca. pH 10. The single peak observed at most pH

TABLE 4-21   Summit Potentials for the AC Polarography of Thiamine Hydrochloride[a]

pH	E_S (V versus SCE)	
	Peak I	Peak II
3.00		−1.273
4.00		−1.275
4.25		−1.299
4.50		−1.316
4.75		−1.320
5.00	−1.275	−1.335
5.25	−1.290	−1.357
5.50	−1.300	−1.370
5.75	−1.307	−1.390
6.00	−1.310	−1.392
7.00	−1.351	−1.445
7.90	−1.320	−1.399
8.95	−1.317	−1.453
10.00		

[a] Data from Okamoto.[139]

TABLE 4-22   Summit Potentials for the AC Polarography of Thiamine Monophosphate[a]

pH	E_S (V versus SCE)
0.90	−1.090
1.75	−1.141
3.00	−1.222
4.00	−1.270
5.00	−1.337
6.00	−1.390
6.80	−1.440
7.85	−1.518
8.85	−1.500, −1.623
9.95	ca. −1.517
10.65	

[a] Data from Okamoto.[140]

values for thiamine monophosphate was primarily under diffusion control. The number of electrons involved in the electrode reaction was determined by a millicoulometric technique. At pH values less than 3, nine electrons were apparently involved. At pH 4, eight electrons were involved. The reduction mechanism could not be elucidated in detail in Okamoto's work, but it was suggested that the reduction could take place in both the pyrimidine and thiazole rings. The decrease in current with increasing pH was proposed to be due to various ionizations of the thiamine monophosphate leading to the existence in solution of less easily reduced species. The disappearance of the AC polarographic peaks above pH 10 was thought to be due to the opening of the thiazole ring.

Although the mechanisms for electrochemical reduction of thiamine and thiamine monophosphate were not elucidated, they were thought to be identical. However, the data presented in the two papers of Okamoto do not support this conclusion. Apart from a similarity between the summit potential of the second peak of thiamine hydrochloride and the peak of thiamine monophosphate and their decrease with increasing pH, there was little similarity. Although it was stated[139] that thiamine hydrochloride was reduced electrochemically to an unspecified dihydrothiamine, this presumably involving two electrons, no coulometric measurements were taken. Reduction involving eight or nine electrons as proposed for thiamine monophosphate would necessarily result in formation of a multihydrogenated species or in considerable reductive degradation of the thiamine molecule. Further, it is extremely unlikely that the AC peaks would involve eight or nine electrons since the peak alternating currents would have been very many times greater than those reported, and in any case the involvement of so many electrons in an AC polarographic process is highly unlikely.

Okamoto[141] has also examined the AC polarography of *S*-benzoylthiamine monophosphate. This is an interesting compound because it has the same physiological properties as vitamin B_1 itself but is able to maintain a much higher blood level. The DC polarography of this compound has also been reported,[142,143] but the ill-defined form and "irreversibility" of the waves precluded any detailed treatment.[141] In view of the apparent irreversibility of the DC waves, it is somewhat surprising that remarkably well-formed AC polarographic peaks were observed and that heterogeneous rate constants calculated for the first (more positive) peak were close to those expected for a reversible DC process (i.e., ca. 5×10^{-4} cm sec^{-1}). Nevertheless, *S*-benzoylthiamine monophosphate did give rise to two well-formed AC polarographic peaks (Table 4-23), which were predominantly under diffusion control. By calculating an approximate value of the diffusion coefficient for the compound, the number of electrons involved in the electrode reaction of the first DC and AC processes was then calculated to be somewhere between 3 and 4, depending

TABLE 4-23   Summit Potentials for the AC Polarography of S-Benzoyl-thiamine Monophosphate[a]

	E_s (V versus SCE)	
pH	Peak I	Peak II
1.00	−0.994	−1.110
1.95	−1.040	−1.176
3.10	−1.125	−1.335
3.50	−1.143	−1.364
4.00	−1.164	−1.377
4.50	−1.186	−1.404
5.00	−1.212	−1.441
5.50	−1.237	ca. −1.15
6.15	−1.262	
6.50	−1.288	ca. −1.56
7.00	−1.325	ca. −1.60
7.50	−1.350	ca. −1.63
8.00	−1.468	
8.55	−1.615	
9.20	−1.627	
10.00	−1.632	
11.00	−1.644	
12.05	−1.653	

[a] Data from Okamoto.[141]

on the pH. On the basis of this evidence a mechanism was developed. The reducibility of the pyrimidine ring moiety and all other moieties except the acyl bond of S-benzoylthiamine monophosphate were disregarded in the first electrochemical reaction. The mechanism proposed for the first DC or AC processes was a primary $2e-2H^+$ reduction in acidic solution of the acyl group of S-benzoylthiamine monophosphate (I, Fig. 4-26) to give the corresponding secondary alcohol (II, Fig. 4-26). In order to account for the calculated n value of between 3 and 4, two secondary processes were proposed. The first, which was predominant between pH 4.5 and 5.5, was a $2e-2H^+$ reaction to give the —SH form of thiamine monophosphate (III, Fig. 4-26) and the primary alcohol (IV, Fig. 4-26). The second, predominant below pH 2, was a $1e-1H^+$ reaction to again give the thiol form of thiamine monophosphate (III, Fig. 4-26) and the pinacol (V, Fig. 4-26). A mixture of the two secondary processes was thought to occur at around pH 3. The second DC wave or AC peak of S-benzoylthiamine monophosphate seemed to be due to the reduction of the thiol form of thiamine monophosphate (III, Fig. 4-26) formed in the first wave or peak process. The

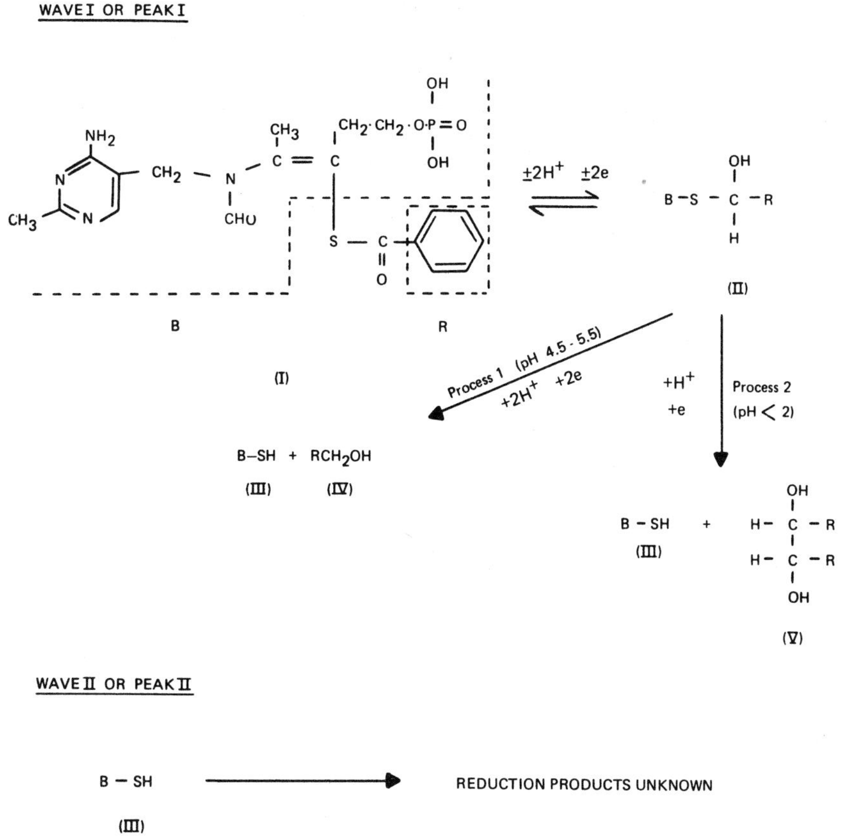

WAVE I OR PEAK I

FIG. 4-26. Possible mechanism for the polarographic reduction of *S*-benzoylthiamine monophosphate.[141]

actual mechanism or products of this process are not known. A large number of papers have appeared that are concerned with the analytical determination of thiamine and its derivatives by polarographic methods.[117,120,129,137,144–153]

B. Electrochemical Oxidation

Reports on the electrochemical oxidation of pyrimidines are relatively rare. Chiang and Chang[154] found that a number of dihydroxypyrimidines dissolved in acetic acid containing 1 M NH_4NO_3 as supporting electrolyte could be oxidized at a rotating platinum electrode. The half-wave potentials for these oxidations are presented in Table 4-24. Under the same conditions uracil and 2-methyl-6-hydroxy-5-benzyloxypyrimidine were not oxidized. The authors[154] concluded

TABLE 4-24   Half-Wave Potentials for Oxidation of Some Pyrimidines at a Rotating Platinum Electrode in Acetic Acid Containing 1 M NH_4NO_3 as Supporting Electrolyte[a]

Compound	$E_{1/2}$ (V)[b]
2-Methyl-4,5,6-trihydroxypyrimidine	0.4
2-Methyl-5,6-dihydroxypyrimidine	0.79
2-Amino-5,6-dihydroxypyrimidine	0.43
2,5,6-Trihydroxypyrimidine	0.47
5,6-Dihydroxypyrimidine	0.8

[a] Data from Chiang and Chang.[154]
[b] Reference electrode not known; presumably SCE.

that pyrimidines without an —OH group at the 5 position could not be oxidized, whereas pyrimidines with an enediol

$$\begin{array}{cc} \text{OH} & \text{OH} \\ | & | \\ -\text{C} & = \text{C}- \end{array}$$

structure could be voltammetrically oxidized to the diketone

$$\begin{array}{cc} \text{O} & \text{O} \\ || & || \\ -\text{C} & -\text{C}- \end{array}$$

Unfortunately, detailed electrode mechanisms are not available. Glicksman[155] has reported the electrochemical oxidation of a number of polyamino- and/or polyhydroxypyrimidines at a graphite electrode in 1.44 M NaOH solution. The unusual mode of studying these compounds, however, makes it difficult to compare the data with those of later workers.

The enzymatic oxidation of phenylalanine to tyrosine in mammalian liver proceeds by a series of reactions that involves the oxidation of a tetrahydropteridine cofactor of structure I (Eq. 30) to an unstable dihydropteridine intermediate (II_a, II_b, Eq. 30). The latter is reduced to I (Eq. 30) in the complete enzymatic system, but in the absence of reductant it quickly tautomerizes to a 7,8-dihydropteridine (III, Eq. 30).[156] The same unstable dihydropteridine (II_a, II_b, Eq. 30) can apparently also be obtained as the primary product of chemical oxidation of I (Eq. 30).[145] The structure of the intermediate dihydropteridine has been proposed to be either a para- or orthoquinoidal structure (II_a and II_b, respectively, (Eq. 30).[156]

Since the basic structural features of I to II_a or II_b (see Eq. 30) are

(30)

reminiscent of 2,5-diaminopyrimidine (**XXIV**) and 5,6-diaminopyrimidine (**XXV**), respectively, Cohen *et al.*[157] investigated the electrochemical oxidation of the latter two compounds and related systems in an attempt to characterize

the electron-transfer processes and mechanisms. Both compounds gave a single voltammetric oxidation wave at a rotating platinum electrode. The current was linearly proportional to concentration up to $2 \times 10^{-4} M$ for **XXIV** and $2 \times 10^{-5} M$ for **XXV**. At higher concentrations the plot of current versus concentration had the appearance of a Langmuir isotherm. The height of the waves for both compounds was invariant with pH. The equations of the half-wave potentials are shown in Table 4-25. The two linear segments of the $E_{1/2}$–pH plots for 5,6-diaminopyrimidine (Table 4-25) intercepted at about pH 6. The pK_a of this compound was found to be 6.05. Thus, below pH 6 it appeared as if the protonated form of the 5,6-diaminopyrimidine was oxidized. In view of the observed current versus concentration behaviour, it was proposed that at low concentration the oxidations of 5,6- and 2,5-diaminopyrimidine were diffusion controlled but that at higher concentration the product of the oxidations was only slowly desorbed, hence blocking the surface somewhat, so that the Langmuir type of isotherm plot of current versus concentration was obtained. By comparison of the currents in the diffusion-controlled region with

TABLE 4-25   Half-Wave Potentials for the Voltammetric Oxidation of Diaminopyrimidines at a Rotating Platinum Electrode[a]

Compound	pH Range	$E_{1/2}$ (V versus SCE)
5,6-Diaminopyrimidine	3–6	$1.28 - 0.08$ pH
	6–10	$0.96 - 0.03$ pH
2,5-Diaminopyrimidine	1–6[b]	$1.09 - 0.077$ pH

[a] Data from Cohen et al.[157]
[b] A wave was observed between pH 6 and 10, but it did not show a linear $E_{1/2}$ versus pH relationship.

those of similar compounds, it was concluded that the electrooxidation involved two electrons. The overall electrochemical mechanisms were therefore proposed to be a 2e oxidation of the appropriately protonated 5,6-diaminopyrimidine (I_a, I_b, Fig. 4-27) to the bisimine (II, Fig. 4-27). 2,5-Diaminopyrimidine (III,

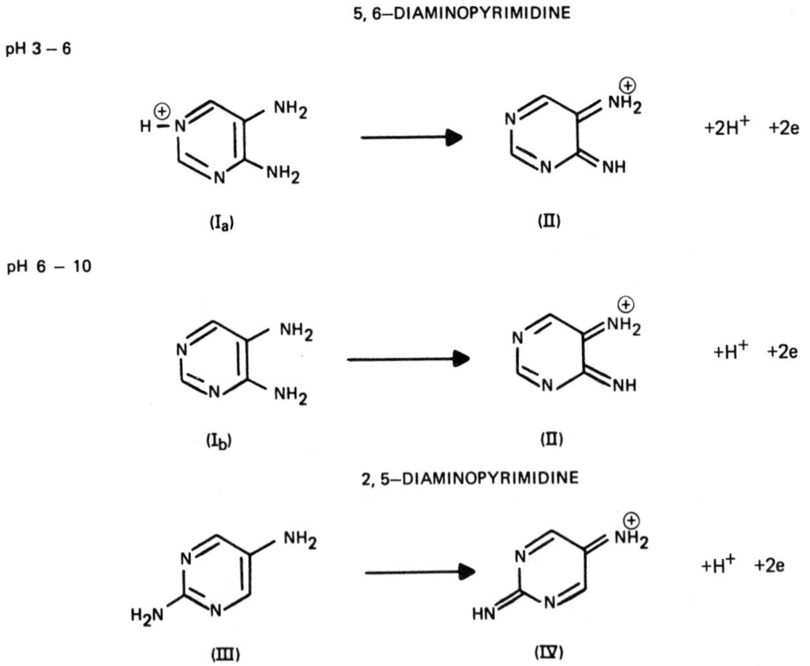

FIG. 4-27.   Proposed mechanism of electrochemical oxidation of diaminopyrimidines at the platinum electrode.[157]

Fig. 4-27) was presumed to behave in a similar way to give the paraquinoidal structure (IV, Fig. 4-27). It appears likely that the ortho- and paraquinoidal structures (II, IV, Fig. 4-27) would further react to give other products.

Cohen and co-workers[157] also measured the half-wave potentials for the oxidation of a number of substituted 2,5- and 5,6-diaminopyrimidines as a function of pH. These data are presented in Table 4-26. The half-wave potentials in this table refer only to the first oxidation wave, although with the exception of 5,6-diamino-4-methylpyrimidine and 2-hydroxy-4,5,6-triaminopyrimidine, all of the compounds gave a further ill-defined wave at more positive potential. The mechanisms responsible for the anodic waves were not studied. A reasonably

TABLE 4-26 Half-Wave Potentials for the Electrochemical Oxidation of Some Polyamino- and Aminohydroxypyrimidines at the Rotating Platinum Electrode[a]

Pyrimidine[b,c]	$E_{1/2}$ (V versus SCE) at pH								
	2.94	3.50	4.12	4.50	4.90	5.89	6.96	7.92	8.94
5,6-Diamino-4-methyl-	1.03		0.91		0.84	0.74	0.66	0.67	0.63
2,5,6-Triamino-	0.69		0.57		0.51	0.42	0.36	0.31	0.28
5,6-Diamino-4-hydroxy-	0.52		0.46		0.45	0.38	0.33	0.29	0.26
4,5,6-Triamino-	0.63		0.54		0.47	0.38	0.37	0.30	0.24
2,6-Dihydroxy-5-methylamino-	0.62	0.58	0.51	0.51	0.45	0.41	0.39	0.35	0.27
5,6-Diamino-2,4-dihydroxy-	0.34		0.22		0.19	0.14	0.09	0.03	0.01

Pyrimidine[b,c]	$E_{1/2}$ (V versus SCE) at pH								
	3.00	3.50	4.10	4.50	5.08	6.31	7.31	8.53	9.68
5-Amino-2,6-dihydroxy-	0.69		0.60		0.57	0.48	0.44	0.40	0.31
5-Amino-2,6-dihydroxy-4-methyl-	0.61		0.55		0.52	0.44	0.43	0.40	0.30
2,6-Diamino-5-methylamino-4-hydroxy-	0.30	0.27	0.22	0.22	0.14	0.10	0.06	0.02	−0.01
2,5,6-Triamino-4-hydroxy-	0.30	0.28	0.22	0.20	0.15	0.09	0.05	0.02	—

Pyrimidine[b,c]	$E_{1/2}$ (V versus SCE) at pH								
	2.81	3.50	4.08	4.50	4.92	5.91	6.84	7.99	8.92
2,5,6-Triamino-4-methoxy-	0.42	0.38	0.30	0.27	0.24	0.19	0.18	0.15	0.12
4,5,6-Triamino-2-hydroxy-	0.43		0.33		0.27	0.19	0.14	0.07	0.0

[a] Data taken from Cohen et al.[157]

[b] Concentration of pyrimidines 6×10^{-5} M.

[c] Numbering of compounds changed from original reference to accord with system employed in this chapter.

good straight-line relationship between the half-wave potentials of these compounds and the Hückel energies of the highest occupied molecular orbitals was obtained.

1. Barbituric Acid and Its Derivatives

Barbituric acid gives an anodic wave at the DME between about pH 3.5 and 13.[158,159] The variation of $E_{1/2}$ with pH is shown in Fig. 4-28. The height of the polarographic wave is proportional to concentration at low concentrations but is constant at higher concentrations. The anodic wave is apparently due to reaction of barbituric acid with the mercury electrode to give a compound that is adsorbed at the electrode surface. Zuman and co-workers[160,161] found that the sleep-inducing agent barbital or veronal (5,5-diethylbarbituric acid) (**XXVI**)

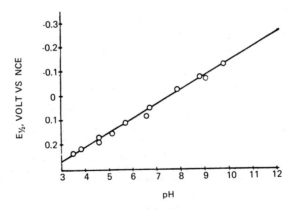

(**XXVI**)

produced an anodic wave in borate buffer, pH 9.3, which was formed as a result of reaction of two atoms of mercury with three molecules of barbital to give an insoluble complex or compound, which was adsorbed on the surface of the DME. Similarly, the 5-alkyl-substituted thiobarbiturates also gave an anodic

FIG. 4-28.   Variation of $E_{1/2}$ with pH for the anodic wave of barbituric acid.[158,159] (Reprinted with the permission of Academia Publishing House, Prague.)

wave at the DME which was complicated by adsorption effects. In alkaline media the reaction that took place is represented as

$$\text{RSH} + \text{Hg} \rightleftharpoons \text{RSHg} + \text{H}^+ + e \tag{31}$$

Fischer and Dracka[162] studied kinetic phenomena in the formation of adsorbable mercury barbiturate salts on mercury. Armstrong and co-workers[163] studied the anodic formation of a number of mercury 5,5-substituted barbiturates at a stationary mercury drop electrode using potentiostatic methods. It was found in all cases that one monomolecular layer of salt was formed, having a structure Hg(II) barbiturate(II). In the region of the electrocapillary maximum (ecm) neutral barbituric acid molecules were adsorbed. At potentials anodic of the ecm these molecules were desorbed and the divalent anions of the acids were adsorbed.

Johnson and Stride[164] electrochemically oxidized barbituric acid in 0.2% sodium hydroxide solution at a platinum electrode at a constant current density of 1 A/cm^2. The main electrochemical reaction appeared to be oxidation of the methylene group in barbituric acid (I, Fig. 4-29) since the products claimed to be formed were hydurilic acid, 4% (II, Fig. 4-29); alloxantin, 25% (III, Fig. 4-29); murexide, 3% (IV, Fig. 4-29); and alloxan, 50% (V, Fig. 4-29). That these products were actually formed should be viewed with considerable skepticism because alloxan, for example, does not exist in solutions of pH > ca. 5, being very rapidly hydrolyzed in alkaline solution to alloxanic acid.[64,65] Alloxantin is also probably unstable at high pH and is very readily oxidized electrochemically.[63,68]

Kato and Dryhurst[165] reported that barbituric acid, 1-methylbarbituric acid, and 1,3-dimethylbarbituric acid are electrochemically oxidized at the PGE at ca. pH 1 in the presence of chloride ion by way of a single voltammetric peak. The voltammetric peak potentials for all the latter barbituric acid derivatives were essentially identical ($E_p = 1.02 \pm 0.02$ V at pH 1.3). Coulometry revealed that barbituric acid and its N-methyl derivatives are oxidized (at pH ca. 1.0 in the presence of chloride ion) in a process involving about 3.2–3.4 electrons. Three major products of the electrode reaction were isolated and identified, namely, the appropriately methylated 5,5′-dichlorohydurilic acids, 5,5-dichlorobarbituric acids, and alloxans. By analysis of the voltammetric peak it was concluded that the initial step in the electrooxidation of barbituric acids (I, Fig. 4-30) is probably a 1e–1H$^+$ process to give a barbiturate radical (II, Fig. 4-30). The radical species was proposed[165] to undergo two possible reactions: dimerization to hydurilic acid (III, Fig. 4-30) or further 1e oxidation to a carbonium ion (IV, Fig. 4-30) followed by attack of Cl$^-$ to give 5-chlorobarbituric acid (V, Fig. 4-30) or perhaps attack by water to give 5-hydroxybarbituric acid (XI, Fig. 4-30), also known as dialuric acid. Dimerization of the radical (II, Fig. 4-30)

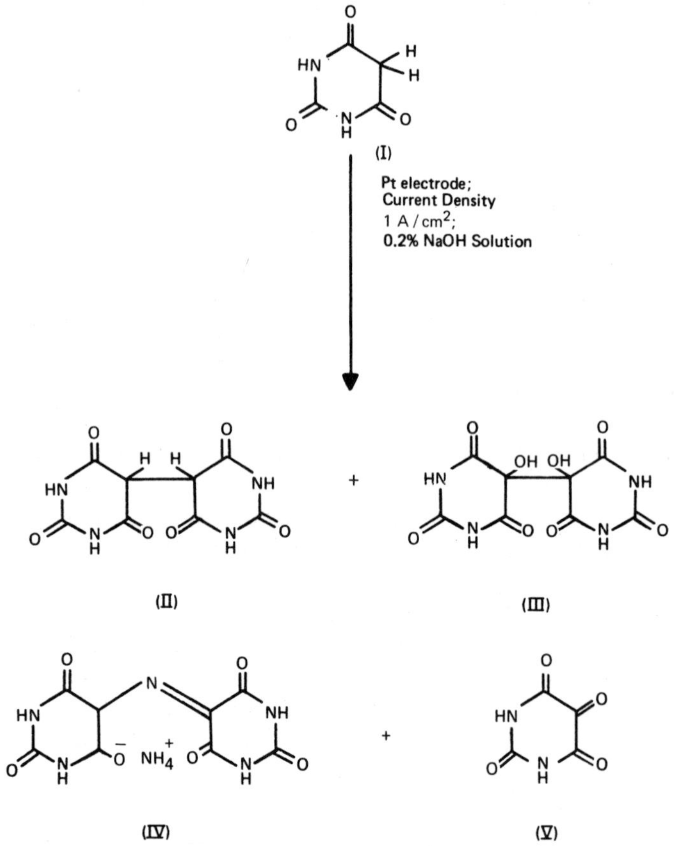

(I)

Pt electrode;
Current Density
1 A / cm^2;
0.2% NaOH Solution

(II)   +   (III)

(IV)   +   (V)

FIG. 4-29.   Products of oxidation of barbituric acid in 0.2% NaOH solution at a platinum electrode according to Johnson and Stride.[164]

to hydurilic acid (III, Fig. 4-30) was proposed to be the less likely (minor) route because although electrochemical oxidation of the latter, at the same potential and under identical conditions employed for oxidation of barbituric acid, gave the same products as were observed from the latter species, the yields were considerably different. On the other hand, electrooxidation of 5-chloro-barbituric acid, which would be formed by attack of Cl$^-$ on the carbonium ion (IV, Fig. 4-30), gave the same products as and in similar yields to those obtained from barbituric acid. Thus, on the basis of the product yields obtained on electrooxidation of barbituric acids, hydurilic acid, and 5-chlorohydurilic acid, it was concluded that the initial radical formed on electrooxidation of the barbituric acids is preferentially further oxidized to a carbonium ion rather than dimerizing. The 5-chlorobarbituric acid (V, Fig. 4-30) was then proposed to be

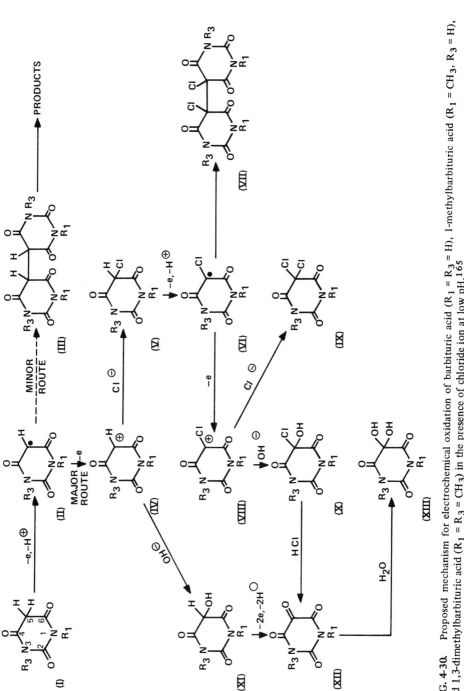

FIG. 4-30. Proposed mechanism for electrochemical oxidation of barbituric acid ($R_1 = R_3 = H$), 1-methylbarbituric acid ($R_1 = CH_3$, $R_3 = H$), and 1,3-dimethylbarbituric acid ($R_1 = R_3 = CH_3$) in the presence of chloride ion at low pH.165

further electrochemically oxidized in a $1e-1H^+$ process first to the radical VI (Fig. 4-30). Again, this radical could dimerize to 5,5'-dichlorohydurilic acid (VII, Fig. 4-30) or be further electrooxidized to carbonium ion VIII (Fig. 4-30). Attack of the latter species by nucleophiles in solution (Cl^- and H_2O) would give 5,5-dichlorobarbituric acid (IX, Fig. 4-30) or 5-hydroxy-5-chloro-barbituric acid (X, Fig. 4-30). The latter species would then lose HCl to give alloxan (XII, Fig. 4-30) and hence alloxan hydrate (XIII, Fig. 4-30). As mentioned previously, the barbituric acid carbonium ion (IV, Fig. 4-30) could be attacked by water to give dialuric acid (XI, Fig. 4-30). This species would be very rapidly electrooxidized to alloxan (XII, XIII, Fig. 4-30) in a $2e-2H^+$ process.[68]

A subsequent study of the electrochemical oxidation of hydurilic acid, 1,1'-dimethylhydurilic acid, and 1,1',3,3'-tetramethylhydurilic acid at the PGE[166] revealed that these compounds give rise to several voltammetric oxidation peaks (Table 4-27, Fig. 4-31). The electrochemistry involved in the peak I_a process was studied in greatest detail, using voltammetry, coulometry, product identification and analysis, and double-potential step chronoampero-metry.[167] These investigations supported the view that the peak I_a process is a $2e$ reaction. At pH values below the pK_a of the hydurilic acids (4.1, 4.4, and 3.4 for hydurilic acid, 1,1'-dimethyl-, and 1,1',3,3'-tetramethylhydurilic acids, respectively[166]) it was concluded that the peak I_a process was a $2e-1H^+$

TABLE 4-27   Voltammetric Peak Potentials for Electrochemical Oxidation of Hydurilic Acids at the PGE[a]

Compound	Peak	pH Range	E_p (V versus SCE)
Hydurilic acid[b]	I_a	1–5	$0.552 - 0.032$ pH
		5–12	$0.595 - 0.032$ pH
	II_a	1–5	$1.140 - 0.060$ pH
		5–11	$1.380 - 0.059$ pH
	III_a	3.5–9	$1.38 - 0.033$ pH
	IV_a	3.4–4.5	1.41 ± 0.02
1,1'-Dimethyl-hydurilic acid	I_a	1–6	$0.525 - 0.031$ pH
		6.5–11	$0.578 - 0.032$ pH
	II_a	1–3.5	$1.115 - 0.077$ pH
	III_a	4.5–10.5	$1.445 - 0.053$ pH
1,1',3,3'-Tetra-methylhydurilic acid	I_a	1–4.5	$0.518 - 0.038$ pH
		4.5–13	0.35 ± 0.01
	II_a	1–4	$1.224 - 0.112$ pH
		4–10	0.77 ± 0.025
	III_a	3.5–11.5	$1.343 - 0.029$ pH

[a] Data from Kato *et al.*[166]
[b] Structures for these compounds are shown in Fig. 4-31.

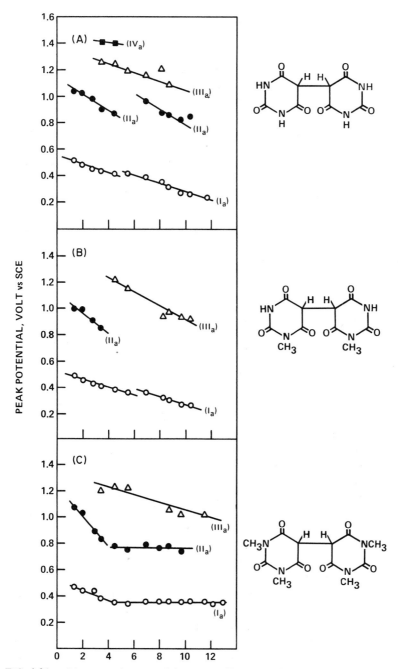

FIG. 4-31. Linear peak potential versus pH relationships for the voltammetric oxidation peaks of (A) 0.669 mM hydurilic acid, (B) 0.3412 mM 1,1′-dimethylhydurilic acid, and (C) 0.4671 mM 1,1′,3,3′-tetramethylhydurilic acid at the pyrolytic graphite electrode at a sweep rate of 0.005 V sec^{-1}.[166]

oxidation to give a carbonium ion (I → IV, Fig. 4-32). Rapid loss of a proton from the latter species then gives dehydrohydurilic acid (V, Fig. 4-32). At pH values greater than the pK_a, peak I_a of tetramethylhydurilic acid became independent of pH (Table 4-27, Fig. 4-31). Hence, the monoanion of this compound (II, Fig. 4-32) is electrooxidized in a $2e$ process to the carbonium ion (IV, Fig. 4-32), which then deprotonates to the corresponding dehydrohydurilic acid (V, Fig. 4-32). However, hydurilic acid and dimethylhydurilic acid are still oxidized in a $2e-1H^+$ process (i.e., $dE_p/d(pH) \approx 30$ mV) at $pH > pK_a$. Accordingly, the anion of the latter two compounds (III, Fig. 4-32) is oxidized directly to the corresponding dehydrohydurilic acid (V, Fig. 4-32). Cyclic voltammetry of hydurilic acids at pH values greater than pK_a indicated that reaction III ⇌ V (Fig. 4-32) is a reversible couple. The apparent heterogeneous rate constant for the latter process was measured for hydurilic acid by the cyclic voltammetric method of Nicholson[168] and had values ranging from 2.8×10^{-3} cm sec^{-1} at pH 7.0 to 1.0×10^{-3} cm sec^{-1} at pH 9.77.[166]

Cyclic voltammetry and double-potential step chronoamperometry however, indicated that dehydrohydurilic acid (V, Fig. 4-32) rapidly hydrates to 5-hydroxyhydurilic acid (VI, Fig. 4-32). The rate constant for the first-order hydration of V to VI (Fig. 4-32) was measured and found to be ca. 8.8, 1.4, and 0.2 liters/mole sec at pH 7.0, 8.15, and 8.70, respectively. Clearly, therefore, the hydration reaction is acid catalyzed. In view of the fact that 5-methoxy-1,1',3,3'-tetramethylhydurilic acid (X, Fig. 4-32) could be isolated from the peak I_a reaction product mixture obtained from tetramethylhydurilic acid after treatment with methanol, it is apparent that dehydrohydurilic acids can be attacked by methanol as well as water. Following formation of 5-hydroxyhydurilic acid (VI, Fig. 4-32), further slow decomposition occurs to give barbituric acids (VII, Fig. 4-32) or alloxans (VIII, Fig. 4-32).

Electrochemical oxidation of hydurilic acids at peak II_a potentials and pH 1 in the presence of chloride ion (only peaks I_a and II_a are observed below pH 3) gave the same products that are observed on electrooxidation of the corresponding barbituric acids[165,166] (i.e., the appropriately methylated 5,5'-dichlorohydurilic acids, 5,5-dichlorobarbituric acids, and alloxans; see Fig. 4-30 for structures). In the absence of chloride ion the sole product observed on electrooxidation of hydurilic acid was alloxan. The detailed mechanism for the peak II_a process could not be evaluated because 5-hydroxyhydurilic acid (VI, Fig. 4-32) and barbituric acid (VII, Fig. 4-32) formed in the peak I_a process gave peaks at peak II_a potentials. The additional voltammetric peaks observed for the hydurilic acids (e.g., peaks III_a, IV_a, and V_a, Table 4-27, Fig. 4-31), which are generally very close to background discharge, were also thought to be due to electrochemical oxidations of 5-hydroxyhydurilic acid. Because of the relative instability of the latter compound in aqueous solution the reactions associated with these peaks were not studied.[166]

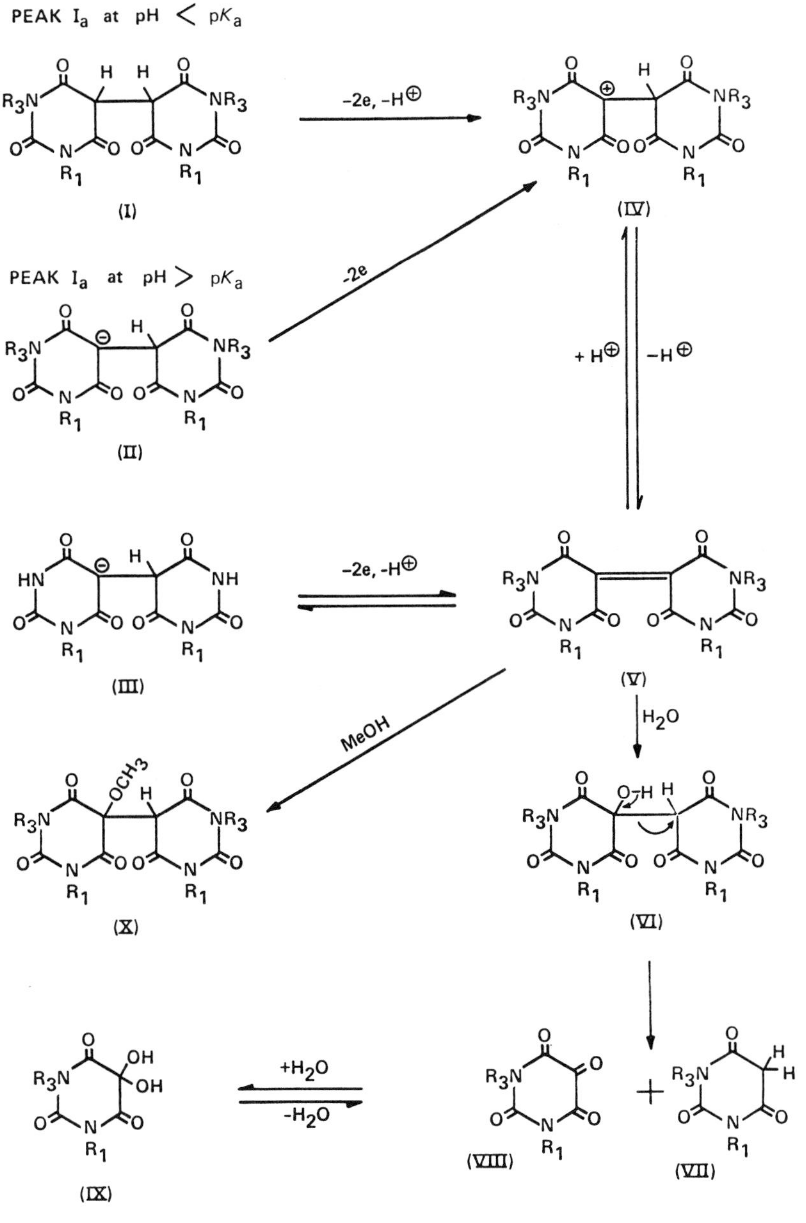

PEAK I_a at pH $<$ pK_a

FIG. 4-32. Proposed mechanism for the peak I_a electrochemical oxidation of hydurilic acids.[166] For structure (II) R_1 and R_3 must be CH_3. For all other structures R_1 and R_3 may be CH_3 or H.

During the course of some investigations of the electrochemistry of barbituric acid derivatives, Dryhurst *et al.*[169] discovered a very simple electrochemical method for the synthesis of a new cyclic barbiturate, 5,6-dihydro-1,3-dimethyl-5,6-di(1′,3′-dimethyl-2′,4′,6′-trioxopyrimid-5′,5′-yl)furo[2,3-*d*]uracil   (**XXVII**).

(XXVII)

This compound was prepared by oxidizing 1,3-dimethylbarbituric acid in 1 M acetic acid at 1.00 V versus SCE (E_p for barbituric acid in 1 M HOAc[169]) at a large pyrolytic graphite electrode. Electrolysis of 700 mg of 1,3-dimethyl-barbituric acid in 150 ml of 1 M HOAc involved transfer of close to 2.5 faradays/mole. The cyclic barbiturate (**XXVII**) precipitates as the electrolysis proceeds in a yield of 46.6%. The other major identified product was 1,3-dimethyl alloxan. The structure of the cyclic barbiturate was determined by NMR and IR spectroscopy and by X-ray crystallography. The mechanism of formation of this and other oligomeric derivatives formed on electrooxidation of various barbituric acids[170] has not yet been determined. A fairly large number of reports have appeared outlining the use of direct and indirect polarography[171–177] and oscillopolarography[178–180] for analysis of barbiturates.

2. Uracil and Its Derivatives

Manoušek and Zuman[181] found that in slightly alkaline solution uracil gave one anodic wave, which is presumably due to compound formation with the DME. Other substituted uracils gave similar behavior. Some typical polarographic data are presented in Table 4-28.

Meinert and Cech[182] have reported that electrooxidation of uracil at platinum electrodes in various nonaqueous solvents containing halide ion (Cl⁻, Br⁻, I⁻, CN⁻) leads to halogenation of uracil at the 5 position. The process, however, appears to proceed by attack of electrogenerated halogen (Cl_2, Br_2, etc.) with uracil.

TABLE 4-28 Polarographic Half-Wave Potentials for the Anodic Waves of Uracil and Some Derivatives at the DME at pH 9.3[a]

	$E_{1/2}$ (V versus SCE) at concentration	
Compound	8×10^{-5} M	5×10^{-4} M
Uracil	0.019	−0.010
5-Methyluracil	0.025	−0.002
6-Methyluracil	0.040	0.015
6-Carboxyuracil	−0.020	−0.047
3-Methyl-6-carboxyuracil	−0.018	−0.070

[a] Data from Manoušek and Zuman.[181]

IV. SUMMARY

A. Electrochemical Reductions

Consideration of the available electrochemical data on the pyrimidines makes it possible to make some generalized statements regarding the ease of electrochemical reduction of these compounds.

1. The ease of electrochemical reduction of a pyrimidine decreases with the number of amino and hydroxy substituents. The principal reason for this appears to be that the hydroxy group in particular seems to favor saturation of the pyrimidine ring by means of tautomeric shifts, thereby removing possible sites of reduction, i.e., −N=C− double bonds, within the ring. It should be particularly noted that experimental evidence favors the fact that hydroxypyrimidines exist principally in the keto form.[15,183−185] Pyrimidines polysubstituted with tautomeric groups are apt to be electrochemically nonreducible or not easily reducible within the available potential range. The very noticeable exception to this is alloxan. However, the evidence available indicates that the electrochemical reduction of alloxan differs from other, less oxidized pyrimidines in that with alloxan the reduction involves carbonyl groups, whereas in the case of most other pyrimidines ring carbon−nitrogen double bonds are reduced.

2. Pyrimidines having 4-amino or 4-hydroxy substituents are less easily reduced than the corresponding 2-substituted compounds. The reason for this is probably that the hydroxy group effectively removes the N-3=C-4 double bond, which is the most susceptible site for reduction. In the case of 4-amino substituents it is presumably the electron-releasing effect of the amino group[186] that decreases the electron-acceptor properties of the N-3=C4 double bond.

3. When the N-3=C-4 double bond is not available for reduction, by virtue of tautomeric shifts of the C-4 position substituents or because the bond has been reduced, further electrochemical reduction occurs at the N-1=C-2 double bond, although this is energetically more difficult than reduction of the former bond.

B. Electrochemical Oxidation

Since relatively few detailed studies of the electrochemical oxidation of pyrimidines have been carried out, it is difficult to generalize about the processes. The relative complexity of electrooxidation reactions of pyrimidines, however, suggests that further studies are warranted in this area, which could be a fruitful source of mechanistic, synthetic, and analytical information. It will also be of considerable interest to discover the correlations that may exist between electrochemical and biological reactions of these compounds.

REFERENCES

1. S. F. Mason, *in* "The Pyrimidines" (D. J. Brown, ed.), Chapter XIII. Wiley (Interscience), New York, 1962.
2. J. R. Marshall and J. Walker, *J. Chem. Soc.* p. 1004 (1951).
3. D. J. Brown, E. Hoerger, and S. F. Mason, *J. Chem. Soc.* p. 4035 (1955).
4. D. J. Brown, ed., "The Pyrimidines." Wiley (Interscience), New York, 1962.
5. G. W. Kenner and A. Todd, *Heterocycl. Compd.* **6**, 234 (1957).
6. V. Meyer and P. Jacobson, "Lehrbuch der organischen Chemie," Vol. II, Part 3, p. 1172. Walter De Gruefter and Co. Berlin, 1923.
7. T. B. Johnson and D. A. Hahn, *Chem. Rev.* **13**, 193 (1933).
8. H. Gilman, "Organic Chemistry," 1st ed., Vol. II, p. 948. Wiley, New York, 1938.
9. E. E. Conn and P. K. Stumpf. "Outlines of Biochemistry," 2nd ed. Wiley, New York, 1967.
10. R. R. Williams and T. D. Spies, "Vitamin B_1 and Its Use in Medicine." Macmillan, New York, 1938.
11. H. R. Rosenberg, "Chemistry and Physiology of the Vitamins." Wiley (Interscience), New York, 1942.
12. F. A. Robinson, "The Vitamin B Complex." Wiley, New York, 1951.
13. K. Lohmann and P. Schuster, *Naturwissenschaften* **25**, 26 (1937).
14. K. Lohmann and P. Schuster, *Angew. Chem.* **50**, 221 (1937).
15. A. Baeyer, *Ann. Chem. Pharm.* **127**, 199 (1863).
16. A. Baeyer, *Ann. Chem. Pharm.* **130**, 129 (1864).
17. W. J. Doran, *Med. Chem.* **4**, 102 (1959).
18. J. A. Russell, *Annu. Rev. Biochem.* **14**, 322 (1945).
19. C. F. Cori and G. T. Cori, *Annu. Rev. Biochem.* **15**, 203 (1946).
20. D. Stetten, *Annu. Rev. Biochem.* **16**, 136 (1947).
21. J. J. Fox, K. A. Watanabe, and A. Bloch, *Prog. Nucleic Acid Res. Mol. Biol.* **5**, 272 (1966).
22. S. Farber, R. Toch, E. M. Sears, and D. Pinkel, *Adv. Cancer Res.* **4**, 2–54 (1956).

23. R. M. S. Smellie, *in* "The Nucleic Acids" (E. Chargaff and J. N. Davidson, eds.), Vol. 2, p. 393. Academic Press, New York, 1955.
24. J. O. Heath, *Nature (London)* **158**, 23 (1946).
25. L. F. Cavalieri and B. A. Lowy, *Arch. Biochem. Biophys.* **35**, 83 (1952).
26. D. Hamer, D. M. Waldron, and D. L. Woodhouse, *Arch. Biochem. Biophys.* **47**, 272 (1953).
27. K. Sugino and K. Shirai, *Nippon Kagaku Zasshi* **70**, 111 (1949).
28. K. Sugino, K. Shirai, and K. Odo, *J. Electrochem. Soc.* **104**, 667 (1957).
29. V. H. Smith and B. D. Christensen, *J. Org. Chem.* **20**, 829 (1955).
30. D. L. Smith and P. J. Elving, *J. Am. Chem. Soc.* **84**, 2741 (1962).
31. A. Albert, R. Goldacre, and J. Phillips, *J. Chem. Soc.* p. 2240 (1948).
32. P. J. Elving, S. J. Pace, and J. E. O'Reilly, *J. Am. Chem. Soc.* **95**, 647 (1973).
33. D. Thévenot, G. Hammouya, and R. Buvet, *J. Chim. Phys. Phys.-Chim. Biol.* **66**, 1903 (1969).
34. D. Thévenot, G. Hammouya, and R. Buvet, *C. R. Hebd. Seances Acad. Sci., Ser. C* **268**, 1488 (1969); D. Thévenot, *J. Electroanal. Chem.* **46**, 89 (1973).
35. G. Dryhurst and P. J. Elving, *Talanta* **16**, 855 (1969).
36. B. Janik and E. Paleček, *Arch. Biochem. Biophys.* **105**, 225 (1964).
37. B. Czochralska and D. Shugar, *Biochim. Biophys. Acta* **281**, 1 (1972).
38. J. W. Webb, B. Janik, and P. J. Elving, *J. Am. Chem. Soc.* **95**, 991 (1972).
39. H. T. Miles, *Proc. Natl. Acad. Sci. U.S.A.* **47**, 791 (1961); B. I. Suchorukov, *Biofizika* **7**, 664 (1962).
40. J. C. M. Henning, *J. Chem. Phys.* **44**, 2139 (1966).
41. E. W. Stone and A. H. Maki, *J. Chem. Phys.* **39**, 1635 (1963).
42. A. Carrington and J. dos Santos-Veiga, *Mol. Phys.* **5**, 21 (1962).
43. A. Carrington, *Q. Rev., Chem. Soc.* **17**, 67 (1963).
44. D. van der Meer and D. Feil, *Recl. Trav. Chim. Pays-Bas* **87**, 746 (1968).
45. D. van der Meer, *Recl. Trav. Chim. Pays-Bas* **88**, 1361 (1969); **89**, 51 (1970).
46. E. van't Land and D. van der Meer, *Recl. Trav. Chim. Pays-Bas* **92**, 409 (1973).
47. J. E. O'Reilly and P. J. Elving, *J. Am. Chem. Soc.* **93**, 1871 (1971).
48. M. L. Olmstead, R. G. Hamilton, and R. S. Nicholson, *Anal. Chem.* **41**, 260 (1969).
49. R. S. Nicholson, J. M. Wilson, and M. L. Olmstead, *Anal. Chem.* **38**, 542 (1966).
50. V. Vetterl, *Experientia* **21**, 9 (1965).
51. V. Vetterl, *Collect. Czech. Chem. Commun.* **31**, 2105 (1966).
52. V. Vetterl, *Abh. Dtsch. Akad. Wiss. Berlin, Kl. Med.* p. 493 (1966).
53. V. Vetterl, *Collect. Czech. Chem. Commun.* (in press) (obtained through Palecek[54]).
54. E. Palecek, *Prog. Nucleic Acid Res. Mol. Biol.* **9**, 31 (1969).
54a. J. W. Webb, B. Janik, and P. J. Elving, *J. Am. Chem. Soc.* **95**, 991 (1972).
55. S. Ono, M. Takagi, and T. Wasa, *Bull. Chem. Soc. Jap.* **31**, 364 (1958).
56. D. Nightingale, *Org. Synth.* **23**, 6 (1943).
57. A. Baeyer, *Ann. Chem. Pharm.* **127**, 1 (1863).
58. R. S. Tipson, *Org. Synth.* **33**, 3 (1953).
59. R. S. Tipson and L. H. Cretcher, *J. Org. Chem.* **16**, 1091 (1951).
60. E. S. Hill and L. Michaelis, *Science* **78**, 485 (1933).
61. E. Biilmann and J. Bentzon, *Ber. Dtsch. Chem. Ges.* **51**, 522 (1918).
62. G. M. Richardson and R. R. Cannan, *Biochem. J.* **23**, 68 (1929).
63. W. A. Struck and P. J. Elving, *J. Am. Chem. Soc.* **86**, 1229 (1964).
64. G. M. Richardson and R. R. Cannon, *Biochem. J.* **23**, 68 (1929).
65. D. Seligson and H. Seligson, *J. Biol. Chem.* **190**, 647 (1951).
66. B. H. Hansen and G. Dryhurst, *J. Electroanal. Chem.* **32**, 405 (1971).
67. B. H. Hansen and G. Dryhurst, *J. Electroanal. Chem.* **30**, 407 (1971).

68.   B. H. Hansen and G. Dryhurst, *J. Electrochem. Soc.* **118**, 1747 (1971).
69.   J. Patterson, A. Lazarow, and S. Levey, *J. Biol. Chem.* **177**, 187 (1949).
70.   S. G. Mairanovskii, "Catalytic and Kinetic Waves in Polarography," p. 37. Plenum, New York, 1968.
71.   P. J. Elving and J. S. Leone, *J. Am. Chem. Soc.* **80**, 1021 (1958).
72.   F. Bergmann and S. Dikstein, *J. Am. Chem. Soc.* **77**, 691 (1955).
73.   H. R. Jacobs, *Proc. Soc. Exp. Biol. Med.* **37**, 407 (1937–1938).
74.   J. S. Dunn, H. L. Sheehan, and N. G. B. McLetchie, *Lancet* **1**, 484 (1943).
75.   J. S. Dunn and N. G. B. McLetchie, *Lancet* **1**, 384 (1943).
76.   J. S. Dunn, J. Kirkpatrick, N. G. B. McLetchie, and S. D. Telfer, *J. Pathol. Bacteriol.* **55**, 245 (1943).
77.   G. Brückmann and E. Wertheimer, *J. Biol. Chem.* **168**, 241 (1947).
78.   O. Koref, L. Vargas, F. H. Rodriguez, and A. Telchi, *Endocrinology* **35**, 391 (1944).
79.   C. C. Bailey, O. T. Bailey, and R. S. Leech, *Bull. N. Engl. Med. Cent.* **7**, 59 (1945).
80.   P. A. Herbut, J. S. Watson, and E. Perkins, *Arch. Pathol.* **42**, 214 (1946).
81.   G. Brückmann and E. Wertheimer, *Nature (London)* **155**, 267 (1945).
82.   M. G. Goldner and G. Gomoni, *Endocrinology* **35**, 241 (1944).
83.   L. Laszt, *Experientia* **1**, 234 (1945).
84.   C. C. Bailey, O. T. Bailey, and R. S. Leech, *Proc. Soc. Exp. Biol. Med.* **63**, 502 (1946).
85.   M. G. Goldner and G. Gomori, *Proc. Soc. Exp. Biol. Med.* **63**, 502 (1946).
86.   G. R. Barker, N. G. Luthy, and M. M. Dhar, *J. Chem. Soc.* p. 4206 (1954).
87.   N. G. Luthy and B. Lamb, *Anal. Chem.* **29**, 1454 (1957).
88.   T. J. Bardes, R. R. Herr, and T. Enkoji, *J. Am. Chem. Soc.* **77**, 960 (1955).
89.   Y. Okazaki, *Bunseki Kagaku* **11**, 1142 (1962); *Chem. Abstr.* **58**, 7786 (1963).
90.   Y. Okazaki, *Bunseki Kagaku* **11**, 986 (1962); *Chem. Abstr.* **58**, 2323 (1963).
91.   T. Uno and Y. Okazaki, *Yakugaku Zasshi* **81**, 1292 (1961); *Chem. Abstr.* **56**, 11711 (1962).
92.   Y. Okazaki, *Bunseki Kagaku* **11**, 1239 (1962); *Chem. Abstr.* **58**, 8854 (1963).
93.   M. Březina and P. Zuman, "Polarography in Medicine, Biochemistry and Pharmacy." Wiley (Interscience), New York, 1958.
94.   P. Zuman, *Collect. Czech. Chem. Commun.* **20**, 649, 876, and 883 (1953).
95.   O. Manoušek and P. Zuman, *Cesk. Farm* **4**, 193 (1956).
96.   O. Manoušek and P. Zuman, *Pharmazie* **11**, 530 (1956).
97.   W. F. Smyth, G. Svehla, and P. Zuman, *Anal. Chim. Acta* **51**, 463 (1970).
98.   W. F. Smyth, G. Svehla, and P. Zuman, *Anal. Chim. Acta* **52**, 129 (1970).
99.   W. F. Smyth, P. Zuman, and G. Svehla, *J. Electroanal. Chem.* **30**, 101 (1971).
100. P. Zuman, *Collect. Czech. Chem. Commun.* **15**, 839 (1950).
101. H. Lund, *Acta Chem. Scand.* **13**, 249 (1959).
102. Y. P. Kitayev and T. V. Troyepolskaya, *Prog. Electrochem. Org. Compd.* **1**, 63 (1969).
103. W. Kahl and W. Pasek, *Rocz. Chem.* **44**, 2425 (1970).
103a. W. Kahl and W. Pasek, *Rocz. Chem.* **46**, 865 (1972).
104. M. Wrona and B. Czochralska, *Acta Biochim. Pol.* **17**, 351 (1970).
105. K. Berens and D. Shugar, *Acta Biochim. Pol.* **10**, 25 (1963).
106. G. K. Budnikov, *Zh. Obshch. Khim.* **38**, 2431 (1968); *Chem. Abstr.* **70**, 53408r (1969).
107. P. C. Jain and R. C. Kapoor, *J. Polarogr. Soc.* **14**, 27 (1968).
108. P. T. Kissinger and C. N. Reilley, *Anal. Chem.* **42**, 12 (1970).
109. K. Wiesner, *Cas. Cesk. Lek.* **57**, 76 (1944) (see Lund[101]).
110. I. Tachi and S. Koide, *J. Agric. Chem. Soc. Jap.* **26**, 255 (1952).
111. I. Tachi and S. Koide, *J. Agric. Chem. Soc. Jap.* **25**, 145 (1951).
112. I. Tachi and S. Koide, *Proc. Int. Polarogr. Congr., 1st, 1951* Part I, Proc. 469 (1951–1952).

113. C. Rubina, *Byull. Eksp. Biol. Med.* **37**, 47 (1954).
114. I. Tachi and S. Koide, *J. Agric. Chem. Soc. Jap.* **26**, 249 (1952).
115. E. Gunther, *Pharmazie* **6**, 577 (1951).
116. Y. Asahi, *J. Vitaminol.* **4**, 118 (1958); *Chem. Abstr.* **52**, 20892 (1958).
117. J. J. Lingane and O. L. Davis, *J. Biol. Chem.* **137**, 567 (1941).
118. R. Pleticha, *Chem. Listy* **47**, 806 (1953).
119. R. Portillo and G. Varela, *An. Bromatol.* **2**, 251 (1950).
120. G. Sartori and C. Catteneo, *Gazz. Chim. Ital.* **74**, 166 (1944).
121. A. M. Shkodin and G. P. Tikhomirova, *Biokhimiya* **18**, 184 (1953); *Chem. Abstr.* **47**, 8547 (1953).
122. I. Tachi and S. Koide, *J. Agric. Chem. Soc. Jap.* **25**, 195 (1951).
123. I. Tachi and S. Koide, *J. Agric. Chem. Soc. Jap.* **26**, 243 (1952).
124. A. Wollenberger, *Science* **101**, 386 (1945).
125. A. Watanabe and Y. Asahi, *J. Pharm. Soc. Jap.* **77**, 153 (1957); *Chem. Abstr.* **51**, 10493 (1957).
126. I. Tachi and S. Koide, *J. Agric. Chem. Soc. Jap.* **25**, 330 (1951); *Chem. Abstr.* **47**, 11035 (1953).
127. I. Yamanouchi, *Bitamins* **7**, 251 (1954).
128. I. Yamanouchi, *Bitamins* **1**, 199 (1948).
129. R. Pleticha, *Cesk. Farm.* **2**, 149 (1953); *Chem. Abstr.* **48** 9835 (1954).
130. Y. Asahi, *Yakugaku Zasshi* **80**, 1222 and 1226 (1960); *Chem. Abstr.* **55**, 4884 (1961).
131. I. Tachi and S. Koide, *J. Agric. Chem. Soc. Jap.* **26**, 330 (1952).
132. G. Tikhomirova and S. L. Belen'kaya, *Ukr. Khim. Zh.* **28**, 1048 (1962); *Chem. Abstr.* **59**, 3544 (1963).
133. J. V. Palomino, A. Serna, and J. Sancho, *An. Quim.* **66**, 311 (1970); *Chem. Abstr.* **73**, 98258v (1970).
134. J. V. Palomino, J. B. V. Abarca, and J. Sancho, *An. Quim.* **66**, 317 (1970); *Chem. Abstr.* **73**, 109104v (1970).
135. J. V. Palomino, J. B. V. Abarca, and A. Serna, *An. Quim.* **67**, 125 (1971); *Chem. Abstr.* **75**, 29273a (1971).
136. G. P. Tikhomirova and S. L. Belen'kaya, *Ukr. Khim. Zh.* **29**, 97 (1963); *Chem. Abstr.* **59**, 5467 (1963).
137. Y. Asahi, *Takeda Kenkyusho Nempo* **16**, 1 (1957); *Chem. Abstr.* **52**, 10265 (1958).
138. I. Tachi, M. Senda, S. Shibabe, and T. Maruyama, *Adv. Polarogr., Proc. Int. Congr., 2nd, 1959* Vol. 3, p. 1099 (1960).
139. K. Okamoto, *Bull. Chem. Soc. Jap.* **34**, 1063 (1961).
140. K. Okamoto, *Bull. Chem. Soc. Jap.* **36**, 366 (1963).
141. K. Okamoto, *Bull. Chem. Soc. Jap.* **36**, 371 (1963).
142. K. Okamoto, *Bitamins* **22**, 364 (1961).
143. I. Tachi, *Bitamins* **21**, 423 (1960).
144. R. Pleticha, *Pharmazie* **15**, 108 and 478 (1960); *Chem. Abstr.* **54**, 16517 (1960); **57**, 2333 (1962).
145. R. Pleticha, *Anal. Chim. Acta* **18**, 146 (1958); *Chem. Abstr.* **53**, 21277 (1959).
146. Y. Asahi, *Bitamins* **13**, 496 (1957); *Chem. Abstr.* **54**, 1802 (1960).
147. J. Sancho and P. Salmeron, *An. R. Soc. Esp. Fis. Quim, Ser. B* **53**, 117 (1957); *Chem. Abstr.* **53**, 9859 (1959).
148. E. Kevei, M. Kiszel, and F. Simek, *Elelmez. Ipar* **9**, 370 (1955); *Chem. Abstr.* **52**, 8401 (1958).
149. E. Kevei, M. Kiszel, and F. Simek, *Acta Chim. Acad. Sci. Hung.* **6**, 345 (1955); *Chem. Abstr.* **51**, 2193 (1957).

150. R. Pleticha, *Pharmazie* **12**, 675 (1957); *Chem. Abstr.* **52**, 9292 (1958).
151. O. Enriquez and V. Kubac, *Rev. Fac. Farm., Univ. Cent. Venez.* **3**, 249 (1962); *Chem. Abstr.* **61**, 10535 (1964).
152. M. Sterescu, S. Arizan, M. Dobrovici, and R. Talmucui, *Rev. Chim. (Bucharest)* **8**, 376 (1957).
153. A. M. Shkodin and G. P. Tikhomirova, *Ukr. Khim. Zh.* **21**, 265 (1955); *Chem. Abstr.* **50**, 426 (1956).
154. K. C. Chiang and P. Chang, *Hua Hsueh Hsueh Pao* **24**, 300 (1958); *Chem. Abstr.* **53**, 20080 (1959).
155. J. Glicksman, *J. Electrochem. Soc.* **109**, 352 (1962).
156. S. Kaufman, *J. Biol. Chem.* **236**, 804 (1961).
157. D. Cohen, M. Koenigsbuch, and M. Sprecher, *Isr. J. Chem.* **6**, 615 (1968).
158. J. Koryta and P. Zuman, *Collect. Czech. Chem. Commun.* **18**, 197 (1953).
159. J. Koryta and P. Zuman, *Chem. Listy* **46**, 389 (1952).
160. P. Zuman, J. Koryta, and R. Kalvoda, *Collect. Czech. Chem. Commun.* **18**, 350 (1953).
161. P. Zuman, J. Koryta, and R. Kalvoda, *Chem. Listy* **47**, 345 (1953).
162. O. Fischer and O. Dracka, *Collect. Czech. Chem. Commun.* **27**, 2727 (1962).
163. R. D. Armstrong, M. Fleischmann, and J. W. Oldfield, *Trans. Faraday Soc.* **65**, 3053 (1969).
164. K. M. Johnson and J. D. Stride, *Chem. Ind. (London)* **24**, 783 (1969).
165. S. Kato and G. Dryhurst, *J. Electroanal. Chem.* **62**, 415 (1975).
166. S. Kato, B. Visinski, and G. Dryhurst, *J. Electroanal. Chem.* **66**, 21 (1975).
167. W. V. Childs, J. T. Malloy, C. P. Keszthelyi, and A. J. Bard, *J. Electrochem. Soc.* **118**, 874 (1971).
168. R. S. Nicholson, *Anal. Chem.* **37**, 1351 (1965).
169. S. Kato, M. Poling, D. van der Helm, and G. Dryhurst, *J. Am. Chem. Soc.* **96**, 5255 (1974).
170. S. Kato and G. Dryhurst, in progress.
171. R. Kalvoda and J. Zyka, *Cas. Cesk. Lek.* **63**, 36 (1950); *Chem. Abstr.* **46**, 4172 (1952).
172. R. Kalvoda and J. Zýka, *Proc. Int. Polarogr. Congr. 1st, 1951* Part I, p. 735, (1951); Part III, p. 550 (1952).
173. R. Kalvoda and J. Zýka, *Sb. Celostatni Konf. Anal. Chem., 1st, 1952* p. 224 (1953).
174. R. Kalvoda and J. Zýka, *Cesk. Farm.* **2**, 154; *Chem. Abstr.* **48**, 9622 (1954).
175. A. Heyndrickx, *J. Pharm. Belg.* **9**, 132 (1954).
176. T. Isshiki, *Yakugaku Saikin No Shimpo* p. 113 (1958); *Chem. Abstr.* **53**, 130 (1959).
177. Sasongko and J. A. S. Sasongko-Ponamon, *Suora Pharm. Madjalah* **5**, No. 4, 101 (1960); *Chem. Abstr.* **58**, 12370 (1963).
178. F. Vorel and J. Prokes, *Soudni Lek.* **2**, 129 (1957); *Chem. Abstr.* **54**, 24980 (1960).
179. J. Prokes, F. Vorel, and V. Dolezal, *Chem. Zvesti* **16**, 411 (1962); *Chem. Abstr.* **58**, 7285 (1963).
180. J. Prokes and F. Vorel, *Chem. Zvesti* **14**, 818 (1960); *Chem. Abstr.* **55**, 18017 (1961).
181. O. Manoušek and P. Zuman, *Chem. Listy* **49**, 668 (1955); *Collect. Czech. Chem. Commun.* **20**, 1340 (1955); *Chem. Abstr.* **49**, 11459 (1955).
182. H. Meinert and D. Cech, *Z. Chem.* **12**, 291 (1972).
183. H. T. Miles, *Biochim. Biophys. Acta* **22**, 247 (1956).
184. L. F. Cavalieri, A. Bendich, J. F. Tinker, and G. B. Brown, *J. Am. Chem. Soc.* **70**, 3875 (1948).
185. L. F. Cavalieri and A. Bendich, *J. Amer. Chem. Soc.* **72**, 2589 (1950).
186. J. Volke, *Phys. Methods Heterocycl. Chem.* **1**, 260 (1963).

5

Purine and Pyrimidine Nucleosides and Nucleotides, Polyribonucleotides, and Nucleic Acids

I. INTRODUCTION, NOMENCLATURE, AND STRUCTURE

The structure and nomenclature of the purine and pyrimidine nucleosides and nucleotides have been outlined in Chapters 3 and 4, respectively. Polynucleotide is a rather broad term that includes not only nucleic acids, i.e., a nucleic acid is a natural polynucleotide, but also synthetic substances containing only specified nucleotides or a particular sequence of nucleotides.

There are two types of nucleic acid. If the sugar associated with the constituent nucleotides is deoxyribose and the pyrimidine bases are cytosine and thymine, the nucleic acid is deoxyribonucleic acid, DNA. If the sugar is ribose and the pyrimidine bases are (mostly) cytosine and uracil, the nucleic acid is ribonucleic acid, RNA. The constituent purine bases in both types of nucleic acid are adenine and guanine. At first sight the sequences of bases in nucleic acids appear to be somewhat random, but in fact the sequence of purine and pyrimidine bases in DNA contains the information for their own replication and for the synthesis of ribosomal, messenger, and transfer RNA molecules.[1] Part of the DNA sequence is copied exactly during synthesis of messenger RNA, except that uracil replaces thymine. Messenger RNA acts as a template for protein synthesis since each three-base sequence in the messenger RNA directs the incorporation of a particular amino acid into a growing peptide chain (protein). As mentioned in earlier chapters, in a similar manner the nucleic acids contain the information for the transfer of genetic information.

In double-stranded nucleic acids the amount of adenine always equals the amount of thymine (DNA) or uracil (RNA), and the amount of guanine always equals the amount of cytosine. The now famous Watson–Crick[1] model of DNA consists of two polynucleotide chains wound into a helix. The chains consist of deoxyribotide phosphates joined together by phosphate diesters, with the bases projecting perpendicularly from the chain to a center axis. For each adenine projecting toward the central axis, one thymine must project toward adenine from the second parallel chain and be held by hydrogen bonding to adenine. Similarly, cytosine and guanine can associate together via hydrogen bonding. The specificity of the hydrogen bonding processes, however, does not permit cytosine and adenine or guanine and thymine to associate. The result is that DNA consists of two polynucleotide chains coiled around a common axis and held together by the selective hydrogen bonding between adenine and thymine, and cytosine and guanine. The chains, however, do not run in the same direction with respect to their internucleotide linkages, but are antiparallel. Thus, considering the structure shown in Fig. 5-1,[2] if the two adjacent deoxyribosides,

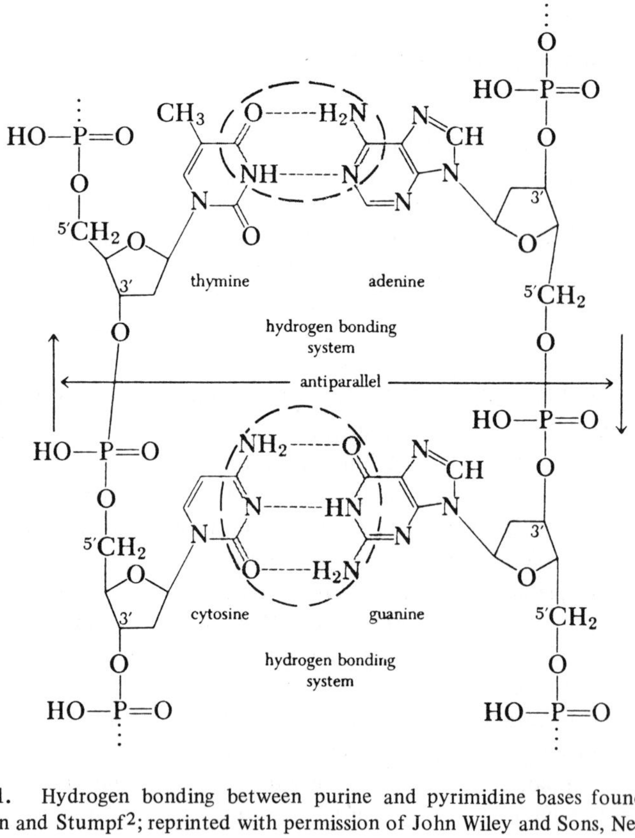

FIG. 5-1.   Hydrogen bonding between purine and pyrimidine bases found in DNA. (From Conn and Stumpf[2]; reprinted with permission of John Wiley and Sons, New York.)

thymine and cytosine, are linked through a phosphate grouping via the 3' and 5' positions in the deoxyribose rings, then the complementary deoxyribosides, adenine and guanine, in the other chain will be linked via the 5' and 3' positions. The generalized DNA structure shown in Fig. 5-2 illustrates the probable double-helix structure. The structure of RNA has not been determined to the same extent as that of DNA, but it is likely that a similar type of base pairing occurs in both macromolecules.

A term frequently referred to in subsequent discussions is *denatured* DNA. It is found that not only does DNA exist in the double-stranded helix, but under certain conditions it may exist as single, flexible strands. The latter condition is referred to as denatured DNA. An increase in temperature or a change in pH may bring about denaturation of DNA, or chemicals that can interfere with the hydrogen bonding between the base pairs, such as urea, can bring about the same effect.

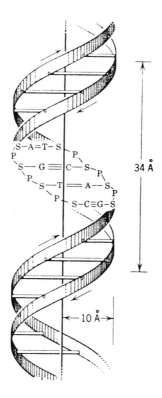

FIG. 5-2. Double helix of DNA. Key: P, phosphate; S, deoxyribose; A, adenine; T, thymine; G, guanine; C, cytosine. (Reprinted with permission from Conn and Stumpf,[2] Wiley, New York.)

Polyribonucleotides are very often synthetic macromolecules that are a long chain made up of one or sometimes two types of nucleotides. A long chain of cytidylic acid molecules is called polycytidylic acid or, more simply, poly(C). Similarly, other polyribonucleotides are poly(A) (polyadenylic acid), poly(I) (polyinosinic acid), poly(U) (polyuridylic acid), etc.

II. ELECTROCHEMISTRY OF NUCLEOSIDES AND NUCLEOTIDES

In 1946 Heath[3] reported, incorrectly, that of all nucleic acid constituents only adenine and its nucleosides and nucleotides yielded any polarographic reduction wave, at least in acidic solution. Luthy and Lamb[4] confirmed these findings, as did McGinn and Brown.[5] However, in 1960 Palaček[6] studied the oscillopolarographic behavior of nucleic acid components and found that all of the bases and their nucleosides and nucleotides gave indentations on the dE/dt versus $f(E)$ curves. This led to a more detailed examination of the polarographic reduction of these compounds. In order to simplify the discussion in this section the nucleosides and nucleotides of each parent base will be discussed separately.

A. Adenine Series

The polarographic half-wave potentials for reduction of a number of adenine nucleosides and nucleotides and related derivatives are presented in Table 5-1.

A very detailed study of the electrochemical reduction of adenine nucleosides and nucleotides has been carried out by Janik and Elving.[7,8] Each adenine derivative examined (Table 5-1) showed one generally well-defined cathodic polarographic wave, the overall behavior pattern being fundamentally similar to that of adenine itself[9] (see Chapter 3). As with adenine in high ionic strength solutions at $25°C$[10] the upper portion of the wave of adenine nucleosides and nucleotides was often distorted by a second, more negative wave whose behavior resembled that of a maximum of the second kind. Also similar to adenine, it was found[7] that this anomalous wave did not appear at low solution ionic strength (i.e. $\leqslant 0.1\ M$) at $1.5°C$. A discussion of the nature of the anomalous wave is presented in Chapter 3. Up to pH 4–5 all adenine nucleosides and nucleotides gave a diffusion-controlled wave. The magnitude of the diffusion current for this wave was, within 10%, the same as that for adenine. On the basis of this and the pH dependence of the half-wave potentials, the polarographic reduction of the adenine nucleosides and nucleotides was assumed to be a $4e$ process identical to that of adenine (see Chapter 3 for a detailed discussion). Plots of limiting current versus pH exhibited a sigmoidal decrease centered at 1.2–2.2 pH units higher than the pK_a for the protonated adenine derivative. The wave was kinetically

TABLE 5-1   Half-Wave Potentials of Adenine and Its Nucleosides and Nucleotides

Compound	$pK_a{}^a$	pH Range	$E_{1/2}$ (V)	Reference
Adenine	4.2	0.05 M HClO$_4$	-1.1^b	5
		0.1 M HClO$_4$	-1.46^c	4
		1.0–6.5	$-0.975 - 0.084$ pH^b	7,8
Adenosine	3.6	0.05 M HClO$_4$	-1.17^b	5
		0.1 M HClO$_4$	-1.42^c	4
		2.0–4.5	$-1.040 - 0.070$ pH^b	7,8
		4.5–6.0	$-1.180 - 0.041$ pH^b	
Deoxyadenosine	3.7	2.5–4.6	$-1.060 - 0.069$ pH^b	7,8
		4.6–6.5	$-1.205 - 0.037$ pH^b	
Adenosine 5'-monophosphate (AMP or adenylic acid)	3.8	0.1 M HClO$_4$	-1.34^c	4
		1.0–4.3	$-1.015 - 0.083$ pH^b	7,8
		4.3–5.5	$-1.115 - 0.060$ pH^b	
Deoxyadenylic acid	4.4	2.0–6.5	$-0.985 - 0.080$ pH^b	7,8
Adenosine 5'-diphosphate (ADP)		3.7	-1.357^b	9
		4.7	-1.415^b	
Adenosine 5'-triphosphate (ATP)	4.1	2.5–4.5	$-1.035 - 0.083$ pH^b	7,8
		4.5–5.5	$-1.175 - 0.052$ pH^b	
Adenosine 1-N-oxide		0.05 M HClO$_4$	(I) -0.84^b	5
			(II) -1.14^b	

^{a} Basic pK_a (i.e., BH$^+$ ⇌ B + H$^+$).

^{b} Potential versus SCE.

^{c} Potential versus mercury pool electrode (anode).

controlled in pH regions where it decreased. These two facts coupled with the observation that plots of $E_{1/2}$ usually changed slope at 0.1–0.3 pH unit higher than pK_a (Table 5-1) indicated that the adenine nucleosides and nucleotides were polarographically reducible only in a protonated form. The proton would probably be added at the N-1 position, which is known to be the most likely protonation site in adenine,[11,12] although it has been suggested[13,14] that equilibria involving monoprotonated species at N-1, N-3, or N-7 are possible.

By subtracting $dE_{1/2}/d$(pH) values for adenine from those for the nucleosides and nucleotides in corresponding pH regions, Janik and Elving[7] concluded that attachment of a sugar or sugar phosphate moiety decreased the ease of reducibility, at least below pH 5. The order of ease of reducibility at hypothetical zero concentration at pH 4.1 was adenine > deoxyadenosine, adenosine, dAMP > AMP > ADP > ATP. This order changed somewhat with increasing concentration so that at the 1 mM level the order was adenine > adenosine, deoxyadenosine > dAMP, AMP > ADP > ATP. The fact that substitution of adenine with ribose or ribose phosphate had such a small effect was considered to be the result of at least two opposing factors. First, substitution by ribose or ribose phosphate should increase the ease of reduction of adenine because of the electron-withdrawing effect of the former groups.[15-17] On the other hand, adsorption[18,19] and intermolecular association[17,20-22] of the electroactive species should decrease the ease of reducibility (*vide infra*). It was also considered probable[7] that the presence of negatively charged phosphate groups in the nucleotides, which could result in electrostatic repulsions between the compound and the negatively charged electrode, could decrease reducibility. The condensed structure of ATP and possibly ADP in aqueous media[23] could also decrease reducibility.

As briefly indicated earlier, all of the adenine nucleosides and nucleotides, and indeed adenine itself, were found to be adsorbed at the DME, as shown by depressions of the base current in AC polarography.[7,24] The AC polarograms of dilute solutions of adenine nucleosides and nucleotides (0.25 m*M*) at pH 2–5 are shown in Fig. 5-3. The deep depression of the base currents around −0.5 V (Fig. 5-3), which is close to the DME potential of zero charge, indicated that an uncharged portion of the molecule was adsorbed. Based on the magnitude of the depression of the base current, the order of uncharged-site-controlled adsorbability was adenine < dAMP < deoxyadenosine. This order agrees nicely with the order of surface activities calculated from differential capacity data in 1 *M*

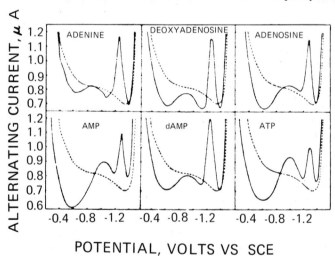

POTENTIAL, VOLTS VS SCE

FIG. 5-3. Alternating current polarograms for 0.25 m*M* solutions of adenine, its nucleosides, and nucleotides in pH 2.5 McIlvaine buffer (0.5 *M* ionic strength) at 25°C. The dotted line is the polarogram for supporting electrolyte alone. The AC polarograms obtained at 50 Hz and 4 mV peak-to-peak.[7] [Reprinted with the permission from B. Janik and P. J. Elving, *J. Am. Chem. Soc.* **92**, 235 (1970). Copyright (1970) by the American Chemical Society.]

NaCl by Vetterl.[19] The low, rounded peak and second base current depression observed for adenine and its derivatives at more negative potential (Fig. 5-3) have been interpreted[7,19] as indicating that the molecules adsorbed with an uncharged site toward the electrode are desorbed or reorient and become readsorbed with a positively charged site toward the electrode (i.e., the protonated N-1 position). Alternating current polarographic data of Vetterl[25] indicate that at pH 7 adenine nucleosides are also adsorbed in the potential region of the faradaic peak. Alternating current polarographic measurement[7] did not provide unequivocable evidence for adsorption at potentials more negative than that of the faradaic peak. However, the summit potential, E_s, of the AC faradaic reduction peak was found to be more negative than the corresponding $E_{1/2}$ (Table 5-2). With increasing pH, $E_{1/2} - E_s$ became smaller, which was interpreted to mean that the reduced form of the adenine species becomes more strongly adsorbed relative to the oxidized form in the region of the summit potential. The AC polarography of nucleosides will be further discussed in Section II, D.

Janik and Elving[7] also noted a marked decrease in the magnitude of experimental diffusion coefficients with concentration. The fact that the value of these diffusion coefficients was 3 to 4 times greater than those calculated on the basis of the Stokes–Einstein relation[26] was interpreted in terms of association, preferential orientation, and conformation of the species diffusing to the electrode. The latter was proposed to involve planar arrangement of the adenine rings and vertical stacking, which results in the effective barrier to diffusional transport being essentially the minimal cross-sectional area of the planar purine moiety.

Skulachev and Denisovich[27] have also reported half-wave potential data for the polarographic reduction of adenine nucleosides and nucleotides.

TABLE 5-2 **Alternating Current Polarographic Data for 0.25 mM Adenine Nucleosides and Nucleotides at pH 2.5[a]**

Compound	E_s[b] (V versus SCE)	$E_{1/2}$ (V versus SCE)	$E_{1/2} - E_s$ (mV)
Adenine	−1.260	−1.185	75
Adenosine	−1.285	−1.220	65
Deoxyadenosine	−1.270	−1.230	40
AMP	−1.295	−1.225	70
dAMP	−1.250	−1.185	65
ATP	−1.295	−1.240	55

[a] Data from Janik and Elving.[7]

[b] Alternating current polarographic data taken at a frequency of 50 Hz and an amplitude of 4 mV peak-to-peak.

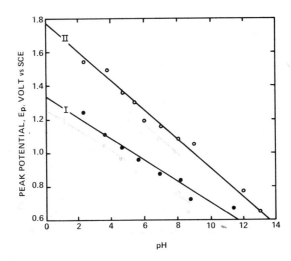

FIG. 5-4. Variation of peak potential with pH for the voltammetric oxidation of adenine (I) and adenosine (II) at the pyrolytic graphite electrode at the 0.25 mM concentration level.[28] (Reprinted with permission of Pergamon Press, New York.)

Adenosine is electrochemically oxidizable at the pyrolytic graphite electrode in aqueous solution[28] at more positive potentials than adenine (Fig. 5-4). No details of the products or mechanism associated with the electrochemical oxidation have been published. An analytical method has been developed which allows adenine and adenosine in mixtures to be determined by electrochemical methods.[28]

B. Cytosine Series

The first detailed study of the electrochemical reduction of cytosine and cytidine and its derivatives was that of Janik and Paleček.[29] Half-wave potential data reported by these authors are presented in Table 5-3.[29,30] Consideration of these data reveals that the half-wave potential becomes more negative in the order nucleotide < nucleoside < base. As with the adenine series, the waves decrease in height and become kinetic in nature at moderately high pH and, in view of the pK_a of the compounds, it has been concluded[29] that the protonated form of the pyrimidine nucleus is the electroactive species. By comparing the height of the polarographic wave of cytidine with that of similar compounds having known n values, Janik and Paleček[29] concluded that the reduction of cytidine involved four electrons. Controlled potential electrolysis of cytidine gave a product that had a UV spectrum similar to that observed on reduction of cytosine. In addition, ammonia was found to be liberated during the electrolysis.

TABLE 5-3 Half-Wave Potentials for Cytosine and Its Nucleosides and Nucleotides

Compound	pK_a	pH Range	$E_{1/2}$ (V versus SCE)	Reference
Cytosine	4.6	2.5–7	$-1.170 - 0.084$ pH	29
		4–6	$-1.125 - 0.073$ pH	30
Cytidine	4.2	2.5–7	$-1.105 - 0.072$ pH	29
Deoxycytidine	4.6	2.5–7	$-1.154 - 0.068$ pH	29
Deoxycytidylic		3.5–7.3	$-0.908 - 0.110$ pH	29
acid		7.3–8.7	$-1.350 - 0.050$ pH	
5-Methyldeoxy-	4.4	4–7.3	$-0.775 - 0.118$ pH	29
cytidine		7.3–10	$-1.325 - 0.042$ pH	
Cytidylic acid		6.5	-1.68	29
5-Hydroxymethyl-		5.6	-1.61	29
cytidine		6.5	-1.70	29

Subsequent work by Elving and co-workers[31] using coulometry indicates, however, that the reduction of cytidine involves three electrons, not four, as proposed by Janik and Paleček.[29] In view of this fact and the very close parallelism between the electrochemical behavior of cytidine and its parent base cytosine, which has been studied in considerable detail[30−32] (see Chapter 4, Section III, A, 4), the original reaction scheme proposed by Janik and Paleček[29] is probably incorrect.

Elving and co-workers[31] have more recently studied the electrochemical behavior of cytosine nucleosides and nucleotides in somewhat more detail. Some typical polarographic results are presented in Table 5-4. The electrochemical behavior of each of the cytosine species supported the conclusion that reduction of the cytosine moieties occurs by a similar mechanism for all of the species, except for CpC [cytidylyl-(3′,5′)-cytidine] above pH 7. In agreement with the earlier report of Janik and Paleček,[29] it was concluded that the protonated form of the cytosine species is the reducible moiety; i.e., the limiting current at a pH slightly greater than the pK_a decreased and assumed increasing kinetic character with increasing pH. By use of DC and AC polarography, cyclic voltammetry, and coulometry it was concluded that the reaction mechanisms for cytosine, cytidine, and cytidine 5′-monophosphate (CMP) were essentially identical. Thus, the appropriate monomeric cytosine nucleoside or nucleotide (I, Fig. 5-5) appears to undergo a rapid protonation at N-3 to form the electroactive species (II, Fig. 5-5). A 2e reduction of the latter species then gives a carbanion (III, Fig. 5-5), which protonates to give the corresponding 3,4-dihydro derivative (IV, Fig. 5-5). Deamination of the dihydro derivative then gives the appropriate nucleoside or nucleotide of 2-oxypyrimidine (V, Fig. 5-5). A rapid protonation of the latter gives a further electroactive species (VI, Fig. 5-5), which is reduced

TABLE 5-4 Typical DC Polarographic Behavior of Cytosine and Its Nucleosides and Nucleotides[a]

Compound	pH	Medium[b]	μ (M)[c]	Compd concn (mM)	Temp (°C)	Wave	−E_{1/2} (V versus SCE)	I[d]
Cytidine	4.2	Ac	0.5	0.97	25		1.44	6.3
	4.2	Ac	0.5	0.97	0		1.45	4.4
	5.0	Mc	0.13	0.1	0		1.38	4.1
CMP[e]	4.1	Ac	0.5	1.00	25		1.38	6.3
	5.0	Mc	0.13	0.1	0		1.46	3.2
	4.1	Ac	0.5	0.99	25		1.37	4.6
				0.49	0		1.37	3.0
CpC[f]	5.0	Mc	0.13	0.018–0.89	0	I	1.308[g]	5.5–0.4
				0.053–0.89		II	1.39–1.49	1.6–5.9
				0.36–0.71		III	1.61	1.3–1.7
	2.5–5.5	Mc	0.5	0.05	25	I	1.040 + 0.041 pH	10[h]
	2.5–5.9	Mc				II	1.085 + 0.056 pH	
	5.9–8.1	Mc + Ca				II	0.800 + 0.104 pH	
	8.1–10.6	Ca				II	1.225 + 0.051 pH	
CpU[i]	4.5	Mc	0.13	0.094			1.32	2.8

[a] Data from Webb et al.[31]
[b] Ac, acetate buffer; Mc, McIlvaine buffer; Ca, carbonate buffer.
[c] μ: Ionic strength.
[d] $I = i_1 / C m^{2/3} t^{1/6}$.
[e] Cytosine 5'-monophosphate.
[f] Cytidylyl-(3',5')-cytidine.
[g] Average value for all concentrations.
[h] $E_{1/2}$ in pH 9.1 ammonium buffer is 65 mV more positive.
[i] Cytidylyl-(3',5')-uridine.

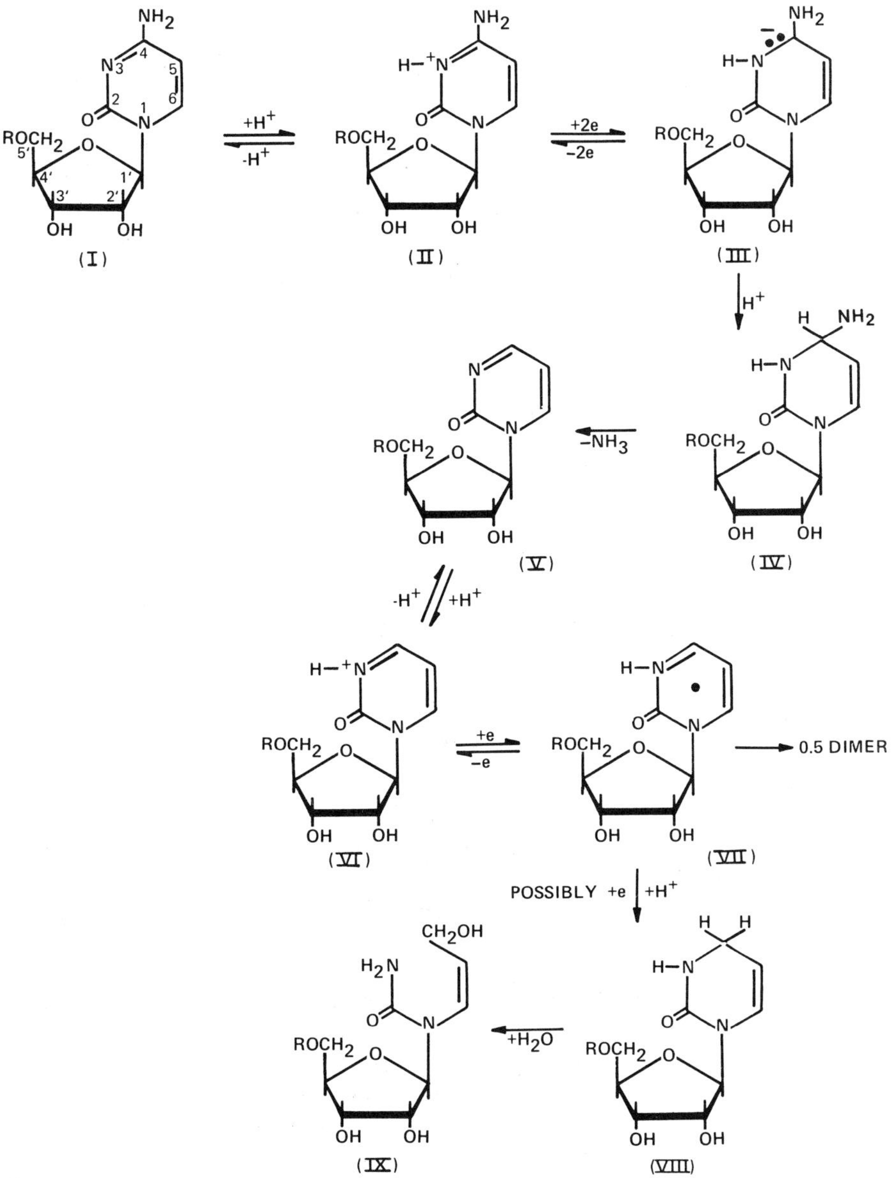

FIG. 5-5. Probable mechanism for the electrochemical reduction of cytosine nucleosides and nucleotides. When R = H, I is a nucleoside. When

$$R = \left(HO - \overset{\overset{\displaystyle O}{\|}}{\underset{\underset{\displaystyle OH}{|}}{P}} - \right)_x$$

I is a nucleotide, where x may be 1, 2, or 3. Steps I–VII adapted from reference 31. Steps VII–IX might occur under conditions of controlled potential electrolysis as suggested in Janik and Paleček.[29]

in a 1e reaction to a free radical (VII, Fig. 5-5). This radical then presumably dimerizes, although the dimer has never been isolated and properly characterized. Indeed, the work of Janik and Paleček[29] on various cytosine nucleosides and nucleotides seems to support the fact that under conditions of controlled potential electrolysis some additional 1e–1H$^+$ reduction of the free radical (VII, Fig. 5-5) might occur, giving the corresponding nucleoside or nucleotide of 3,4-dihydro-2-oxypyrimidine (VIII, Fig. 5-5), which then hydrolyzes to the appropriate 3-ureidoallyl alcohol derivative (IX, Fig. 5-5).

The reduction of CMP apparently differs slightly from that observed for cytosine and cytidine in that the deamination reaction (i.e., IV → V, Fig. 5-5) is slower. The rate constant for this reaction in the case of cytosine has been estimated to be 10 sec^{-1}, whereas in the case of CMP it has a value of ca. 3 sec^{-1}.

In the case of CpC, coulometry indicated that four electrons are involved in

CpC

its overall electrochemical reduction.[31] To Elving *et al.*[31] this implied that each cytosine ring of CpC is reduced in a 2e process and that the deamination step (i.e., equivalent to IV → V, Fig. 5-5) occurs very slowly or not at all. Unfortunately, this conclusion was not proved by evidence from product analysis. However, at 25°C in neutral or slightly alkaline solution, CpC exists in an equilibrium between stacked single-stranded and disordered conformations.[33] In addition, charge-transfer complexes stabilized by van der Waals–London forces have been observed[34] in frozen solutions (77°K) of CpC at pH values around its pK_a. Elving *et al.*[31] concluded that similar complexes and stacking might occur at 25°C, shielding the reduction sites of CpC and, consequently, hindering deamination. It was also proposed that higher pH would stimulate the deamination. Cytidylyl-(3',5')-cytidine shows one generally moderately well-defined cathodic polarographic wave, which in certain pH and concentration ranges may split into two waves (see, e.g., Table 5-4). The first wave (wave I), which occurs at the more positive potentials, showed many of the characteristics of an adsorption prewave, indicating adsorption of the product of CpC reduction at the electrode/solution interface (see Chapter 2). Some preliminary data on the polarography of CpU and CpG have been reported.[31]

Elving and co-workers[31] have also measured the diffusion coefficients for various cytosine derivatives (Table 5-5). These results indicated that the ribose group had little effect on the diffusing species, while the relatively bulky phosphate group decreased the diffusion coefficient by about a factor of 2. It was concluded,[31] therefore, that the cytosine species diffuse to the electrode with the plane of the ring perpendicular to the plane of the electrode surface; i.e., the ribose group is behind the cytosine ring and adds little to the effective cross-sectional area of the molecule. The phosphate group, however, probably therefore sticks out to one side of the molecule, increasing the diffusional cross-sectional area.

The fact that the diffusion coefficient of CpC is about the same as that for cytosine and cytidine indicates that their effective cross-sectional areas may be similar. This suggests that CpC is oriented differently to cytosine as it diffuses to the electrode. However, no specific suggestions regarding the orientation of CpC diffusing toward the electrode were made.[31]

C. Guanine Series

Several groups of workers have confirmed that guanine, guanosine, and guanylic acid do not give a polarographic reduction wave at the DME.[3,9,35] Using alternating current oscillographic polarography, Paleček[36–39] and co-workers,[40–42] however, have shown that guanine and its nucleosides and

TABLE 5-5 Diffusion Coefficients for Cytosine and Its Nucleosides and Nucleotides[a]

Compound	Temperature (°C)	Diffusion coefficient $\times 10^5$ (cm^2 sec^{-1})[b]
Cytosine	25	1.18
	0	0.58
Cytidine	25	1.20
	0	0.51
CMP	25	0.65
	0	0.31
CpC	25	1.69
	0	0.53

[a] Data from Webb *et al.*[31]
[b] Obtained by DC polarography.

nucleotides and indeed nucleic acids (see subsequent Section III) give an anodic indentation on the dE/dt versus $f(E)$ curve. Prolonged reduction of deoxy-guanosine at a mercury pool electrode at potentials beyond background electrolyte discharge potentials gives a product that yields an anodic wave in conventional DC polarography.[40] Studies of guanine derivatives by Janik[43] revealed that the reduction probably took place in the imidazole ring at the N-7=C-8 double bond to give directly or indirectly the electrochemically oxidizable product (see Chapter 3). This phenomenon of reduction of guanine residues at or beyond background discharge potentials will be mentioned frequently in subsequent discussions of nucleic acids. The electrochemical reduction of guanosine in aqueous solution could not be observed at platinum, silver, or carbon electrodes.[8]

Guanosine is electrochemically oxidized in aqueous solution at the pyrolytic graphite electrode (PGE) by way of four voltammetric oxidation peaks, all of which occur at more positive potential than the single oxidation peak of guanine (Fig. 5-6).[44] Both guanine and guanosine are very strongly adsorbed at the PGE. A simple and rapid voltammetric method for the analysis of guanine in the presence of guanosine has been developed.[44]

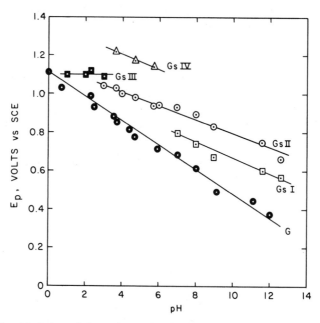

FIG. 5-6.   Variation of E_p with pH for the single voltammetric oxidation peak of guanine (G) and for the four voltammetric oxidation peaks of guanosine (Gs). Guanine concentration, 0.300 mM; guanosine, 0.302 mM. Sweep rate, 5 mV sec^{-1}.[44] (Reprinted with permission of Elsevier Publishing Company, Amsterdam.)

D. Adsorption of Purine and Pyrimidine Nucleosides and Nucleotides

By measuring the differential capacity of the DME in solutions of deoxyribonucleosides and deoxyribonucleotides in 1 M NaCl solution, Vetterl[19] found that the adsorbability (or surface activity) of the compounds increased in the order deoxycytidine < thymidine, deoxyadenosine < deoxyguanosine for the deoxynucleosides and deoxycytidilic acid < thymidylic acid < deoxyadenylic acid < deoxyguanylic acid for the deoxynucleotides. Increasing adsorbability for the parent bases was in the same order, i.e., cytosine < thymine < adenine < guanine, which has been interpreted[8] as indicating that the base, which is presumably the hydrophobic moiety of the nucleoside or nucleotide, is the portion of the molecule adsorbed at the electrode surface. By comparing the relative surface activity it appears that in the case of the cytosine, thymine, and adenine series the base is less strongly adsorbed than the nucleotide, which in turn is less strongly adsorbed than the nucleoside. The guanine series, however, exhibits the reverse order.

In a later study Vetterl[45] studied the AC polarography of biologically important nucleosides in 0.5 M NaF with 0.1 M phosphate buffer, pH 7. The AC polarograms of uridine, thymidine, and cytidine (Fig. 5-7) showed a minimum at around -0.4 to -0.6 V and a cathodic peak or desorption maximum at about -0.9 to -1.5 V. The minima indicated extensive adsorption of the three compounds. The large peak for cytidine and deoxycytidine at ca. -1.6 V was due to the faradaic reduction of the compound and was not associated with any adsorption—desorption phenomena. At higher concentrations of deoxycytidine there was association of the adsorbed molecules, which was characterized by formation of a pit on the AC polarograms (Fig. 5-7D). Alternating current polarograms of adenosine (Fig. 5-8A) up to concentrations of 1.1 mM gave a normal minimum at about -0.4 V due to adsorption of the compound. At higher concentrations association took place, as evidenced by pits at about -1.1 V, and at concentrations exceeding 3.3 mM a further pit was observed at ca. -0.4 V. The existence of two areas in which association of adenosine occurred was thought to indicate that, when the potential of the electrode changes, the association capacity of the adsorbed molecules alters through a change in their orientation at the electrode surface. When the temperature was increased, the tendency for adsorption and association decreased.

Deoxyadenosine showed adsorption in the region of -0.4 V and association pits at around -1.2 V (Fig. 5-8B). Alternating current polarograms of guanosine (Fig. 5-9A) indicated that guanosine molecules were adsorbed at about -0.5 V, and at higher concentrations association occurred at about the same potential region. The sharp peak or maximum observed at about -1.2 V was due to desorption of guanosine. With deoxyguanosine, association of the adsorbed

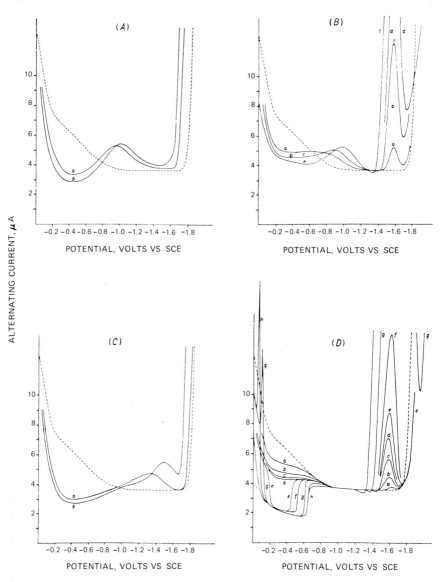

FIG. 5-7.   Alternating current polarograms of nucleoside solutions in 0.5 M NaF with 0.1 M phosphate buffer, pH 7.0, at 25°C. (*A*) Uridine: a, 1.0 mM; b, 5.5 mM. (*B*) Cytidine: a, 0.12 mM; b, 0.23 mM; c, 0.46 mM; d, 0.92 mM; e, 17 mM. (*C*) Thymidine: a, 8.0 mM; b, 42 mM. (*D*) Deoxycytidine: a, 0.044 mM; b, 0.24 mM; c, 0.46 mM; d, 0.68 mM; e, 0.90 mM; f, 1.8 mM; g, 7.3 mM; h, 10 mM; i, 50 mM. The AC polarograms were obtained at 78 Hz and an amplitude of 18 mV.[45] (Reprinted with permission of Elsevier Publishing Company, Amsterdam.)

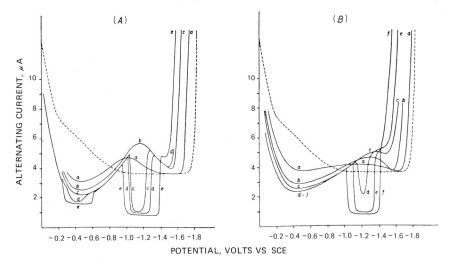

FIG. 5-8. Alternating current polarograms of (*A*) adenosine and (*B*) deoxyadenosine in 0.5 *M* NaF with 0.1 *M* phosphate buffer, pH 7. (*A*) Concentrations of adenosine: a, 0.55 m*M*; b, 1.1 m*M*; c, 3.3 m*M*; d, 4.4 m*M*; e, 11 m*M*. (*B*) Concentrations of deoxyadenosine: a, 0.084 m*M*; b, 0.66 m*M*; c, 4.6 m*M*; d, 4.7 m*M*; e, 6.2 m*M*; f, 11 m*M*. The AC polarograms were obtained at 78 Hz and an amplitude of 18 mV.[45] (Reprinted with permission of Elsevier Publishing Company, Amsterdam.)

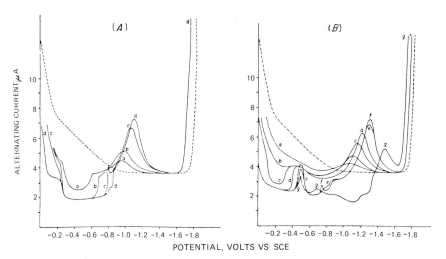

FIG. 5-9. Alternating current polarograms of (*A*) guanosine and (*B*) deoxyguanosine in 0.5 *M* NaF with 0.1 *M* phosphate buffer, pH 7. (A) Concentration of guanosine: a, 0.10 m*M*; b, 0.19 m*M*; c, 0.52 m*M*; d, 0.78 m*M*. (B) Concentration of deoxyguanosine: a, 0.056 m*M*; b, 0.11 m*M*; c, 0.26 m*M*; d, 0.56 m*M*; e, 1.1 m*M*; f, 1.8 m*M*; g, 9.5 m*M*. The AC polarograms were obtained at 78 Hz and an amplitude of 18 mV.[45] (Reprinted with permission of Elsevier Publishing Company, Amsterdam.)

molecules appeared to occur at about -0.2 V. However, at concentrations above 1.1 mM the AC polarograms were more complicated (Fig. 5-9B), possibly due to reorientation of the molecules on the electrode surface as the potential changed and by aggregation of the deoxyguanosine in solution.[46-48]

Elving *et al.*[31] have reported that cytosine and various cytosine nucleosides and nucleotides are adsorbed at the DME in pH 4 acetate buffer at 25°C. In agreement with the earlier findings of Vetterl,[19] at pH 7 depressions of the AC base current below background at potentials around the electrocapillary maximum (ecm) were observed. The depth of these depressions increased in the order cytosine $<$ cytidine, CMP, the same as that observed by Vetterl.[19] In view of this order it has been concluded that the ribose and ribose phosphate groups strongly influence the adsorption of these molecules through production of additional adsorption sites in the molecule as a result of the added ribose moiety and/or through the effect of the ribose group on adsorption sites in the cytosine ring.

At potentials more negative than the ecm a broad peak was observed for compounds other than cytosine itself. These broad peaks, which were similar to those observed by Vetterl (see, for example, Fig. 5-7B), were again proposed to be due to a gradual reorientation of the molecule on the electrode surface. Since cytosine did not exhibit this orientation peak, the influence of the ribose and ribose phosphate on the interfacial properties of these molecules is apparent. A depression of the AC base current, immediately prior to the faradaic reduction process, was observed by Elving *et al.*[31]; this depression was similar to that observed by Vetterl[19] at higher pH (see Fig. 5-7B). The former workers have suggested that the depression observed by Vetterl indicates that the molecules are probably oriented so that the protonated ring nitrogen (see Fig. 5-5 and associated discussion) is closest to the electrode surface.

The dinucleoside phosphates of cytosine, CpC, CpU, and CpG, are adsorbed much more strongly than cytosine, cytidine, and CMP,[31] which is probably due to the presence of more adsorption sites in the dinucleoside phosphate.

The so-called reorientation peaks observed for various cytosine derivatives at potentials negative of the ecm are much larger for the dinucleoside phosphates, which may indicate more complex changes at the electrode/solution interface than for cytosine, cytidine, and CMP. As mentioned earlier (Section II,B) the reduced product of CpC is probably adsorbed at the DME, as evidenced by the presence of a typical adsorption prewave under DC polarographic conditions at certain pH values and concentrations.[31]

Cytosine-containing dinucleoside phosphates have been shown to associate, in the adsorbed state at the DME, much more than the cytosine nucleoside and nucleotide. For example, CpC associates (i.e., forms a well-defined pit or well in the AC base current) at pH 7.3 at bulk solution concentrations of less than 0.05 mM. Cytidine does not exhibit association effects at bulk concentrations of

17 m*M*. By examination of the association pits observed on AC polarography of various cytosine dinucleoside phosphates it has been concluded[31] that the extent of association at the DME surface increases in the order CpC < CpU < CpA, ApC, ApG. This order parallels that for the tendency of nucleosides and nucleotides to associate in solution, i.e., pyrimidine–pyrimidine < purine–pyrimidine < purine–purine,[49] and hence might indicate similar modes of association at the interface and in solution, i.e., vertical overlap or stacking of bases.[50]

E. Oscillopolarographic Studies

It appears that a very large number of purine and pyrimidine nucleosides give indentations on the *dE/dt* versus *f(E)* traces obtained in alternating current oscillopolarography. This behavior seems to have been used primarily for analytical purposes. Thus, the guanine plus cytosine content of DNA has been determined oscillopolarographically.[51] Paleček has employed the technique for following the course of thermal denaturation of DNA[52,53] and for the determination of denatured DNA.[52] It has also been applied to defining the genetic relationships of bacterial DNA.[54]

As has been outlined in the discussion of electrochemical techniques (Chapter 2), the position of an indentation formed on oscillopolarography is defined by the *Q* value, where

$$Q = \frac{\text{linear distance of the indentation peak from the potential of anodic mercury dissolution}}{\text{linear distance between potential of anodic mercury dissolution and that of background electrolyte discharge}}$$

Almost all purines and pyrimidines and their nucleosides and nucleotides of interest in nucleic acid chemistry have been studied oscillopolarographically.[36–39,55–57] A complete tabulation of the *Q* values in appropriate media has been prepared by Janik and Elving[8] and in slightly modified form is presented in Table 5-6. Although the resolution of oscillopolarography is considerably less than that of conventional DC polarography, it can be seen from the data presented in Table 5-6 that the addition of sugar or sugar phosphate moieties to adenine has very little effect on the potential or Q_c value for reduction. In the case of the cytosine series the *Q* value is influenced rather more, although no definite trend is obvious. The anodic indentations for the guanine and xanthine series are, as mentioned earlier, due to oxidation of a product of reduction of the compound at background discharge potentials. For further details on the oscillopolarography of these systems the review of Janik and Elving[8] and the references cited therein should be consulted.

TABLE 5-6   Indentations on Oscillopolarographic Traces of Purines and Pyrimidines and Their Nucleosides and Nucleotides[a]

Compound	1 N H_2SO_4	1 N $HCOONH_4$	2 N $HCOONH_4$	1 N NaOH	KCl
Adenine	Q_c 0.85	—	—	R	C
Adenosine	Q_c 0.85[c]	—	—	R	C
Deoxyadenosine	Q_c 0.85[c]	—	—	R	C
Adenylic acid	Q_c 0.85[c]	C	C	R	C
Deoxyadenylic acid	Q_c 0.85[c]	C	C	R	C
Guanine	—	Q_a 0.17	Q_a 0.17, C	R	C
Guanosine	—	Q_a 0.17[c]	Q_a 0.17, C[c]	R	C
Deoxyguanosine	—	Q_a 0.17[c]	Q_a 0.17, C[c]	R	C
Guanylic acid	A, C	Q_a 0.17[c]	Q_a 0.17, C[c]	R	C
Deoxyguanylic acid	A, C	Q_a 0.17[c]	Q_a 0.17, C[c]	R	C
Xanthine	—	Q_a 0.22	—		
Xanthosine	—	Q_a 0.22[c]	—		
Cytosine	C[d]	Q_c 0.93	Q_c 0.92	R	C
Cytidine	Q_c 0.91	C[e]	C[e]	R	C
Deoxycytidine	Q_c 0.91	C[e]	C[e]	R	C
Cytidylic acid	Q_c 0.90	C[e]	Q_c 0.85	R	C
Deoxycytidylic acid	Q_c 0.90	C[e]	C[e]	R	C
5-Methylcytosine	Q_c 0.97	Q_c 0.93	Q_c 0.90	R	C
5-Methylcytidine		Q_c 0.88	C[e]		
Uracil	X	—	C	Q_c 0.12	C
Uridine	X	—	C	Q_c 0.12[c]	C
Deoxyuridine	X	—	C	X	C
Uridylic acid	X	—		X	
Thymine	X	—	Q_c 0.11	X	
Thymidine	X	—		X	
Thymidylic acid	X	—	X	X	
6-Azauridine	Q_c 0.67	Q_c 0.73		Q_c 0.14	

[a] All data taken from Paleček[6] and Paleček and Kalab[57] except for those on 6-azauridine, which are taken from Humlova.[55] Table adapted with minor modification from Janik and Elving.[8]

[b] For definition of Q see text; Q_c is Q for a cathodic indentation; Q_a is Q for an anodic indentation; C denotes cathodic indentation; A is anodic indentation; R is reversible indentation or deformation close to anodic mercury dissolution; X is no indentation. Indentations denoted by C, A, or R were not suitable or have not been employed for analytical purposes. Dots indicate no indentation or one not suitable for analysis; a blank indicates that data are not available.

[c] The Q value does not differ from that of the corresponding base by more than 0.02.

[d] Indentation very close to background discharge.

[e] Well-developed indentation, but Q values not available.

III. ELECTROCHEMISTRY OF DEOXYRIBONUCLEIC ACIDS

Paleček[58,59] has authoritatively reviewed the applications of electrochemical techniques in nucleic acid research through 1970 in great detail. For details of experimental methodology the review articles by Paleček[58,59] and co-workers[60] should be consulted.

A. Native DNA

Native DNA does not give a detectable reduction wave under DC polarographic conditions,[61,62] although this technique has been employed to study the interaction of DNA with daunomycin,[63] Cu^{2+} and Cu^+,[64] and CuEDTA.[65] Under AC polarographic conditions, however, native DNA yields a peak in 0.5 M ammonium formate with 0.1 M sodium phosphate, pH 7, solution at about -1.2 V versus SCE[61,62,66-69] (Fig. 5-10). This peak will be referred to as wave I in subsequent discussions.* Wave I is nonfaradaic and is apparently formed as a result of desorption of DNA that is adsorbed at more positive potential.[70] The adsorption of native DNA has been studied by Miller, who employed a differential capacity method.[70] Miller found that at pH 6 native DNA is adsorbed at potentials more positive than -1.2 V versus SCE and then desorbed at ca. -1.3 V and is probably not further adsorbed at more negative potential. Particularly significant is the fact that by measuring the electrode area covered per nucleotide at full surface coverage it was found that, at potentials where the mercury electrode carried a negative charge and where DNA was adsorbed, the area covered per nucleotide was 35 $Å^2$. When the mercury surface was positively charged, the value was 86 $Å^2$ per nucleotide. These data were interpreted to indicate that native DNA preserves its double-helical structure (Fig. 5-2) when adsorbed on a negatively charged surface, but the double helix unfolds on a positively charged surface. These findings of Miller,[70] however, have been seriously criticized by Fleming,[71] who did not believe that unwinding of native DNA takes place on a DME even under the conditions of Miller's experiments.

Under conditions of alternating current oscillopolarography native DNA exhibits a cathodic indentation (CI-1, Fig. 5-11) at approximately the same potential as AC polarographic wave I.

Using differential pulse polarography, Paleček and Frary[72] observed, at very high sensitivities and high concentrations of native DNA at pH 7, two small waves (Fig. 5-12a), which were designated waves I and II, wave I being that

---

* Alternating current polarography and differential pulse polarography normally give peaks. It is a common practice of workers studying nucleic acids to refer to these peaks as waves. This practice has been observed throughout this chapter.

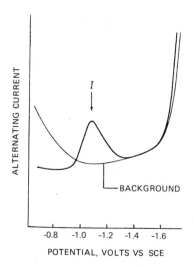

FIG. 5-10. Alternating current polarogram of native DNA at a concentration of 500 μg/ml in 0.5 *M* ammonium formate with 0.1 *M* sodium phosphate, pH 7. Amplitude of alternating voltage, 18 mV; frequency, 78 Hz.[58] (Reprinted with permission of Academic Press, Inc., New York.)

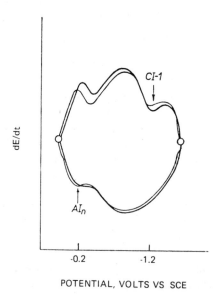

POTENTIAL, VOLTS VS SCE

FIG. 5-11. Oscillopolarogram of native calf thymus DNA at a concentration of 100 μg/ml in 0.5 *M* ammonium formate with 0.1 *M* sodium phosphate, pH 6.8.[58] Frequency of alternating current, 50 Hz; CI, cathodic indentation; AI_n, anodic indentation. (Reprinted with permission of Academic Press, Inc., New York.)

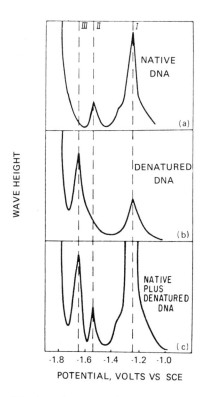

FIG. 5-12. Differential pulse polarograms of native and denatured DNA. (a) Native calf thymus DNA at a concentration of 500 μg/ml in 0.3 M ammonium formate plus 0.05 M sodium phosphate, pH 7.0; (b) denatured calf thymus DNA at a concentration of 10 μg/ml in 0.3 M ammonium formate plus 0.05 M sodium phosphate, pH 7.0; (c) 1 mg of native calf thymus DNA plus 15 μg of denatured calf thymus DNA per milliliter in 0.3 M ammonium formate with sodium phosphate, pH 7.0.[72] (Reprinted with permission of Academic Press, Inc., New York.)

which occurred at most positive potential. The actual nature of the differential pulse polarographic wave I has not been studied in detail, but it appears to correspond to that of the nonfaradaic AC polarographic wave I (Fig. 5-10) and oscillographic indentation CI-1 (Fig. 5-11). At pH 6 native DNA yields a third wave (wave III); the potential of this third wave corresponds to a wave observed for denatured DNA (*vide infra*). This wave grows slightly to about pH 4 and then increases very sharply with further decrease in pH.

Under conditions of multiple sweep alternating current oscillopolarography native DNA in an ammonium formate medium exhibits an anodic indentation, AI_n, at ca. -0.2 V (Fig. 5-11).[37,73] In order for this anodic indentation to appear the electrode must first be polarized to potentials at which background electrolyte is reduced. This behavior is typical of that observed for guanine (see Chapter 3) and its nucleosides and nucleotides (Section II, C). Accordingly, it has been concluded that the guanine residues in DNA are responsible for this anodic indentation. Paleček[54] found that, if only a single alternating current cycle was applied to the DME, the anodic indentation of DNA observed upon multiple cycles did not appear. However, after more polarization cycles the indentation did appear. This behavior indicated that on the first polarization cycle the guanine residues contained in DNA are essentially inaccessible to the electrode surface, so that no electron transfer could occur. However, after one or more polarization cycles guanine residues do become accessible. Paleček[58] considered that a reorientation of the DNA molecule to make the guanine residues accessible might take place at positive potentials, since Miller[70] had previously suggested that at such potentials DNA unravels. Alternatively, it could also take place at very negative potentials where, in any case, the guanine residues are reduced. It is the reoxidation of the product of this latter reduction that probably gives rise to the anodic indentation.

Further studies on native DNA[74] have shown that in neutral and alkaline media the anodic indentation is not observed on the first oscillographic polarization cycle, in agreement with the previous discussion. However, in acidic solution the anodic indentation can be observed on the first polarization cycle, which indicates that the guanine residues of native DNA are accessible to the electrode. Paleček[58] has suggested that this effect might be due to changes in the conformation of DNA owing to protonation of at least some of the purine and pyrimidine bases. Paleček[54] has investigated native DNA's from various *Bacillus* bacteria, with approximately the same guanine plus cytosine residues, and has shown a correlation between the depth of the anodic indentation and the genetic relations of the bacteria. The differences between the oscillopolarographic behavior were probably associated with differences in the base sequences of the various DNA's.

B. Denatured DNA

Unlike native DNA, denatured DNA shows a small but distinct DC polarographic reduction wave,[61,62] which in a 0.5 M ammonium formate plus 0.1 M sodium phosphate medium, pH 7.0, is of approximately the size expected for a diffusion-controlled wave. The half-wave potential occurs at about -1.4 V versus SCE. Interestingly, it has been found that, as the concentration of ammonium formate is increased, the height of the DC wave increases and the $E_{1/2}$ tends to shift toward more negative potential.[62] Other salts have a similar effect, but divalent cations produced the effects at much lower concentrations. In any particular background electrolyte the current for the polarographic wave increases linearly with concentration.[62]

Denatured DNA shows two AC polarographic waves (waves I and II) (Fig.

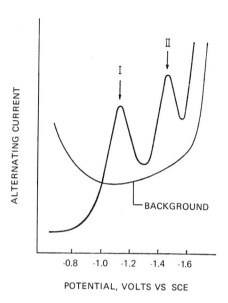

FIG. 5-13. Alternating current polarogram of denatured DNA. Concentration, 500 μg/ml; background solution, 0.5 M ammonium formate with 0.1 M sodium phosphate, pH 7. Alternating voltage, 78 Hz; amplitude, 18 mV.[58] (Reprinted with permission of Academic Press, Inc., New York.)

5-13).[61,62,66–69] Wave I, which occurs at about −1.2 V versus SCE at pH 7, is the same as the nonfaradaic wave I of native DNA and is generally thought to be formed as a result of desorption of adsorbed DNA.[70] Valenta and Nürnberg,[75] however, have proposed that the latter wave I is not a desorption peak but is due to some unspecified alteration of the structure of the adsorbed denatured DNA layer. The second wave (wave II) occurs at about −1.4 V at pH 7; it is at least partly due to a faradaic electron-transfer process[62] and corresponds to the single observed DC polarographic wave. The adsorption studies on denatured DNA[70] reveal that it is adsorbed at potentials more positive than −1.2 V versus SCE and is not adsorbed above about −1.3 V. The area per nucleotide at full surface coverage has been found to be 93 Å^2 per nucleotide for denatured DNA, regardless of the potential of the DME (at potentials where it is adsorbed).

Under conditions of alternating current oscillopolarography denatured DNA gives rise to two cathodic indentations (Fig. 5-14).[58] The indentation designated CI-1 (Fig. 5-14) corresponds to the nonfaradaic AC polarographic wave I of native and denatured DNA. The indentation designated CI-2 is faradaic in nature and corresponds to AC polarographic wave II and the single DC polarographic wave of denatured DNA.

Using differential pulse polarography[74] it is found that denatured DNA shows two waves (Fig. 5-12b). The most positive of these waves (wave I), which

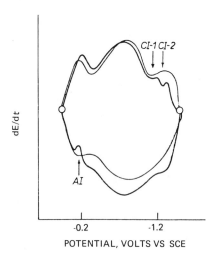

dE/dt

POTENTIAL, VOLTS VS SCE

-0.2   -1.2

FIG. 5-14. Oscillopolarogram of denatured calf thymus DNA at a concentration of 100 μg/ml in 0.5 *M* ammonium formate plus 0.1 *M* sodium phosphate, pH 6.8. Frequency of alternating current, 50 Hz; CI, cathodic indentation; AI, anodic indentation.[58] (Reprinted with permission of Academic Press, Inc., New York.)

occurs at about −1.2 V at pH 7, is at the same potential as wave I observed for native DNA; it is probably due to a nonfaradaic process and corresponds to AC polarographic wave I. Wave III, which occurs at about −1.65 V at pH 7, is specific for denatured DNA and is probably due to the same processes responsible for the DC polarographic wave and AC polarographic wave II.

One can therefore characterize and, under suitable situations, analyze for native and denatured DNA by observation of the DC, AC, differential pulse, and oscillopolarographic behavior of the compound. A summary of this behavior is presented in Table 5-7. The data appear to indicate that the processes responsible for AC polarographic wave I, oscillopolarographic indentation CI-1, and differential pulse polarographic wave I for both denatured and native DNA are the same and are nonfaradaic in nature. Native DNA is characterized by the faradaic differential pulse polarographic wave II, while denatured DNA is characterized by AC polarographic wave II, oscillopolarographic indentation CI-2, and differential pulse polarographic wave III. All of the latter peaks for denatured DNA appear to be essentially faradaic in nature.

The detailed nature of the electrode reactions responsible for the faradaic electron-transfer processes of denatured or indeed native DNA is not known. Until recently it was generally assumed that the most probable electroactive species were the cytosine residues.[61,62] This seemed reasonable since adenine and its nucleosides and nucleotides are reducible only at low pH and are not normally reducible at pH 7,[8] where the vast bulk of electrochemical work on

TABLE 5-7   Summary of the Response of Native and Denatured DNA to Various Electrochemical Techniques

Experiment	Response	Nature electrode process
Native DNA		
DC polarography	None	
AC polarography	Wave I	Desorption peak
Oscillopolarography	Indentation CI-1	Nonfaradaic cathodic capacitive process
	Indentation $AI_n{}^a$	Anodic faradaic process
Differential pulse polarography	Wave I	Probably nonfaradaic capacitive process
	Wave II	Probably faradaic due to breaks in DNA structure
Denatured DNA		
DC polarography	Wave I	Predominantly faradaic
AC polarography	Wave I	Nonfaradaic capacitive process
	Wave II	Predominantly faradaic
Oscillopolarography	Indentation CI-1	Nonfaradaic cathodic capacitive process
	Indentation CI-2	Cathodic faradaic process
	Indentation AI	Anodic faradaic process
Differential pulse polarography	Wave I	Probably nonfaradaic capacitive process
	Wave III	Faradaic reduction step

^{a} Observed only on multiple-cycle experiments.

DNA has been carried out. In order to investigate this point Brabec and Paleček[76] have specifically studied the polarographic reduction of adenine and cytosine moieties in denatured DNA. These investigators examined the polarographic reduction currents yielded by samples of denatured DNA's differing in guanine plus cytosine content in a medium of 1 M ammonium formate buffered with Britton–Robinson buffers. It was conclusively demonstrated that both cytosine and adenine residues are reduced in denatured DNA over the pH range of 6.0–8.7. However, the contribution to the total current by adenine reduction decreases with increasing pH over this pH range. That is, for any given sample of denatured DNA the total faradaic reduction current, measured by differential pulse polarography (i.e., wave III, Table 5-7), decreased with increase in pH because, presumably, the protonation of the adenine moieties to give the electroactive form of these residues[8,31] becomes less and less favorable. At pH 6 in 1 M ammonium formate medium protonation of both cytosine and adenine residues is effectively complete since both residues yield their maximum expected currents based on the adenine residues undergoing a 4*e* reaction and the cytosine residues undergoing a 3*e* reaction.[76] Valenta and Nürnberg[75] have agreed, in essence, with these conclusions, i.e., that the reduction sites in DNA are protonated adenine and cytosine residues. However, they have proposed that both protonated adenine and cytosine residues are reduced in $4e-4H^+$ processes. Although this is probably true of adenine moieties under polarographic conditions, the preponderance of evidence on a number of cytosine derivatives[31,32] clearly suggests that cytosine residues in DNA are reduced either by 3*e* or, in view of the data obtained earlier on CpC,[31] 2*e* processes. That the various polarographic waves of denatured DNA designated as faradaic in Table 5-7 are in fact faradaic in nature seems to have been confirmed by the fact that controlled potential electrolysis at potentials corresponding to this process results in their disappearance.[77]

The faradaic waves of denatured DNA (e.g., wave I observed in DC polarography at $E_{1/2}$ = ca. −1.4 V versus SCE) give a limiting current that is independent of pH in acid solutions and decreases to a very low value in weakly alkaline solutions.[78] The S-shaped pH dependence of the limiting current in DC polarography or peak current in, for example, differential pulse polarography (Fig. 5-15) is shifted to higher pH values on going to more concentrated

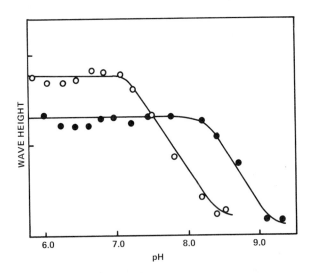

FIG. 5-15. Dependence of the differential pulse polarographic wave height of denatured DNA on pH. The DNA concentration is the same for both curves. Open circles are for DNA in 0.3 *M* ammonium formate with McIlvaine buffers. Closed circles are for DNA in 1.0 *M* ammonium formate with McIlvaine buffers.[78] (Reprinted with permission of Elsevier Publishing Company, Amsterdam.)

supporting electrolyte.[78] As indicated earlier, the position of the pH dependence and the height of the faradaic reduction wave also depend on the nature of the salt. In acid solutions the denatured DNA reduction current is said to be followed by a catalytic hydrogen ion reduction process[75] which, particularly in relatively low pH solutions, limits the range of observable faradaic waves of DNA. For this reason, ammonium formate is commonly employed as a supporting electrolyte since the S-shaped pH dependence of the current is shifted to more alkaline regions with increasing concentration of the latter salt (see Fig. 5-15) so that the interference due to catalytic currents is diminished.

Denatured DNA gives a DC polarographic wave having a conventional shape (Fig. 5-16a),[79] particularly in pH regions on the upper part of the S-shaped pH versus current plots (see Fig. 5-15). However, sometimes at higher pH values it appears that a so-called double wave[79,80] is produced (Fig. 5-16b)[75,79,80] because of the appearance of a peak or maximum at more negative potentials than those of the main wave. With increasing pH the latter peak or maximum shifts to more positive potentials and merges with the main wave, the net effect being formation of a single peak (Fig. 5-16c). At even higher pH (e.g., pH 7.7, Fig. 5-16d) the faradaic wave disappears and only a small nonfaradaic capacity wave is observed. This behavior has been interpreted in several ways. Paleček and Brabec[79,81] have concluded that, under conditions in which denatured DNA yields a faradaic polarographic reduction wave, the reduction of only the adsorbed DNA takes place. Desorption of denatured DNA takes place only at potentials more negative than the $E_{1/2}$ value. It is this desorption of the electroactive species that limits the supply of DNA to the surface of the electrode at potentials negative of $E_{1/2}$ that results in the peak-shaped polarogram shown, for example, in Fig. 5-16c. The appearance of a single, normal-shaped wave of denatured DNA at relatively low pH (Fig. 5-16a) has

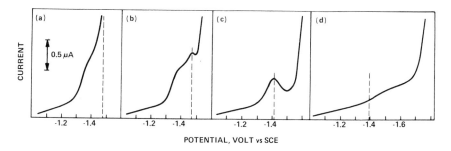

FIG. 5-16. Direct current polarograms of denatured calf thymus DNA (concentration, 200 μg/ml) in 0.1 M NaCl with McIlvaine buffer at pH (a) 5.3, (b) 5.9, (c) 6.3, and (d) 7.7.[79] (Reprinted with permission of Elsevier Publishing Company, Amsterdam.)

been explained[79] by desorption of the electroactive species at potentials substantially more negative than that where the reduction current reaches its limiting value. At these potentials the large increase in background discharge currents obscures any current decrease due to denatured DNA desorption. The appearance of the so called double wave at slightly higher pH values (Fig. 5-16b) has been explained by the fact that the desorption of the denatured DNA and consequent current decrease take place at potentials only slightly more negative than the potential at which the faradaic reduction current reaches its limiting value. The "double wave" is thus composed of a more positive wave due to denatured DNA reduction followed by a peak caused by depression on the curve of background electrolyte discharge. The shape of the DC polarographic curve of denatured DNA therefore depends on the potential difference between the foot of the faradaic DC wave and the desorption of the denatured DNA, which inhibits the current. The difference in the latter potentials can be varied by change in pH, ionic strength, or the concentration of other substances.[79]

Filipski *et al.*[80] explained the appearance of the DC polarographic "double wave" of denatured DNA by the existence of two different faradaic reduction processes. The faradaic reduction of the electrochemically reducible protonated cytosine and adenine residues of denatured DNA was proposed to be responsible for the first (least negative) wave, in agreement with Paleček and Brabec.[79,81] However, the second, more negative wave or peak was proposed to result from a catalytic process, presumably due to hydrogen ion reduction. It is worth mentioning that the possibility that the "double wave" is due to sequential reduction of the adenine and cytosine residues is unlikely since single-stranded homogeneous polynucleotides such as poly(C) and poly(A) also yield the "double wave" (see Sections IV, B and IV, C).[79,82,83]

Valenta and Nürnberg[75] have provided a third suggestion regarding the DC polarographic behavior of denatured DNA. According to these workers, formation of a peak-shaped DC polarogram of denatured DNA such as that shown in Fig. 5-16c is due to strong adsorption of the reduction products on the DME surface, hence blocking the electrode surface and inhibiting the electrode process. The net result, therefore, is that as soon as appreciable quantities of denatured DNA are reduced, the electrode surface becomes blocked and the current consequently decreases, giving a peak rather than a normal-shaped DC polarographic wave.

Valenta and Nürnberg[75] have also proposed that the effect of ammonium ions (i.e., ammonium formate) in shifting the S-shaped pH dependence of the polarographic limiting current to more negative potentials (Fig. 5-15) is due to the presence of NH_4^+ ions in the double layer which behave as proton donors and enhance formation of protonated DNA near the electrode surface. In other words, in the vicinity of the electrode surface NH_4^+ ions behave as proton donors and protonate the adenine and cytosine residues of DNA to give their electroactive forms to an extent not expected on the basis of the bulk pH of the solution. Increasing NH_4^+ ion concentration hence allows the DNA faradaic polarographic process to be observed at increasingly high bulk solution pH. The effect of other cations on the pH dependence of the polarographic reduction current and $E_{1/2}$ of denatured DNA has been attributed to the influence of the various salts on the protonation rate of DNA.[75] A further explanation of the effect of cations on the height and shape of DC polarographic waves of DNA and other polynucleotides has been suggested[58,62,94] in terms of the adsorbability of the polynucleotide being affected by the cation screening the negative charges of the polynucleotide phosphate group, which otherwise are electrostatically repelled from the negatively charged electrode. The greater effect of NH_4^+ compared to that of Na^+ and K^+ was proposed to be connected either with the higher screening effectiveness of NH_4^+ and/or with its proton-donor ability.

A number of other interpretations of the various electrochemical waves, peaks, and indentations of native and denatured DNA initially advanced by Paleček and co-workers[58,59] have been challenged. For example, Berg and co-workers[84] have indicated that the height of the DC polarographic wave I of denatured DNA increases with decreasing pH while the height of the corresponding AC polarographic wave II decreases. It was concluded from this that the faradaic process associated with the DC polarographic wave is not connected with the process responsible for the AC polarographic wave. In other words, Berg and co-workers[84] assert that AC polarographic wave II is in fact a purely capacitive wave. Brabec and Paleček[85] have rejected this argument because when polarograms were taken in a medium containing ammonium ions, which Berg *et al.*[84] did not use, then with decreasing pH both the AC wave II and DC wave I increased. However, AC polarographic wave I decreased with decreasing pH. Since AC wave I is undoubtedly capacitive in nature, it was therefore argued[85] that AC wave II cannot be totally capacitive. Berg *et al.*[84] also found that the phase angle dependence of the height of AC polarographic wave II of denatured DNA was similar to that of the AC polarographic wave of cyclohexanol, the latter being a true capacitive or tensammetric process. Brabec and Paleček,[85] however, have contended that to deduce from this that the wave II of denatured DNA is totally capacitive is erroneous since an AC wave resulting from simultaneous adsorption–desorption (i.e., capacitive processes) and faradaic processes would probably behave with respect to the phase angle to some extent like the purely capacitive processes responsible for the cyclohexanol wave.

The process responsible for the anodic indentation AI on oscillographic polarography of denatured DNA is apparently oxidation of the product of reduction of guanine residues. This reduction reaction takes place at potentials very close to background discharge. Unlike native DNA, denatured DNA gives the indentation AI even on single-cycle oscillopolarography, indicating the immediate accessibility of the guanine residues in denatured DNA.

C. Single-Strand Breaks in DNA

It will be recalled that at high sensitivity native DNA gives rise to two differential pulse polarographic waves, waves I and II. In the discussion of native DNA the origin of wave II was not mentioned. It has been found that wave II is present to some extent in all sources of native DNA and that the height of the wave could be increased markedly under the influence of γ rays and DNase I, which are known to introduce single-strand breaks into the DNA molecule.[86] It appears likely, therefore, that wave II is a faradaic wave associated with reduction of one of the bases in DNA, probably cytosine, in the vicinity of broken phosphate—diester bonds, which appear to allow the base to become accessible to the electrode surface.

D. Electrochemical Studies on the Denaturation of DNA

Normally, the thermal denaturation of DNA is followed by an increase in absorbance at 260 nm. The same data can be obtained by warming the native DNA sample at some predetermined temperature for ca. 10 min, then cooling very rapidly and running a differential pulse polarogram. The height of pulse polarographic wave III is used to estimate the concentration of the denatured DNA produced.[41,47,52,54,61,87,88] Interestingly, it has been found,[61,87−89] using oscillopolarography in ammonium formate plus sodium phosphate buffer, pH 7, that at temperatures 20 deg. below that at which UV absorption spectrophotometry indicates any denaturation very characteristic changes do occur. Thus, at such temperatures oscillographic polarography reveals that CI-2 indentation is present and that its depth increases with temperature. Such characteristic evidence for some denaturation below the so-called *melting temperature* has also been observed by DC polarography,[62] AC polarography,[62,69] and differential pulse polarography,[90] the responses obtained by each technique being compatible with the expected behavior of the appropriate denatured DNA. By use of the presently most sensitive polarographic technique, differential pulse polarography, it is possible to determine both native and denatured DNA during the denaturation process, native DNA via wave II and denatured DNA via wave III.[60]

Paleček[61] has explained the polarographic behavior of DNA at *premelting* temperatures as a release of reducible groups that are not accessible to the electrode at lower temperatures. This release occurs in the bulk of solution, not at the electrode due to some adsorption type of effects. It was further thought[58,61,90] that the conformational changes required to make the reducible bases (cytosine and adenine) accessible to the electrode at premelting temperatures occurred at the most thermally labile parts of the DNA molecule. The most probable regions of this type are at broken phosphate—diester bonds, where the bases loop out of the double-helical structure, or in adenine- plus thymine-rich regions, which are relatively weakly bonded regions in DNA. This is verified by the fact that the behavior of native DNA at elevated temperatures but below the melting temperature (UV) is very similar to that of native DNA irradiated with small doses of γ rays or treated with DNase I, i.e., where single-stranded breaks are introduced into the DNA, probably at phosphate— diester bonds. Thus, it appears that native DNA's normally contain single-stranded breaks in their double-helical structure. Upon raising the temperature, conformational changes could occur at these breaks, making polarographically reducible groups more accessible to the electrode yet not disrupting the overall structure sufficiently to affect the absorption of ultraviolet radiation.

Filipski and co-workers[77] have agreed that the fact that the faradaic waves for DNA can be observed by various polarographic techniques at temperatures lower than the beginning of the spectrophotometrically observed melting temperature may indicate some early stages of unwinding. However, they have suggested, on the basis of their experiments, that although premelting may take place in the bulk solution at such spectrophotometrically premelting temperatures, adsorption of DNA on the mercury electrode might be involved in the observed polarographic behavior. In other words, the effect of the charged mercury surface on the structure of the DNA double helix could be crucial, and the observed polarographic melting curves could exclusively reflect changes in the structure of adsorbed native DNA. In view of this, Filipski and co-workers[77] have proposed that the polarographic behavior of native DNA at premelting temperatures might simply reflect a surface-induced unwinding of the double helix giving rise to accessible and reducible purine and pyrimidine bases. Accordingly, they suggest that polarographic methods should not be considered adequate to prove the existence of a premelting state of DNA molecules in the bulk of solution.

Some recent reports, however, appear to verify Paleček's interpretations of the polarographic behavior of DNA at premelting temperatures. Thus, it was found[91] that not only differential pulse polarography indicates that conformational changes occur making polarographically reducible groups accessible, but circular dichroism confirms the same effect.[91,92] This would indicate that the conformational changes do occur in the bulk of the solution.

E. Adsorption of Native and Denatured DNA

In the preceding discussion the adsorption/desorption behavior of native and denatured DNA has been referred to repeatedly. Some controversy clearly exists regarding the interpretation of various capacitive waves observed for various DNA species. Brabec and Paleček[93] have reported some further observations and interpretations of the observed adsorption behavior of native and denatured

DNA. Their results suggest that all three constituents of nucleic acids – bases, sugar, and phosphate groups – are involved in the adsorption. However, the extent of their participation depends on the ionic strength of the solution, the electrode potential, and the conformation of the polynucleotide in solution. In relatively high ionic strength solutions (Britton–Robinson buffer at pH 8 containing 0.3–1 M KCl) native DNA gave AC polarograms similar to those reported previously, i.e., a single capacitive wave I at about –1.2 V versus SCE[58] preceded by a depression of the base current (below background) at less negative potentials (see Fig. 5-10). Alternating current polarography of denatured DNA under similar conditions at pH 8 yields waves I and III. The latter wave should not be confused with faradaic AC wave II observed at lower pH. At pH 8 in the medium mentioned above, denatured DNA is not reducible; i.e., wave III is capacitive in nature. Again, the base current was depressed at potentials less negative than wave I. Native DNA does not normally give AC polarographic wave III at pH 8 and room temperature, although at elevated temperatures (but lower than the temperature at which denaturation appears to begin by UV spectral studies) a wave appears in the AC polarogram of native DNA which increases with increasing temperature.[93] This wave, which will be referred to subsequently as wave III', does not occur at the same potential as wave III of denatured DNA but is slightly more positive than wave III. For example, native DNA at pH 8 and 77°C exhibits waves I and III' (Fig. 5-17c). Denatured DNA under identical conditions shows waves I and III (Fig. 5-17d). Significantly, lowering the molecular weight of DNA by sonication caused native DNA to exhibit wave III' even at room temperature (Fig. 5-17a). Sonicated denatured DNA, however, exhibited only waves I and III under the latter conditions. Also, damage of DNA by relatively low doses of X radiation or digestion with DNase I caused an increase of wave III', while wave I remained practically unchanged. Finally, decrease in the ionic strength of the supporting electrolyte from 1.0 to 0.3 M KCl caused a decrease in the base current depression around the potential of the electrocapillary maximum.

At lower ionic strength (e.g., ≤0.3 M KCl with Britton–Robinson buffer), native DNA yields a new capacitive wave (wave 0) around –0.5 V (Fig. 5-18). With decreasing KCl concentrations the height of wave 0 increased and the base current depression of DNA decreased at potentials more negative than wave 0. On the other hand, at more positive potentials the depression of base current increased.

From this dependence on ionic strength it was concluded[93] that double-helical DNA is adsorbed mostly on a positively charged electrode at low ionic strength (ca. 0.1 M in Na$^+$, K$^+$, Cs$^+$). Addition of neutral salts to native DNA solutions in 0.1 M sodium acetate, pH 6.7, led to a suppression of wave 0. The salts of divalent cations were about 20 times more effective than the salts of monovalent cations. The presence of single-stranded breaks in DNA (made by

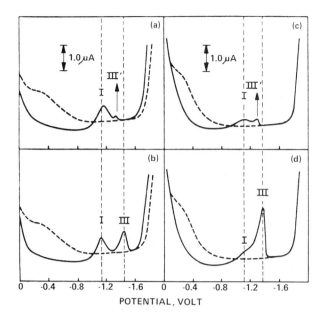

FIG. 5-17. Alternating current polarograms of sonicated and nonsonicated DNA. Sonicated DNA (a, b) in 0.3 M ammonium formate with McIlvaine buffer, pH 7.9, at 23°C. Nonsonicated DNA (c, d) in 0.5 M KCl with Britton–Robinson buffer, pH 8.0, at 77°C. The DNA concentration, 300 μg/ml. (a, c) Double-helical DNA; (b, d) denatured DNA. Dashed line is background electrolyte. Potential measured versus SCE (a, b) or mercury pool electrode (c, d).[93] (Reprinted by permission of John Wiley and Sons, Inc., New York.)

DNase I or ionizing radiation) did not cause measurable changes in the height of wave 0.

Since at room temperature and an ionic strength greater than ca. 0.1 single-stranded polynucleotides [i.e., denatured DNA, poly(C), poly(A)] contain a large number of helical regions, the single-stranded regions should be adsorbed through purine and pyrimidine bases, while the helical regions should be adsorbed, less firmly, through sugars and perhaps a few unpaired bases (loop-out bases, bases in the region of bends in hairpinlike structures, etc.). Brabec and Paleček[93] have proposed that the more positive wave I of single-stranded polynucleotides corresponds to desorption of the helical regions that are adsorbed less firmly through sugar residues and exposed bases. The more negative wave III was accordingly proposed to be due to desorption of the more firmly adsorbed bases from nonhelical regions. At lower ionic strengths, it was suggested that electrostatic forces stimulate adsorption of sugar residues of native DNA at a positively charged mercury surface. At the negatively charged electrode adsorption of sugars would be difficult owing to the repulsion between

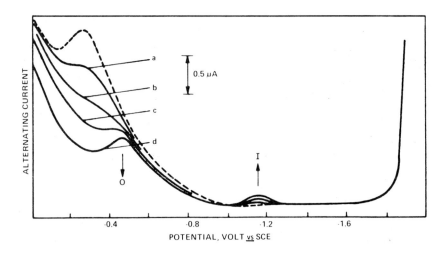

FIG. 5-18.   Alternating current polarograms of native DNA in a medium of 0.1 M KCl, pH 6, at various concentrations of DNA: (a) 100 μg/ml, (b) 200 μg/ml, (c) 300 μg/ml, (d) 500 μg/ml. Dashed line is background electrolyte.[93] (Reprinted by permission of John Wiley and Sons, Inc., New York.)

negatively charged phosphate groups. The capacitive wave 0 (Fig. 5-18) was hence proposed to be due to marked changes in the surface concentration of adsorbed DNA as the sign of the electrode charge changes. That is, the deep depression shown at ca. −0.2 V (i.e., potentials positive of the ecm, where the electrode is positively charged) in Fig. 5-18 indicates extensive adsorption of native DNA principally through its sugar residues. On passing to potentials negative of the ecm, electrostatic repulsion of phosphate residues caused considerable desorption of the DNA. The net result, at appropriate DNA concentrations, was wave 0. Wave III′ of native DNA was thought to be due to desorption of native DNA adsorbed through open or intermediate regions of the double helix, i.e., in the regions of single-strand breaks, regions of the molecule ends, and in (adenine–thymine)-rich regions. The more positive potential of wave III′ of native DNA compared to wave III of denatured DNA was suggested to be the result of easier desorption of the former due to the fact that the intermediate regions of native DNA are probably shorter than the single-stranded regions in denatured DNA.

IV. ELECTROCHEMISTRY OF POLYRIBONUCLEOTIDES

Paleček[58] found that polycytidylic acid [poly(C)], polyadenylic acid [poly(A)], and polyinosinic acid [poly(I)] are electrochemically reducible and that they behave in many respects like denatured DNA. Polyuridylic acid [poly(U)], polythymidylic acid [poly(rT)], polyguanylic acid [poly(G)], and [poly(UG)] are not electrochemically reducible. However, as has been observed for all guanine residues, poly(G), poly(UG), etc., show an oscillopolarographic anodic indentation once the potential has been swept to the cathodic background discharge region where the guanine residues are reducible to an oxidizable species. To facilitate further discussion, the electrochemical behavior of each polynucleotide will be dealt with separately.

A. Polycytidylic Acid

In an ammonium formate with sodium phosphate medium of pH 7, poly(C) gives a reasonably well-defined DC polarographic reduction wave[58,94] at about -1.4 V versus SCE. The height and shape of the wave are dependent, however, on the concentration of ammonium formate. For example, at ammonium formate concentrations below 1 M, the DC polarographic wave of poly(C) has the appearance of a peak or maximum (Fig. 5-19A) but, at the 1 M ammonium formate concentration level or above, a well-defined normal wave is obtained. With increasing concentration of ammonium formate up to about 1 M, the wave height for poly(C) increases. However, poly(C) is polarographically reducible even in the absence of ammonium formate.

The dependence of the polarographic wave height on pH for poly(C) in the region where it possesses a more or less single-stranded structure (pH > 6) has been found to be a characteristic S-shaped polarographic dissociation curve (Fig.

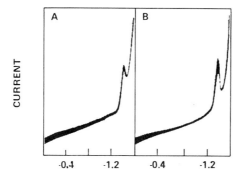

POTENTIAL, VOLTS VS SCE

FIG. 5-19. Direct current polarogram of 0.6 mM poly(C) (A) and poly(A) (B) in 0.5 M ammonium formate with 0.1 M sodium phosphate, pH 6.9.[94] (Reprinted with permission of Elsevier Publishing Company, Amsterdam.)

5-20). This, of course, suggests that poly(C) is reduced only in its protonated state in accord with its parent base and nucleoside.[30-32] Ammonium formate has the effect of shifting the S-shaped curve to higher pH values (Fig. 5-20). Similar, but less pronounced effects are brought about by other salts such as NH_4Cl, NaCl, and KCl. The pH dependence of the half-wave potential of poly(C) is also shown in Fig. 5-20, where it is seen that $E_{1/2}$ shifts to more negative potential with increasing pH. The wave height of poly(C) is a nonlinear function of concentration, the wave height versus concentration curves having the apperance of a Langmuir isotherm,[58,94] which indicates the involvement of adsorption processes (*vide infra*).

Controlled potential electrolysis of poly(C) at potentials corresponding to its DC polarographic reduction wave results in disappearance of the characteristic UV spectrum of the compound, confirming that a faradaic reduction of the compound takes place.

At pH values below about 6, poly(C) can form a double-helical structure consisting of two strands of poly(C) stabilized by protons.[95,96] In this double-helical form of poly(C), which will be referred to as poly($C \cdot C^+$), one proton is bound between two cytosine residues and forms a third hydrogen bond, as is shown in Fig. 5-21.[97] Paleček[98] has reported that at a pH (e.g., pH 5.0) where poly(C) is in its double-helical form an abrupt decrease in the height of the DC polarographic wave is noted compared to the value at which the single-stranded form predominates. In addition, in the presence of double-stranded poly($C \cdot C^+$) the AC polarographic current at positive potentials (see

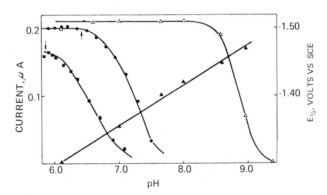

FIG. 5-20.   Dependence of $E_{1/2}$ and wave height on pH for 0.1 mM polycytidylic acid in (■) 0.3 M NaCl with Britton–Robinson buffer (current); (●) 0.3 M ammonium formate with Britton–Robinson buffer (current); (△) 1.0 M ammonium formate with Britton–Robinson buffer (current); (▲) $E_{1/2}$. To the right of the arrows the waves had the appearance of a maximum.[94] (Reprinted by permission of Elsevier Publishing Company, Amsterdam.)

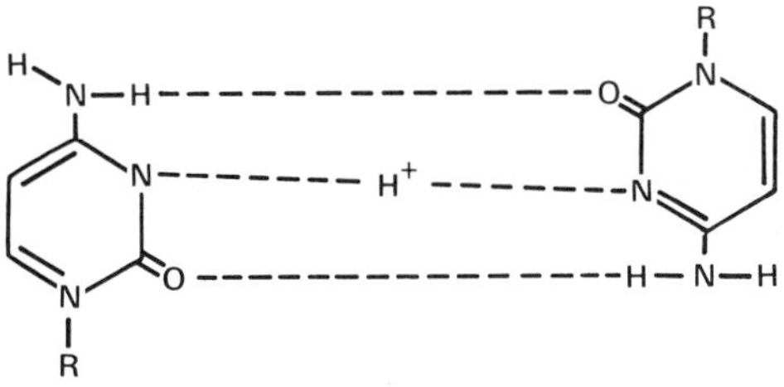

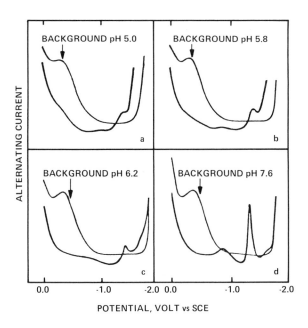

FIG. 5-21. Scheme of hydrogen bonds between bases in double-helical protonated poly(C),[97] i.e., poly(C·C$^+$).

Fig. 5-22a) is lowered to a smaller extent than with the single-stranded poly(C). This effect was interpreted as a decrease in adsorbability of the protonated polynucleotide at positive potentials due to partial neutralization of the negative charges of phosphate groups by the positive charges of the cytosine residues.[98] At negative potentials there is a marked difference between the curve of poly(C) at a pH where it is not reducible (i.e., Fig. 5-22d) and that obtained at lower pH values where it is reducible in the single-stranded form (Fig. 5-22b and c). That

FIG. 5-22. Alternating current polarograms of 1 mM poly(C) in 0.5 M NaCl plus citrate buffer.[98] (Reprinted with permission of Academia Publishing House, Czechoslovakia.)

is, a large capacitive AC wave is observed at pH 7.6 for nonreducible poly(C) at about the same potential where it is reduced at lower pH, i.e., pH 5.8 and 6.2. Clearly, a capacitive AC wave is also observed for poly(C·C$^+$) (Fig. 5-22a) at around the same potential (ca. −1.4 V) at which poly(C) is reduced (Fig. 5-22b and c). Accordingly, unlike DC polarography, AC polarography cannot really distinguish the transition of poly(C) to poly(C·C$^+$) based on the observed behavior around −1.4 V.

On the other hand, differential pulse polarography of poly(C) shows an abrupt decrease in current corresponding to the poly(C) → poly(C·C$^+$) transition.[98] However, the differential pulse polarography is a little more complex than might at first be imagined. Thus, a solution of poly(C) in 0.5 M NaCl plus citrate buffer at pH 6.6 gives a single differential pulse polarographic peak (Fig. 5-23A). With decreasing pH down to 6.0 a linearly increasing inflection occurs on the rising portion of this peak as indicated by an arrow in Fig. 5-23B and C. With a further decrease in pH both the peak and the inflection decrease in height, particularly from pH 5.5 to 5.0. Near pH 5.0, where poly(C·C$^+$) exists, the inflection alone is observed. A further decrease in pH causes a change of the inflection into an increasing peak whose growth ceases at pH 4. The

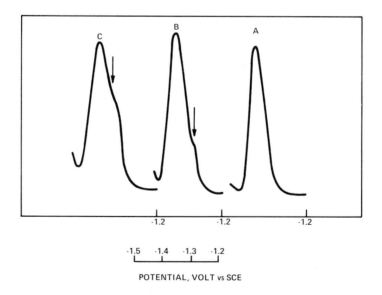

POTENTIAL, VOLT vs SCE

FIG. 5-23.   Differential pulse polarograms of 5 x 10^{-5} M poly(C) in 0.5 M NaCl plus citrate buffer plus 10^{-3} M EDTA. A, pH 6.6; B, pH 6.4; C, pH 6.0.[98] Each polarogram begins at −1.20 V with increasingly negative potential from right to left. (Reprinted with permission of Academia Publishing House, Czechoslovakia.)

investigations of Paleček[98] suggest that the inflection is due to reduction of cytosine residues that are protonated in the bulk of the solution, i.e., reduction of poly(C·C$^+$). The initial peak (Fig. 5-23A) corresponds to reduction of cytosine residues in poly(C). The increase in the differential pulse polarographic current as the pH changes from 5.0 to 4.0 has been suggested[98] to be due to a structural change of poly(C·C$^+$), allowing better accessibility of the cytosine residues to the electrode surface.

It will be recalled (Section III, B) that denatured DNA under certain pH conditions gives a normal-shaped DC polarographic reduction wave, while at somewhat higher pH a "double wave" or a peak is observed. Poly(C) exhibits very similar behavior, which has led Brabec and Paleček[99] to conclude that the polarographic reduction of poly(C) takes place only in the adsorbed state. The "double wave" and DC polarographic peak are apparently identical in nature to those proposed in the case of denatured DNA, i.e., the peculiar shape of the waves is due to inhibition of the reduction process owing to polynucleotide desorption from the negatively charged electrode.

The effect of the cation of the supporting electrolyte on the height and shape of the DC polarographic wave and on the shift of the S-shaped dependence of current on pH is thought[94] to be related to the cation influencing the adsorbability of poly(C). The cation has been proposed to influence the polynucleotide adsorbability by its screening of the negative charges associated with the phosphate groups and through its presence in the electrode double layer. The greater effect of NH_4^+ compared to K^+ and Na^+ is thought to be connected with the higher screening efficiency of NH_4^+ and/or with its proton-donor ability.[94,100]

Differential pulse polarography of poly(C) in 0.3 M ammonium formate with 0.1 M sodium phosphate, pH 7.0, gives rise to two waves (Fig. 5-24) compared to only one DC wave.[94] The more negative differential pulse polarographic wave is probably capacitive.

B. Polyadenylic Acid

In ammonium formate with sodium phosphate medium, pH 7, poly(A) gives a relatively well-defined DC polarographic wave at about -1.4 V versus SCE (Fig. 5-19B).[58,94] In many respects this wave is similar to that produced by poly(C). For example, the height and shape of the DC polarographic wave of poly(A) depends on the concentration of ammonium formate. The wave height for poly(A) increases up to a concentration of 0.6 M in ammonium formate. Poly(A), however, is polarographically reducible in the absence of ammonium formate. The dependence of the DC polarographic wave height on pH for poly(A) in the region where it possesses a more or less single-stranded structure (pH $>$ 6) is a characteristic S-shaped polarographic dissociation curve (Fig. 5-25).

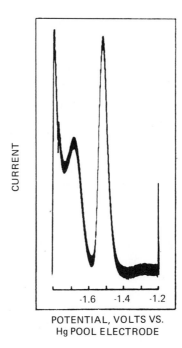

POTENTIAL, VOLTS VS.
Hg POOL ELECTRODE

FIG. 5-24.   Differential pulse polarogram of $5 \times 10^{-5}\,M$ poly(C) in $0.3\,M$ ammonium formate plus $0.1\,M$ sodium phosphate, pH 7.5[58,94] (Reprinted with permission of Elsevier Publishing Company, Amsterdam.)

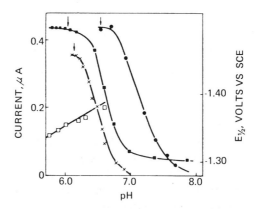

FIG. 5-25.   Dependence of $E_{1/2}$ and wave height on pH for 0.1 mM polyadenylic acid (x) in Britton–Robinson buffer (current); (■) in $0.3\,M$ NaCl with Britton–Robinson buffer (current); (□) $E_{1/2}$; (●) in $0.3\,M$ Britton–Robinson buffer (current). To the right of the arrows the waves had the appearance of a maximum.[94] (Reprinted with permission of Elsevier Publishing Company, Amsterdam.)

Accordingly, it is probable that the protonated form of poly(A) is the reducible form of the molecule. Ammonium formate and other salts such as NH_4Cl, $NaCl$, and KCl have the effect of shifting the S-shaped curve to higher pH values (see Fig. 5-25). The pH dependence of the half-wave potential of poly(A) is shown in Fig. 5-25, where it is clear that $E_{1/2}$ shifts to more negative potentials with increasing pH. Similar to poly(C), poly(A) gives a DC polarographic wave the height of which is a nonlinear function of concentration, indicating the involvement of adsorption processes (see below).

Controlled potential electrolysis of poly(A) at potentials corresponding to its DC polarographic wave results in the loss of the characteristic UV absorption spectrum, hence confirming that the compound is reduced. Poly(A) gives a differential pulse polarographic wave, the pH dependence of which is similar to that of the DC wave[94] shown in Fig. 5-25. Alternating current polarography of poly(A) reveals that it causes a pronounced base current depression at potentials positive of where the reduction wave is observed, indicating that the compound is strongly adsorbed.

It is now quite well-established that poly(A) has a helical, single-stranded structure with partially stacked bases at neutral pH and a helical, double-stranded structure with parallel chains and stacked protonated bases at acidic pH.[96,101] There is no internucleotide hydrogen bonding in the neutral form of poly(A).[102] In the double-helical protonated form of poly(A), which will be referred to as poly(A·A⁺), the protons do not participate in the formation of hydrogen bonds but stabilize the structure by their charge. The amino group, N-7 nitrogen, and phosphate oxygen atom are, however, involved in hydrogen bonding. A schematic diagram of the hydrogen bonds between bases in double-helical poly(A·A⁺) is shown in Fig. 5-26.

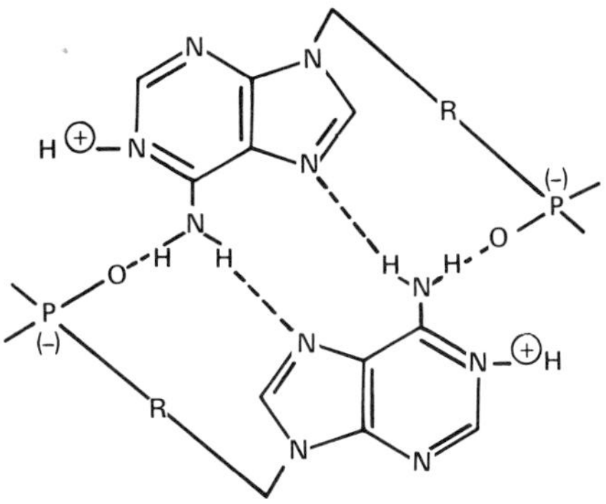

FIG. 5-26. Scheme of hydrogen bonds between nucleotides in double-helical protonated polyadenylic acid, poly(A·A⁺).[96]

The conversion of single-stranded poly(A) into its double-helical protonated form, poly(A·A$^+$), has been studied by Paleček[98] using DC and differential pulse polarography. Both techniques showed a decrease in the polarographic current as the pH decreased due to formation of poly(A·A$^+$). The pH range in which the drop in current was observed depends on the ionic strength of the solution, which influences the poly(A) $\rightleftharpoons$ poly(A·A$^+$) transition. Thus, in 0.1 M NaCl plus citrate buffer the decrease of the polarographic current occurs at about pH 5.5. However, in the presence of 0.5 M NaCl plus the same citrate buffer the decrease of polarographic current did not occur until ca. pH 5.0. The decrease of the polarographic current was only ca. 25–30% on passing from a medium where single-stranded poly(A) exists (pH > 6) to a medium where double-helical poly(A·A$^+$) exists. This decrease in current is far less than was observed for the poly(C) → poly(C·C$^+$) transition discussed earlier.[98] Also, in contrast to the observed behavior of poly(C), the differential pulse polarograms of poly(A) did not exhibit any new peaks or inflections with change in pH.[98]

Janik and co-workers,[103] however, subsequently reported that DC polarography of poly(A) at pH values where it exists in its single-stranded form gives a normal-shaped wave, for example, at pH 5.64 at an ionic strength of 0.65 (Fig. 5-27a). With increasing pH, however, the wave becomes distorted, first forming a "double wave" of the type discussed with respect to poly(C) and denatured DNA (see above) and, at higher pH, forming a peak or maximum (see Fig. 5-27). With increasing ionic strength the depression of the current at negative potentials (with consequent peak formation) decreases and eventually disappears.[103] Formation of the "double wave" and peak under DC polarographic conditions has been explained[103,104] in terms of desorption and repulsion of poly(A) from the electrode at potentials slightly more negative than the limiting portion of the DC wave with a consequent current decrease. Poly(A) is therefore probably reduced only in its adsorbed state.[105] The repulsive effect is due to the polyanionic character of poly(A). The suppression of current giving rise to the observed effects can be enhanced by increasing negative potential of the electrode and by increasing the exposure of the negative charges of the phosphate groups, e.g., by increasing pH and temperature and by decreasing ionic strength and buffer capacity. The current suppression may be at least partially eliminated by reversing the latter conditions. Polyamines, which appear to shield the negative phosphate groups from the negatively charged electrode, are very effective in eliminating the current suppression (see, for example, Fig. 5-27b′ and c′).

Although Paleček[98] has reported that the poly(A) → poly(A·A$^+$) transition can be followed by DC and pulse polarography (see earlier discussion) in a rather simple manner, Janik *et al.*[103] have noted far more dramatic decreases in current as the pH is decreased to the point where the latter transition occurs. Further decrease in pH then gave more large changes in the observed

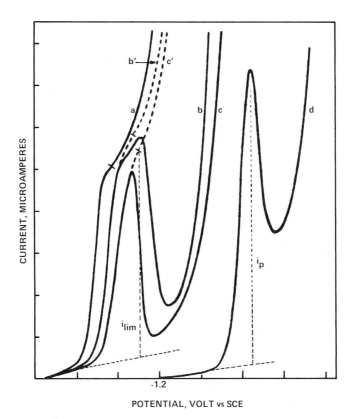

FIG. 5-27.   Direct current (a–c) and differential pulse (d) polarograms of 0.2 mM poly(A) at an ionic strength of 0.65 M at pH 5.64 (a, d), 6.11 (b), and 6.39 (c). The dashed lines show the curve shape in the presence of 14 μM (b') and 33 μM (c') spermidine.[103] Each curve starts at −1.2 V; the potential scale is 100 mV per division. (Reprinted with permission of Elsevier Publishing Company, Amsterdam.)

polarographic reduction current. The latter workers have concluded that the changes they observed indicate that poly(A·A⁺) can exist in three different forms at acidic pH values. Each of these forms appeared to have a different conformation; hence, the accessibility of the adenine moieties to the electrode surface was changed, causing the observed polarographic current to alter. Brabec and Paleček,[105,106] however, have disputed the explanation that the decrease in polarographic current observed, when the poly(A) → poly(A·A⁺) transition occurs, is due to hiding of the adenine moieties inside the double helix of the poly(A·A⁺) molecule. They have studied the polarographic behavior of poly(A) having different molecular weights and have shown that the decrease of current depends on the length (MW) of the poly(A) molecule. Shorter molecules (i.e., having a MW considerably less than 10^6, measured by Janik *et al.*[103]) show almost no decrease when the poly(A) → poly(A·A$^+$) transition occurs. This result seems to confirm that the formation of poly(A·A$^+$) does not cause the adenine bases to become inaccessible. The lower currents observed on formation of poly(A·A$^+$) have been proposed to be due to a decrease in the diffusion coefficient of the molecule and perhaps by shielding of a limited number of reduction sites as a result of molecular aggregation. Longer poly(A·A$^+$) molecules appear to aggregate more extensively than shorter molecules.[105,106] Vetterl and Guschlbauer[107] have also confirmed that poly(A·A$^+$) aggregates in solution, and such aggregation is facilitated by increasing ionic strength and temperature, which is manifested by a decrease in the height of the polarographic wave.

Reynaud and Leng[108] have reported that poly(A·A$^+$) gives two DC polarographic waves in acidic solution. The first, more positive wave was thought to be due to reduction of the adenine residues, while the second, larger, more negative wave was due to catalytic reduction of protons. Furthermore, it was found that the potential of the first reduction wave depended very strongly on the method used to prepare the double-stranded poly(A·A$^+$) from a neutral solution of the single-stranded poly(A). The variations observed in the potentials of the first reduction wave of poly(A) were thought to be related to differences in the degree of overlap of the two strands.

C. Polyuridylic Acid

Poly(U) is not polarographically reducible.[58,94] Janik and Sommer,[109] however, have examined in some detail the adsorption properties of poly(U) using AC polarographic techniques. In a McIlvaine buffer of pH 5.45 poly(U) is adsorbed at the DME between about 0 and −1.1 V versus SCE. At about −1.3 V in the same medium a small tensammetric wave is observed. The area A occupied by the monomeric units of adsorbed poly(U) molecules of different chain lengths was determined. The values of A were around 77–87 Å^2 per nucleotide, although an upward trend of A with decreasing chain length was noted. This variation of A with chain length was thought to suggest that the orientation of poly(U) molecules in the electrode double layer depends on the chain length and that at lower chain lengths the area adsorbed per adsorbed monomer is larger. It has been suggested,[110,111] in general, that adsorption of a linear, flexible macromolecule forming a random coil (which is a reasonable description of poly(U) above 15°C[112]) brings about a change in conformational topology. The molecule becomes a structure of two-dimensional adsorbed segments alternating with three-dimensional standing loops. It has been proposed[110] that the fraction of segments in contact with the surface tends to increase as the chain length decreases. A similar model has been assumed to occur with poly(U) adsorbed at the DME surface[109]; i.e., with increasing chain length, looping of longer poly(U) chains out from the surface occurs.

D. Other Polynucleotides

Poly(I) yields two DC polarographic waves in acidic solution, but at pH 5 it appears to be electrochemically inactive.[58] Poly(U), poly(U-G), and poly(rT) are not electrochemically reducible, although at relatively high concentrations they do give rise to tensammetric waves under AC polarographic conditions.[58]

The polarographic reducibility of a number of double-helical synthetic polynucleotides has been studied by differential pulse polarography.[113] Thermally stable polynucleotides such as poly(rG-rC) and poly(dG-dC) are not reducible at premelting (double-helical) temperatures. However, less stable poly(rI-rC) and poly[d(A-T)·d(A-T)] are polarographically reducible at about −1.45 V versus a mercury pool electrode at pH 6.6–7.0 even at room temperature. The polarographic wave height of the latter molecules increases linearly with increasing temperature. Such behavior suggests that, in the thermally less stable polynucleotides, conformational changes occur as a result of temperature changes in the premelting region.

Very little work has been carried out on the electrochemical behavior of RNA, although some preliminary data of Paleček[58] suggest that some type of activity can be expected.

V. CONCLUSIONS

There are several important conclusions that can be drawn from the published work on nucleic acids and polynucleotides.

1. As straightforward analytical methods, many of the electrochemical techniques have very great potential for the study of nucleic acids. In this respect differential pulse polarography should prove to be particularly useful since it has an analytical sensitivity for electroactive materials several orders of magnitude greater than more conventional (e.g., spectrophotometric) methods. It clearly has the potential of detecting changes in nucleic acid systems which are quite inaccessible by other techniques.

2. Randomly coiled polynucleotides containing polarographically reducible bases yield faradaic currents.

3. Polynucleotides having the double-helical structure such as poly(C·C$^+$) and native DNA generally have the majority of the reduction sites hidden within the helix, so that to a large extent they are not accessible for the electron-transfer process at the electrode. In the case of poly(A·A$^+$), however, aggregation

appears to be more important in decreasing the extent of electroreducibility of this compound.

4. In native DNA having the double-helix structure a small number of the reducible bases are accessible to the electrode and can be detected by differential pulse polarography. The pulse polarographic wave for these bases occurs at a different potential than the wave of the parent random-coiled polynucleotide. These electrochemically accessible bases are probably located in the labile regions of the double helix, i.e., at broken phosphate–diester bonds, at loops of the double-helical structures, etc. The extent or length of these labile regions seems to be increased with increasing temperature. Irradiation with γ rays or treatment with DNase I can increase the number of the labile regions.

It is unfortunate in many respects that so little work has been done on the details of the electrode mechanisms. Although it is likely that the primary electrode reactions of nucleosides, nucleotides, polynucleotides, and nucleic acids might be superficially the same as for the parent purines and pyrimidines, it has yet to be positively established that this is so. There is also a vast amount of research to be done on the electrochemical oxidation of these types of compounds.

From a biological viewpoint it is likely that some of the most valuable electrochemical studies on polynucleotides and nucleic acids are those concerned with the adsorption and related interfacial behavior of these molecules. It is becoming increasingly apparent that a large number of processes that occur in biological systems are related directly or indirectly to the behavior of various molecules at charged interfaces such as the surfaces of cells, bones, other cellular components, endothelial lining of blood vessels, and protheses. It is also clear that our knowledge of such surface processes in biological systems is extremely limited.[114]

It is generally thought that RNA is adsorbed at the surface of many different cells, possibly with their phosphate groups directed away from the electrokinetic surface.[115] Schell[116] has demonstrated, for example, that living mammalian cells recognize double- and triple-stranded polynucleotides, and poly(I)·poly(C) has been shown to be adsorbed at the cell surface. In the adsorbed state strand separation occurs followed by breakage of poly(C). It may be a general phenomenon that interaction of polynucleotides and nucleic acids with cells may involve, as the initial step, adsorption of the macromolecule. Janik and Sommer[109] have suggested that interaction of polynucleotides with mammalian cell interfaces is probably a prerequisite for manifestation of interferon induction, adjuvant effects, and enzyme inhibition.

Although an electrode cannot be regarded as equivalent to a biological surface, it can in a relatively crude sense be employed as a model of such a surface. Interactions of polynucleotides observed at the electrode surface such as adsorption, stacking effects, and structural changes brought about by the electrical potential at the interface can be studied and perhaps ultimately applied to an overall understanding of the behavior of these compounds in biological systems.

REFERENCES

1. J. D. Watson, "Molecular Biology of the Gene." Benjamin, New York, 1965.
2. E. E. Cohn and P. K. Stumpf, "Outlines of Biochemistry," 2nd ed. Wiley, New York, 1967.
3. J. C. Heath, *Nature (London)* **158**, 23 (1946).
4. N. G. Luthy and B. Lamb, *J. Pharm. Pharmacol.* **8**, 410 (1956).
5. F. A. McGinn and G. B. Brown, *J. Am. Chem. Soc.* **82**, 3193 (1960).
6. E. Paleček, *Collect. Czech. Chem. Commun.* **25**, 2283 (1960).
7. B. Janik and P. J. Elving, *J. Am. Chem. Soc.* **92**, 235 (1970).
8. B. Janik and P. J. Elving, *Chem. Rev.* **68**, 295 (1968).
9. D. L. Smith and P. J. Elving, *J. Am. Chem. Soc.* **84**, 1412 (1962).
10. B. Janik and P. J. Elving, *J. Electrochem Soc.* **117**, 457 (1970).
11. C. A. Dekker, *Annu. Rev. Biochem.* **29**, 463 (1963).
12. J. H. Lister, *Adv. Heterocycl. Chem.* **6**, 1 (1966).
13. J. M. Read and J. H. Goldstein, *J. Am. Chem. Soc.* **87**, 3440 (1965).
14. S. I. Chan and J. H. Nelson, *J. Am. Chem. Soc.* **91**, 168 (1969).
15. C. D. Jardetzky and O. Jardetzky, *J. Am. Chem. Soc.* **82**, 222 (1960).
16. J. Clauwaert and J. Stockx, *Z. Naturforsch., Teil B* **23**, 25 (1968).
17. M. P. Schweizer, A. D. Broom, P. O. P. Ts'o, and D. P. Holis, *J. Am. Chem. Soc.* **90**, 1042 (1968).
18. I. R. Miller, *J. Mol. Biol.* **3**, 229 and 357 (1961).
19. V. Vetterl, *Collect. Czech. Chem. Commun.* **31**, 2105 (1966).
20. P. O. P. Ts'o, S. A. Rapaport, and F. J. Bollum, *Biochemistry* **5**, 4153 (1966).
21. A. D. Broom, M. P. Schweizer, and P. O. P. Ts'o, *J. Am. Chem. Soc.* **89**, 3612 (1967).
22. P. O. P. Ts'o, *in* "Molecular Association in Biology" (B. Pullman, ed.), p. 39. Academic Press, New York, 1968.
23. B. Pullman and A. Pullman, "Quantum Biochemistry," pp. 364 and 374–376. Wiley, New York, 1963.
24. G. Dryhurst and P. J. Elving, *Talanta* **16**, 855 (1969).
25. V. Vetterl, *J. Electroanal. Chem.* **19**, 169 (1968).
26. L. Meites, "Polarographic Techniques," p. 133. Wiley, New York, 1965; I. M. Kolthoff and J. J. Lingane, "Polarography," pp. 56–59. Wiley (Interscience), New York, 1952.
27. V. P. Skulachev and L. I. Denisovich, *Biokhimiya* **81**, 132 (1966); *Chem. Abstr.* **64**, 14467 (1966).
28. G. Dryhurst, *Talanta* **19**, 769 (1972).
29. B. Janik and E. Paleček, *Arch. Biochem. Biophys.* **105**, 225 (1964).
30. D. L. Smith and P. J. Elving, *J. Am. Chem. Soc.* **84**, 2741 (1962).
31. J. W. Webb, B. Janik, and P. J. Elving, *J. Am. Chem. Soc.* **95**, 991 (1973).
32. B. Czochralska and D. Shugar, *Biochim. Biophys. Acta* **281**, 1 (1972).
33. J. Brahms, J. C. Maurizot, and A. M. Michelson, *J. Mol. Biol.* **25**, 465 (1967).
34. T. M. Garestier and C. Hélène, *Biochemistry* **9**, 2865 (1970).
35. D. Hamer, D. M. Waldron, and D. L. Woodhouse, *Arch. Biochem. Biophys.* **47**, 272 (1953).

36. E. Paleček, *Biochem. Biophys. Acta* **51**, 1 (1961).
37. E. Paleček, *Biokhimiya* **25**, 803 (1960).
38. E. Paleček, *Nature (London)* **188**, 656 (1960).
39. E. Paleček, *Chem. Zvesti* **14**, 798 (1960).
40. B. Janik and E. Paleček, *Z. Naturforsch., Teil B* **21**, 1117 (1966).
41. B. Janik and E. Paleček, *Abh. Dsch. Akad. Wiss. Berlin, Kl. Med.*, p. 513 (1966).
42. E. Paleček and B. Janik, *Chem. Zvesti* **16**, 406 (1962).
43. B. Janik, *Z. Naturforsch., Teil B* **24**, 539 (1969).
44. G. Dryhurst, *Anal. Chim. Acta* **57**, 137 (1971).
45. V. Vetterl, *J. Electroanal. Chem.* **19**, 169 (1968).
46. O. Jardetzky, *Biopolym. Symp.* **1**, 501 (1964).
47. R. V. Ravidranathan and H. T. Miles, *Biochim. Biophys. Acta* **94**, 603 (1965).
48. M. P. Schweizer, S. I. Chan, and P. O. P. Ts'o, *J. Am. Chem. Soc.* **87**, 5241 (1965).
49. P. O. P. Ts'o, J. S. Melvin, and A. C. Olsen, *J. Am. Chem. Soc.* **85**, 1289 (1963).
50. C. E. Bugg and J. M. Thomas, *Biopolymers* **10**, 175 (1971).
51. J. Bohacek and E. Paleček, *Collect. Czech. Chem. Commun.* **30**, 3456 (1965).
52. E. Paleček, *Biochim. Biophys. Acta* **94**, 293 (1965).
53. E. Paleček, *J. Mol. Biol.* **11**, 839 (1965).
54. E. Paleček, *Collect. Czech. Chem. Commun.* **31**, 2360 (1966).
55. A. Humlova, *Collect. Czech. Chem. Commun.* **29**, 182 (1964).
56. E. Paleček, *Naturwissenschaften* **45**, 186 (1958).
57. E. Paleček and D. Kalab, *Chem. Listy* **57**, 13 (1963).
58. E. Paleček, *Prog. Nucleic Acid Res. Mol. Biol.* **9**, 31 (1969).
59. E. Paleček, in "Methods in Enzymology" (L. Grossman and K. Moldave, eds.), Vol. 21, Part D, p. 3. Academic Press, New York, 1971.
60. M. Vorlíčková, G. Ježková, V. Brabec, Z. Pechan, and E. Paleček, *Stud. Biophys.* **24–25**, 131 (1970).
61. E. Paleček, *J. Mol. Biol.* **20**, 263 (1966).
62. E. Paleček and V. Vetterl, *Biopolymers* **6**, 917 (1968).
63. E. Calendi, A. Di Marco, R. Reggiani, B. Scarpinato, and L. Valentini, *Biochim. Biophys. Acta* **103**, 25 (1965).
64. D. Bach and I. R. Miller, *Biopolymers* **5**, 161 (1967).
65. I. R. Miller and D. Bach, *Biopolymers* **4**, 705 (1966).
66. H. Berg and F. A. Gollmick, "Vorträge des I. Seminars über Molekularbiologie, Reinhardsbrunn," p. 74. Inst. Mikrobiol. Exp. Ther., Jena, 1965.
67. H. Berg, H. Bar, and F. A. Gollmick, *Biopolymers* **5**, 61 (1967).
68. H. Berg and F. A. Gollmick, *Abh. Dsch. Akad. Wiss., Berlin, Kl. Med.* p. 533 (1966).
69. H. Berg, *Monatsber. Dsch. Akad. Wiss., Berlin* **7**, 210 (1965).
70. I. R. Miller, *J. Mol. Biol.* **3**, 229 (1961).
71. J. Flemming, *Stud. Biophys.* **8**, 209 (1968); *Biopolymers* **6**, 1967 (1968).
72. E. Paleček and B. D. Frary, *Arch. Biochem. Biophys.* **115**, 431 (1966).
73. E. Paleček, *Nature (London)* **188**, 4751 (1960).
74. E. Paleček, *Abh. Dsch. Akad. Wiss, Berlin, Kl. Med.* 501 (1966).
75. P. Valenta and H. W. Nürnberg, *J. Electroanal. Chem.* **49**, 55 (1974).
76. V. Brabec and E. Paleček, *Z. Naturforsch., Teil C* **29**, 323 (1974).
77. J. Filipski, J. Chmielowski, and M. Chorazy, *Biochim. Biophys. Acta* **232**, 451 (1971).
78. E. Paleček and V. Brabec, *Biochim. Biophys. Acta* **262**, 125 (1972).
79. V. Brabec, *J. Electroanal. Chem.* **50**, 235 (1974).
80. J. Filipski, J. Chmielowski, and M. Chorazy, *Biochim. Biophys. Acta* **232**, 451 (1971).
81. V. Brabec and E. Paleček, *Biophysik* **6**, 290 (1970).

82. B. Janik, R. G. Sommer, and A. M. Bobst, *Biochim. Biophys. Acta* **281**, 152 (1972).
83. B. Janik and R. G. Sommer, *Biochim. Biophys. Acta* **269**, 15 (1972).
84. H. Berg, D. Tresselt, J. Fleming, H. Bär, and G. Horn, *J. Electroanal. Chem.* **21**, 181 (1969).
85. V. Brabec and E. Paleček, *J. Electroanal. Chem.* **27**, 145 (1970).
86. E. Paleček, *Biochim. Biophys. Acta* **145**, 410 (1967).
87. E. Paleček, *Z. Chem.* **2**, 260a (1962).
88. E. Paleček, *Abh. Dsch. Akad. Wiss., Berlin, Kl. Med.* p. 280 (1964).
89. E. Paleček, *J. Mol. Biol.* **11**, 839 (1965).
90. E. Paleček, *Arch. Biochem. Biophys.* **125**, 142 (1968).
91. E. Paleček and I. Fric, *Biochem. Biophys. Res. Commun.* **47**, 1262 (1972).
92. J. Brahms and W. H. F. M. Mommaerts, *J. Mol. Biol.* **10**, 73 (1964).
93. V. Brabec and E. Paleček, *Biopolymers* **11**, 2577 (1972).
94. E. Paleček, *J. Electroanal. Chem.* **22**, 347 (1969).
95. R. F. Steiner and R. F. Beers, "Polynucleotides." Elsevier, Amsterdam, 1961.
96. A. M. Michelson, J. Massoulié, and W. Guschlbauer, *Prog. Nucleic Acid Res. Mol. Biol.* **6**, 83 (1967).
97. R. Langrindge and A. Rich, *Nature (London)* **198**, 725 (1963).
98. E. Paleček, *Collect. Czech. Chem. Commun.* **37**, 3198 (1972).
99. V. Brabec and E. Paleček, *Stud. Biophys.* **42**, 1 (1974).
100. A. Katchalsky, *Biophys. J.* **4**, 9 (1964).
101. J. T. Yang and T. Samejima, *Prog. Nucleic Acid Res. Mol. Biol.* **9**, 223 (1969).
102. C. L. Stevens and A. Rosenfeld, *Biochemistry* **5**, 2714 (1966).
103. B. Janik, R. G. Sommer, and A. M. Bobst, *Biochim. Biophys. Acta* **281**, 152 (1972).
104. B. Janik and R. G. Sommer, *Biophys. J.* **13**, 449 (1973).
105. V. Brabec and E. Paleček, *Z. Naturforsch., Teil C*, **28**, 685 (1973).
106. E. Paleček, V. Vetterl, and J. Sponar, *Nucleic Acids Res.* **1**, 427 (1974).
107. V. Vetterl and W. Guschlbauer, *Arch. Biochem. Biophys.* **148**, 130 (1972).
108. J. A. Reynaud and M. Leng, *C. R. Hebd. Seances Acad. Sci., Ser. D* **271**, 854 (1970).
109. B. Janik and R. G. Sommer, *Biopolymers* **12**, 2803 (1973).
110. A. Silberberg, *J. Chem. Phys.* **48**, 2835 (1968).
111. A. Silberberg, *J. Polym. Sci., Part C* **30**, 393 (1970).
112. L. D. Inners and G. Felsenfeld, *J. Mol. Biol.* **50**, 373 (1970).
113. A. Bezděkova and E. Paleček, *Stud. Biophys.* **34**, 141 (1972).
114. G. Poste and C. Moss, *Prog. Surf. Sci.* **2**, 198 (1972).
115. L. Weiss, *in* "The Chemistry of Biosurfaces" (M. L. Hair, ed.), Vol. II, p. 378. Dekker, New York, 1972.
116. P. L. Schell, *Biochim. Biophys. Acta* **240**, 472 (1971).

6

Pteridines

I. INTRODUCTION, NOMENCLATURE, AND STRUCTURE

The pteridine ring system is formed by fusion of a pyrimidine ring with a pyrazine ring. The structure of the pteridine ring and the numbering system normally used in this country are shown in Fig. 6-1. As with most heterocyclic compound groups there is considerable confusion and contradiction in the literature regarding the numbering systems and nomenclature of the pteridines; these have been illustrated by Elderfield and Mehta.[1] Fairly extensive surveys of the chemical and biological properties of pteridine derivatives have appeared in several review articles.[2-7]

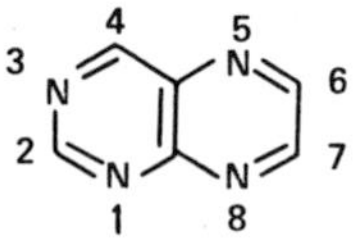

FIG. 6-1.   Structural and numbering system of pteridine.

Pteridines have been under intensive investigation for only a relatively short time. Although they were investigated by Hopkins[8,9] in the last century, his source being the wings of certain butterflies, it was not until the early 1940's that the structures of some pteridines were elucidated.[10-14]

Pteridines are often known by trivial names or as derivatives of compounds having trivial names, much as are the purines and pyrimidines. In order to facilitate subsequent discussion a summary of the trivial names, chemical names, and structural formulas of some common pteridines is presented in Table 6-1. This table is not meant to be complete and reflects only some of the compounds that are discussed in this chapter.

Details of the properties, synthesis, and some reactions of the pteridines are presented elsewhere.[1-8] Solubilities have been discussed by Albert, Lister, and Pederson,[15] ionization constants by Albert,[16] ultraviolet absorption spectra by Albert[4] and Mason,[17] infrared spectra by Mason,[18] Brown and Mason,[19] and Waller *et al.*,[20] NMR spectra by Matsuura and Goto,[21] and mass spectra by Goto *et al.*[22]

II. BIOLOGICAL SIGNIFICANCE OF PTERIDINE DERIVATIVES

As mentioned earlier, pteridine derivatives are found extensively in butterfly wings. Other naturally occurring sources of pteridines are the eyes and scales of fish, amphibians, and reptiles as well as other diverse insect sources.[6] The biological implications of many pteridines are not yet completely understood, but the fact that pteridines are so widely dispersed, albeit in generally very low concentrations, has led to the suspicion that they are implicated in biogenetic phenomena.[6,23] Xanthopterin, leucopterin, and isoxanthopterin are by far the most widespread of the naturally occurring pteridines. Xanthopterin occurs at low concentration as a normal constituent of animal and human urine,[24] as does biopterin.[25,26]

The pteridine derivative subjected to the most intensive study has been folic acid and related compounds. Generally, folic acid occurs as its mono-, tri-, and heptaglutamyl peptide derivatives (Fig. 6-2) In folic acids containing three or seven glutamic acid residues, the residues are combined in a γ-glutamyl linkage.

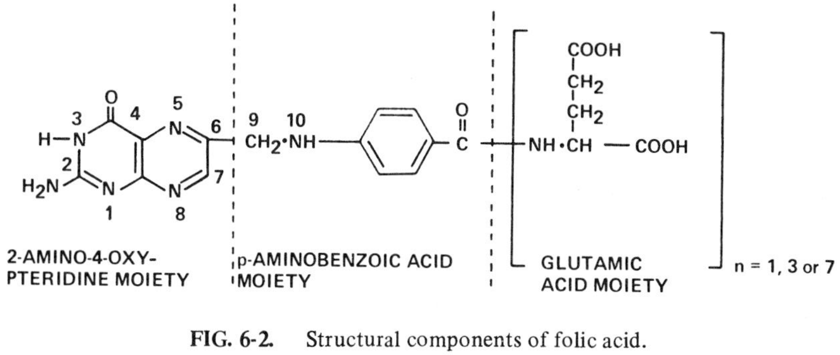

FIG. 6-2. Structural components of folic acid.

TABLE 6-1   Trivial Names, Structural Formulas, and Chemical Names of Some Important Pteridines

Trivial name	Structure	Chemical name
Pteridine		Pyrazino[2,3-d]pyrimidine
Leucopterin		2-Amino-4,6,7-trioxypteridine; 2-amino-4,6,7-pteridinetriol
Xanthopterin		2-Amino-4,6-dioxypteridine; 2-amino-4,6-pteridinediol
Isoxanthopterin		2-Amino-4,7-dioxypteridine; 2-amino-4,7-pteridinediol
Folic acid		Pteroyl-L-glutamic acid* (or pteroylglutamic acid)

Pteroic acid

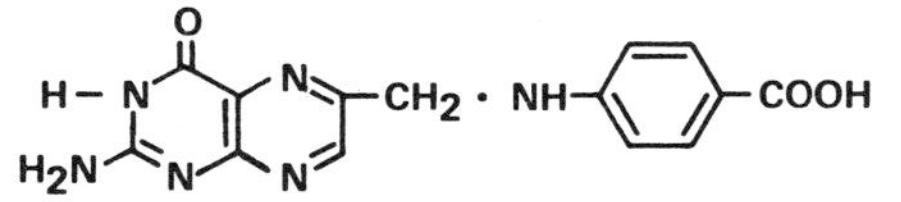

Rhizopterin

Biopterin

Lumazine

Pterorhodin

p-[(2-Amino-4-hydroxy-6-pteridinylmethyl) amino] benzoic acid

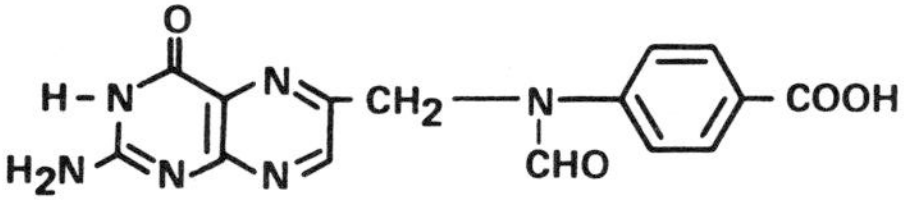

10-Formyl pteroic acid

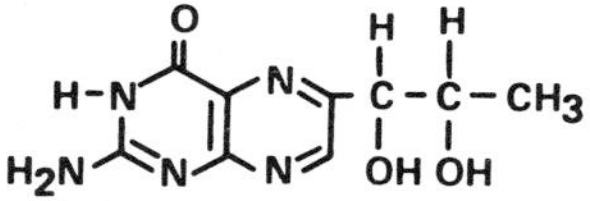

6-L-*erythro*-1',2'-(Dihydroxypropyl) pterin

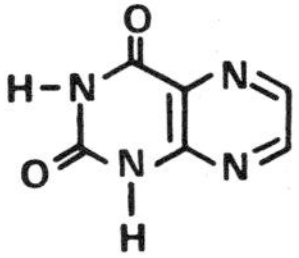

2,4-Dioxypteridine; 2,4-pteridinediol

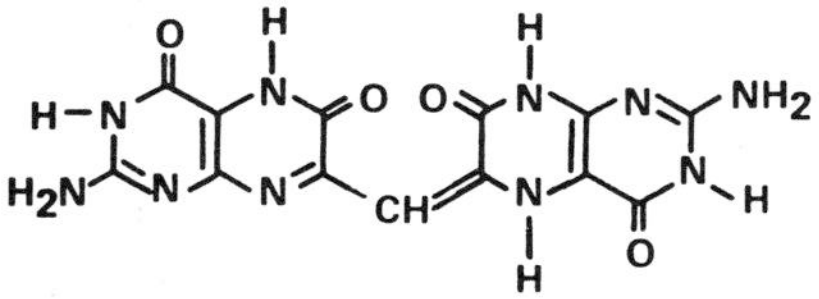

Folic acid has long been recognized as a part of the vitamin B complex[27] and as such is a constituent essential for the growth of many organisms. Derivatives of folic acid play an important but unknown role in the formation of normal red blood cells.[28] Folic acid is a vitamin, but in fact its reduction products are the actual coenzymes. In the presence of the enzyme folic reductase and the reduced form of coenzyme II, nicotinamide adenine dinucleotide phosphate (NADPH), folic acid (I, Fig. 6-3) is reduced across the C-7=N-8 double bond to give dihydrofolic acid (FH_2) (II, Fig. 6-3). This compound can be reduced at the N-5=C-6 double bond to tetrahydrofolic acid (FH_4) (III, Fig. 6-3) in the presence of the enzyme dihydrofolic reductase and NADPH. Tetrahydrofolic acid is important because it can act as a carrier for a formate unit, which subsequently is utilized in the biosynthesis of purines, serine, and glycine. Thus, FH_4 (I,

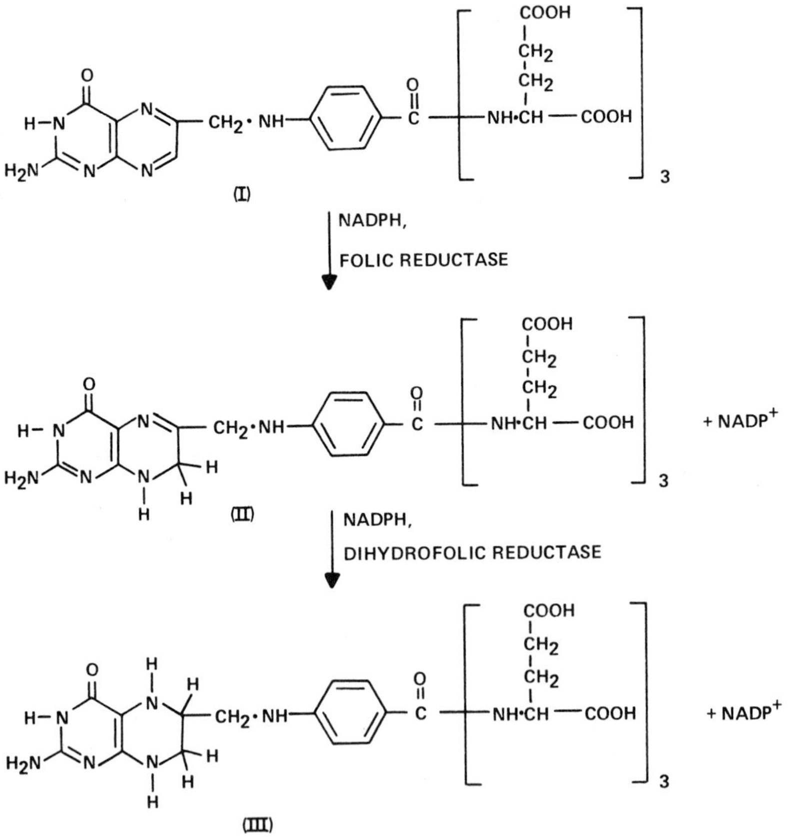

FIG. 6-3.   General mode of biochemical reduction of folic acid (I) to dihydrofolic acid (FH_2, II) and tetrahydrofolic acid (FH_4, III).

Fig. 6-4) reacts with formic acid in the presence of ATP (see Chapters 3 and 4) to give formyl-N^{10}-FH$_4$ (II, Fig. 6-4). This then goes on further into the biosynthetic processes.[28]

In view of the fact that formyl-N^{10}-tetrahydrofolic acid is vitally involved in the biosynthesis of purines and hence nucleic acids, which themselves are the primary constituents of living cells, whether normal or malignant, considerable work has been devoted to developing analogs of folic acid that could *selectively* prevent the development of malignant or neoplastic cells but not normal cells. In other words, a search has developed for pteridines that could act as selective antimetabolites for neoplastic cells.[29] Among the pteridine antimetabolites that have achieved some success are aminopterin[30,31] (I, Fig. 6-5) and A-methopterin (II, Fig. 6-5) Both of these compounds are used widely, often in conjunction with purine antimetabolites such as 6-thiopurine (see Chapter 3), for the successful treatment of acute leukemia.[31,32] They appear to function by blocking the conversion of folic acid to formyl-N^{10}-tetrahydrofolic acid, which, as might be expected, prevents or blocks purine and hence nucleic acid biosynthesis.

Fuller and Nugent[33] proposed that 2-amino-4-hydroxy-6-substituted pteridines could play a primary role in photosynthetic electron transport. It was found that these types of unconjugated pteridines, which are known to occur in association with the photosynthetic apparatus of green plants and photosynthetic bacteria, can be reduced by light from the dihydro to the tetrahydro form in the presence of a bacterial chromophore fraction. Accordingly, it was

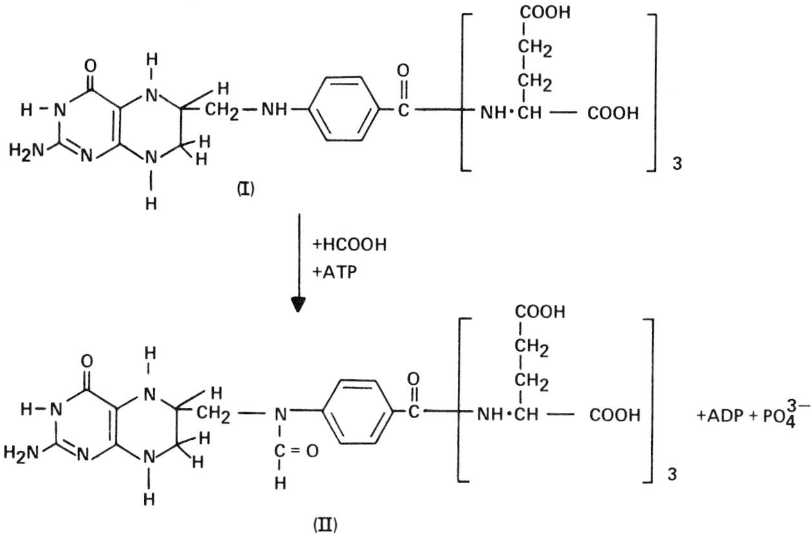

FIG. 6-4. Biosynthesis of formyl-N^{10}-FH$_4$ from FH$_4$.

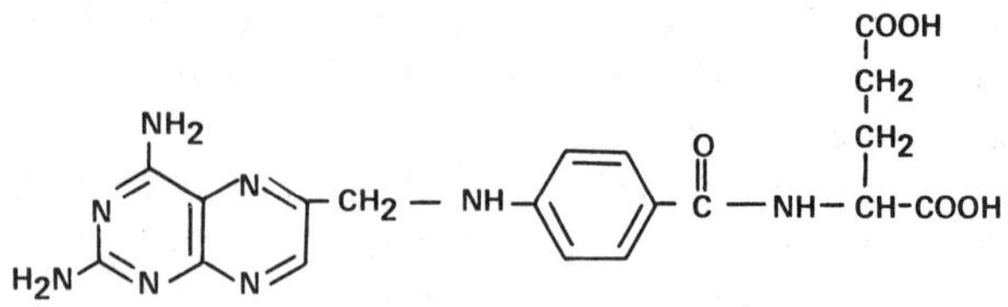

FIG. 6-5.   Structures of the pteridine antimetabolites aminopterin (I) and A-methopterin (II).

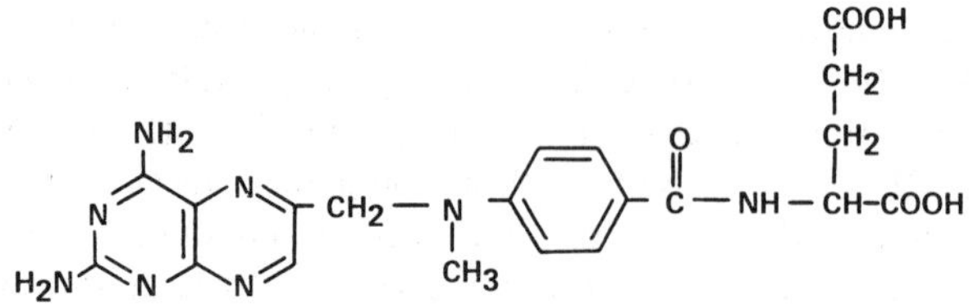

proposed that the dihydropteridine acts as the (previously unknown) primary electron acceptor in photosynthesis. The mechanism of the acceptance and consequent stabilization of the primary electron produced in the photochemical act is not completely clear. However, theoretical and experimental observations require that a single electron transfer be involved in the production of the primary photochemical reductant. The tentative mechanism proposed is that the dihydropteridine (I, Fig. 6-6), via an unspecified biochemical reaction, yields the free-radical species (II, Fig. 6-6). It is this species that acts as the primary electron acceptor generated by the action of light on the light-harvesting chlorophyll or bacteriochlorophyll. After protonation the resultant product is

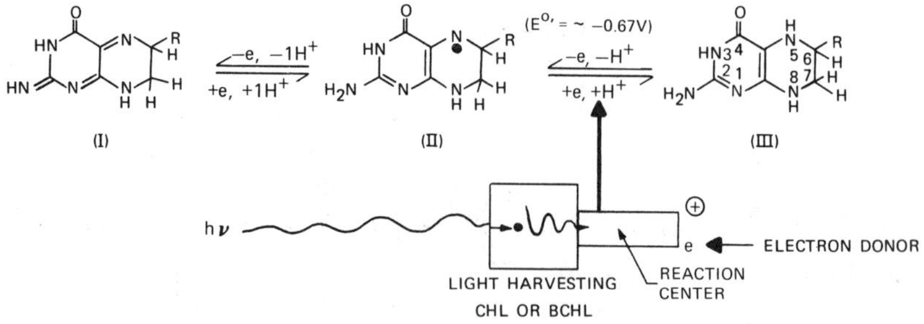

FIG. 6-6.   Proposed initial photochemical electron-transfer steps in photosynthesis (CHL, chlorophyll; BCHL, bacteriochlorophyll).[33]

the tetrahydropteridine (III, Fig. 6-6). Ferredoxin (an iron-containing protein) was also shown to react very readily with tetrahydropteridines to give the reduced ferredoxin and a dihydropteridine. It appears that this type of oxidation–reduction sequence might be responsible for the initial steps in the electron-transfer sequences involved in photosynthesis.

III. ELECTROCHEMISTRY OF PTERIDINE DERIVATIVES

There have been relatively few studies of the electrochemistry of pteridine derivatives, and most of these have been concerned with polarographic reduction processes at the dropping mercury electrode.

A. Pteridine

The first report on the electrochemistry of pteridine is that of Komenda and Laskafeld,[34] who observed that it gives three polarographic reduction waves in aqueous solution. The half-wave potentials reported by these workers for the three waves are shown in Fig. 6-7. A more recent report by McAllister and Dryhurst[35] has confirmed that three polarographic waves are indeed observed for pteridine, although the apparent breaks in the pH dependence of $E_{1/2}$ for waves I and II at around pH 8 observed by Komenda and Laskafeld[34] (Fig. 6-7)

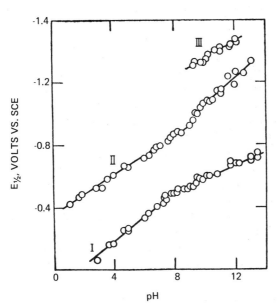

FIG. 6-7. Polarographic $E_{1/2}$ versus pH plots for pteridine waves I, II, and III.[34] (Reprinted with permission of Academia Publishing House, Prague, Czechoslovakia.)

could not be confirmed. Between pH 1 and 13 the equations of the pH dependence of $E_{1/2}$ for pteridine reported by McAllister and Dryhurst[35] are as follows:

$$
\begin{aligned}
\text{wave I} \quad (\text{pH } 3{-}13) \quad & E_{1/2} = 0.04 - 0.064 \text{ pH} \\
\text{wave II} \quad (\text{pH } 1{-}13) \quad & E_{1/2} = -0.36 - 0.070 \text{ pH} \\
\text{wave III} \,(\text{pH } 11{-}13) \quad & E_{1/2} = -0.54 - 0.080 \text{ pH}
\end{aligned}
$$

all potentials being versus the SCE.

The single wave observed below pH 3 (wave II, Fig. 6-7) was proposed by Komenda and Laskafeld[34] to be due to a $2e$ reduction of the cation of N-(3-formyl-2-pyrazinyl)formamidine

formed by hydrolysis of pteridine in acidic solution. With increasing pH wave II decreased and wave I appeared at less negative potentials (Fig. 6-7), which was proposed to be due to a $2e$ reduction of pteridine. Wave III, which appears only at pH values greater than 9, was proposed[34] to be due to reduction of an anionic form of pteridine. Because of their failure to properly identify the polarographically reducible pteridine species and to isolate and identify products of the reduction, the polarographic processes suggested by Komenda and Laskafeld[34] are incorrect.

In order to understand the electrochemistry of pteridine it is necessary to review the rather complex equilibria of this compound in aqueous solution. These equilibria have been studied extensively by Perrin[36] and Albert *et al.*[37] using UV and NMR spectroscopy. When pteridine is initially dissolved in neutral aqueous solution, it exists almost entirely as the nonhydrated neutral molecule (I, Fig. 6-8) but over the course of several minutes attains equilibrium with the neutral 3,4-hydrate (II, Fig. 6-8). In acidic solutions (pH $<$ 3), pteridine exists initially as the cation of the 3,4-hydrate (III, Fig. 6-8), which slowly equilibrates with the cation of the 5,6,7,8-dihydrate (IV, Fig. 6-8). In basic solution of pH $\geqslant$ 11 the anion of the 3,4-hydrate (V, Fig. 6-8) is formed.

The pteridine species responsible for each polarographic wave is readily established by comparison of the changes in polarography of pteridine at various pH values as a function of time with the slow equilibria processes. This can be seen by reference to Fig. 6-9.[35] Thus, a freshly prepared solution of pteridine at pH 7.0 shows a well-defined wave at $E_{1/2} = -0.44$ V (wave I) and a very small wave at $E_{1/2} = -0.82$ V (wave II) (Fig. 6-9A). After about 15 min, however, wave I has decreased in height while wave II is larger. This behavior implies,[35] therefore, that wave I is due to the reduction of nonhydrated pteridine and that wave II is due to the reduction of the 3,4-monohydrate. Similarly, in 1 M acetic

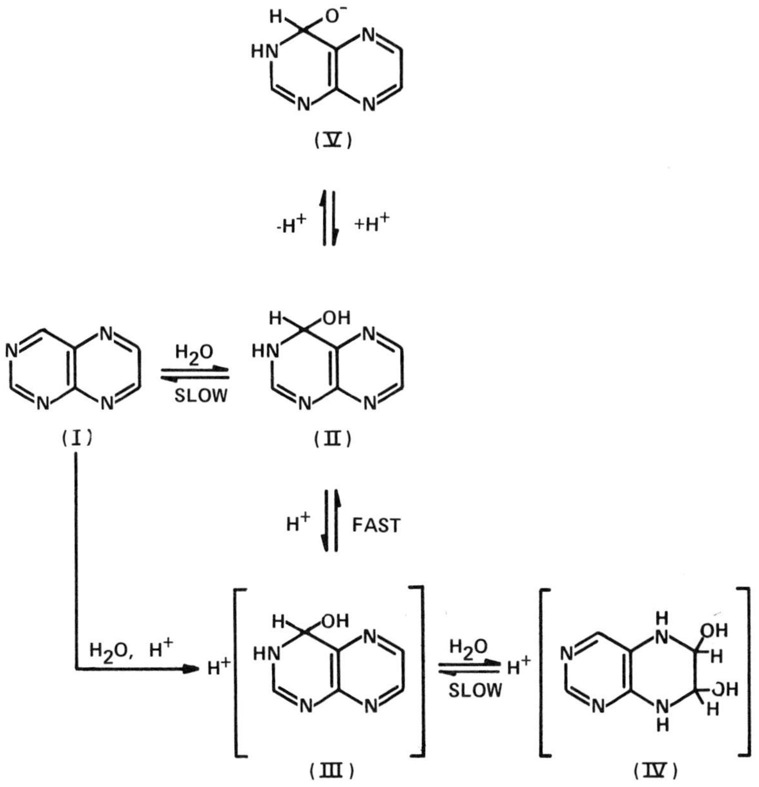

FIG. 6-8. Equilibrium between the neutral, nonhydrated form of pteridine (I), neutral 3,4-hydrate (II), the cation of the 3,4-hydrate (III), the cation of the 5,6,7,8-dihydrate (IV), and the anion of the 3,4-hydrate (V).[35] (Reprinted by permission of Elsevier Publishing Company, Amsterdam.)

acid (pH 2.3), pteridine shows a very small wave I and a much larger wave II (Fig. 6-9B). After a few minutes, wave II decreases in height owing to formation of the nonelectroactive 5,6,7,8-dihydrate. Wave III, by use of similar arguments,[35] has been ascribed to reduction of the anionic form of the 3,4-hydrate.

Using DC polarography and cyclic voltammetry it has been demonstrated[35] that the wave I process is a $2e-2H^+$ reversible reduction of pteridine (I, Fig. 6-10) to 5,8-dihydropteridine (II, Fig. 6-10). The latter compound could not be isolated, however, because it reacts rapidly with pteridine in a base-catalyzed Michael reaction to give a dihydro dimer, probably 7,7',8,8'-tetrahydro-7,7'-dipteridyl (III, Fig. 6-10). The latter species was identified by mass, infrared, and NMR spectrometry. The heterogeneous rate constant, k_s, for the electron-transfer reaction (I → II, Fig. 6-10), determined by the cyclic voltam-

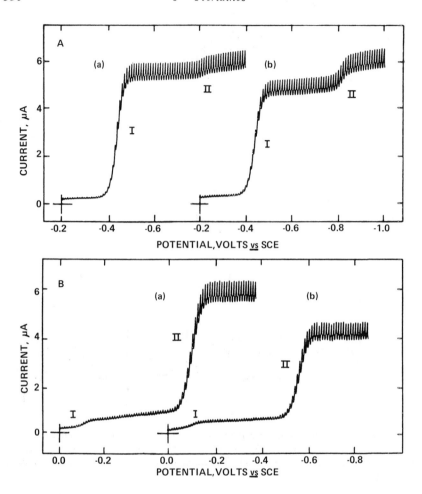

FIG. 6-9. Direct current polarograms of 1 m*M* pteridine (A) in pH 7.0 McIlvaine buffer and (B) in 1 *M* acetic acid, pH 2.3. (a) Immediately after dissolution; (b) after 15 min.[35] (Reprinted with permission of Elsevier Publishing Company, Amsterdam.)

metric methods of Nicholson,[38] is ca. 3×10^{-2} cm sec^{-1} at the hanging mercury drop electrode and ca. 1.3×10^{-2} cm sec^{-1} at the pyrolytic graphite electrode. The second-order rate constant, k_2 (Fig. 6-10), for the reaction of pteridine with 5,8-dihydropteridine was determined by a double-potential step chronoamperometric method[39] and had values ranging from 1.1×10^2 liters mole^{-1} sec^{-1} at pH 7.0 to 4.1×10^3 at pH 11.0.[35]

The wave II process involves reduction of the monohydrated form of pteridine, 3,4-dihydro-4-hydroxypteridine (IV, Fig. 6-10), which exists in major

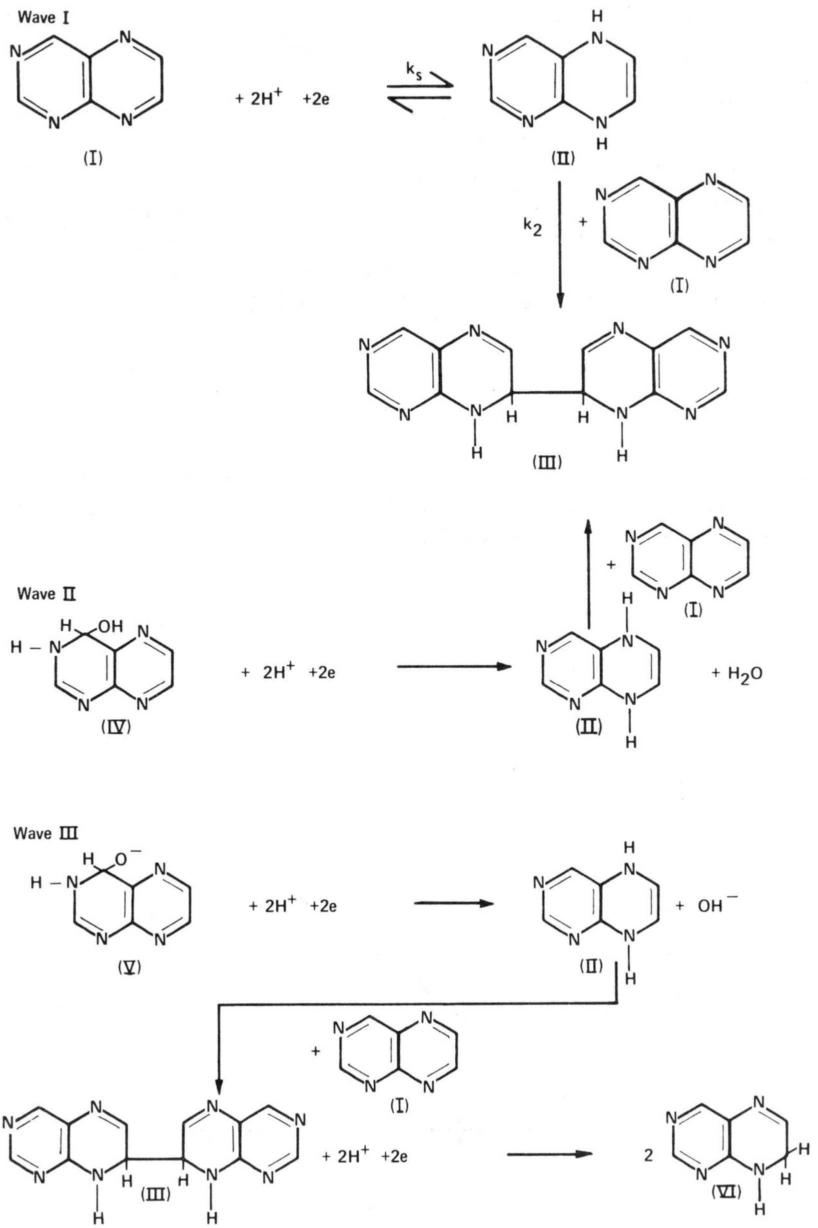

FIG. 6-10. Proposed electrochemical reduction scheme for pteridine.[35] (Reprinted with permission of Elsevier Publishing Company, Amsterdam.)

amounts at low pH, in a $2e-2H^+$ irreversible process to 5,8-dihydropteridine (II, Fig. 6-10). Again, the latter compound reacts with pteridine to give the dihydro dimer (III, Fig. 6-10).

Wave III is due to reduction of the anion of 3,4-dihydro-4-hydroxypteridine (V, Fig. 6-10) in a $2e-2H^+$ irreversible process to 5,8-dihydropteridine (II, Fig. 6-10). which reacts with pteridine to give the dihydro dimer (III, Fig. 6-10). This dihydro dimer is also electrochemically reducible at wave III potentials $[E_{1/2}$ (pH 2.3–12) = $-0.48 - 0.068$ pH in volts versus SCE] so that it is further reduced in a $2e-2H^+$ process to 7,8-dihydropteridine (VI, Fig. 6-10). The nature of this product was confirmed by UV, mass, IR, and NMR spectroscopy.[35]

B. 4-Hydroxypteridine

4-Hydroxypteridine shows a single, diffusion-controlled, pH-dependent polarographic wave (Fig. 6-11) between pH 0 and ca. 10. Komenda and Laskafeld[34] have proposed that it is due to a $2e$ process.

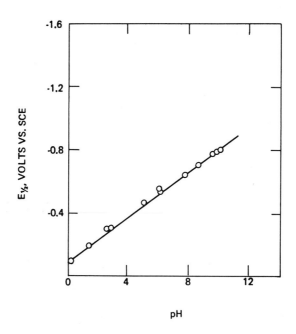

FIG. 6-11.   Polarographic $E_{1/2}$ versus pH relationship for 4-hydroxypteridine.[34] (Reprinted with premission of Academia Publishing House, Prague, Czechoslovakia.)

C. Substituted 4-Hydroxypteridines [4(3*H*)-Pteridones]

Asahi[40] and Komenda[41,42] first reported that 2-amino-4-hydroxy-6-methylpteridine is polarographically reducible. Both authors observed three polarographic waves, although their interpretations of the processes involved were different. Asahi[40] has reported that this compound gives three pH-dependent polarographic reduction waves at the DME (Table 6-2).

The first wave (wave I), which occurred up to about pH 9, was proposed to be due to a $2e-2H^+$ reduction of neutral 2-amino-4-hydroxy-6-methylpteridine (I, Fig. 6-12) to its 7,8-dihydro derivative.[40] The second wave (wave II), which replaced wave I above pH 9, was proposed to be due to a $2e$ reduction of the anionic form of the pteridine (I_a, Fig. 6-12) to the same 7,8-dihydro derivative (II, Fig. 6-12). The third wave (wave III) observed in acidic solution (Table 6-2) was thought to be due to reduction of the protonated form of 7,8-dihydro-2-amino-4-hydroxy-6-methylpteridine (III, Fig. 6-12) (site of protonation unknown) to the 5,6,7,8-tetrahydro derivative (IV, Fig. 6-12). The 7,8-dihydro derivative (II, Fig. 6-12) was prepared by controlled potential electrolysis and, in support of the nature of the wave III process, it gave a single polarographic reduction wave corresponding to wave III. The 5,6,7,8-tetrahydro derivative (IV, Fig. 6-12), also prepared by controlled potential electrolysis, did not show any cathodic polarographic waves, but it did exhibit a pH-dependent anodic wave ($E_{1/2} = 0.27 - 0.060$ pH) which is apparently due to oxidation of the 5,6,7,8-tetrahydro derivative to the 7,8-dihydro compound. Wave I at pH values above 9 and wave III at about pH 5 became kinetic in nature, the kinetic step being the protonation of I_a to give I (Fig. 6-12) or of II to give III (Fig. 6-12). Komenda[41,42] has proposed that wave I of 2-amino-4-hydroxy-6-methyl-

TABLE 6-2   Linear $E_{1/2}$ versus pH Relationships for the Polarographic Reduction of 2-Amino-4-hydroxy-6-methylpteridine[a]

Wave	pH Range	$E_{1/2}$ (V versus NCE[b])
I	0 to ca. 9	$-0.25 - 0.070$ pH
II	>9	$-c$
III	0 to ca. 5	$-0.93 - 0.064$ pH

[a] Data from Asahi.[40]

[b] Normal calomel electrode; add -0.038 V to convert to SCE.

[c] The pH dependence not given in abstract of Asahi.[40] Presumably wave II occurs at more negative potential than wave I.

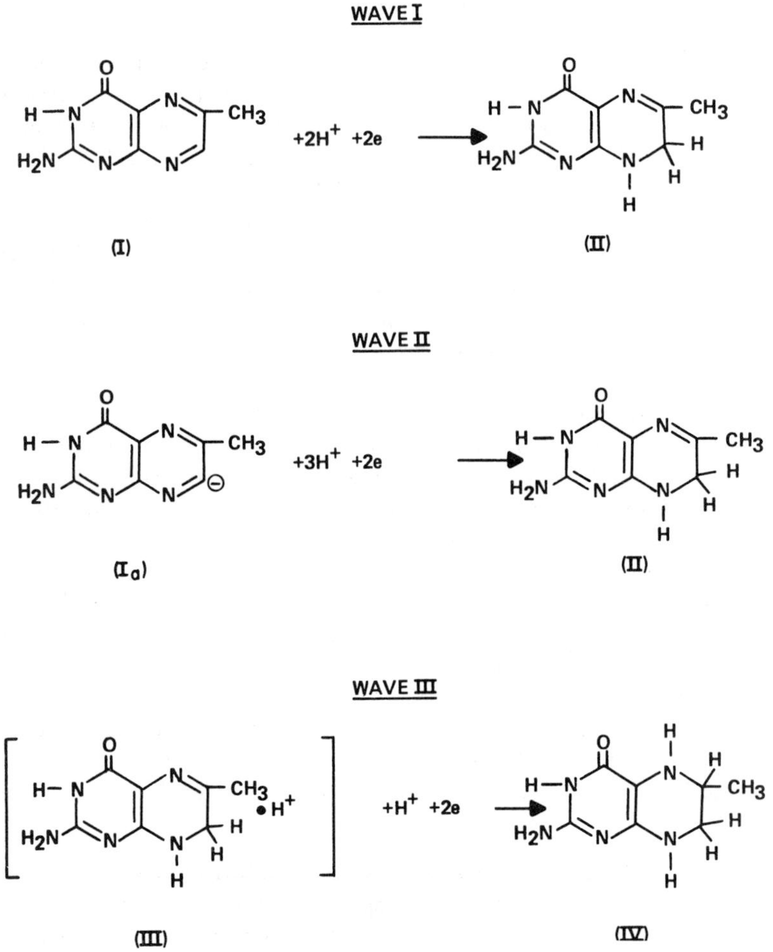

WAVE I

(I)   (II)

WAVE II

(I$_a$)   (II)

WAVE III

(III)   (IV)

FIG. 6-12.   Interpretation of the mechanism of polarographic reduction of 2-amino-4-hydroxy-6-methylpteridine according to Asahi.[40]

pteridine (I, Fig. 6-13) is due to its $2e-2H^+$ reduction to the corresponding 5,8-dihydro derivative (IV, Fig. 6-13), not the 7,8-dihydro compound proposed by Asahi.[40] The wave I reaction scheme proposed by Komenda is shown in Fig. 6-13. A number of other derivatives of 2-amino-4-hydroxypteridine have been studied by Asahi.[43]

Some of the details of the wave I and wave II processes of 2-amino-4-hydroxy-6-methylpteridine have been considerably clarified by Kwee and Lund,[44] who studied the electrochemistry of this compound and a variety of

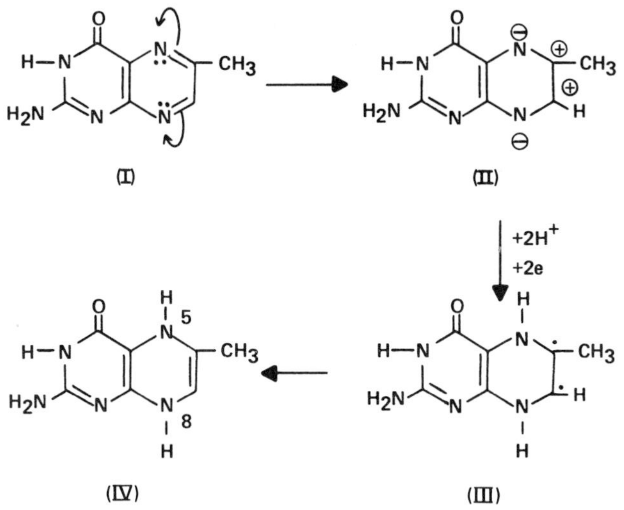

FIG. 6-13. Interpretation of the mechanism of polarographic wave I of 2-amino-4-hydroxy-6-methylpteridine according to Komenda.[41,42]

other related compounds (structures I_{a-f}, II, III, and IV_{g-i}, Table 6-3) by DC polarography, cyclic voltammetry, and controlled potential electrolysis. Unfortunately, complete details of polarographic $E_{1/2}$ versus pH data or voltammetric E_p versus pH data were not published. Some typical data, however, were reported regarding the half-peak potentials observed on cyclic voltammetry of these compounds at a hanging mercury drop electrode (HMDE) (Table 6-3). Considering compound I_a, 2-amino-4-hydroxy-6-methylpteridine, cyclic voltammetry at pH 9 gives rise to two cathodic peaks: A_c at $E_{p/2} = -0.90$ V and B_c at $E_{p/2} = -1.55$ V (Table 6-3). After the sweep is reversed at potentials more negative than peak B_c, an anodic peak (peak A_a, $E_{p/2} = -0.93$ V) is observed that forms a reversible couple with peak A_c. At even more positive potentials a further anodic peak is observed (peak C_a, $E_{p/2} = -0.30$ V) which forms a reversible couple with cathodic peak C_c, $E_{p/2} = -0.27$ V. Controlled potential reduction and coulometry of I_a at potentials corresponding to peak A_c resulted in transfer of two electrons and in formation of 5,8-dihydro-2-amino-4-hydroxy-6-methylpteridine. This species is not stable but, in a pH-dependent reaction, is transformed into another tautomer, 7,8-dihydro-2-amino-4-hydroxy-6-methylpteridine. The tautomerism of the 5,8-dihydro derivative to the 7,8-dihydro derivative is more rapid at high pH (i.e., at pH 9 compared to pH 7).[44]

Controlled potential electrolysis and coulometry of a number of 4-hydroxy-

TABLE 6-3 Half-Peak Potentials Observed on Cyclic Voltammetry of Various Pteridines at the HMDE[a]

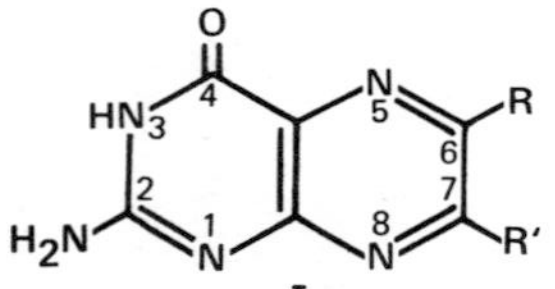

I

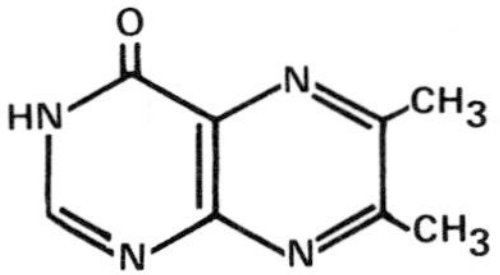

Ia: R = CH$_3$, R' = H
Ib: R = H, R' = CH$_3$
Ic: R = R' = CH$_3$
Id: R = H, R' = C$_6$H$_5$
Ie: R = R' = C$_6$H$_5$
If: R = R' = H

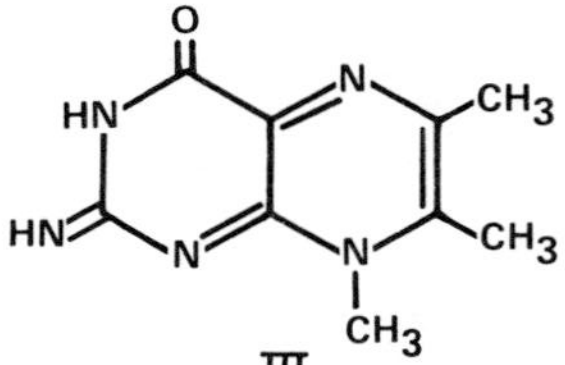

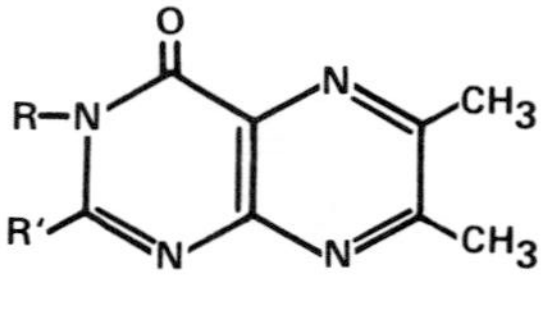

II

III

IV

IVg: R = H, R' = OH
IVh: R = CH$_3$, R' = NH$_2$
IVi: R = CH$_3$, R' = CH$_3$CONH

Compound	Substituents						$E_{p/2}$, (V versus SCE)[c] Peak						
	2	3	4	6	7	8	A_c	A_a	B_c	C_a	C_c	pH
Ia	NH$_2$	H	O	CH$_3$	H		−0.90	−0.93	−1.55	−0.27	−0.30	9
									−1.35	−0.10	−0.14	7
Ib	CH$_3$CONH	H	O	CH$_3$	H		−0.92	−0.96	−1.30	−0.25	−0.25	9
	NH$_2$	H	O	H	CH$_3$							9

Compound													
I$_c$	NH$_2$	H	O	CH$_3$	CH$_3$		-0.96	-0.96	-1.55	-0.22	-0.26		9
I$_d$	CH$_3$CONH	H	O	CH$_3$	CH$_3$		-0.90	-0.94	-1.50	-0.22	-0.22		9
I$_e$	NH$_2$	H	O	H	C$_6$H$_5$		-0.85	-0.83	-1.45		-0.00		9
I$_f$	NH$_2$	H	O	C$_6$H$_5$	C$_6$H$_5$		-0.94	-0.91					9
				H	H		-0.94	-0.78					8
II	H	H	O	CH$_3$	CH$_3$		-1.00	-0.77	-1.45	-0.25	-0.25		9
III	NH$_2$	H	O	CH$_3$	CH$_3$	CH$_3$	-0.77	-0.81	-1.55	-0.45	-0.22		9
IV$_g$	O	H	O	CH$_3$	CH$_3$		-0.82	-0.87	-1.50	-0.24	-0.27		9
IV$_j$	NH$_2$		NH$_2$	CH$_3$	CH$_3$		-0.95	-0.97	-1.55	-0.16	-0.20		9
IV$_h$	NH$_2$	CH$_2$	O	CH$_3$	CH$_3$		-0.87	-0.92					9
IV$_i$	CH$_3$CONH	CH$_3$	O	CH$_3$	CH$_3$		-0.97	-0.93	-1.58	-0.22	-0.23		9
							-0.77	-0.80	-1.36	-0.07	-0.12		7
	CH$_3$CONH	H	O	CH$_3$	H		-0.95	-0.94	-1.30	-0.22	-0.22		9
							-0.85	-0.82	-1.47				8

[a] Data from Kwee and Lund.[44]
[b] See above for structures.
[c] At a voltage sweep rate of 154 mV sec^{-1}

TABLE 6-4   Coulometric n Values and Products Observed on Electrolysis of Various Pteridines[a]

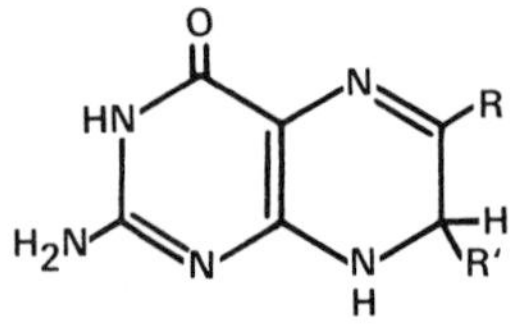

V_a: R = CH$_3$, R' = H
V_b: R = H, R' = CH$_3$
V_c: R = R' = CH$_3$
V_d: R = H, R' = C$_6$H$_5$
V_e: R = R' = C$_6$H$_5$
V_f: R = R' = H

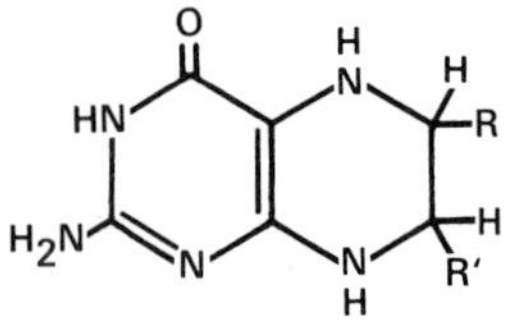

VI_a: R = CH$_3$, R' = H
VI_b: R = H, R' = CH$_3$
VI_c: R = R' = CH$_3$
VI_d: R = H, R' C$_6$H$_5$
VI_e: R = R' = C$_6$H$_5$
VI_f: R = R' = H

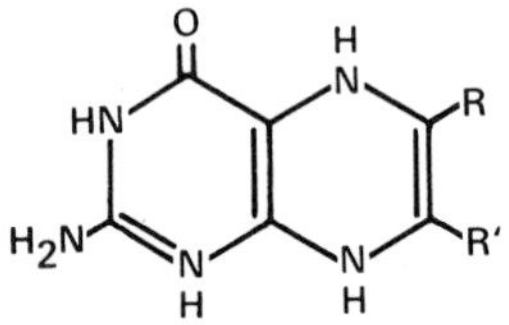

VII_a: R = CH$_3$, R' = H
VII_b: R = H, R' = CH$_3$
VII_c: R = R' = CH$_3$
VII_d: R = H, R' = C$_6$H$_5$
VII_e: R = R' = C$_6$H$_5$
VII_f: R = R' = H

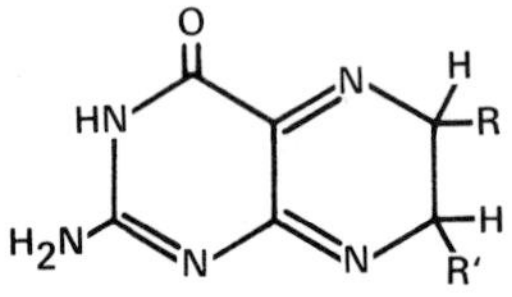

VIII_a: R = CH$_3$, R' = H
VIII_b: R = H, R' = CH$_3$
VIII_c: R = R' = CH$_3$
VIII_d: R = H, R' = C$_6$H$_5$
VIII_e: R = R' = C$_6$H$_5$
VIII_f: R = R' = H

Compound[b]	Medium[c]	Controlled potential[d] (V versus SCE)		n Value[e]	Intermediate[b]	Product[b]
		Reduction	Oxidation			
I_a	A	−1.00		4		VI_a
I_a	B	−0.80		2	VII_a	V_a
I_b	A	−1.00		4		VI_b
I_c	A	−1.00		4		VI_c
I_b	B	−0.80		2	VII_b	I_b
I_b	B	−1.60		4		VI_b
I_c	B	−1.20		2	VII_c	V_c
I_c	B	−1.60		4		VI_c
I_d	B	−0.80		2	VII_d	I_d
I_e	B	−1.20		2	VII_e	V_e
II	B	−1.00		2	5,6-Dihydro-II	7,8-Di-hydro-II
II	B	−1.50		4		Tetra-hydro-II

TABLE 6-4 *Continued*

Compound[b]	Medium[c]	Controlled potential[d] (V versus SCE)		n Value[e]	Intermediate[b]	Product[b]
		Reduction	Oxidation			
Tetra- hydro-II	B		−0.10	2		6,7-Di- hydro-II
III	B	−0.80		2	5,8-Dihydro- III	7,8-Di- hydro-III
7,8-Di- hydro-III	B	−1.60		2		Tetra- hydro-III
Tetra- hydro-III	B		−0.10			7,8-Di- hydro-III
IV$_g$	B	−1.00		2	5,8-Dihydro- IV$_g$	7,8-Di- hydro-IV$_g$
IV$_h$	B	−1.20		2	5,8-Dihydro- IV$_h$	7,8-Di- hydro-IV$_h$
IV$_i$	B	−1.00		2	5,8-Dihydro- IV$_i$	7,8-Di- hydro-IV$_i$
IV$_j$	B	−1.00		2	5,8-Dihydro- IV$_j$	7,8-Di- hydro-IV$_j$
7,8-Di- hydro-IV$_j$	B	−1.60		2		Tetra- hydro-IV$_j$
Tetra- hydro-IV$_j$	B		−0.30	2	6,7-Dihydro- IV$_j$	7,8-Di- hydro-IV$_j$
V$_a$	B	−1.40		2		VI$_a$
VI$_a$	B		−0.10	2	VIII$_a$	V$_a$ + I$_a$
VI$_b$	B		−0.10	2	VIII$_b$	I$_b$
VI$_c$	B		−0.10	2		VIII$_c$

[a] Data from Kwee and Lund.[44]
[b] See above and Table 6-3 for structures.
[c] A, 1 *M* HCl; B, borate buffer, pH 9.0
[d] Mercury pool working electrode.
[e] From text of Kwee and Lund.[44]

pteridine derivatives was carried out by Kwee and Lund.[44] Typical *n* values, intermediates, and products are shown in Table 6-4. Based on the information presented in Tables 6-3 and 6-4 a general reaction scheme has been proposed for the electrochemical behavior of these 4-hydroxypteridine derivatives in neutral or alkaline solution. Using 2-amino-4-hydroxy-6-methylpteridine (I$_a$, Fig. 6-14) at pH 9 as an example, the first polarographic reduction wave or voltammetric peak is a reversible $2e-2H^+$ process to give the corresponding 5,8-dihydro derivative (VII$_a$, Fig. 6-14). The 5,8-dihydro compound is difficult to isolate,

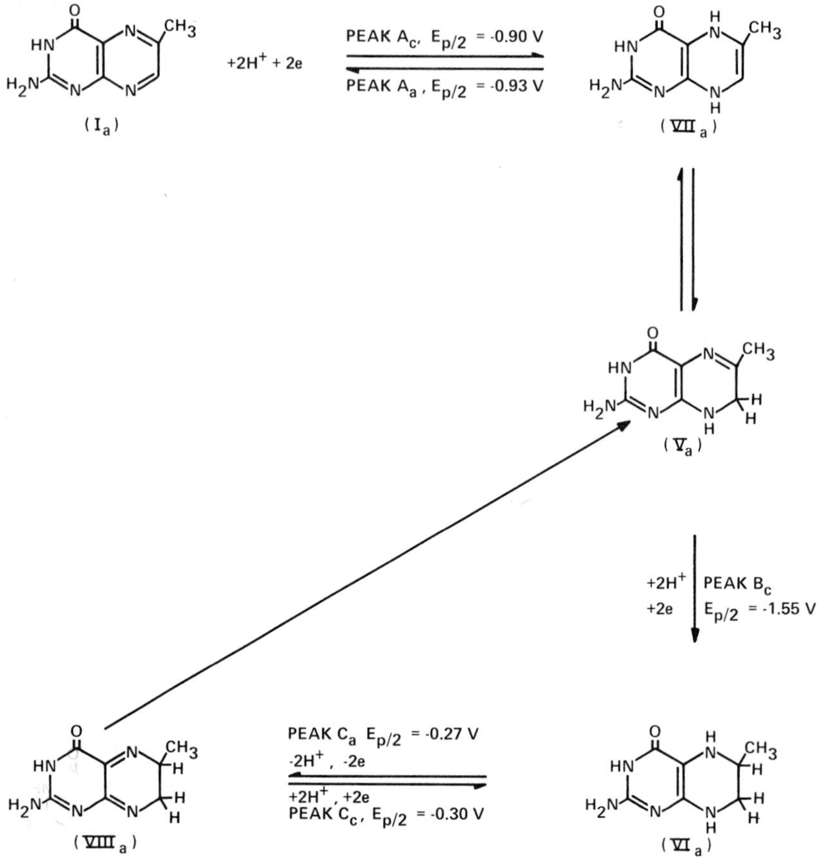

FIG. 6-14. General reaction scheme proposed for the electrochemical reduction and related behavior of 2-amino-4-hydroxy-6-methylpteridine in borate buffer, pH 9.[44]

principally because it tautomerizes to the corresponding 7,8-dihydro derivative (V_a, Fig. 6-14). The tautomerization step is apparently reversible, but for most compounds the equilibrium is displaced strongly in favor of the 7,8-dihydro derivative. In the case of 7-substituted compounds such as 2-amino-4-hydroxy-7-methylpteridine (I_b, Table 6-3) the 5,6-dihydro derivative would be expected, not the 7,8-dihydro derivative. However, the 5,8-dihydro derivative was found to be the most abundant form, and the presence of the 5,6- or 7,8-dihydro derivative was not detected.[44]

The 7,8-dihydro derivatives formed in the tautomerization step (e.g., V_a, Fig. 6-14) are further reducible at more negative potential (peak B_c, Table 6-3) in a $2e-2H^+$ step to the corresponding 5,6,7,8-tetrahydro derivative (VI_a, Fig. 6-14). It is the latter compound that, on cyclic voltammetry, is oxidized by way of peak C_a ($E_{p/2} = -0.27$ V at pH 9, Table 6-3). This process involves two electrons and two protons and is due to oxidation of the tetrahydro derivative (VI_a, Fig. 6-14) to a quinoid dihydropteridine (VIII_a, Fig. 6-14), not to the 7,8-, 5,6-, or 5,8-dihydro derivative. The quinoid form is not stable but rapidly transforms to the 7,8-dihydro tautomer.

The rates of transformation of the 5,8-dihydro derivative to the 7,8-dihydro derivative (i.e., equivalent to $VII_a \rightarrow V_a$, Fig. 6-14) show the following approximate decreasing order: $III \sim IV_i > IV_h > IV_j > IV_g > I_a > I_c \sim II \gg I_b \sim I_d$.[44]

Kwee and Lund[44] have pointed out that the oxidation of the tetrahydropteridine cofactor during the phenylalanine hydroxylase reaction is equivalent to the peak C_a process[45] (i.e., $VI_a \rightarrow VIII_a$, Fig. 6-14). In addition, they suggest that the rate of this reaction, the potential at which it occurs, and the rate of tautomerization of the primarily formed quinoid dihydro derivative are probably important for the effectiveness of the cofactor. To justify this statement they indicate that the tetrahydro derivatives of 7-substituted 4-hydroxypteridines (e.g., I_b, Table 6-3) are oxidized at rather positive potentials which are not attainable at a mercury electrode (see compound I_b, Table 6-3) and their rate of tautomerization of the quinoid dihydro derivative is low. In accordance with these observations, 2-amino-4-hydroxy-7-methylpteridine has been reported to be a less effective cofactor in the enzymatic hydroxylation process than the corresponding 6-substituted derivative.[46]

D. 6- and 7-Hydroxypteridine

1. Electrochemical Reduction

6-Hydroxypteridine exists in aqueous solution below a pH of about 7, primarily as a hydrated species (I, Fig. 6-15A) with water added covalently across the 7,8 bond (i.e., 7,8-dihydro-6,7-dihydroxypteridine). At higher pH the principal form of 6-hydroxypteridine is the anhydrous anion (III, Fig. 6-15A).[47] The nonhydrated neutral molecule (II, Fig. 6-15A) does not exist to any appreciable extent in aqueous solution. Between pH 1.2 and 6.4, 7-hydroxypteridine exists predominantly as the nonhydrated neutral species (II, Fig. 6-15B). At lower pH a nonhydrated cation is formed (I, Fig. 6-15B), while at higher pH a nonhydrated anion is formed (III, Fig. 6-15B).

6-Hydroxypteridine gives a single, pH-dependent polarographic reduction wave above pH 7 (Table 6-5). Coulometry reveals that the polarographic process involves two electrons, and the observed product is 7,8-dihydro-6-

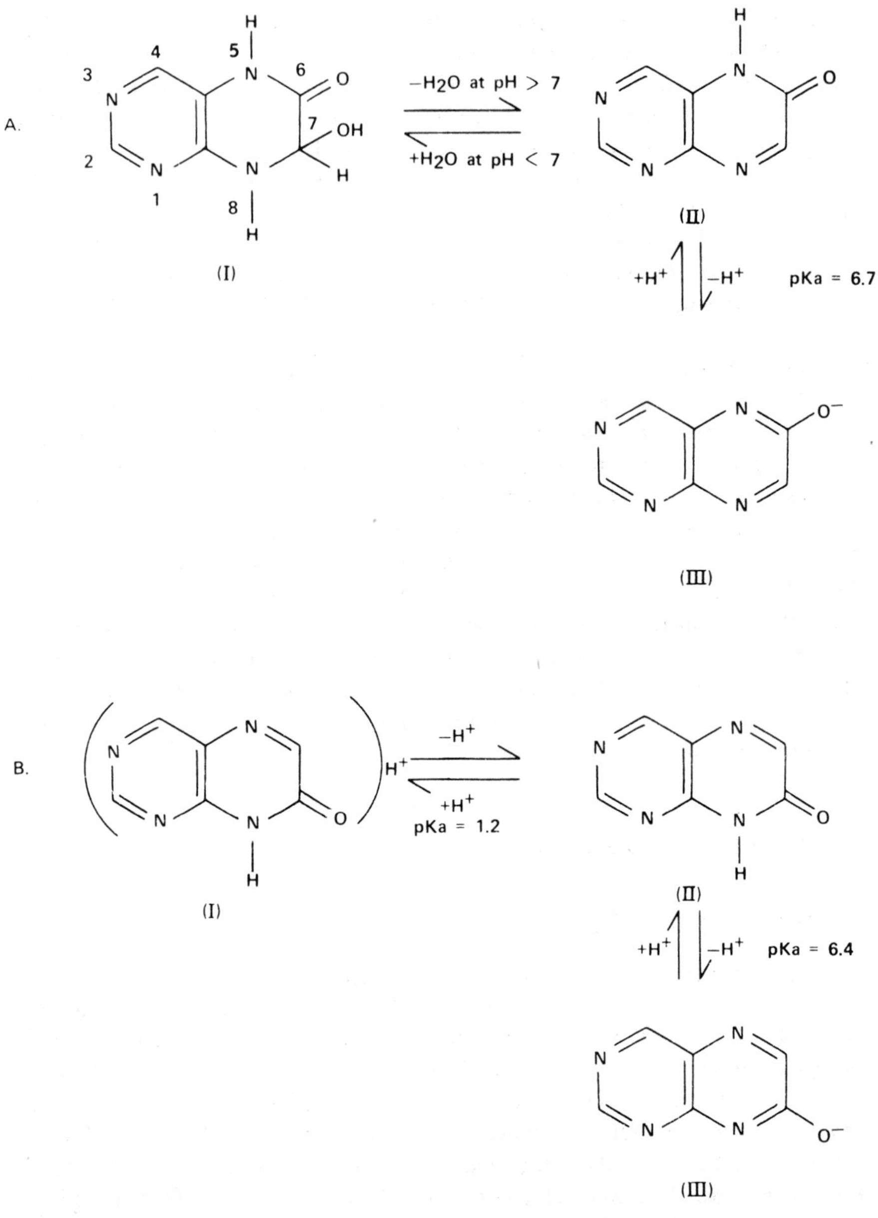

FIG. 6-15. (A) Equilibrium between the hydrated neutral form of 6-hydroxypteridine (I), nonhydrated 6-hydroxypteridine (II), and the nonhydrated anion of 6-hydroxypteridine (III). (B) Cationic (I), nonhydrated neutral (II), and anionic forms of 7-hydroxypteridine (III).[48] (Reprinted with permission of Elsevier Publishing Company, Amsterdam.)

TABLE 6-5   Polarographic Half-Wave Potentials for Reduction of 6- and 7-Hydroxypteridine[a]

Compound	pH Range	$E_{1/2}$ (V versus SCE)
6-Hydroxypteridine	7.1–12	$-0.16 - 0.064$ pH
7-Hydroxypteridine	0–12	$-0.23 - 0.070$ pH

[a] Data from McAllister and Dryhurst.[48]

hydroxypteridine. However, AC polarography revealed that 7,8-dihydro-6,7-dihydroxypteridine (i.e., the monohydrate of 6-hydroxypteridine) below pH 7 and 7,8-dihydro-6-hydroxypteridine (the reduction product of 6-hydroxypteridine) above and below pH 7 are strongly adsorbed at mercury and that they associate at the electrode surface as evidenced by pronounced pits or wells in the AC polarogram. It has been proposed[48] that the latter compounds when adsorbed and associated on the electrode surface form a continuous polymeric film. A possible form of the polymeric film is shown in Fig. 6-16A, where chains of 7,8-dihydro-6,7-dihydroxypteridine are formed by hydrogen bonding between the pyrimidine and pyrazine ring nitrogens. A two-dimensional polymer can be formed by further hydrogen bonding between the $>$C-6=O and the $-$OH group at position 7 of 7,8-dihydro-6,7-dihydroxypteridine molecules. By such hydrogen bonding between two $>$N $\cdots$ H $\cdots$ N$<$ hydrogen-bonded antiparallel chains, a large polymeric chain could be formed. Almost identical polymeric hydrogen-bonded structures can be drawn for 7,8-dihydro-6-hydroxypteridine.[48] An alternative hydrogen-bonded structure, of a dimeric form,[48] has also been proposed (Fig. 6-16B).

In view of the fact that 6-hydroxypteridine is electrochemically reduced only above pH 7, where it exists as a nonhydrated anion (see Fig. 6-15A), a reaction scheme has been proposed whereby the latter species (I, Fig. 6-17) is reduced in a $2e-2H^+$ reaction to the anion of 7,8-dihydro-6-hydroxypteridine (II, Fig. 6-17). Protonation of the latter gives the neutral dihydro species (III, Fig. 6-17). This dihydro species is strongly adsorbed (see earlier discussion) at the DME, and above a critical concentration association of the adsorbed molecules occurs to give a polymeric species (IV, Fig. 6-17).[48]

7-Hydroxypteridine gives a single well-defined polarographic reduction wave (Table 6-5) between pH 0 and 12. Although over this pH range 7-hydroxy pteridine exists in three different ionic forms, no discontinuities in the $E_{1/2}$ versus pH relationship were noticed.[48] Although both DC and AC polarography gave evidence for adsorption of 7-hydroxypteridine at the DME, no association of the adsorbed molecules was observed. Coulometry indicated that two electrons are involved in the electrochemical reduction of 7-hydroxypteridine (I,

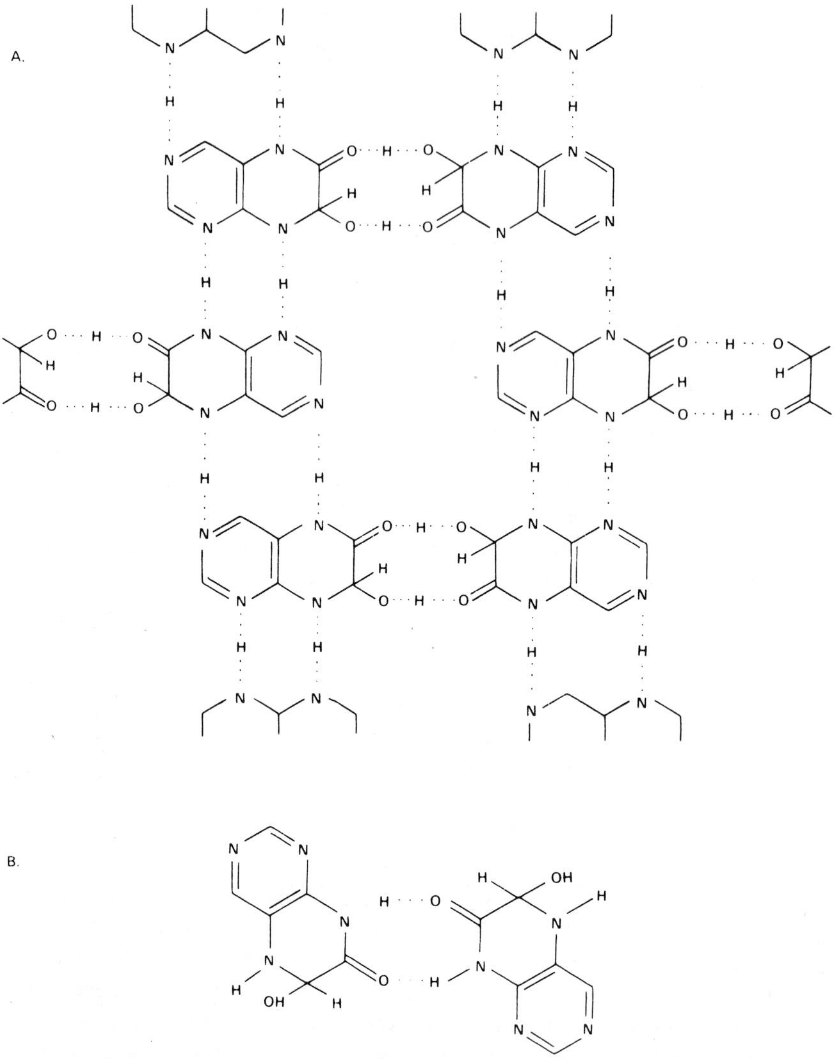

FIG. 6-16. (A) Possible two-dimensional hydrogen-bonded polymer of 7,8-dihydro-6,7-dihydroxypteridine. (B) Possible two-dimensional hydrogen-bonded dimeric form of 7,8-dihydro-6,7-dihydroxypteridine.[48] (Reprinted with permission of Elsevier Publishing Company, Amsterdam.)

Fig. 6-18), and the product, formed in almost quantitative yield, was 5,6-dihydro-7-hydroxypteridine (II, Fig. 6-18). A simple DC polarographic method has been proposed for analysis of 6- and 7-hydroxypteridine mixtures.[48]

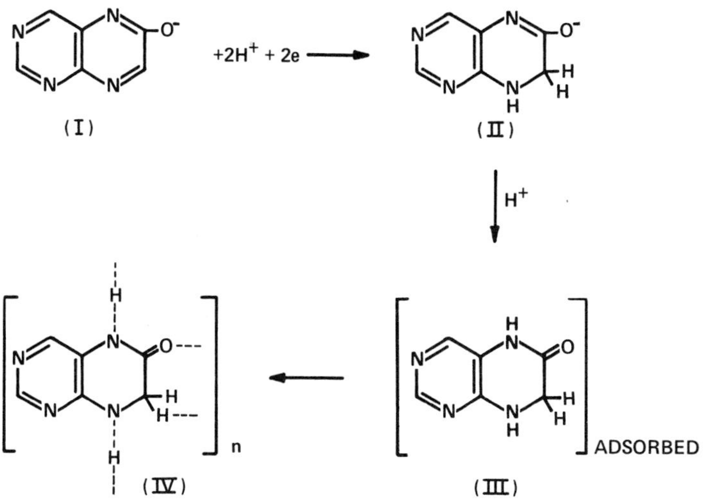

FIG. 6-17. Reaction scheme proposed for the electrochemical reduction of 6-hydroxypteridine at the DME.[48]

2. Electrochemical Oxidation

6-Hydroxypteridine exhibits two pH-dependent voltammetric oxidation peaks at the pyrolytic graphite electrode (PGE) (Table 6-6).[49] In view of the variation of peak current with pH it has been concluded that only the neutral, hydrated form of 6-hydroxypteridine is oxidized (I, Fig. 6-19). 7-Hydroxypteridine is oxidized only in very acidic solutions, i.e., $1-2\,M\,H_2SO_4$ (Table 6-6).[49]

Coulometry of both 6- and 7-hydroxypteridine at peak I_a (Table 6-6) indicated that the electrode reaction is a $2e$ process. In both cases the identified product in 95–100% yield was 6,7-dihydroxypteridine (II, IV, Fig. 6-19).

TABLE 6-6 Voltammetric Peak Potentials for Oxidation of 6- and 7-Hydroxy-pteridine and 6,7-Dihydroxypteridine at the PGE[a]

Compound	Peak	pH Range	Peak potential (V versus SCE)[b]
6-Hydroxypteridine	I_a	2.2–10	$1.13 - 0.071$ pH
	II_a	2.2–9	$1.40 - 0.064$ pH
7-Hydroxypteridine	I_a	$1\,M\,H_2SO_4$	1.08
6,7-Dihydroxypteridine		2.2–9	$1.36 - 0.056$ pH

[a] Data from McAllister and Dryhurst.[49]
[b] Voltage sweep rate, 10 mV sec^{-1}.

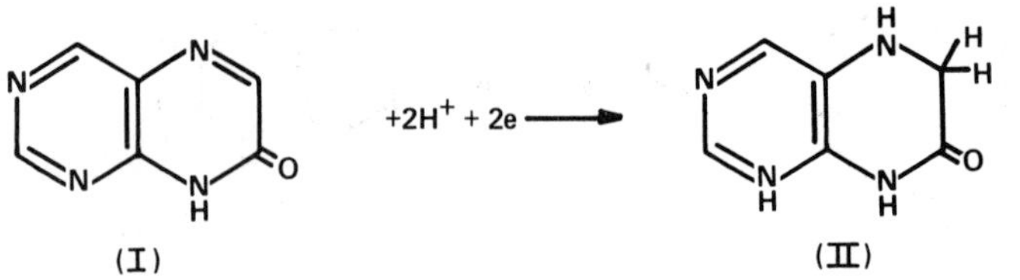

FIG. 6-18. Reaction scheme proposed for the electrochemical reduction of 7-hydroxypteridine at the DME.[48]

Accordingly, the reaction schemes for the peak I_a voltammetric processes of 6-hydroxypteridine (I, Fig. 6-19) and 7-hydroxypteridine (III, Fig. 6-19) are as shown in Fig. 6-19.

The peak II_a process of 6-hydroxypteridine, which is observed between pH 2.2 and 9 (Table 6-6), involves electrochemical oxidation of the 6,7-dihydroxypteridine formed in the peak I_a process. Under voltammetric conditions the single oxidation peak of 6,7-dihydroxypteridine (I, Fig. 6-20) is a

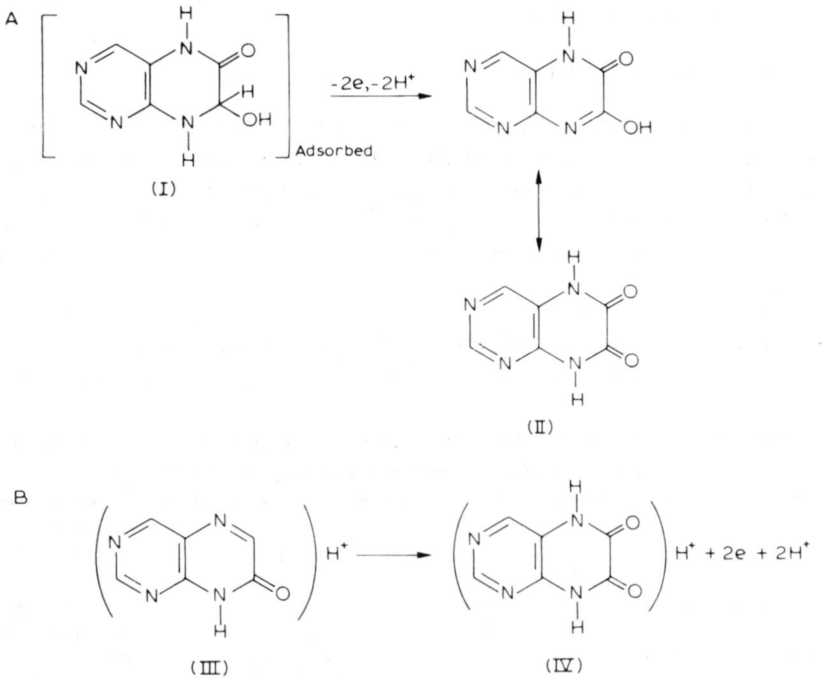

FIG. 6-19. Proposed reaction scheme for the electrochemical oxidation of (A) 6-hydroxypteridine and (B) 7-hydroxypteridine at the PGE at peak I_a.[49] (Reprinted with permission of Elsevier Publishing Company, Amsterdam.)

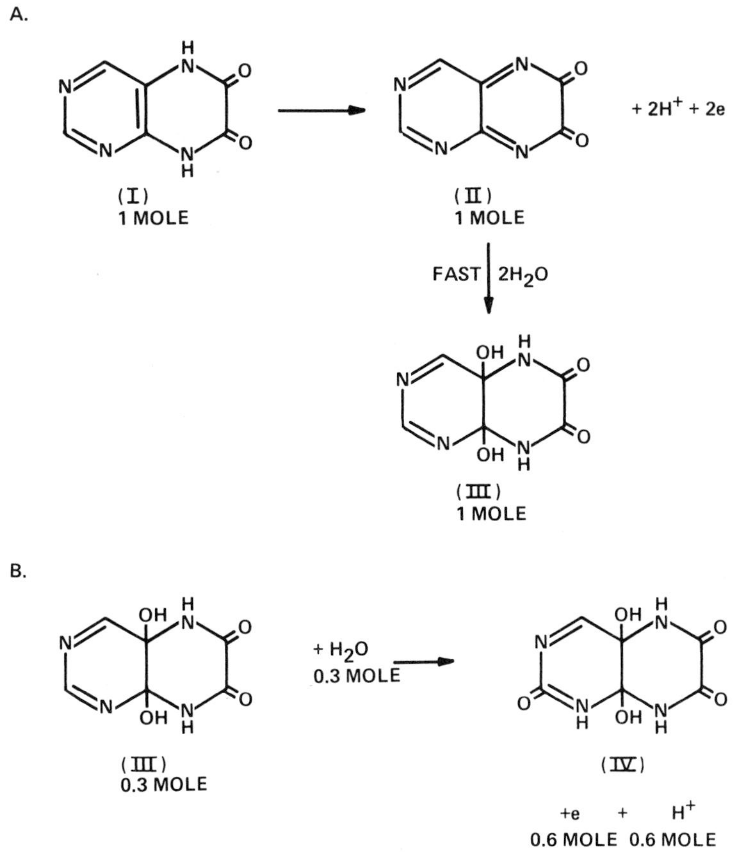

A.

(I)
1 MOLE

(II)
1 MOLE

+ 2H$^+$ + 2e

FAST | 2H$_2$O

(III)
1 MOLE

B.

(III)
0.3 MOLE

+ H$_2$O
0.3 MOLE

(IV)

+e   +   H$^+$
0.6 MOLE 0.6 MOLE

FIG. 6-20.   (A) Proposed initial electrochemical oxidation of 6,7-dihydroxypteridine and (B) possible secondary electrochemical oxidation.[49] (Reprinted with permission of Elsevier Publishing Company, Amsterdam.)

$2e-2H^+$ reaction giving[49] a diimine species (II, Fig. 6-20) that very rapidly hydrates to a diol (III, Fig. 6-20). Under conditions of controlled potential electrolysis, however, the electrooxidation involves transfer of approximately six electrons. It has therefore been proposed[49] that under the latter conditions the bridgehead diol of 6,7-dihydroxypteridine (III, Fig. 6-20) undergoes a partial further oxidation to the bridgehead diol of 2,6,7-trihydroxypteridine (IV, Fig. 6-20). Proposed electrochemical and chemical reactions for formation of the ultimate products of the oxidation from the bridgehead diols of 6,7-dihydroxypteridine and 2,6,7-trihydroxypteridine are shown in Figs. 6-21 and 6-22, respectively. The ultimate identified products of these reactions are tetraketopiperazine (XV, Fig. 6-22), oxamide (XVII, Figs. 6-21 and 6-22), urea (XII,

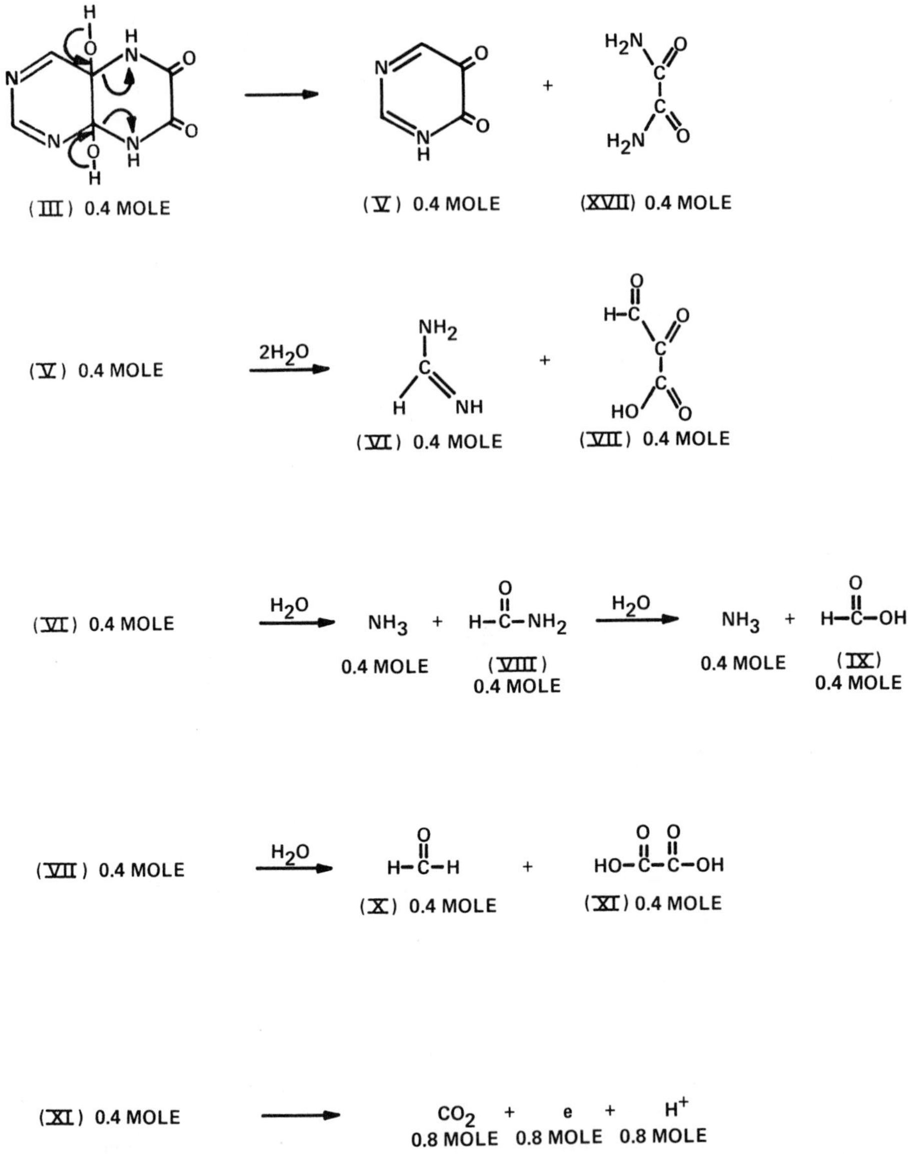

(III) 0.4 MOLE   (V) 0.4 MOLE   (XVII) 0.4 MOLE

(V) 0.4 MOLE $\xrightarrow{2H_2O}$ (VI) 0.4 MOLE   +   (VII) 0.4 MOLE

(VI) 0.4 MOLE $\xrightarrow{H_2O}$ NH_3 + $H-\overset{O}{\underset{}{C}}-NH_2$ $\xrightarrow{H_2O}$ NH_3 + $H-\overset{O}{\underset{}{C}}-OH$

0.4 MOLE   (VIII)   0.4 MOLE   (IX)
                     0.4 MOLE   0.4 MOLE

(VII) 0.4 MOLE $\xrightarrow{H_2O}$ $H-\overset{O}{\underset{}{C}}-H$ + $HO-\overset{O}{\underset{}{C}}-\overset{O}{\underset{}{C}}-OH$

(X) 0.4 MOLE   (XI) 0.4 MOLE

(XI) 0.4 MOLE $\longrightarrow$ CO_2 + e + H^+
                                   0.8 MOLE   0.8 MOLE   0.8 MOLE

FIG. 6-21.   Proposed reaction scheme for rearrangement of the bridgehead diol of 6,7-dihydroxypteridine (III) to oxamide (XVII) and 4,5-diketopyrimidine (V) followed by fragmentation of 4,5-diketopyrimidine to ammonia, formaldehyde, formic acid, oxalic acid, and CO_2.[49]

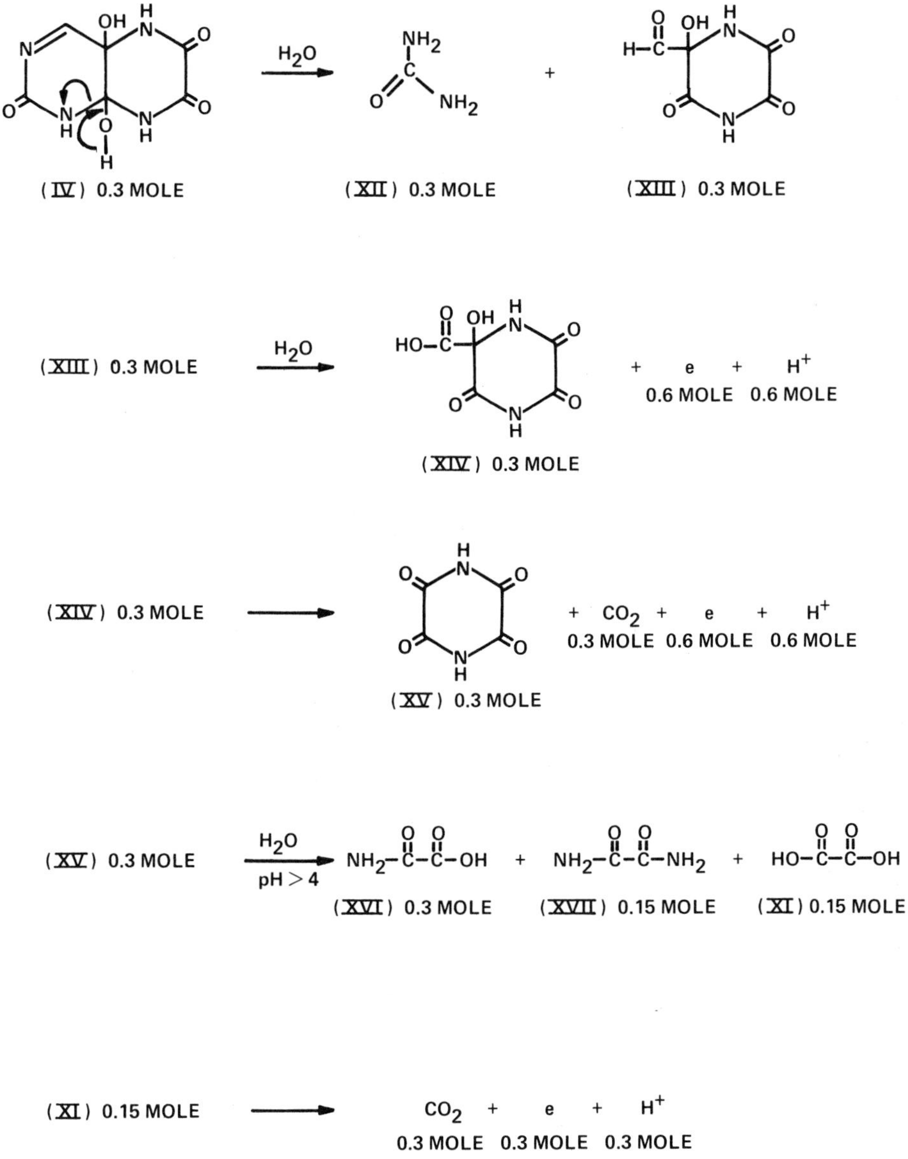

(IV) 0.3 MOLE → H$_2$O → (XII) 0.3 MOLE + (XIII) 0.3 MOLE

(XIII) 0.3 MOLE → H$_2$O → (XIV) 0.3 MOLE + e (0.6 MOLE) + H$^+$ (0.6 MOLE)

(XIV) 0.3 MOLE → (XV) 0.3 MOLE + CO$_2$ (0.3 MOLE) + e (0.6 MOLE) + H$^+$ (0.6 MOLE)

(XV) 0.3 MOLE → H$_2$O, pH > 4 → NH$_2$–C(=O)–C(=O)–OH (XVI) 0.3 MOLE + NH$_2$–C(=O)–C(=O)–NH$_2$ (XVII) 0.15 MOLE + HO–C(=O)–C(=O)–OH (XI) 0.15 MOLE

(XI) 0.15 MOLE → CO$_2$ (0.3 MOLE) + e (0.3 MOLE) + H$^+$ (0.3 MOLE)

FIG. 6-22. Proposed reaction scheme for the rearrangement and oxidation of the bridgehead diol of 2,6,7-trihydroxypteridine (IV) to tetraketopiperazine (XV), urea (XII), and CO$_2$.[49]

Fig. 6-22), oxamic acid (XVI, Fig. 6-22), ammonia, formaldehyde, formic acid, and CO_2. The electrochemical oxidation of oxalic acid (XI, Figs. 6-21 and 6-22) to CO_2 has subsequently been shown to proceed quantitatively in a $2e-2H^+$ reaction at the PGE.[50]

E. Xanthopterin

Asahi[51] has found that xanthopterin undergoes a $2e-2H^+$ reduction to form a dihydro derivative. In view of the various dissociation equilibria of xanthopterin, its polarographic reduction wave was split into many steps and many waves were kinetic owing to the protonation equilibria involved.

Below pH 8 the reduction of the neutral xanthopterin (pK_a 8.65,[52] 9.99[53]) was accompanied by an adsorption wave; this was confirmed by AC polarography and electrocapillary studies. From the height of the adsorption wave it was calculated that the area of the electrode occupied by one molecule was 55 Å^2. This seemed to indicate that the plane of the ring was parallel to the electrode surface. Data on the half-wave potentials and pH ranges over which waves are observed were not specified.

F. 2-Amino-4-hydroxypteridine 6-Carboxaldehyde

Both Asahi[54] and Hrdý[55] have studied the DC polarography of 2-amino-4-hydroxypteridine 6-carboxaldehyde, which shows a total of three polarographic reduction waves. The $E_{1/2}-$pH relationships of Hrdý are shown in Fig. 6-23.

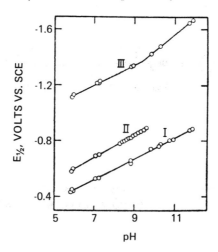

FIG. 6-23.   Variation of $E_{1/2}$ with pH for the polarographic reduction waves of 2-amino-4-hydroxypteridine 6-carboxaldehyde.[55] (Reprinted with permission of Academia Publishing House, Prague, Czechoslovakia.)

According to both Asahi[54] and Hrdý[55] the first two waves (wave I and II) represent, overall, a $2e-2H^+$ reduction process. Since the height of the first wave increased relative to the second wave with increasing pH and was kinetic in nature at low pH, it was proposed[54] that the first wave was due to reduction of the nonhydrated 2-amino-4-hydroxypteridine 6-aldehyde (I, Fig. 6-24), the second wave to reduction of the hydrated aldehyde (IV, Fig. 6-24). However, the reduction process, which involved two electrons and two protons, did not involve an aldehyde moiety but was apparently due to reduction of the pteridine nucleus[54,55] to give an unspecified dihydro derivative. The third wave was proposed to be due to reduction of the aldehyde moiety, presumably to give the

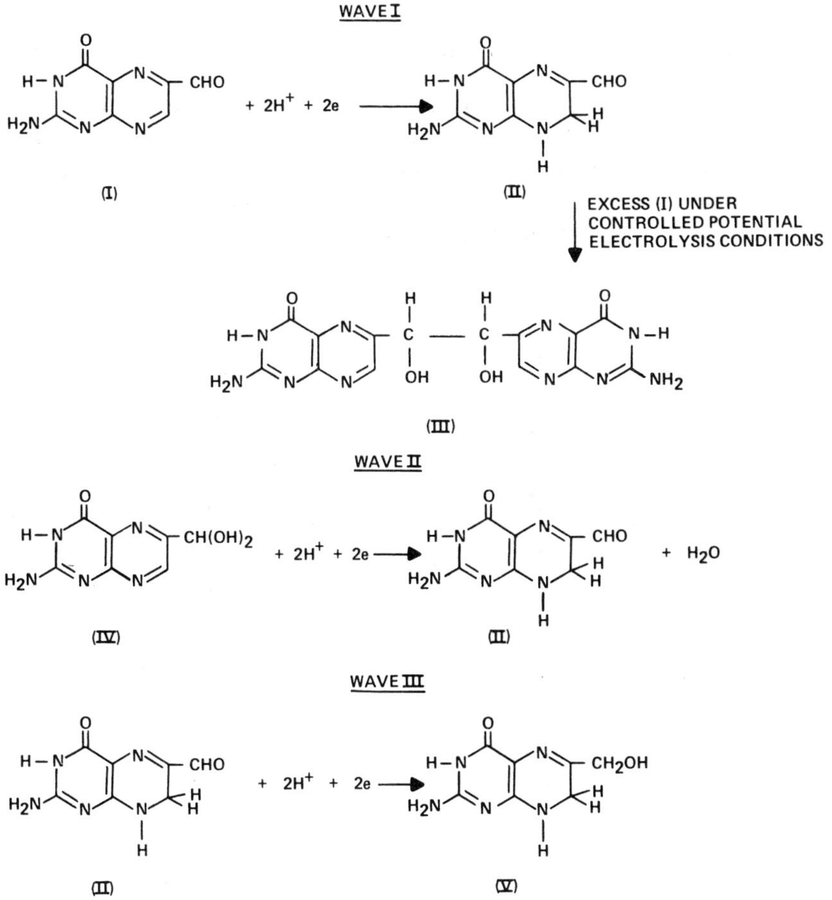

FIG. 6-24. Interpretation of the electrochemical reduction of 2-amino-4-hydroxy-pteridine 6-aldehyde.[54]

corresponding alcohol (V, Fig. 6-24). Asahi[54] found that electrolysis of 2-amino-4-hydroxypteridine 6-aldehyde (I, Fig. 6-24) at potentials corresponding to wave I resulted in formation of a dihydro derivative (II, Fig. 6-24), possibly the 7,8-dihydro compound, although this is not explicit (in the abstract). The dihydro in turn reacted with the unreduced parent compound in the bulk of the solution to give 1,2-bis(2-amino-4-hydroxy-6-pteridinyl)ethylene glycol, which probably has the structure III (Fig. 6-24).

Asahi[54] also found that 2-amino-4-hydroxy-6-hydroxymethylpteridine gave a one-step reduction at potentials close to those of the second wave of 2-amino-4-hydroxypteridine 6-aldehyde and considered it to be a $2e-2H^+$ reduction to a dihydro derivative.

G. 2-Amino-4-hydroxy-6-pteridinecarboxylic Acid

2-Amino-4-hydroxy-6-pteridinecarboxylic acid gives a single, pH-dependent (Fig. 6-25) polarographic reduction wave.[55] The single wave produced was diffusion controlled, involved two electrons, and accordingly would be expected to give rise to a dihydro derivative. 2-Amino-4-hydroxy-6-methoxypteridine behaved in a very similar manner (Fig. 6-25).

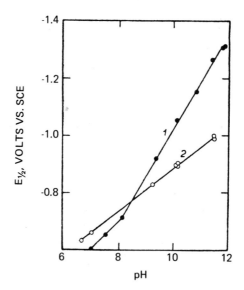

FIG. 6-25.   The $E_{1/2}$ versus pH dependence for the polarographic reduction of (1) 2-amino-4-hydroxy-6-pteridinecarboxylic acid and (2) 2-amino-4-hydroxy-6-methoxypteridine.[55] (Reprinted with permission of Academia Publishing House, Prague, Czechoslovakia.)

H. 7-Methylxanthopterin

Komenda *et al.*[56] found that 7-methylxanthopterin gives three pH-dependent polarographic reduction waves at the DME (Fig. 6-26). All of these waves appeared to be straightforward $2e-2H^+$ reductions of the pteridine nucleus, complicated by the various ionization states of 7-methylxanthopterin.

I. 6-Methylisoxanthopterin

Komenda *et al.*[56] reported that 6-methylisoxanthopterin exhibits polarographic activity at $pH \geqslant 8.5$. Three waves were observed. Wave I appeared between about pH 8.5 and 10.5 and corresponded to a $2e$ reduction of the cationic form of 6-methylisoxanthopterin. As wave I disappeared, wave II appeared and was due to a $2e$ reduction of the neutral compound. Above pH 12 only the third wave was observed due to a $2e$ reduction of the anionic species. The pH dependence of the half-wave potentials is shown in Fig. 6-27.

J. Leucopterin (2-Amino-4,6,7-trihydroxypteridine)

Leucopterin shows two pH-dependent polarographic reduction waves in alkaline solution[56] (Fig. 6-28). In view of the fact that at pH greater than about

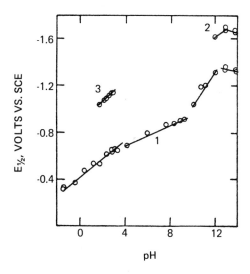

FIG. 6-26. The $E_{1/2}$ versus pH relationships for 7-methylxanthopterin.[56] (Reprinted with permission of Academia Publishing House, Prague, Czechoslovakia.)

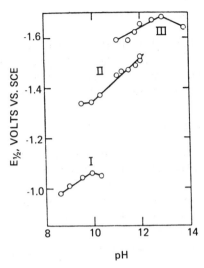

FIG. 6-27.   The $E_{1/2}$ versus pH relationships for the polarographic reduction of 6-methylisoxanthopterin.[56] (Reprinted with permission of Academia Publishing House, Prague, Czechoslovakia.)

9.5 wave I decreased and wave II appeared and both gave currents that varied with pH according to a dissociation curve, the process responsible for wave I was a $2e-2H^+$ reduction of the monoanionic leucopterin (pK_a for proton loss 7.56, 9.78, 13.6[57]), while wave II was probably a $2e-2H^+$ reduction of the dianionic species.

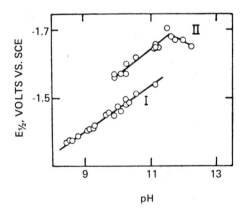

FIG. 6-28.   The $E_{1/2}$ versus pH relationships for the polarographic reduction of leucopterin.[56] (Reprinted with permission of Academia Publishing House, Prague, Czechoslovakia.)

K. Aminopterin, Pteroic Acid, 10-Formylpteroylglutamic Acid, and A-Methopterin

Asahi[58] found that all of these compounds give rise to three polarographic reduction waves in acidic solution and one or two reduction waves in alkaline solution. The first electrochemical process to occur was a $2e-2H^+$ reduction of the pteridine rings of these compounds to give an unspecified dihydro derivative. In the case of pteroic acid and 10-formylpteroylglutamic acid, this first process was split into two waves because of ionization steps. The second wave was proposed to be due to reductive cleavage of the CH_2N: grouping in the 6 position, and the third wave to further reduction of the dihydropteridine formed in the first electrochemical step to a tetrahydropteridine. Unfortunately, detailed half-wave potential and mechanistic data are not available.

L. Folic Acid

The first reports of the electrochemical reduction of folic acid are those of Hrdý[55] and Asahi.[59] However, the data reported by these authors are often contradictory. More recently, Kretzschmar and Jaenicke[60,61] reexamined the electrochemical reduction of folic acid which, according to these workers, occurs by way of three polarographic waves (I_c, II_c, and III_c) between pH 1 and 12 (Fig. 6-29). The electrochemical behavior of folic acid was examined most carefully at pH 5–7, where waves I_c, II_c, and III_c appear, and pH 9, where waves I_c and III_c appear (Fig. 6-29). It has been proposed that in the former pH range the wave I_c process is a reversible $2e-2H^+$ reduction of folic acid (I, Fig. 6-30) to 5,8-dihydrofolic acid (II, Fig. 6-20). Waves II_c and III_c are under kinetic control between pH 5 and 7 because they are dependent on the chemical rearrangement of 5,8-dihydrofolic acid (II, Fig. 6-30) to 7,8-dihydrofolic acid (III, Fig. 6-30). Wave II_c is due to a $2e-2H^+$ reductive cleavage of the latter dihydro derivative between the C-9 and N-10 positions to give 7,8-dihydro-2-amino-4-hydroxy-6-methylpteridine (IV, Fig. 6-30) and RNH_2. Wave III_c is then due to a $2e-2H^+$ reduction of 7,8-dihydro-6-methylpterin (IV, Fig. 6-30) to the corresponding 5,6,7,8-tetrahydro derivative (V, Fig. 6-30).

In basic solution the mechanism changes somewhat.[60,61] The wave I_c process remains a reversible $2e-2H^+$ reduction of folic acid (I, Fig. 6-31) to the 5,8-dihydro derivative (II, Fig. 6-31). Tautomerization of the latter species, however, has been proposed to involve the 7,8-dihydro derivative (III, Fig. 6-31) and the 6,7-dihydro derivative (IV, Fig. 6-31). Wave III_c is proposed to involve a $2e-2H^+$ reversible reduction of the 6,7-dihydro species to 5,6,7,8-tetrahydrofolic acid (V, Fig. 6-31). Controlled potential electrolysis of folic acid at its wave III_c potentials at pH 9 supported the involvement of the 7,8- and 6,7-dihydro species (III and IV, Fig. 6-31) since some side reactions appeared to

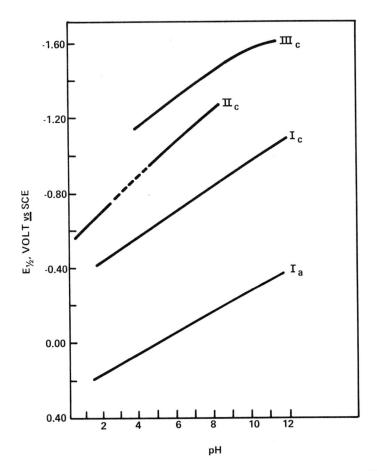

FIG. 6-29.   Variation of $E_{1/2}$ with pH for the polarographic waves of folic acid.[60]

occur. The 7,8-dihydro derivative (III, Fig. 6-31) apparently undergoes a $2e-2H^+$ reductive cleavage across the C-9—N-10 bond, while the 6,7-dihydro-derivative (IV, Fig. 6-31) undergoes a reductive cleavage across the C-6—C-9 bond.

Kretzschmar and Jaenicke[60,61] have also investigated the process responsible for the anodic wave observed for 5,6,7,8-tetrahydrofolic acid between pH 1 and 12 (wave I_a, Fig. 6-29). At pH 6.8 evidence seemed to suggest that tetra-hydrofolic acid (I, Fig. 6-32) is oxidized in an initial $1e-1H^+$ reaction to give a radical (II, Fig. 6-32), which could dimerize to III (Fig. 6-32) or be further oxidized in a $1e-1H^+$ reaction to 6,7-dihydrofolic acid (IV, Fig. 6-32). The

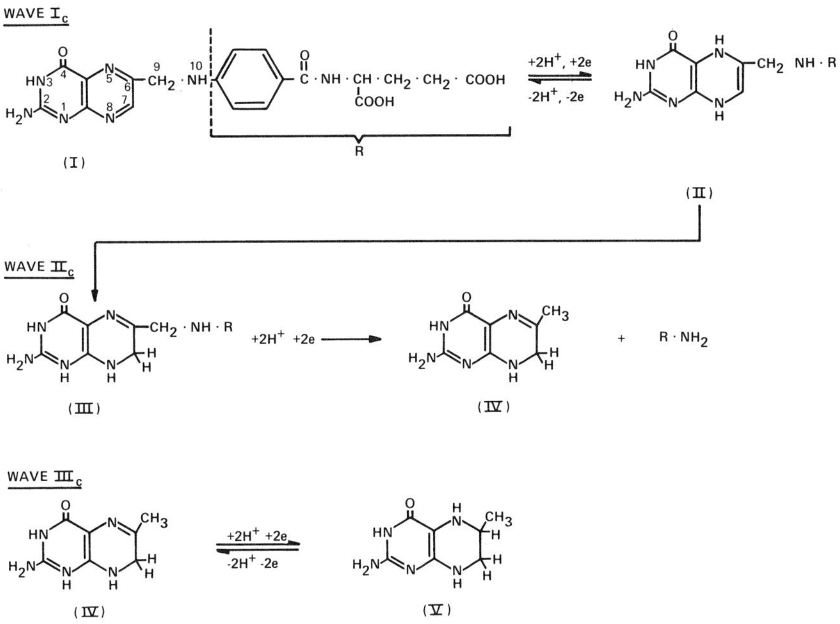

WAVE I_c

(I)

(II)

WAVE II_c

(III)   (IV)

WAVE III_c

(IV)   (V)

FIG. 6-30. Proposed reaction scheme for polarographic reduction of folic acid between pH 5 and 7.[60,61]

latter species is unstable in aqueous solution and gives 7,8-dihydropterin (V, Fig. 6-32) and the alcohol VI (Fig. 6-32). In view of the complexity of this proposed reaction it is not surprising that rather variable n values were obtained on controlled potential electrolysis.

Some additional anodic polarographic waves for 5,6,7,8-tetrahydrofolic acid and 7,8-dihydrofolic acid were observed between about pH 8 and 12.[60,61] The original papers should be consulted for details.

Asahi[59] observed up to five polarographic reduction waves for folic acid, the additional waves being apparently due to dissociation of folic acid or its various reducible reduction products. Although cathodic wave I_c is proposed by Kretzschmar and Jaenicke[60,61] to be a reversible process, oscillopolarographic studies by Hrdý[55] suggest that the process is irreversible.

In view of the apparent discrepancies between the various reports of the polarographic reduction of folic acid and the fact that only a very limited number of electrochemical techniques have been applied to study the reaction, it is to be hoped that a much more detailed mechanistic picture will be forthcoming.

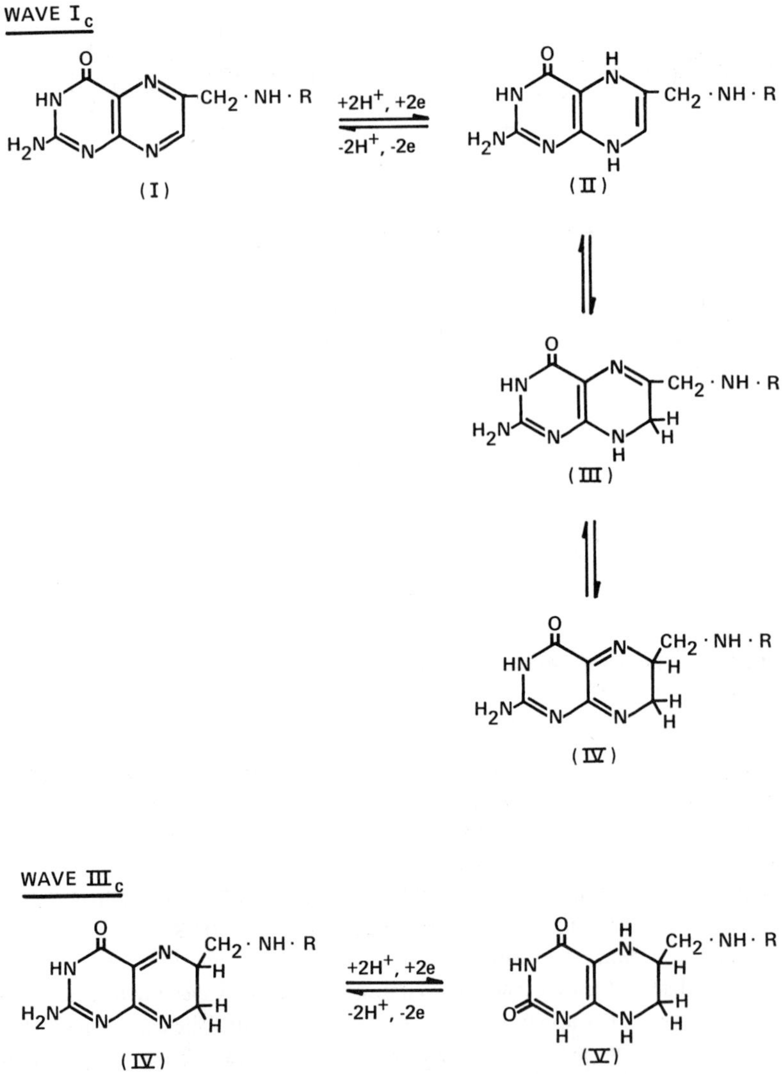

WAVE I_c

WAVE III_c

FIG. 6-31.   Proposed reaction scheme for the polarographic reduction of folic acid at pH 9.[60,61]

M. Other Studies

Komenda[62] and Komenda and Kišová[63] have investigated the electrochemical reductions of other pteridines, especially by use of oscillopolarography; the original papers should be consulted for details. Asahi[64] and Solunina *et al.*[65]

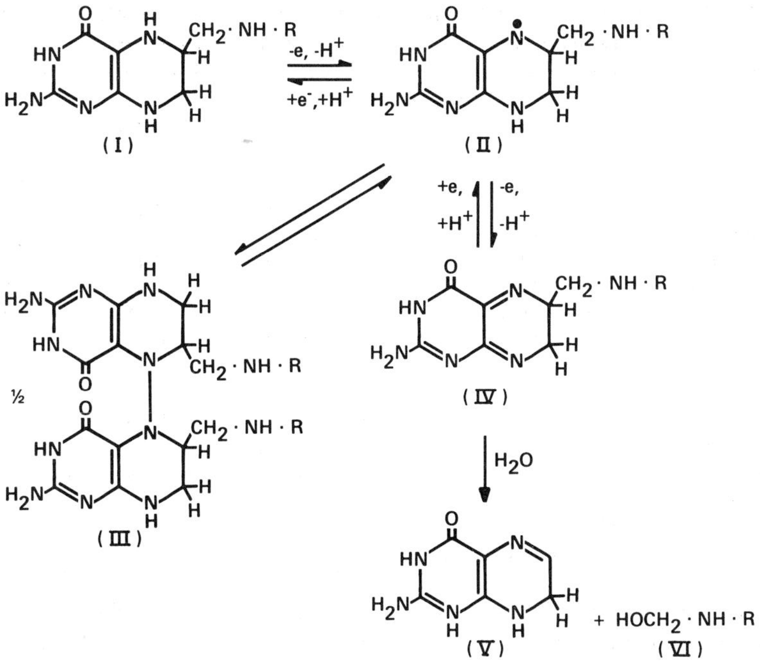

FIG. 6-32. Proposed reaction scheme for the electrochemical oxidation of 5,6,7,8-tetrahydrofolic acid at wave I_a potentials at pH 6.8.[60,61]

have outlined DC polarographic methods for the analytical determination of folic acid and other pteridine derivatives. Rickes and co-workers[66] briefly examined the polarographic behavior of a number of pteridine derivatives, including xanthopterin, rhizopterin, aporhizopterin, and folic acid. In lithium borate medium, pH 9.12, all apparently gave a single wave with $E_{1/2}$ of -0.82 ± 0.03 V versus SCE.

N. Correlations

Komenda[67] has reviewed data for a number of pteridines and has concluded that the presence of the hydroxyl (keto) group at C-4 results in an increase in the chemical stability of the compounds. He also prepared a tabulation of the half-wave potentials for a number of pteridines (presumably for their first waves), which in effect demonstrated the effect of substituents, particularly in the 6 position, of 2-amino-4-hydroxypteridines (pterins) (Table 6-7).[34,42,55,56,67,68]

Komenda[67] also found an approximately linear relationship between the $E_{1/2}$ value at pH 5 of the compounds in Table 6-7 and the logarithm of the

TABLE 6-7   Effect of Substitution on the Half-Wave Potentials for the First Polarographic Reduction Wave of Some Pteridines[a]

Compound	Substituted in position				$E_{1/2}$ at pH 5 (V versus SCE)	Reference
	2	4	6	7		
Pteridine	H	H	H	H	−0.26	42
2-Aminopteridine	NH_2	H	H	H	−0.28	42
4-Hydroxypteridine	H	OH	H	H	−0.46	42
6-Methylpterin	NH_2	OH	CH_3	H	−0.52	34
Pterin 6-alcohol	NH_2	OH	CH_2OH	H	−0.51[b]	55
Pterin 6-aldehyde	NH_2	OH	CHO	H	−0.38[b]	55
Folic acid	NH_2	OH	−[c]	H	−0.56	55,68
6-Methylisoxanthopterin	NH_2	OH	CH_3	OH	−0.91	56
7-Methylxanthopterin	NH_2	OH	OH	CH_3	−0.74	56
Leucopterin	NH_2	OH	OH	OH	−1.13	56

[a] Data from Komenda.[67]

[b] Extrapolated values of $E_{1/2}$.

[c] See Table 6-1 for structure of folic acid.

dissociation constants corresponding to dissociation of the 4-hydroxyl group. A linear relationship between the $E_{1/2}$ and the sum of the polar constants of substituents of pteridine molecules was also obtained. The relationship between the $E_{1/2}$ and the wave number (cm^{-1}) of the second $\pi \to \pi^*$ band of the ultraviolet absorption spectrum of pteridines was also found to be linear. Asahi[69] has similarly examined the relationships between the half-wave potentials and substituents of some 4-hydroxypteridines. Among the compounds that he examined were 2-amino-4-hydroxypteridine-6-carboxylic acid, 2-amino-4-hydroxypteridine-7-carboxylic acid, 2-amino-4-hydroxy-7-methylpteridine, 2-amino-4-hydroxy-6-(D-*arabino*-tetrahydroxybutyl)pteridine, leucopterin, lumazine, and 6,7-dimethyl-8-ribityllumazine. Asahi found that all of these compounds showed a primary electrochemical process that involved a $2e-2H^+$ reduction of the compound to a dihydro derivative. Owing to dissociations of the 4-hydroxyl group this primary electrochemical process was often in the form of two waves, the more positive being due to reduction of the neutral pteridine, the more negative being due to reduction of the anion. 2-Amino-4-hydroxy-7-methylpteridine and lumazine also exhibited further polarographic waves due to further reduction of the dihydro derivatives to the tetrahydro derivatives. Asahi[69] compared the $E_{1/2}$ values of dozens of pteridines and concluded that −CHO and −COOH groups shifted the $E_{1/2}$ to more positive potentials, while −OH, −O−, −CH_3, and alkyl groups shifted $E_{1/2}$ to more negative potentials. It was also concluded that the effect of substituents in the 7 position was greater than in the 6 position.

O. Conclusions

Pteridines, in general, appear to be rather readily reduced electrochemically. Reducible pteridines are reduced in the pyrazine ring and not in the pyrimidine ring. The more recent, detailed investigations of pteridine,[35] various 2-amino-4-hydroxypteridines,[44] folic acid,[60,61] and a related family of compounds (the quinoxalines)[70-72] appear to support the view that these compounds are reducible in an initial $2e-2H^+$ process to an unstable 5,8-dihydro derivative, when formation of such a product is structurally possible. For pteridines that are able to form a 5,8-dihydro derivative, the reduction process is reversible. Once formed, the 5,8-dihydro derivative rapidly rearranges to the more stable 7,8-dihydro derivative. Pteridines such as 6-hydroxypteridine (pteridine-6-one) are structurally unable to form the 5,8-dihydro derivatives upon $2e-2H^+$ reduction and hence are reduced in irreversible processes giving 7,8-dihydro and 5,6-dihydro derivatives, respectively.[48] Thus, the reversibility of the initial $2e-2H^+$ reduction of pteridines appears to depend on the ability of the compound to form a 5,8-dihydro derivative. Clearly a great deal more work is needed to properly define the precise mechanisms and products of both the electrochemical reduction and oxidation of pteridines.

One rather interesting fact in the biochemistry of tetrahydrofolic acid is that metabolic degradation apparently proceeds first by a reductive cleavage at the C-9–N-10 bond to give a 2-amino-4-hydroxy-5,6,7,8-tetrahydropteridine and *p*-aminobenzoylglutamic acid.[73] Such processes are not unlike some of the reductive cleavages of folic acid observed electrochemically between about pH 5 and 7 (see Fig. 6-30 and associated discussion). The substituted tetrahydropteridine formed in the biological cleavage of tetrahydrofolic acid apparently is converted rapidly to a more stable form, possibly as biopterin or isoxanthopterin. Pullman and Pullman[73] have suggested that the isolation of isoxanthopterin (2-amino-4,7-dihydroxypteridine) from biological systems (e.g., human urine[74]), along with the fact that this compound is to some extent the analog in the pteridine series of uric acid in the purine series, indicates an oxidative mechanism for the degradation of the pteridine ring of folic acid that is similar to the mechanism of degradation of purines by xanthine oxidase. Blair[74,75] found that folic acid (II, Fig. 6-33) could be degraded first by a dehydrogenation reaction of the C-9–N-10 bond to give an imide (III, Fig. 6-33) which would readily hydrolyze to the pteridine 6-aldehyde (IV, Fig. 6-33) and *p*-aminobenzoylglutamic acid (V, Fig. 6-33). Xanthine oxidase can oxidize the pteridine 6-aldehyde to the corresponding 6-carboxylic acid (VI, Fig. 6-33), which upon decarboxylation gives 2-amino-4-hydroxypteridine (VII, Fig. 6-33). This in turn can be oxidized to isoxanthopterin (VIII, Fig. 6-33). It is not unlikely that the electrochemical oxidation and reduction processes associated with folic acid and dihydro- and tetrahydrofolic acid could shed considerable

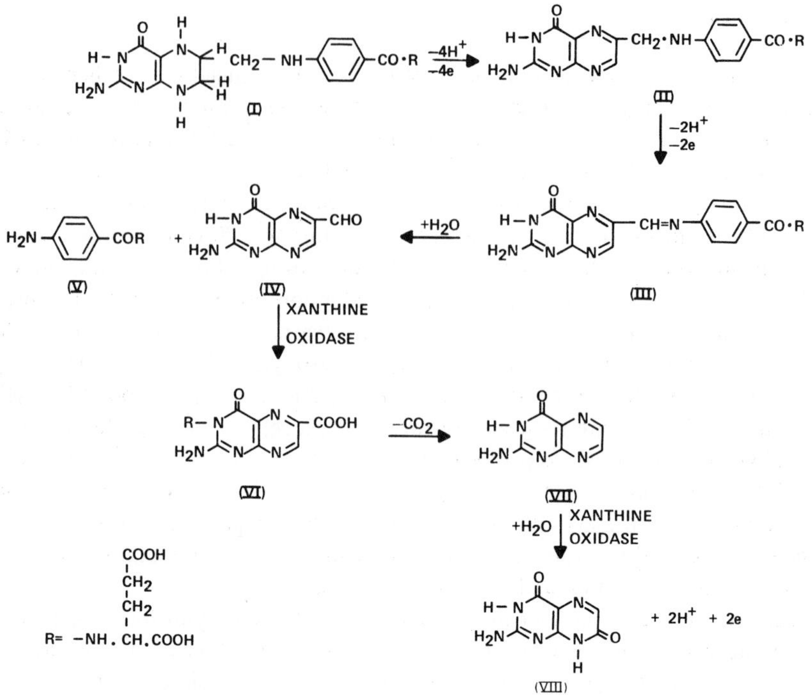

FIG. 6-33.   Possible interpretation of the biological oxidation of folic acid.[74,75]

light on the mechanisms of these processes, intermediates, and stability of products. Molecular orbital calculations[73] certainly indicate that many of the species outlined in Fig. 6-33 should be electrochemically active. The apparent superficial similarity between the modes of xanthine oxidase oxidations of pteridines[76-78] compared to the purines (see Chapter 3 for discussion of purine oxidation and reduction) and the degree of parallelism that is evident between the electrochemical and biological reactions of the latter group of compounds suggest that considerably more detailed effort and study should be devoted to the pteridine family.

REFERENCES

1. R. C. Elderfield and A. C. Mehta, *Heterocycl. Compd.* 9, 1 (1967).
2. M. Gates, *Chem. Rev.* 41, 63 (1947).
3. G. E. W. Wolstenholme and M. Cameron, eds., "Chemistry and Biology of Pteridines," Ciba Found. Symp. Little, Brown, Boston, Massachusetts, 1954.

4. A. Albert, *Q. Rev., Chem. Soc.* **6**, 197 (1952).
5. A. Albert, *Fortschr. Chem. Org. Naturst.* **11**, 350 (1954).
6. W. Pfleiderer, *Angew. Chem.* **75**, 993 (1963); *Angew. Chem., Int. Ed. Engl.* **3**, 114 (1964).
7. W. Pfleiderer and E. C. Taylor, eds., "Pteridine Chemistry," 3rd Int. Symp., 1962. Pergamon, Oxford, 1964.
8. F. G. Hopkins, *Nature (London)* **45**, 581 (1891).
9. F. G. Hopkins, *Phil. Trans. R. Soc. London, Ser. B* **186**, 661 (1893).
10. C. Schopf and H. Wieland, *Ber. Dtsch. Chem. Ges.* **59**, 2067 (1926).
11. H. Wieland and C. Schopf, *Ber. Dtsch. Chem. Ges.* **58**, 2178 (1925).
12. H. Wieland and P. Decker, *Justus Liebigs Ann. Chem.* **547**, 180 (1941).
13. R. Purrmann, *Justus Liebigs Ann. Chem.* **544**, 182 (1940).
14. R. Purrmann, *Justus Liebigs Ann. Chem.* **546**, 98 (1940).
15. A. Albert, J. H. Lister, and C. Pederson, *J. Chem. Soc.* p. 4621 (1956).
16. A. Albert, *Phys. Methods Heterocycl. Chem.* **1**, 71, 77, 81–88, and 92 (1963).
17. S. F. Mason, *Chem. Biol. Pteridines, Ciba Found. Symp., 1954* p. 74 (1954).
18. S. F. Mason, *J. Chem. Soc.* p. 2336 (1955).
19. D. J. Brown and S. F. Mason, *J. Chem. Soc.* p. 3443 (1956).
20. C. W. Waller, B. L. Hutchings, J. H. Mowat, E. L. R. Stokstad, J. H. Boothe, R. B. Angier, J. Semb, Y. SubbaRow, D. B. Cosulich, M. J. Fahrenbach, M. E. Hultquist, E. Kuh, E. H. Northey, D. R. Seeger, J. P. Sickels, and J. M. Smith, *J. Am. Chem. Soc.* **70**, 19 (1948).
21. S. Matsuura and T. Goto, *J. Chem. Soc.* p. 1773 (1963).
22. T. Goto, A. Tatematsu, and S. Matsuura, *J. Org. Chem.* **30**, 1844 (1965).
23. S. H. Hutner, H. A. Nathan, and H. Baker, *Vitam. Horm. (N.Y.)* **17**, 1 (1959).
24. W. Koschara, *Hoppe-Seyler's Z. Physiol. Chem.* **240**, 127 (1936).
25. E. L. Patterson, H. P. Broquist, A. M. Albrecht, M. H. von Saltza, and E. L. R. Stokstad, *J. Am. Chem. Soc.* **77**, 3167 (1955).
26. E. L. Patterson, M. H. von Sultza, and E. L. R. Stokstad, *J. Am. Chem. Soc.* **78**, 5871 (1956).
27. H. K. Mitchell, E. E. Snell, and R. J. Williams, *J. Am. Chem. Soc.* **63**, 2284 (1941).
28. E. E. Conn and P. K. Stumpf, "Outlines of Biochemistry," 2nd ed., p. 175. Wiley, New York, 1967.
29. D. W. Wooley, "A Study of Antimetabolites." Wiley, New York, 1952.
30. S. Farber, *N. Engl. J. Med.* **238**, 787 (1948).
31. C. P. Rhoades, ed., "Antimetabolites and Cancer," p. 199. Am. Assoc. Adv. Sci., Washington, D.C., 1955.
32. W. Jacobson, *in* "Biological Approaches to Cancer Chemotherapy" (R. J. C. Harris, ed.), p. 149. Academic Press, New York, 1961.
33. R. C. Fuller and N. A. Nugent, *Proc. Natl. Acad. Sci. U.S.A.* **63**, 1311 (1969).
34. J. Komenda and D. Laskafeld, *Collect. Czech. Chem. Commun.* **27**, 199 (1962).
35. D. L. McAllister and G. Dryhurst, *J. Electroanal. Chem.* **59**, 75 (1975).
36. D. D. Perrin, *J. Chem. Soc.* p. 645 (1962).
37. A. Albert, T. J. Batterham, and J. J. MacCormack, *J. Chem. Soc. B* p. 1105 (1966).
38. R. S. Nicholson, *Anal. Chem.* **37**, 1351 (1965).
39. W. V. Childs, J. T. Malloy, C. P. Keszthelyi, and A. J. Bard, *J. Electrochem. Soc.* **118**, 874 (1971).
40. Y. Asahi, *Yakugaku Zasshi* **79**, 1554 (1959); *Chem. Abstr.* **54**, 10592 (1960).
41. J. Komenda, *Chem. Listy* **52**, 1065 (1958).
42. J. Komenda, *Collect. Czech. Chem. Commun.* **24**, 903 (1959).

43. Y. Asahi, *Abh. Dtsch. Akad. Wiss., Berlin, Kl. Chem., Geol. Biol.* p. 74 (1964); *Chem. Abstr.* **62**, 14181 (1965).
44. S. Kwee and H. Lund, *Biochim. Biophys. Acta* 297, 285 (1973).
45. S. Kaufman, *J. Biol. Chem.* 239, 332 (1964).
46. D. B. Fisher and S. Kaufman, *Biochem. Biophys. Res. Commun.* 38, 663 (1970).
47. Y. Inone and D. D. Perrin, *J. Chem. Soc.* p. 2600 (1962).
48. D. L. McAllister and G. Dryhurst, *J. Electroanal. Chem.* 47, 479 (1973).
49. D. L. McAllister and G. Dryhurst, *J. Electroanal. Chem.* 55, 69 (1974).
50. G. Dryhurst and D. L. McAllister, *Anal. Chim. Acta* 72, 209 (1974).
51. Y. Asahi, *Yakugaku Zasshi* 79, 1559 (1959); *Chem. Abstr.* 54, 10592 (1960).
52. A. Albert, D. J. Brown, and G. W. H. Cheeseman, *J. Chem. Soc.* p. 4219 (1952).
53. Y. Inoue and D. D. Perrin, *J. Chem. Soc.* p. 2600 (1962).
54. Y. Asahi, *Yakagaku Zasshi* 79, 1565 (1959); *Chem. Abstr.* 54, 10593 (1960).
55. O. Hrdý, *Collect. Czech. Chem. Commun.* 24, 1180 (1959).
56. J. Komenda, L. Kišová, and J. Koudelka, *Collect. Czech. Chem. Commun.* 25, 1020 (1960).
57. W. Pfleiderer and M. Rukweid, *Chem. Ber.* 94, 118 (1961).
58. Y. Asahi, *Yakugaku Zasshi* 79, 1570 (1959); *Chem. Abstr.* 54, 10593 (1960).
59. Y. Asahi, *Yakugaku Zasshi* 79, 1548 (1959); *Chem. Abstr.* 54, 10592 (1960).
60. K. Kretzschmar and W. Jaenicke, *Z. Naturforsch., Teil B* 26, 225 (1971).
61. K. Kretzschmar and W. Jaenicke, *Z. Naturforsch., Teil B* 26, 999 (1971).
62. J. Komenda, *Pteridine Chem., Proc. Int. Symp. 3rd, 1962* p. 511 (1964); *Chem. Abstr.* 62, 11667 (1965).
63. J. Komenda and L. Kišová, *Chem. Zvesti* 16, 368 (1962); *Chem. Abstr.* 59, 1286 (1963).
64. Y. Asahi, *Sankyo Kenkyusho Nempo* 17, 1 (1958); *Chem. Abstr.* 53, 6535 (1959).
65. I. A. Solunina, V. A. Devyatnin, and T. N. Kuznetsova, *Farmatsiya (Moscow)* 17, 45 (1968); *Chem. Abstr.* 69, 5249 (1968).
66. E. L. Rickes, N. R. Trenner, J. B. Conn, and J. C. Keresztesy, *J. Am. Chem. Soc.* 69, 2751 (1947).
67. J. Komenda, *Collect. Czech. Chem. Commun.* 27, 212 (1962).
68. P. Zuman, *Collect. Czech. Chem. Commun.* 15, 1107 (1950).
69. Y. Asahi, *Yakugaku Zasshi* 79, 1574 (1959); *Chem. Abstr.* 54, 10592 (1960).
70. T. Goto, A. Tatematsu, and S. Matsuura, *J. Org. Chem.* 30, 1844 (1965).
71. M. Strier and J. Cavagnol, *J. Am. Chem. Soc.* 79, 4331 (1957).
72. J. Pinson and J. Armand, *Collect. Czech. Chem. Commun.* 36, 385 (1971).
73. B. Pullman and A. Pullman, "Quantum Biochemistry," pp. 390–391, and references quoted therein. Wiley (Interscience), New York, 1963.
74. J. A. Blair, *Biochem. J.* 68, 385 (1958).
75. J. A. Blair, *Biochem. J.* 65, 209 (1957).
76. F. Bergmann and M. Kwietny, *Biochim. Biophys. Acta* 28, 613 (1958).
77. F. Bergmann and M. Kwietny, *Biochim. Biophys. Acta* 33, 29 (1959).
78. F. Bergmann and M. Kwietny, *J. Chromatogr.* 2, 162 (1959).

7

Isoalloxazines, Flavins, and Flavin Nucleotides

I. INTRODUCTION, NOMENCLATURE, AND STRUCTURE

The terms flavin and isoalloxazine can be considered interchangeable. The flavin or isoalloxazine ring can be best described as the structure formed by condensation of o-phenylenediamines with pyrimidines. The basic isoalloxazine ring structure and the usual numbering system are illustrated in structure **I**.

(I)

The principal interest in the chemistry of the isoalloxazine systems stems quite simply from the fact that vitamin B_2 or riboflavin is a member of this family of compounds. The structures of riboflavin and several other biologically important flavins are shown in Table 7-1. Clearly, riboflavin consists of the sugar alcohol, D-ribitol, attached at the N-10 position to the dimethylated isoalloxazine ring. Riboflavin occurs in nature almost exclusively as a constituent of the two flavin coenzymes, flavin mononucleotide (FMN) and flavin adenine dinucleotide (FAD) (Table 7-1).

TABLE 7-1   Some Biologically Important Flavins

Name	Structure
Lumichrome	
Lumiflavin	
Riboflavin	
Flavin mononucleotide (FMN)	

TABLE 7-1 *Continued*

Name	Structure

Flavin adenine dinucleotide (FAD)

II. BIOLOGICAL SIGNIFICANCE OF FLAVINS

Flavin is a trivial name given to the redox-active prosthetic group of respiratory enzymes, known as flavoproteins, which occur widely in plants and animals. The flavin itself constitutes part of the so-called flavocoenzyme, which does not contain protein. There are two main groups of flavoproteins, one of which contains iron or iron and molybdenum, and the other of which contains no metals.

Flavins or, more correctly, flavoproteins are enzymes that catalyze oxidation–reduction reactions in biological systems. Both groups of flavoproteins (those with and without metals) as well as the free coenzymes[1] react with suitable substrates in an overall two-electron process, the flavin being reduced and the substrate being oxidized. The electrons accepted by the flavin are then generally transferred very rapidly to some other suitable electron acceptor, often oxygen. The product is then hydrogen peroxide and the regenerated protein—FAD. For example, a typical oxidative reaction of a flavoprotein enzyme that is important in intermediary metabolism involves the enzyme succinic dehydrogenase. This enzyme contains FAD as a prosthetic group and catalyzes the oxidation of succinate (I, Fig. 7-1) to fumarate (II, Fig. 7-1). The flavoprotein enzymes are also often found in the redox systems involving pyridine nucleotides. Often they serve to reoxidize reduced pyridine nucleotides.

Typical of other substrates of flavoprotein-containing enzymes are dihydrolipoamide, glucose, purines, amino acids, and, as mentioned, dihydropyridine nucleotides and succinate. Quite clearly, therefore, flavo-proteins are vitally important constituents of the metabolic processes of respiratory organisms. Hemmerich, Veeger, and Wood[2] have described some of the chemistry and biochemistry of flavins. This article should be consulted for further, detailed information.

A considerable amount of evidence, much of it via electrochemical techniques (*vide infra*) and some via ESR spectroscopy,[3,4] has indicated that the $2e$ reduction of a flavin occurs by way of two separate $1e$ steps. After the addition of a single electron (and proton) to the flavin (I, Fig. 7-2), a partially reduced flavin is formed which is generally known as a semiquinone (II, Fig. 7-2). Addition of a further electron and proton results in the formation of the dihydroflavin (III, Fig. 7-2). The latter species is often referred to as a flavohydroquinone. For more details on the biochemistry of the flavoprotein enzymes other more authoritative and extensive reviews should be consulted.[5-8]

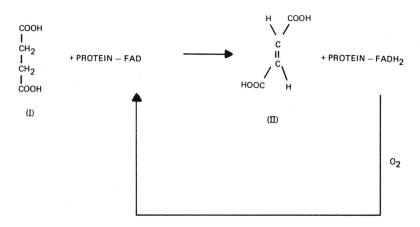

FIG. 7-1.   Oxidation of succinate to fumarate by FAD present in succinic dehydrogenase.

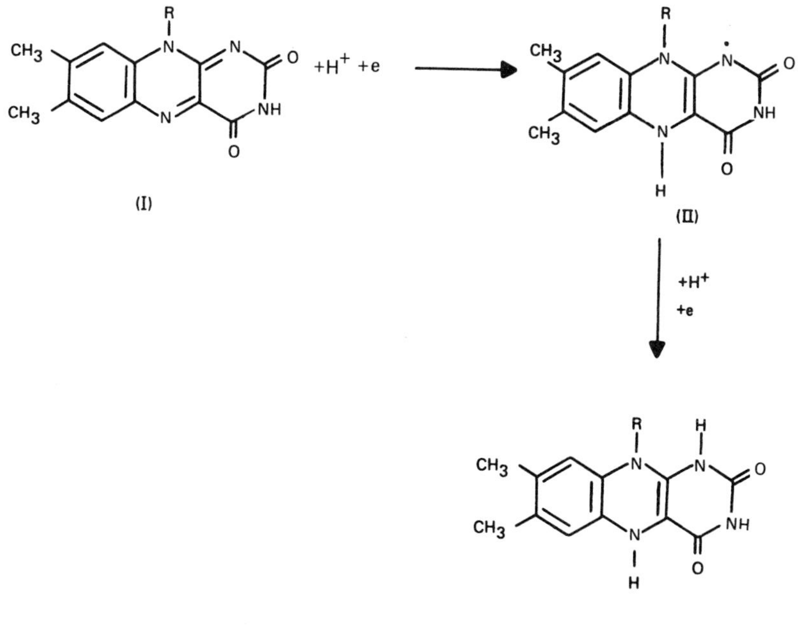

FIG. 7-2. Probable course of reduction of flavins (I) to flavin semiquinones (II) and dihydroflavins (III).

III. ELECTROCHEMISTRY OF FLAVINS

A. Potentiometric Studies

It appears that all of the substituted isoalloxazine derivatives considered in this chapter form thermodynamically reversible redox systems. For this reason they have been studied quite extensively by potentiometric techniques.[9-17] The potentiometric data indicate the formation of semiquinones and in some cases, particularly at higher concentrations, dimeric intermediates for riboflavin,[9,10,12-14,16] FMN, and FAD.[18,19] Generally, however, potentiometric titration curves do not show two distinct breaks, but rather the two 1e processes overlap. Formation constants for the intermediate semiquinones have been calculated from the potentiometric data.[10,13-15] Clark,[20] however, has pointed out that experimental potentiometric titration curves show only small departures from those expected for a straightforward 2e process. In view of this, and small experimental errors, there are considerable discrepancies in the literature regarding values of standard potentials and semiquinone formation constants. The potentiometric data for substituted isoalloxazines, including riboflavin and other related derivatives, have been critically summarized by Clark.[20]

Rather interestingly, Haas[21] concluded from potentiometric studies that the attachment of riboflavin to its protein carrier in enzymes favored semiquinone formation.

B. Polarographic and Related Studies on Riboflavin

Studies of riboflavin in aqueous solution are complicated by the existence of a number of ionized forms of the oxidized, semiquinone (free-radical), and reduced forms at different pH values. Thus, a total of at least nine monomolecular flavin species can be distinguished, i.e., one cationic, one neutral, and one anionic species in each redox state. Hemmerich and co-workers[2] have represented the flavin redox system as shown in Fig. 7-3.[2,22]

A large number of workers have found that riboflavin gives rise to a reversible 2e polarographic reduction wave at the DME over the normal pH range. Brdička[23–27] first observed the polarographic reduction of riboflavin, and his observations have been essentially verified by many other studies.[28–31] Careful analysis of the polarographic wave using the index potential method of Michaelis[14,32] has indicated that the polarographic reduction proceeds by way of two overlapping 1e steps, i.e., with intermediate formation of a semiquinone or free-radical species. Janik and Elving[22] reviewed the electrochemistry of riboflavin and summarized the dependence of the half-wave potential on pH, as shown in Fig. 7-4. The wave that follows the relationships a, b, and c in Fig. 7-4 is the faradaic overall 2e process for reduction of riboflavin to the dihydro species (see Fig. 7-3). (Wave d, which is represented by the solid line in Fig. 7-4, is discussed in Section III, C.) The two changes in slope of the linear $E_{1/2}$–pH relationships occur at pH values that correspond closely to the acid dissociation constants of the reduced and oxidized forms of riboflavin, i.e., 6.05 and 9.95[25,33] (Fig. 7-3). Brdička and Knobloch[25] determined the $E_{1/2}$ versus pH relationships at 20°C for riboflavin from pH 1.81 to 11.98 and calculated the equation of $E_{1/2}$ (volts versus NCE) as follows:

$$E_{1/2} = 0.188 + 0.029 \log \frac{5.01 \times 10^{-7} + [H^+]}{6.31 \times 10^{-10} + [H^+]} - 0.058 \text{ pH} \qquad (1)$$

These workers concluded that the polarographic values of $E_{1/2}$ were essentially identical to the potentiometric oxidation–reduction potentials.

Kaye and Stonehill[34] more recently confirmed the overlapping nature of the two 1e polarographic reduction waves of riboflavin. At 25°C and pH 7.38 they

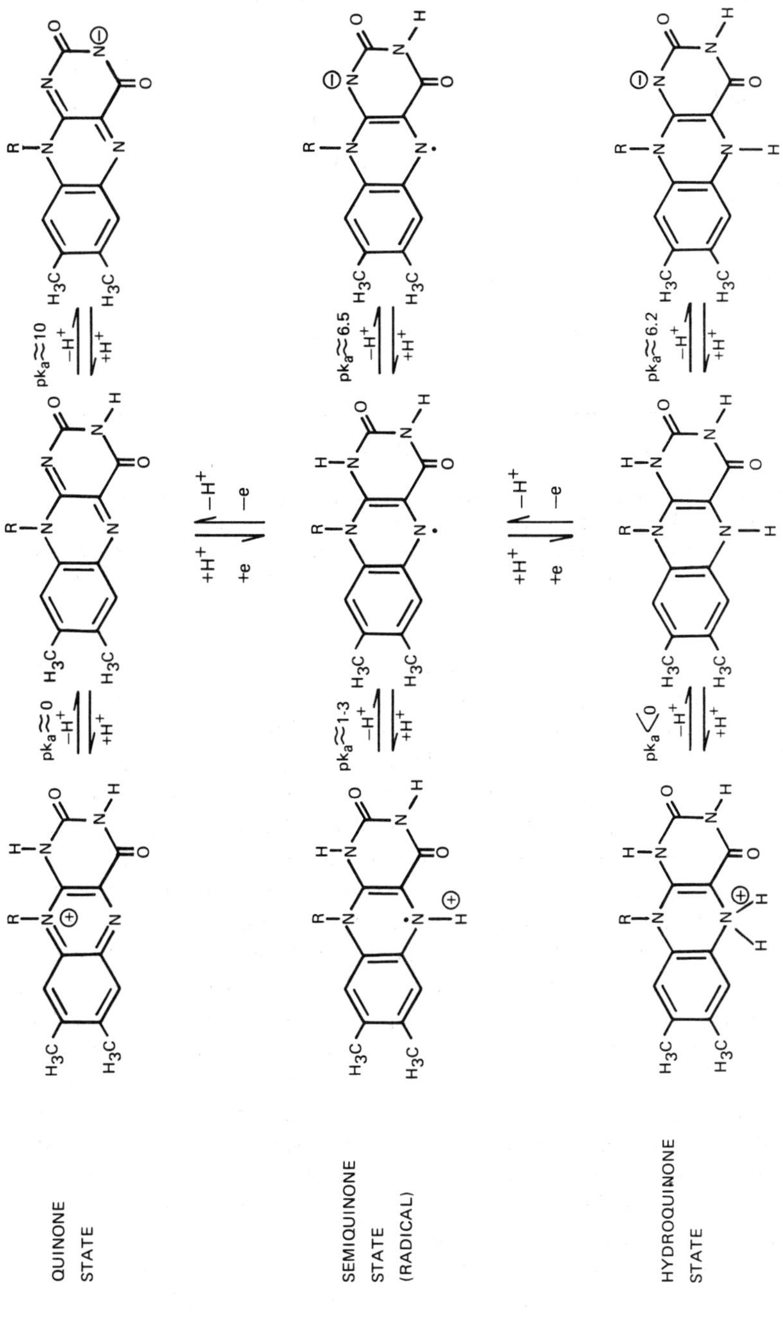

FIG. 7-3. The flavin redox system. Occurrence of the flavin species as a function of pH and redox state. (Adapted from Hemmerich et al.[2] and Janik and Elving.[22])

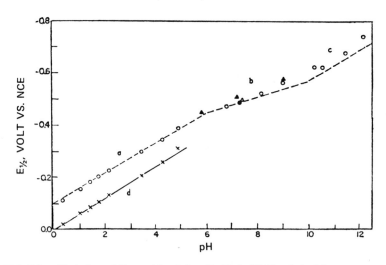

FIG. 7-4.   Variation of $E_{1/2}$ with pH for riboflavin.[22] The dashed line represents the 2e faradaic wave[25,41] with concentration 0.3 mM; open circles (0.475 mM)[33]; solid circle[31]; open triangle (1 mM)[34]; solid triangles.[28] The continuous line represents the adsorption prewave at 0.475 mM.[33] (Reprinted with permission of the American Chemical Society.)

calculated a semiquinone formation constant of 1.49. On the basis of this and related data they developed a tentative theory for the mechanism of action of bacterial respiratory catalysts and anticatalysts.[35] Thus, it was supposed that bacterial respiration was a radical chain reaction of some type and that substances which provide relatively labile semiquinone (free) radicals could participate in the electron-transport reaction, thus acting as catalysts.[36] However, they considered that compounds which readily formed highly stable semiquinones on reduction, especially if no proton was taken up along with the electron, would act as chain breakers (anticatalysts or antibacterials) by reacting with the labile chain radical to form a stable semiquinone. They therefore acted as anticatalysts or antibacterials. In view of the observed potentiometric and polarographic behavior of riboflavin, it appeared that riboflavin should (and indeed does) act as a respiratory catalyst. Merkel and Nickerson[31] also examined the action of riboflavin as a photocatalyst and hydrogen carrier in photochemical reductions. Their polarographic data supported the ease and reversibility of riboflavin oxidation–reduction behavior. Polarography was used to confirm the fact that, when riboflavin absorbs visible light, it becomes reduced, and in turn reduced riboflavin can transfer its electrons to suitable hydrogen acceptors added to the system.

In a rather different vein, Manoušek and Jiřička[37] employed the electrochemical reduction of riboflavin to its rather insoluble dihydro derivative as a means of isolation and purification of riboflavin from diverse biological sources.

Bard and co-workers[38] have reported a study of the electrochemical reduction of riboflavin in dimethyl sulfoxide. Riboflavin is quite readily soluble in this solvent. A nonaqueous solvent was employed in the hope of clearly observing the two 1e reduction processes separately, in much the same way as such processes are observed with aromatic hydrocarbons in nonaqueous (but not in aqueous) media.[39] In dimethyl sulfoxide (DMSO) riboflavin does not exhibit adsorption phenomena at mercury or platinum electrodes. Using sodium perchlorate as the supporting electrolyte, Bard *et al.*[38] observed three DC polarographic reduction waves (Fig. 7-5). Alternating current polarography showed two peaks at potentials corresponding to the first two DC polarographic waves. The diffusion current constant for wave I, at $E_{1/2} = -0.71$ V versus SCE, was that expected for a 1e process. The sum of the heights of waves II and III was about equal to the height of wave I. Cyclic voltammetry of riboflavin at a hanging mercury drop electrode in DMSO (Fig. 7-6) showed, at low concentrations (ca. 1 m*M*), a well-defined reduction peak at $E_p = -0.77$ V and a second broad peak at $E_p = -1.3$ V. However, at higher concentrations of

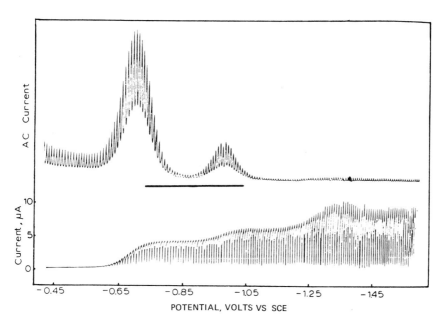

FIG. 7-5. Polarograms for the reduction of riboflavin (3 m*M*) in DMSO containing 0.8 *M* NaClO₄. Upper curve, AC polarogram, with 10 mV, 200 Hz superimposed; lower curve, DC polarogram.[38] (Reprinted with permission of Elsevier Publishing Company, Amsterdam.)

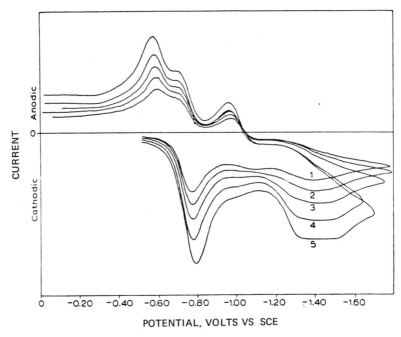

FIG. 7-6.   Cyclic   voltammetry   of   riboflavin   (1.4 m*M*)   in   DMSO   containing
0.1 *M* NaClO₄.   Scan   rates: (1) 152,   (2) 222,   (3) 312,   (4) 476,   (5) 714 mV sec⁻¹.[38]
(Reprinted with permission of Elsevier Publishing Company, Amsterdam.)

riboflavin, e.g., 3.2 m*M*, another peak corresponding to the DC polarographic wave II with $E_p = -1.07$ V appeared. The peak current function (proportional to $i_p/v^{1/2}$, where i_p is the peak current and v is the scan rate) for the first peak was constant with scan rate, indicating a diffusion-controlled process. The current function of the second peak decreased with increasing scan rate, suggesting that the process responsible for this peak was reduction of a substance formed by a homogeneous chemical reaction of a product formed during the first reduction wave. By clipping the cyclic voltammogram after the appearance of the first cathodic peak, two anodic peaks were observed, $E_p = -0.72$ and -0.57 V (Fig. 7-7). The current function for the first anodic peak ($E_p = -0.72$ V) was independent of scan rate, although the height of the anodic peak was slightly less than that of the first cathodic peak. The second anodic peak ($E_p = -0.57$ V) increased in height with increasing scan rate; at very low scan rates it was almost absent. These data again suggested that the substance giving rise to the second anodic peak was decomposing by a homogeneous chemical reaction. Addition of a large excess of proton donor, such as hydroquinone, resulted in the doubling of the height of the first cathodic polarographic wave and the disappearance of

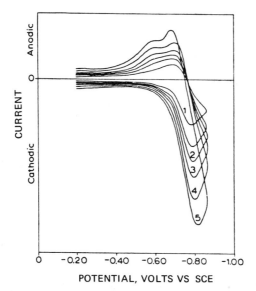

FIG. 7-7. Cyclic voltammetry of riboflavin for reversal following first peak. Scan rates: (1) 67, (2) 222, (3) 312, (4) 476, (5) 714 mV sec[-1].[38] (Reprinted with permission Elsevier Publishing Company, Amsterdam.)

the third cathodic wave. However, the second cathodic wave (or a new wave at about the same potential) remained.

Coulometry at potentials corresponding to the first cathodic polarographic wave revealed that one electron was transferred and that the product showed an ESR spectrum similar to that observed for the riboflavin semiquinone species in aqueous solution.[3,4] The product also showed an anodic polarographic wave at about −0.7 V, although the height of this wave was 30–40% smaller than the original height of the first cathodic wave of riboflavin. Reversal coulometry also revealed, in agreement with the polarographic data, that only ca. 60% of the product of reduction could be reoxidized back to riboflavin. Accordingly, Bard *et al.*[38] proposed a reaction scheme whereby riboflavin (I, Fig. 7-8) in DMSO was first reduced in a $1e$ reaction to the riboflavin anion radical (II, Fig. 7-8). The data, however, indicated that the anion radical was not stable, and it was proposed that the experimental observations could best be explained by assuming that the decomposition of II occurred by parallel or simultaneous reactions to A and B (Fig. 7-8). Species A, which was not identified, was thought to be the substance reduced at potentials of the second cathodic wave ($E_p = -1.07$ V), while B⁻ was the anion radical of an unidentified substance B whose electrochemical properties were very similar to those of riboflavin itself. This implied that the radical observed by ESR was B⁻ rather than II, (Fig. 7-8). The third cathodic polarographic wave was proposed to be composed of the

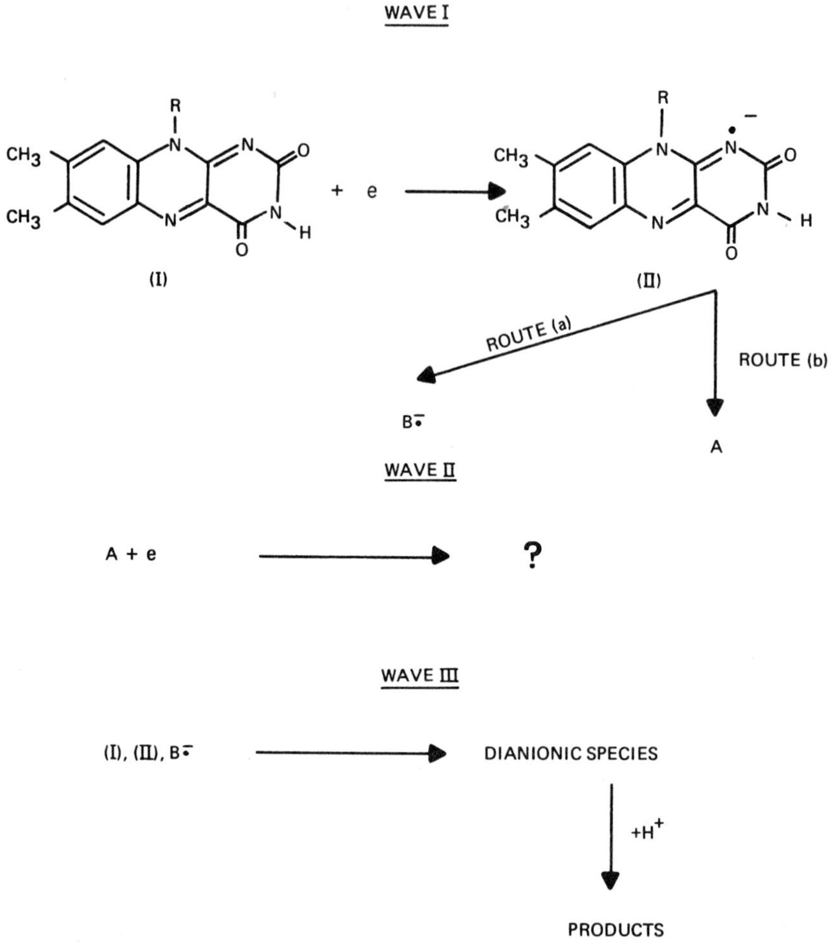

WAVE I

(I)   (II)

ROUTE (a)

ROUTE (b)

B⁻

A

WAVE II

A + e ⟶ ?

WAVE III

(I), (II), B⁻ ⟶ DIANIONIC SPECIES

+H⁺

PRODUCTS

FIG. 7-8.   Proposed reaction scheme for the electrochemical reduction of riboflavin in dimethyl sulfoxide.[38]

reductions of I, II, B⁻, and B (Fig. 7-8), although the source of the latter species is obscure, to give dianions which would then become protonated (Fig. 7-8). The anodic waves observed by cyclic voltammetry were ascribed to the oxidation of B⁻ and II (Fig. 7-8) and perhaps other species. Although some attempts were made to classify the products of the various electrochemical and chemical reactions involved in these processes, no definite conclusions could actually be drawn regarding the details of the mechanism.

Ostrowski and Krawczyk[40] found that the riboflavin proteins that occur in egg white and yolk are not reducible at the DME and that they are incapable of oxidizing reduced pyridine nucleotides.

C. Adsorption of Riboflavin and Dihydroriboflavin

It is evident from Fig. 7-4 that up to about pH 6 riboflavin exhibits two polarographic waves. The more negative of these waves is a normal faradaic wave[41] and has been discussed in Section III, B. The wave occurring at more positive potentials, however, is an adsorption prewave. This prewave or, as it is sometimes called, "anomalous" wave was first observed by Brdička.[25] After studying a similar prewave observed with methylene blue,[42] Brdička[23] concluded that the prewave appearing before the main reduction wave of riboflavin was due to the adsorption of the product of riboflavin reduction, i.e., dihydroriboflavin or, as it is sometimes called, leucoriboflavin. After a careful analysis of the current−voltage curves with due consideration for the formation of semiquinones, Brdička[23] concluded that only the dihydroriboflavin was adsorbed (see later discussion). However, by studying current−time curves, it was found that at certain applied potentials two delayed maxima occurred. These were thought to result from the separate adsorption of both the semiquinone and dihydro forms of riboflavin. Brdička[23] proposed that the delays were due to an incubation period in which the originally nonadsorbable forms of the reduction product are converted into isomers that are adsorbed. The pK_a of dihydroriboflavin is 6.3, so that at pH values greater than this value the principal solution form of the compound is the anionic form. It appeared that the anionic form of the dihydroriboflavin was not adsorbed since no prewave appears above ca. pH 6.

Using AC polarography, Breyer and Biegler[43−45] found depressions of the alternating base current on both sides of the faradaic AC peak of riboflavin except in very acidic solutions (Fig. 7-9). Since the base current depressions at potentials negative of the AC peak persisted at least until pH 9.5, it was apparent that dihydroriboflavin was in fact adsorbed up to that pH. The implication here is that above about pH 6 the anionic form of dihydroriboflavin is adsorbed, although presumably not so strongly as the neutral form since no DC polarographic prewave was observed. The fact that a base current depression occurred before the AC peak indicated that riboflavin itself was also adsorbed. The base current depressions on the positive side of the main AC peak were observed in all but very acidic solutions, which clearly indicated that both the uncharged and anionic forms of riboflavin are adsorbed at the DME. In acid solutions adsorption of the riboflavin molecule disappeared somewhere between pH 1 and 2 (Fig. 7-9), which confirmed the fact that the pK_a of the protonated form of riboflavin lay somewhere within the same range. It was also found that riboflavin gave rise to an AC peak at potentials much more positive than the

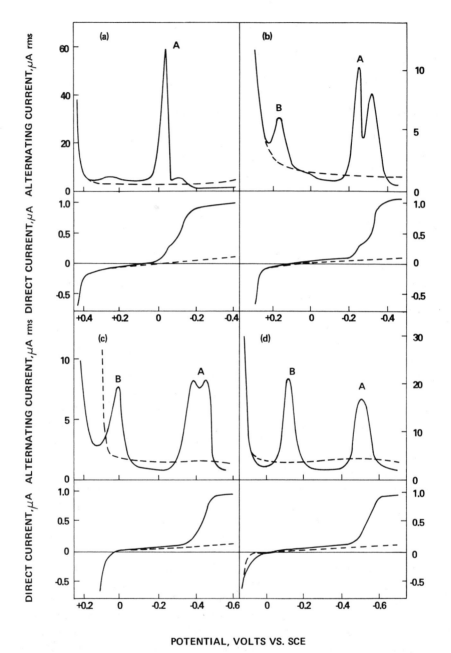

FIG. 7-9. Alternating and direct current polarograms, showing base and residual currents, for $1.5 \times 10^{-4}\,M$ riboflavin in the following supporting electrolytes: (a) M NaClO$_4$ + 0.1 M HClO$_4$; alternating voltage, 15 mV. (b) Acetate buffer, pH 4.7; alternating voltage, 5 mV. (c) M Ammonium acetate, pH 6.9; alternating voltage, 5 mV. (d) M KNO$_3$ + 0.1 M borax; pH 9; alternating voltage, 15 mV.[45]

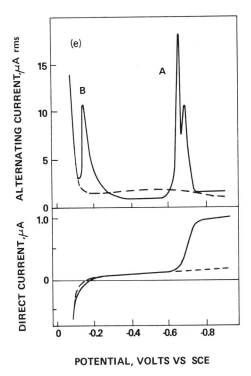

FIG. 7-9. (e) M KNO$_3$ + 0.1 M KOH; alternating voltage, 5 mV.[45]

main faradaic AC peak (peak B, Fig. 7-9). The height of this positive peak increased with increasing riboflavin concentration but only up to a limit of about 0.06 mM. A DC polarographic wave was not generally observed in the same potential region as the latter AC peak. Base current depressions were often observed both to the positive and to the negative sides of the positive AC peak. In addition, the summit potential of the peak did not follow the concentration and temperature dependencies expected for a normal tensammetric (adsorption–desorption) AC peak (see Chapter 2). It was therefore proposed that the process responsible for the positive peak was formation of a compound between riboflavin (R$_f$H) and mercury, according to a reaction of the type shown in Eq. 2.

$$2R_fH + nHg_2^{2+} \rightleftharpoons 2Hg_nR_f + 2H^+ \qquad (2)$$

Some rather nice AC and DC polarograms of riboflavin are presented in Fig. 7-9. It is clear from some of these polarograms that at certain pH values the

DC polarographic prewave and main wave can also be resolved into two well-formed AC peaks.[45]

Over the past five or six years a number of workers have employed chronopotentiometric methods to study the adsorption of riboflavin and its reduction products. Bard and co-workers[46] used a current reversal method and found that dihydroriboflavin was adsorbed at mercury electrodes, the extent of adsorption varying with the magnitude of the current density employed. Tatwawadi and Bard[47] also used a chronopotentiometric method and electrocapillary studies to show that riboflavin itself was adsorbed at mercury, thus confirming the AC polarographic conclusion that both riboflavin and its reduction product are adsorbed. In the latter two reports extensive data on the surface excess values of riboflavin (Γ) were calculated. However, in view of the reports of Lingane,[48,49] it appears that the mathematical and chemical models employed by Bard *et al.*[46,47] were incorrect, so that the values for the extent of adsorption of riboflavin and dihyroriboflavin are also apparently incorrect. Laitinen and Chambers[50] also seem to consider the chronopotentiometric method to be of only qualitative usefulness in the study of adsorption, and indeed with the more recent advent of chronocoulometry[51] it seems pointless continuing with the former method for adsorption studies. Nevertheless, in all fairness it should be pointed out that various workers, using a variety of techniques, have found that the amount of riboflavin and dihydroriboflavin adsorbed at the electrode, Γ, is rather constant at around $0.15-0.20 \times 10^{-9}$ mole/cm[2,23,43,46,52-54] for both compounds.

Khmel'nitskya and co-workers[55] have further studied the mechanism of prewave formation in the riboflavin–dihydroriboflavin system at mercury electrodes using chronopotentiometry and a faradaic impedance method. Both methods confirmed previous findings, i.e., that both riboflavin and dihydroriboflavin are adsorbed. However, these workers proposed that the reduction of riboflavin results in formation of a charge-transfer complex at the electrode surface. This was thought to be a complex of two identical semiquinones or a complex of oxidized and reduced forms of riboflavin situated in parallel layers and stabilized by π-orbital overlap. They therefore assumed that the appearance of a prewave connected with the adsorption of dihydroriboflavin depended on the presence of a layer of riboflavin molecules on the electrode surface which behaved as active centers for the adsorption of one or the other of the reduced forms of riboflavin.

D. Catalytic Hydrogen Reduction Waves

Knobloch[56] found that riboflavin, FAD, and FMN give rise to a very large catalytic hydrogen reduction wave at potentials more negative than the normal reduction wave. The catalytic wave appeared up to about pH 8. Since the height

$$(RH_2)_{ads} + H^+ \longrightarrow (RH_3)^+$$

$$(RH_3)^+ + e \longrightarrow \cdot RH_3$$

$$2 \cdot RH_3 \longrightarrow (RH_3 \cdot RH_3)$$

$$(RH_3 \cdot RH_3) \longrightarrow 2 RH_2 + H_2$$

FIG. 7-10. Schematic representation of the process responsible for the catalytic hydrogen discharge wave observed on polarography of some isoalloxazine derivatives below pH 8; RH_2 is the reduced form of the compound, e.g., dihydroriboflavin.[56]

of the catalytic wave decreased with increasing temperature, it was assumed that the catalyst was adsorbed onto the electrode surface. Surprisingly, isoalloxazine derivatives having no sugar moiety gave a much smaller catalytic wave, while lumichrome gave none. On the basis of his data, Knobloch[56] concluded that the active form of the catalyst at the mercury cathode was the reduced, protonated form of the isoalloxazine derivative. The mechanism proposed by Knobloch[56] is shown schematically in Fig. 7-10. Subsequently, Jambor[57,58] confirmed that in 1 M KCl and phosphate buffers at low pH, riboflavin gave a catalytic polarographic wave that was a thousand times higher than the normal reduction wave. At concentrations of riboflavin greater than about 3 mM, a second catalytic wave was observed. In view of the very large current for the catalytic waves of riboflavin and related derivatives, very low concentrations of these compounds can be determined by ordinary DC polarography.[56-58]

E. Analytical Applications

Apart from the references already quoted, which themselves often contain many analytically useful data, a number of methods have appeared dealing with the polarographic determination of riboflavin,[59-64] its determination in various vitamin preparations,[28,33,65-80] *Butyogenous clostridii* cultures,[81] photochemical reactions[82] in animal tissue and fluids,[87,88] and foods.[89,90] Riboflavin has also been determined by oscillopolarography[91] and AC polarography.[92]

F. Polarography and Related Studies of Flavin Adenine Dinucleotide and Flavin Mononucleotide

Kuhn and Boulanger,[10] many years ago, determined the standard oxidation–reduction potential of FMN by potentiometric titration. At 20°C and pH 7 the $E^{0'}$ value for FMN was found to be -0.185 V versus the normal hydrogen electrode (NHE). Flavin mononucleotide has also been studied potentiometrically by many other workers.[9,14–16,93] Flavin adenine dinucleotide has been studied potentiometrically to a much lesser extent. Ball[94] has reported an $E^{0'}$ value of -0.250 V for FAD at pH 7.8 and 30°C, while Clark[95] has indicated that the FAD system has a standard potential of about -0.22 V at pH 7.

Asahi[33] and Ke[96] independently reported the polarography of FAD. In much the same way as was observed for riboflavin, FMN and FAD show a 2e DC polarographic reduction wave which is due to two overlapping 1e steps, indicating that the first step in the reduction is due to semiquinone formation.[19,33,77,96] The polarographic wave is complicated somewhat by the formation of adsorption prewaves and postwaves (*vide infra*). Janik and Elving[22] have combined much of the polarographic $E_{1/2}$ data for both FMN and FAD from several sources into a single curve (Fig. 7-11). The first inflection point occurs at about pH 6.8 and corresponds to the dissociation constant of the reduced forms of FMN and FAD. The second inflection at pH 9.65 occurs at the dissociation constant of the oxidized form. The $E_{1/2}$ values of the normal wave of FMN and FAD (a, b, c, Fig. 7-11) correspond closely to the variation of the $E^{0'}$ value with pH obtained in the potentiometric study of Kuhn and

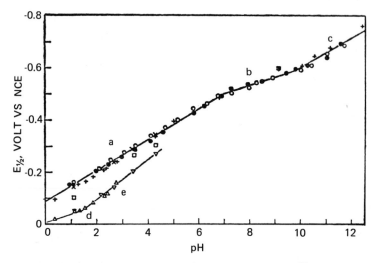

FIG. 7-11. Variation of $E_{1/2}$ with pH for FMN and FAD.[22] Lines a, b, and c correspond to the main reduction wave; lines d and e correspond to the adsorption prewave. The main reduction wave of FAD: closed circles.[96] Pluses, concentration 0.45 mM[33]; crosses, 0.092 mM[33]; squares, 0.0184 mM.[33] Adsorption prewave of FAD: triangles, 0.46 mM[33]; reversed triangles, 0.092 mM.[33] [Reprinted with permission from B. Janik and P. J. Elving, *Chem. Rev.* **68**, 295 (1968). Copyright by the American Chemical Society.]

Boulanger.[10] The semiquinone formation constants for both FMN and FAD have been calculated using the Michaelis index potential method, and it has been found that they increase with increasing concentration of FMN or FAD[96] and with changes in pH.[19] The original papers should be consulted for the values of these formation constants.

G. Adsorption of FMN and FAD

In acid solution FMN and FAD exhibit a DC polarographic adsorption prewave,[19,33,78,97] which according to Brdička's theories indicates preferential adsorption of the product of the polarographic reduction reaction.[23,40] Senda, Senda, and Tachi[19] have further shown that AC polarography gives behavior similar to that observed for riboflavin, i.e., both product and reactant are adsorbed. That both the oxidized and reduced forms of FMN are adsorbed in acid solution was clearly demonstrated by Hartley and Wilson,[97] who found that electrocapillary curves in 0.1 M HClO$_4$ solution were depressed in the presence of FMN (Fig. 7-12). The rather abrupt change in the drop time (surface tension) at about −0.05 V versus the saturated sodium chloride calomel electrode (SSCE) was coincident with the prewave process. However, further studies employing cyclic voltammetry revealed that the adsorption process was somewhat more complex than it superficially appeared to be.[97] If the hanging mercury drop electrode (HMDE) was allowed to equilibrate at +0.1 V versus SSCE in a 0.1 M HClO$_4$ solution of FMN for various times, rather different cyclic voltammetric behavior was observed (Fig. 7-13). Thus, when a scan was begun immediately after the formation of a mercury drop, a sharp cathodic prepeak was observed at E_p = ca. −0.02 V which was well separated from the main process at E_p = −0.13 V (Fig. 7-13A). With increasing equilibration time, the prepeak, however, merged with the main peak and ultimately disappeared (Fig. 7-13B and C). These experiments were therefore thought to support the notion that during the equilibration period formation of a stable film not reducible in the prewave region occurs. The main reduction peak was a reversible process, as evidenced by the appearance of the anodic peak and by the invariance of the peak potentials with scan rate.[98] It was also observed that with very short equilibration times the cyclic voltammogram exhibited two postpeaks after the main anodic peak. The actual nature of these two anodic postpeaks was not clear. Hartley and Wilson[97] also found that equilibration of the HMDE in a solution of FMN, as described earlier, to presumably form a stable adsorbed film of FMN on the electrode surface, followed by removal of the electrode from the solution, washing, and transfer to a second electrochemical cell containing only background electrolyte (usually 0.1 M HClO$_4$), revealed only one set of peaks corresponding closely to the prepeak−postpeak processes, except that an absence of anodic peak splitting was observed on cyclic voltammetry.

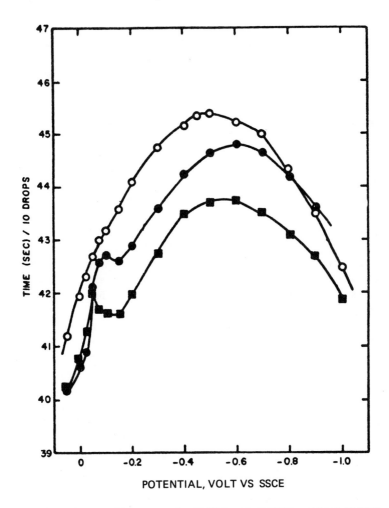

FIG. 7-12.   Electrocapillary curves for FMN in 0.1 *M* HClO₄: (○) 0.1 *M* HClO₄, (●) 1.33×10^{-4} *M* FMN, (■) 3.81×10^{-4} *M* FMN. Potential versus saturated sodium chloride calomel electrode (SSCE).[97] [Reprinted with permission from A. M. Hartley and G. S. Wilson, *Anal. Chem.* 38, 681 (1966). Copyright by the American Chemical Society.]

It would seem that the normal cyclic voltammetric behavior and the film transfer studies are contradictory. After having performed additional chrono-potentiometric and polarographic experiments, these workers concluded that under conditions where a cathodic prewave or prepeak is observed, two reactions occur in the prewave region:

$$(FMN)_{soln} + 2e + 2H^+ \longrightarrow (FMNH_2)_{ads} \tag{3}$$

$$(FMN)_{ads} + 2e + 2H^+ \longrightarrow (FMNH_2)ads \tag{4}$$

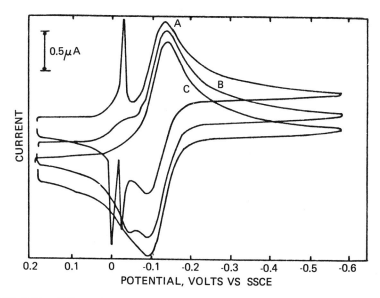

FIG. 7-13. Effect of equilibration time on cyclic scan for 1.90×10^{-4} M FMN in stirred solution. (A) immediately, (B) after 5 min, (C) after 15 min. Scan rate, 0.050 V sec^{-1}. Potential in volts versus the SSCE.[97] [Reprinted with permission from A. M. Hartley and G. S. Wilson, *Anal. Chem.* 38, 681 (1966). Copyright by the American Chemical Society.]

Reaction 3 is the classical Brdička scheme in which diffusion-controlled adsorption occurs. This was verified by the chronopotentiometric and cyclic voltammetric behavior of the prepeak process. The alternative path of reaction 4 was postulated because the oxidized form was not in rapid adsorption equilibrium with the electrode surface. The film-transfer experiments supported the view that reaction 4 could occur in the prewave or prepeak region, although the normal equilibration studies do not. Estimation of the equilibrium surface coverage of oxidized and reduced forms of FMN gave values for the latter (1.22×10^{-10} mole/cm^2) that were almost twice that of the former (0.7×10^{-10} mole/cm^2).

Rather interestingly, calculations revealed that the FMN molecule could readily assume a planar orientation toward the electrode, although the dihydro molecule cannot be coplanar owing to the quinoidal structure of the central isoalloxazine ring. Hartley and Wilson[97] therefore suggested that reduction of an adsorbed layer would necessarily imply a reorientation of the surface layer after reduction. The observed slow adsorption of FMN suggested the possibility of a preferred electroactive orientation resulting from the reorientation of the initially adsorbed material. It was suggested that the initial approach of FMN occurred "edge on," with the plane of the rings normal to the electrode surface to produce a weakly bound unstable state which slowly reoriented to a stable film in which the ring system was coplanar with the electrode. Cyclic voltammetry and chronopotentiometric studies of FMN and FAD have also been reported by Takemori.[99]

Alternating current polarograms of FMN and FAD in acidic solution exhibit two peaks. The summit potentials (E_s) of these peaks agree quite well with the DC polarographic $E_{1/2}$ values of the adsorption prepeak and the normal reduction wave.[19,79,80] In acidic solution the AC adsorption peak of FMN was much higher than the normal AC peak and, as might be expected, at very low FMN concentrations only the AC adsorption peak was observed, its height being abnormally high but directly proportional to the bulk FMN concentration.

In neutral or alkaline solutions of FMN and FAD, Ke[96] observed a small wave at more negative potentials (postwave) than the main DC polarographic reduction wave. The prewave was not observed in neutral or alkaline solutions. Electrocapillary curves for both FMN and FAD at pH 7 revealed that both coenzymes were strongly adsorbed at the DME, as evidenced by marked depressions of the electrocapillary curves in their presence. According to the classical Brdička theory,[41] the appearance of an adsorption postwave is indicative of the strong adsorption of the oxidized form of the flavin coenzymes. The postwave obeyed all of the criteria for a wave under adsorption control. The postwave could not be observed below pH 5.6, either because it was masked by background discharge or, more probably, because the product of the FMN or FAD reductions became more strongly adsorbed than the parent compounds. It is not unlikely that dissociations of the FMN and FAD parent molecules and their reduction products play a very important role in the relative adsorption of each species.

Alternating current polarography of FMN in weakly basic solution indicates that indeed both FMN and its reduction product are adsorbed at the DME.[19,92] Senda, Senda, and Tachi[19] also found that FMN gave rise to an AC wave at potentials close to 0.0 V, which, by analogy to the behavior observed for riboflavin at such potentials (*vide supra*), is probably due to the formation of an insoluble compound of mercury with FMN, involving oxidation of mercury.

Okamoto[100] has studied the oscillographic square-wave polarography of FMN and has reported that at pH 7 in phosphate buffer FMN exhibits a very high peak along with a smaller peak at more negative potential. The summit potentials of the two peaks almost correspond to the summit potentials of the adsorption and normal peak observed by AC polarography.

Takemori[101] has studied the FMN–Fe^{3+} system in neutral oxalate medium and found that two ferric ions combine with one molecule of FMN to give a fairly stable complex species. Polarography of this system gave rise to a kinetic current due to reduction of the Fe^{3+} liberated from the FMN–Fe^{3+} complex.

H. Conclusions

Although it is quite obvious that a considerable amount of work has been carried out on the flavin systems, it appears that not a great deal more is known about the details of the reaction mechanisms than was reported 30 years ago by Brdička using DC polarography and by even earlier workers using potentiometry. Although a number of workers have set out on ambitious projects to study the electrochemistry of flavins, the failure to isolate products, to completely and unambiguously characterize adsorption phenomena, and to carry out a really in-depth study is particularly unfortunate. This is so because the flavin coenzymes comprise two of the five known electron-transferring enzymes [the others being nicotinamide adenine dinucleotide (NAD^+), nicotine adenine dinucleotide phosphate $(NADP^+)$, and thioctic acid] and, consequently, *a priori* constitute systems whose electrochemistry should provide extremely valuable information for an understanding of the *in vivo* electron-transfer and interfacial processes. Thus, further *in vitro* experiments, in which the electrochemistry of these compounds is studied in a medium similar to the biological matrix, could be very informative.

I. Polarography of Other Isoalloxazines

Brdička and Knobloch[25] found that lumichrome (7,8-dimethylisoalloxazine, Table 7-1) is the main product formed on photolysis of riboflavin. It was also found that lumichrome is polarographically reduced at more negative potentials than riboflavin. Lumichrome is an isoalloxazine derivative and differs polarographically from derivatives of isoalloxazine discussed earlier. Thus, apart from the half-wave potentials being at more negative values, the polarographic wave is not reversible.[102] The dependence of $E_{1/2}$ on pH is presented in Fig. 7-14, where it is seen that two discontinuities occur. These correspond closely to the pK_a of the reduced $(pK_a = 6.4)$ and the oxidized form $(pK_a = 7.2)$ of lumichrome. The polarographic waves of lumichrome do not exhibit the typical isoalloxazine adsorption prewave in acid solution. Moore *et al.*[103] have used polarographic methods, to detect lumichrome in photolyzed solutions of riboflavin. The polarographic wave of lumichrome is proposed to be a $2e$ process, although the product of the reduction has not been identified.[102]

Lumiflavin (7,8,10-trimethylisoalloxazine, Table 7-1), which can be formed on photolysis of riboflavin in alkaline solution,[104] apparently behaves polarographically in the same way as riboflavin.[102]

Another product of photolysis of riboflavin is 7,8-dimethyl-10-formylmethylisoalloxazine.[105] Polarographic reduction of this compound takes place by way of two waves.[102] At pH below ca. 2 a single wave appears, but at pH 2 a more negative wave appears that increases in height with increasing

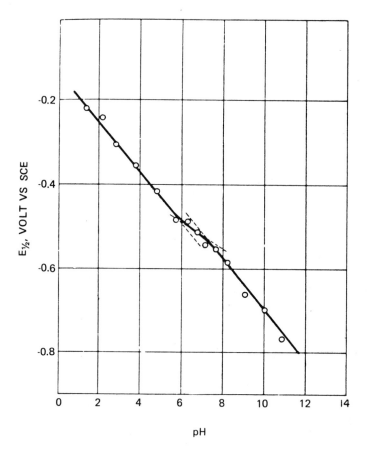

FIG. 7-14.   The pH dependence of the half-wave potential of lumichrome.[102]

pH at the expense of the first, more positive wave. At pH 7 the latter wave disappears. This suggests[102] that at low pH the protonated form of the compound is reduced ($pK_a = 2.9$), and at higher pH the neutral species is reduced. The polarographic process is irreversible, since the reduced form of the compound gives an anodic wave at more positive potentials than the cathodic wave. The polarographic process is not associated with reduction of the aldehydic group.[102]

Knobloch[102] has found that 7,8-dimethyl-10-(2'-hydroxyethyl)isoalloxazine behaves very similarly polarographically to riboflavin, the $E_{1/2}$ pH dependence being essentially identical to that of the latter compound. Knobloch[102] has also used a polarographic method to calculate semiquinone formation constants for a variety of important isoalloxazine derivatives.

REFERENCES

1. C. H. Suelter and D. E. Metzler, *Biochim. Biophys. Acta* **44**, 23 (1960).
2. P. Hemmerich, C. Veeger, and H. C. S. Wood, *Angew. Chem., Int. Ed. Engl.* **4**, 671 (1965).
3. A. Ehrenberg, *in* "Electronic Aspects of Biochemistry" (B. Pullman, ed.), pp. 379–396. Academic Press, New York, 1964.
4. A. V. Guzzo and G. Tollin, *Arch. Biochem. Biophys.* **105**, 380 (1964), and references contained therein.
5. H. Beinert, *in* "The Enzymes" (P. D. Boyer, H. Lardy, and K. Myrbäck, eds.), 2nd ed., Vol. 2, p. 360, 1960; Vol. 7, p. 467, 1963. Academic Press, New York.
6. V. Massey and C. Veeger, *Annu. Rev. Biochem.* **32**, 579 (1963).
7. P. Hemmerich, *in* "Mechanism enzymatischer Reaktionen," Mosbacher Colloq. Ges. Physiol. Chem., p. 183. Springer-Verlag, Berlin and New York, 1964.
8. C. Veeger, *in* "Mechanism enzymatischer Reaktionen," Mosbacher Colloq. Ges. Physiol. Chem., p. 157. Springer-Verlag, Berlin and New York, 1964.
9. E. S. G. Baron and A. B. Hastings, *J. Biol. Chem.* **105**, Proc. vii (1934).
10. R. Kuhn and P. Boulanger, *Ber. Dtsch. Chem. Ges. B* **69**, 1557 (1936).
11. R. Kuhn and G. Moruzzi, *Ber. Dtsch. Chem. Ges. B* **67**, 120 (1934).
12. L. Michaelis, *Chem. Rev.* **16**, 243 (1935).
13. L. Michaelis and G. Schwarzenbach, *J. Biol. Chem.* **123**, 527 (1938).
14. L. Michaelis, M. D. Schubert, and S. V. Smythe, *J. Biol. Chem.* **116**, 587 (1936).
15. F. J. Stare, *J. Biol. Chem.* **112**, 223 (1935).
16. K. G. Stern, *Biochem. J.* **28**, 949 (1934).
17. M. Valentinnuzzi, L. E. Cotino, and M. Portnoy, *An. Soc. Cient. Argent.* **147**, 45 (1949).
18. B. Ke, *Arch. Biochem. Biophys.* **68**, 330 (1957).
19. M. Senda, M. Senda, and I. Tachi, *Rev. Polarog.* **10**, 142 (1962).
20. W. M. Clark, "Oxidation-Reduction Potentials of Organic Systems," p. 442. Williams & Wilkins, Baltimore, Maryland, 1960.
21. E. Haas, *Biochem. Z.* **290**, 291 (1937).
22. B. Janik and P. J. Elving, *Chem. Rev.* **68**, 295 (1968).
23. R. Brdička, *Z. Elektrochem.* **48**, 686 (1942).
24. R. Brdička, *Chem. Listy* **36**, 286 (1943).
25. R. Brdička and E. Knobloch, *Z. Elektrochem.* **47**, 721 (1941).
26. R. Brdička, *Collect. Czech. Chem. Commun.* **14**, 130 (1949).
27. R. Brdička, *Chem. Listy* **36**, 299 (1943).
28. J. J. Lingane and O. L. Davis, *J. Biol. Chem.* **137**, 567 (1941).
29. Y. Mori and K. Murata, *Vitamins* **2**, 14 and 24 (1949).
30. K. Murata and Y. Mori, *Vitamins* **2**, 68 (1949).
31. J. R. Merkel and W. J. Nickerson, *Biochim. Biophys. Acta* **14**, 303 (1954).
32. R. Brdička, *Z. Elektrochem.* **47**, 314 (1941).
33. Y. Asahi, *J. Pharm. Soc. Jpn.* **76**, 378 (1956).
34. R. C. Kaye and H. I. Stonehill, *J. Chem. Soc.* p. 3244 (1952).
35. R. C. Kaye, *J. Pharm. Pharmacol.* **2**, 902 (1950).
36. G. Parravano, *J. Am. Chem. Soc.* **73**, 183 (1951).
37. O. Manoušek and J. Jiřička, *Chem. Tech. (Leipzig)* **13**, 169 (1961); *Chem. Abstr.* **55**, 21217 (1961).

38. S. V. Tatwawadi, K. S. Santhanam, and A. J. Bard, *J. Electroanal. Chem.* **17**, 411 (1968).
39. M. E. Peover, *Electroanal. Chem.* **2**, 1 (1967).
40. Z. Ostrowski and A. Krawczyk, *Acta Chem. Scand., Suppl.* **17**, 241 (1963).
41. R. Brdička, *Collect. Czech. Chem. Commun.* **12**, 522 (1947).
42. R. Brdička, *Z. Elektrochem.* **48**, 278 (1942).
43. B. Breyer and T. Biegler, *Collect. Czech. Chem. Commun.* **25**, 3348 (1960).
44. B. Breyer, *Aust. J. Sci.* **23**, 225 (1961).
45. T. Biegler, Ph.D. Thesis, University of Sidney, 1962.
46. H. Herman, S. V. Tatwawadi, and A. J. Bard, *Anal. Chem.* **35**, 2210 (1963).
47. S. V. Tatwawadi and A. J. Bard, *Anal. Chem.* **36**, 2 (1964).
48. P. J. Lingane, *Anal. Chem.* **39**, 541 (1967).
49. P. J. Lingane, *Anal. Chem.* **39**, 485 (1967).
50. H. A. Laitinen and L. M. Chambers, *Anal. Chem.* **36**, 5 (1964).
51. F. C. Anson, *Anal. Chem.* **38**, 54 (1966).
52. T. Biegler and H. A. Laitinen, *J. Phys. Chem.* **68**, 2374 (1964).
53. H. B. Herman and M. N. Blount, *J. Electroanal. Chem.* **25**, 165 (1970).
54. G. A. Tedoradze, E. Y. Khmel'nitskya, and Y. M. Zolotovitskii, *Elektrokhimiya* **3**, 200 (1967); *Chem. Abstr.* **66**, 110961j (1967).
55. E. Y. Khmel'nitskya, G. A. Tedoradze, and Y. M. Zolotovitskii, *Elektrokhimiya* **4**, 886 (1968); *Sov. Electrochem. (Engl. Transl.)* **4**, 804 (1968).
56. E. Knobloch, *Collect. Czech. Chem. Commun.* **31**, 4503 (1966).
57. B. Jambor, *Magy. Kem. Foly.* **73**, 178 (1967); *Chem. Abstr.* **67**, 39585c (1967).
58. B. Jambor, *Acta Chim. Acad. Sci. Hung.* **48**, 89 (1966); *Chem. Abstr.* **65**, 6741 (1966).
59. W. J. Seagers, *J. Am. Pharm. Assoc., Sci. Ed.* **42**, 317 (1953); *Chem. Abstr.* **47**, 7018 (1953).
60. J. Jilek and M. Liska, *Tech. Publ. Stredisko Tech. Inf. Potravin. Prum.* **139**, 330 (1963); *Chem. Abstr.* **60**, 11293 (1964).
61. O. Enrique and V. Kubac, *Rev. Fac. Farm., Univ. Cent. Venez.* **3**, 249 (1962); *Chem. Abstr.* **61**, 10535 (1964).
62. A. J. Zimmer and C. L. Huyck, *J. Am. Pharm. Assoc.* **44**, 344 (1955).
63. K. H. Tsao, Y. C. Loo, and T. H. Tang, *J. Pharm. Assoc.* **4**, 117 (1956).
64. G. P. Tikhomirova, A. M. Shkodin, and A. I. Yermakova, *Uk. Khim. Zh.* **22**, 687 (1956).
65. M. Sterescu, S. Arizan, and M. Popa, *Rev. Chim. (Bucharest)* **10**, No. 2, 109 (1959); *Chem. Abstr.* **57**, 959 (1962).
66. M. Sterescu, S. Arizan, and R. Talmuciu, *Rev. Chim. (Bucharest)* **8**, 376 (1957).
67. M. Sterescu and S. Negritescu, *Rev. Chim. (Bucharest)* **7**, 159 (1956); *Chem. Abstr.* **52**, 10501 (1958).
68. P. Salmeron, *Ann. Univ. Murcia* **14**, 376 (1955–1956); *Chem. Abstr.* **51**, 18064 (1957).
69. R. Pleticha, *Pharmazie* **13**, 622 (1958); *Chem. Abstr.* **53**, 7291 (1959).
70. E. Knobloch, *Abh. Dtsch. Akad. Wiss. Berlin, Kl. Chem., Geol. Biol.* p. 12 (1964); *Chem. Abstr.* **62**, 5142 (1965).
71. Y. Asahi, *Vitamins* **13**, 490 (1957); *Chem. Abstr.* **54**, 1802 (1960).
72. S. Fowler and R. C. Kaye, *J. Pharm. Pharmacol.* **4**, 748 (1952).
73. G. P. Tikhomirova, S. L. Belen'kaya, R. G. Madievskay, and O. Kurochkina, *Vopr, Pitan.* **24**, 32 (1965); *Chem. Abstr.* **62**, 12978 (1965).

74. A. Maquinay and N. Brouhon, *J. Pharm. Belg.* **12**, 350 (1957); *Chem. Abstr.* **52**, 3259 (1958).
75. Y. Asahi, *J. Vitaminol.* **4**, 118 (1958); *Chem. Abstr.* **52**, 20892 (1958).
76. K. Okamoto, *Rev. Polarog.* **11**, 225 (1964).
77. V. Moret and S. Pinamorti, *Gn. Biochim.* **9**, 223 (1960).
78. Y. Asahi, *Rev. Polarogr.* **8**, 1 (1960).
79. T. Tachi and Y. Takemori, *Vitamins* **9**, 441 (1955).
80. T. Takemori, M. Senda, and M. Senda, *Vitamins* **16**, 492 (1959).
81. O. Sebek, *Chem. Listy* **41**, 238 (1947); *Chem. Abstr.* **42**, 2723 (1948).
82. J. S. Sancho Gomez, P. Salmeron, and J. G. Hurtado, *An. R. Soc. Espan. Fis. Quim.,* *Ser. B.* **53**, 597 (1957); *Chem. Abstr.* **54**, 5293 (1960).
83. K. Enns and W. H. Burgess, *J. Am. Chem. Soc.* **87**, 1822 (1965).
84. A. Kočent, *Chem. Listy* **47**, 195 (1953).
85. A. Kočent, *Chem. Listy* **47**, 652 (1953).
86. K. Enns and W. H. Burgess, *J. Am. Chem. Soc.* **87**, 5766 (1965).
87. M. A. Hoffman, *Int. Z. Vitaminforsch.* **20**, 238 (1948); *Chem. Abstr.* **43**, 1076 (1949).
88. R. Portillo and G. Varela, *An. R. Acad. Farm.* **15**, 787 (1949); *Chem. Abstr.* **44**, 10782 (1950).
89. E. Kevei, M. Kiszel, and M. Simek, *Acta Chim. Acad. Sci. Hung.* **6**, 345 (1955); *Chem. Abstr.* **51**, 2193 (1957).
90. E. Kevei, M. Kiszel, and F. Simek, *Elelmez. Ipar* **9**, 287 (1955); *Hung. Tech. Abstr.* **8**, No. 2, Abstr. 105 (1956); *Chem. Abstr.* **52**, 8403 (1958).
91. G. Dusinsky and L. Faith, *Pharmazie* **22**, 475 (1967); *Chem. Abstr.* **68**, 53280 (1968).
92. B. Breyer and T. Biegler, *J. Electroanal. Chem.* **1**, 453 (1959/1960).
93. B. Bierich and A. Lang, *Hoppe-Seyler's Z. Physiol. Chem.* **223**, 180 (1934).
94. E. G. Ball, *Cold Spring Harbor Symp. Quant. Biol.* **7**, 100 (1939).
95. W. M. Clark, "Topics in Physical Chemistry," 2nd ed., p. 475. Williams & Wilkins, Baltimore, Maryland, 1952.
96. B. Ke, *Arch. Biochem. Biophys.* **68**, 330 (1957).
97. A. M. Hartley and G. S. Wilson, *Anal. Chem.* **38**, 681 (1966).
98. R. S. Nicholson and I. Shain, *Anal. Chem.* **36**, 706 (1964).
99. Y. Takemori, *Rev. Polarogr.* **12**, 63 (1964).
100. K. Okamoto, *Rev. Polarogr.* **11**, 225 (1964).
101. Y. Takemori, *Rev. Polarogr.* **12**, 176 (1964).
102. E. Knoblock, *in* "Methods in Enzymology" (D. B. McCormick and L. D. Wright, eds.), Vol. 18, Part B, p. 305. Academic Press, New York, 1971.
103. W. M. Moore, J. T. Spence, and F. A. Raymond, *J. Am. Chem. Soc.* **85**, 3367 (1963).
104. R. Kuhn and T. W. Jauregg, *Ber. Dtsch. Chem. Ges. B* **66**, 1577 (1931).
105. E. C. Smith and D. E. Metzler, *J. Am. Chem. Soc.* **85**, 3285 (1963).

8

Pyrroles and Porphyrins

I. INTRODUCTION, NOMENCLATURE, AND STRUCTURE

The basic pyrrole skeleton is a five-membered ring containing one heterocyclic nitrogen atom (Fig. 8-1). The use of Greek letters as well as numbers for denoting the various positions in the pyrrole nucleus is very common, especially in the older literature. Unlike many other groups of *N*-heterocycles, the simple pyrroles do not suffer from a proliferation of trivial names; they are simply named using the numbering system shown in Fig. 8-1. The bile pigments, however, which are discussed in Section V, do have trivial names. A summary of the physical properties, synthesis, and reactions of pyrroles has been prepared by Schofield[1] and Corwin.[2]

Porphyrins are intensely colored substances of great importance in the metabolism of animals and plants. Porphyrins are all derived from the fundamental porphyrin skeleton (Fig. 8-2). Quite clearly, the porphyrin molecule is formed from four pyrrole molecules linked together in a ring system by four methene bridges (—CH groups). Some of the chemistry and biochemistry of porphyrins and metalloporphyrins has been reviewed by Falk[3] and others.[4]

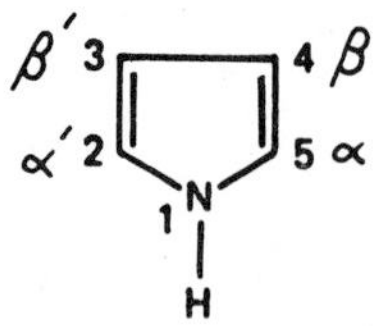

FIG. 8-1.   Basic structure and numbering of pyrrole.

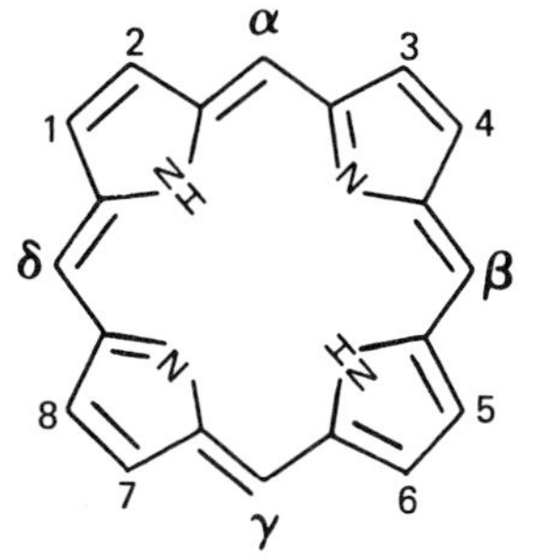

FIG. 8-2. Basic porphyrin ring structure.

II. OCCURRENCE AND BIOLOGICAL IMPORTANCE

Pyrrole itself was probably first recognized by Runge[5] from dry distilled proteins. Pyrrole has also been obtained from bone tar.[6] It is probably true that interest in the chemistry of pyrrole derivatives was stimulated by the discovery that pyrrole is the building block of hemin, bilirubin, and chlorophyll.[2] Simple pyrroles appear to have relatively little biological significance except as intermediates in the biosynthesis of the porphyrin ring.

Only rather small amounts of porphyrins are found in the free state in nature. Most actually occur in the form of metal complexes, i.e., metalloporphyrins, in which a metal atom, usually iron or magnesium, replaces the two central hydrogen atoms. The iron–porphyrin complex represents the prosthetic group of heme proteins. The most important heme proteins are (a) hemoglobin, which is an oxygen-carrying heme protein that consists of the protein globin linked to a hemin prosthetic group, (b) the cytochromes, which are electron-transfer agents involved in the respiratory chain, and (c) several oxidoreducto enzymes such as the peroxidases and catalases. A magnesium–dihydroporphyrin complex is found in chlorophyll, which is employed by plants in the photosynthetic process. Bacteriochlorophyll, a form of chlorophyll found in certain types of green bacteria, is a magnesium–tetrahydroporphyrin complex. For the sake of clarity the structures of some important porphyrins are shown in Fig. 8-3.

Another interesting group of compounds consisting of conjugated pyrrole molecules includes the so-called bile pigments. These are generally open-chain tetrapyrroles and are produced in organisms, including man, by the metabolic degradation of hemoglobin. This degradation appears to result from an initial oxidative cleavage of the protoporphyrin ring (Fig. 8-3) by elimination of the methene group situated between the two pyrrole rings carrying vinyl substituents, i.e., the α-methene grouping. The ultimate product is the green bile pigment

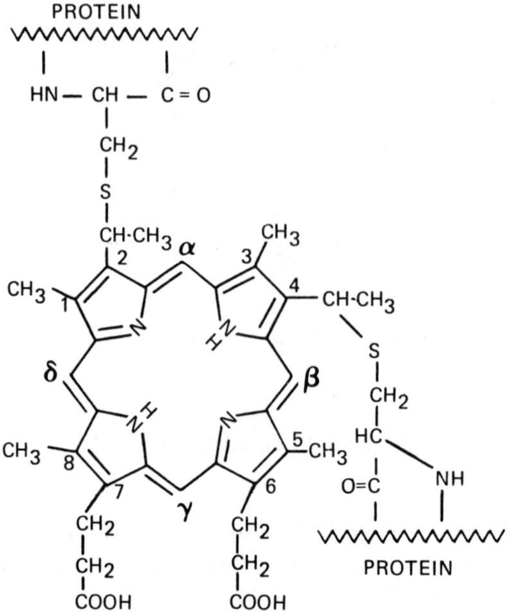

CYTOCHROME C

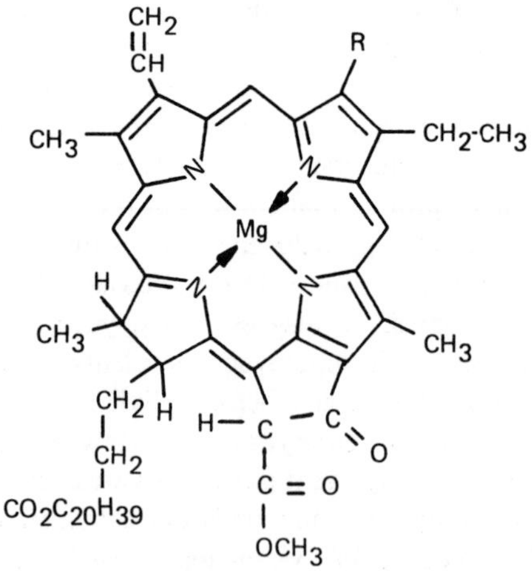

CHLOROPHYLL a when R = —CH$_3$

CHLOROPHYLL b when R = —CHO

FIG. 8-3.   Some biologically important porphyrins.

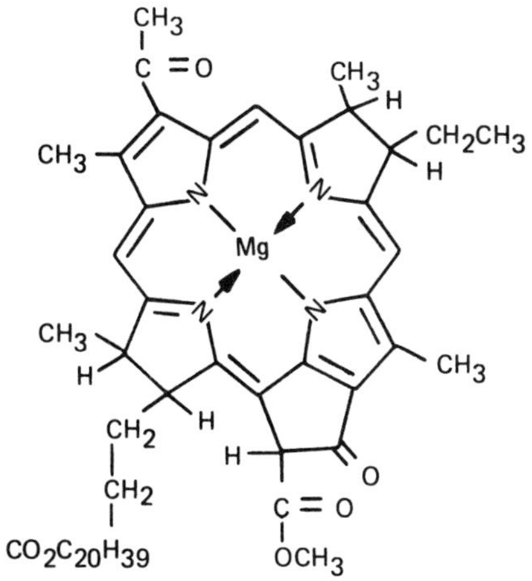

BACTERIOCHLOROPHYLL a

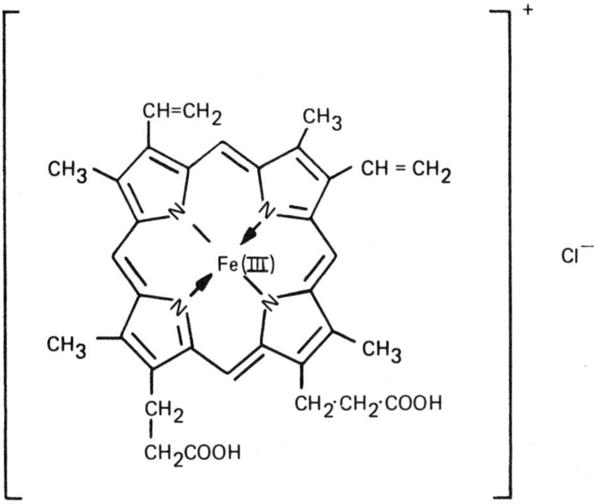

HEMIN
(IRON PROTOPORPHYRIN)

FIG. 8-3. *Continued.*

biliverdin (I, Fig. 8-4). This compound then undergoes a series of metabolic reductions to give orange bilirubin (II, Fig. 8-4) and urobilin (III, Fig. 8-4). Further reductions lead to stercobilin (IV, Fig. 8-4). The chemistry of some of these transformations has been reviewed elsewhere.[7,8]

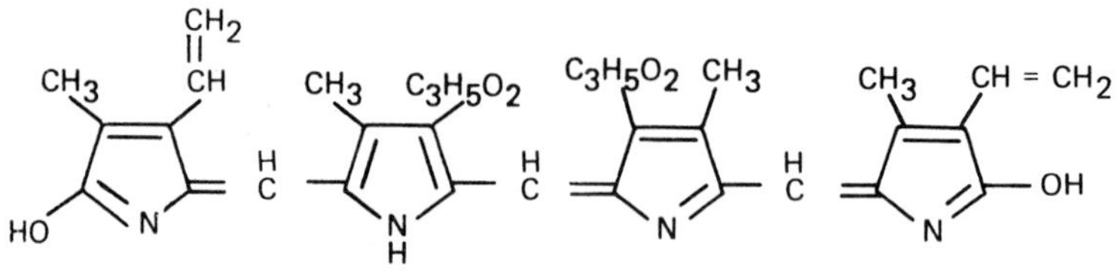

(I) BILIVERDIN

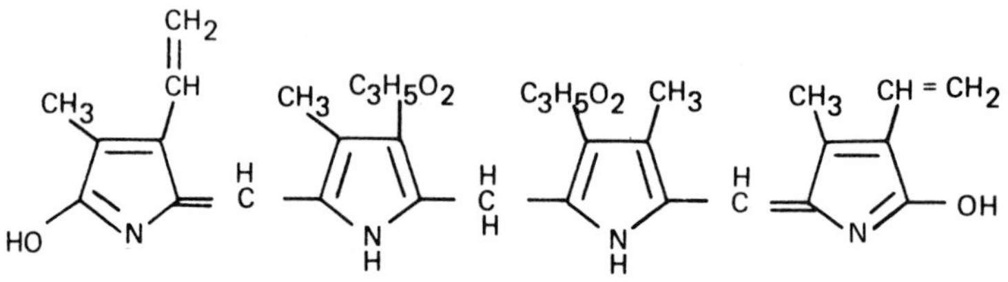

(II) BILIRUBIN

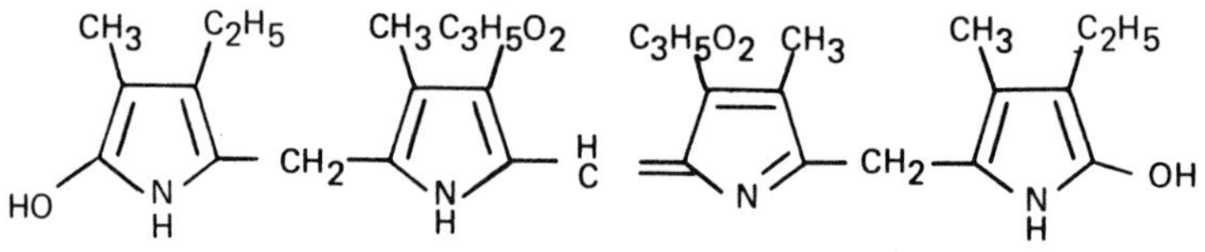

(III) UROBILIN

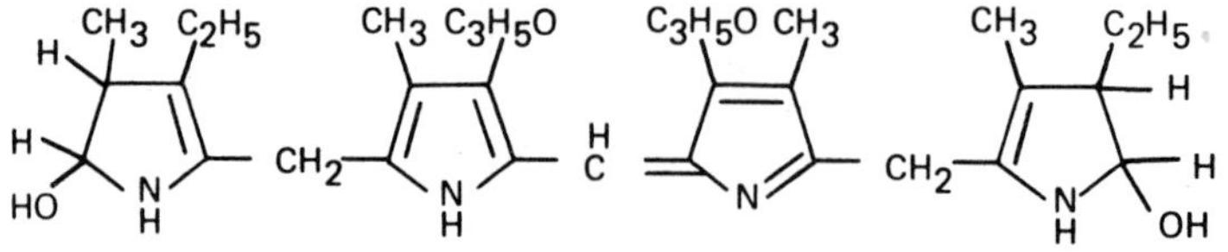

(IV) STERCOBILIN

FIG. 8-4.   Structure of some bile pigments.

III. ELECTROCHEMICAL REDUCTION OF PYRROLES

Unlike all the groups of compounds discussed in earlier chapters, the parent compound in this series is not polarographically reducible.[9] The polarographic nonreducibility of pyrrole is expected from molecular orbital calculations.[10] The apparent general inability of the monomeric pyrrole ring to undergo electrochemical reduction is in a sense peculiar. This peculiarity arises from the fact that the metabolic degradation of other conjugated *N*-heterocyclic compounds such as pteridines, purines, and pyrimidines proceeds primarily by oxidative processes, although the parent compounds are very easily reducible. Pyrroles, on the other hand, are metabolically degraded primarily by reductive processes, but the parent compound is not normally reducible electrochemically.

Much of the work carried out on pyrroles is quite old and unfortunately is reported in very obscure journals which have been abstracted rather poorly; i.e., more often than not half-wave potentials have been given without indication of solvent, supporting electrolyte, pH, or reference electrode. A fairly complete summary of the available literature and findings on pyrrole electrochemistry is presented in Table 8-1.[9,11-31] Very little work has been done on the mechanisms of the electrochemical reductions, although the information available would appear to support the view that substituents are responsible for the electroactivity of the pyrroles and that reduction of the pyrrole ring is not normal. Thus, most data on pyrrole reductions have been obtained on compounds possessing either carbonyl groups in one guise or another or nitro groups, which would be expected to be the electrochemically reducible moieties. However, Nakaya, Kinoshita, and Ono[17] indicate that 2-acetylpyrrole and 2-pyrrylacetic acid are reduced to a dimer in acid medium and to saturated compounds in neutral or basic medium.

IV. ELECTROCHEMICAL OXIDATION OF PYRROLES

Stanienda[32] has shown that pyrrole can be oxidized at a rotating platinum electrode in acetonitrile solution containing $0.1 M$ $LiClO_4$ supporting electrolyte. A single, irreversible oxidation wave is produced that seems to involve about 3.5 electrons per molecule of pyrrole. The half-wave potential for oxidation of pyrrole under these conditions was found to be slightly concentration dependent and was described by

$$E_{1/2} = 0.804 - 0.10 \log C \qquad (1)$$

where C is the concentration of pyrrole (moles per liter x 10^5). The mechanism proposed for the electrochemical oxidation of pyrrole (I, Fig. 8-5) in acetonitrile was a primary one-electron oxidation to give a cation radical (II, Fig. 8-5), which then lost a proton to give a radical (III, Fig. 8-5). Condensation and further electrochemical oxidation of this product gave rise to an unidentified polymeric material (IV, Fig. 8-5). Lund[33,34] has also reported that electrochemical oxidation of pyrrole in acetonitrile at platinum yields a tar that covers the electrode with an insulating layer.

When benzaldehyde was added to solutions of pyrrole in acetonitrile, but in concentrations less than equivalent to pyrrole, tetraphenylporphyrin was obtained in yields $\leqslant 1\%$ (based on benzaldehyde).[32] It was also found that if proton acceptors such as water, pyridine, amines, or even excess benzaldehyde were present in the acetonitrile solution, the yields of tetraphenylporphyrin were drastically reduced. Electrochemical oxidation of pyrrole liberates protons (Fig. 8-5); hence, it was proposed that these protons could protonate unoxidized pyrrole (I → II, Fig. 8-6), which could then condense with benzaldehyde to give the secondary alcohol (III, Fig. 8-6). Further condensation of III (Fig. 8-6) with protonated pyrrole (II, Fig. 8-6) would then lead to the tautomeric bipyrroles IV_a and IV_b (Fig. 8-6), which after further reaction with benzaldehyde and loss of H_3^+O would lead to tetraphenylporphinogen (VI, Fig. 8-6). Electrochemical oxidation of the latter species gives rise to tetraphenylporphyrin (VII, Fig. 8-6).

Weinberg and Brown[35] found that 1-methylpyrrole (I, Fig. 8-7) could be oxidized at a platinum electrode in methanolic solution to 1-methyl-2,2,5,5-tetramethoxypyrroline (IV, Fig. 8-7). This work was done without control of

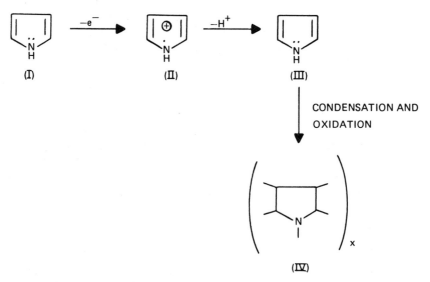

CONDENSATION AND
OXIDATION

FIG. 8-5. Proposed mechanism for the electrochemical oxidation of pyrrole at a platinum electrode in acetonitrile solution.[32]

TABLE 8-1 Summary of the Polarographic Behavior of Pyrrole and Its Derivatives

Compound	pH or solvent and supporting electrolyte	$E_{1/2}$ (V)	Reference electrode	Reference
Pyrrole		NR[a]		9
α-Pyrrole aldehyde	50% Dioxane buffered to pH 7.5	−1.81	NA[b]	11
	Aqueous methanol/ $(CH_3)_4NBr$	−1.70	SCE[c]	12
	0.1 N NH_4Cl in 10% aqueous ethanol	−1.43	SCE	12–14
	Aqueous 0.1 M Et_4NI	Oscillopolaro-graphic peak at −1.75	SCE	31
	0.1 M Et_4NI in DMF^d	Oscillopolaro-graphic peak at −1.65	Hg pool[e]	31
	0.1 M Et_4NI in DMF containing 16% water	Oscillopolaro peaks at (I) −1.50, (II) −1.62	Hg pool	31
1-Methyl-2-pyrrole carboxaldehyde	3.6–7.0	−1.594 to −1.732	SCE	15
	7.0–9.35	−1.732	SCE	15
	NA	−1.224	NA	16
Pyrrole 2-aldehyde[f]	0.1 N NH_4Cl in 50% ethanol at 18–20°C	−1.452	NCE^g	14
2,3,4-Trimethyl-5-formylpyrrole	0.1 N NH_4Cl in 50% ethanol at 18–20°C	−1.534	NCE	14
2-Methyl-3,4-diethyl-5-formylpyrrole	0.1 N NH_4Cl in 50% ethanol at 18–20°C	−1.536	NCE	14

TABLE 8-1 *Continued*

Compound	pH or solvent and supporting electrolyte	$E_{1/2}$ (V)	Reference electrode	Reference
2,4-Dimethyl-3-ethyl-5-formylpyrrole	0.1 N NH$_4$Cl in 50% ethanol at 18–20°C	−1.538	NCE	14
2,4-Dicarbethoxy-3-methyl-5-formylpyrrole	0.1 N NH$_4$Cl in 50% ethanol at 18–20°C	−0.935	NCE	14
2-Carbethoxy-3,5-di-methyl-4-formylpyrrole	0.1 N NH$_4$Cl in 50% ethanol at 18–20°C	−1.524	NCE	14
1-Methyl-2-formylpyrrole	0.1 N NH$_4$Cl in 50% ethanol at 18–20°C	−1.420	NCE	14
2-Acetylpyrrole	0.1 N NH$_4$Cl in 50% ethanol at 20–30°C	−1.646	NCE	14
	NA	NA but apparently reduced to dimer in acid medium and to saturated compounds in neutral or basic medium	NA	17
1-Acetylpyrrole	0.1 N NH$_4$Cl in 50% EtOH at 18–20°C	NR		14
2,3,5-Trimethyl-4-acetylpyrrole	0.1 N NH$_4$Cl in 50% EtOH at 18–20°C	NR		14
2,4-Dimethyl-5-carbethoxy-3-acetyl-pyrrole	0.1 N NH$_4$Cl in 50% EtOH at 18–20°C	NR		14

Compound	Conditions	Value		Ref.
1-Methyl-2-acetyl-pyrrole	0.1 N NH$_4$Cl in 50% EtOH at 18–20°	−1.700	NCE	14
2,5-Diacetylpyrrole	0.1 N NH$_4$Cl in 50% EtOH at 18–20°C	(I) −1.031,^{h} (II) −1.650	NCE	14; see also Ref. 30
1-Methyl-2,5-diacetylpyrrole	0.1 N NH$_4$Cl in 50% EtOH at 18–20°C	(I) −1.053,^{h} (II) −1.685	NCE	14
3,3'-Dimethyl 5,5'-diethyldipyrryl ketone	0.1 N NH$_4$Cl in 50% EtOH at 18–20°C	−1.618	NCE	14
5,5'-Dimethyl 3,3'-diethyldipyrryl ketone	0.1 N NH$_4$Cl in 50% EtOH at 18–20°C	−1.640	NCE	14
3,3',5,5'-Tetramethyl 4,4'-dipropyldipyrryl ketone	0.1 N NH$_4$Cl in 50% EtOH at 18–20°C	−1.680	NCE	14
Pyrrole azomethines	NA	3-Step reduction between pH 8 and 10	NA	18
Pyrrole 2-carbox-aldehyde	Aqueous dioxane at pH 7.4^i	−1.81	NA	19
2-Pyrrylacetic acid	NA	NA but apparently reduced to dimers in acid solution and to saturated compounds in neutral or basic medium	NA	17
syn- and anti-Pyrrole oximes	NA	NA	NA	20
4-Methyl-3,5-di-carbethoxypyrrole	Aqueous dioxane buffered to pH 7.4^i	−1.188	NA	19
3,5-Dimethyl-4-carbethoxy-2-pyrrole carboxaldehyde	Aqueous dioxane buffered to pH 7.4^i	−1.787	NA	19
2,4-Dimethyl-5-carbethoxy-3-pyrrole carboxaldehyde	Aqueous dioxane buffered to pH 7.4^i	−2.00	NA	19

TABLE 8-1 *Continued*

Compound	pH or solvent and supporting electrolyte	$E_{1/2}$ (V)	Reference electrode	Reference
2,4-Dimethylpyrrole carboxaldehyde	Aqueous dioxane buffered to pH 7.4[i]	−2.16	NA	19
1,3,5-Trimethylpyrrole carboxaldehyde	Aqueous dioxane buffered to pH 7.4[i]	NA but reduced at pH 5.5	NA	19
Dipyrryl ketones	NA	NA	NA	21
4-Methyl-3,5-dicarbethoxy-2-pyrrole carboxaldehyde	NA; possibly 0.1 N NH₄Cl	−0.739	NA	22
2,4-Dimethyl-5-carbethoxy-3-pyrrole carboxaldehyde	NA; possibly 0.1 N NH₄Cl	−1.328	NA	22
Nitropyrroles	NA	NA	NA	23, 28
2-Pyrrolesulfonic acid (Ba salt)	0.5 N KCl, 0.003 N HCl	0.022 (A)[j]	0.5 NCE[k]	24, 25
1-Methyl-2-pyrrolesulfonic acid (Ba salt)	0.5 N KCl, 0.003 N HCl	0.018 (A)	0.5 NCE	24, 25
3,5-Dimethyl-2-pyrrolesulfonic acid (Ba salt)	0.5 N KCl, 0.003 N HCl	0.015 (A)	0.5 NCE	24, 25
2,4,5-Trimethyl-3-pyrrolesulfonic acid (Ba salt)	0.5 N KCl, 0.003 N HCl	0.052 (A)	0.5 NCE	24, 25
1,2,5-Trimethyl-3-pyrrolesulfonic acid (Ba salt)	0.5 N KCl, 0.003 N HCl	0.044 (A)	0.5 NCE	24, 25

Substituted α-pyrrole aldehydes	NA	NA	NA	26
2-Pyrrole carboxaldehyde	0.1 M Et$_4$NI	NA; oscillopolarography	NA	27
1-Methyl-2-pyrrole carboxaldehyde	0.1 M Et$_4$NI	NA; Oscillopolarography	NA	27
1-Methyl-2-acetyl-pyrrole	NA	−1.700	NA	29
1-Methyl-2,5-diacetylpyrrole	NA	(I) −1.053, (III) −1.685	NA	29
Substituted pyrrole α-aldehydes	NA	NA	NA	11

[a] Not reducible.

[b] Information not available.

[c] Saturated calomel electrode.

[d] N,N'-Dimethylformamide.

[e] Mercury pool anode employed as reference electrode.

[f] Concentration, 1.5 × 10^{-3} M.

[g] Normal calomel electrode.

[h] Two waves observed.

[i] The solvent system not specified in abstract; however, data from the same laboratory at the same time would support the view that the solvent is buffered aqueous dioxane.

[j] A, anodic wave.

[k] 0.5 N Calomel electrode.

the working electrode potential, electron numbers were not obtained, and none of the usual electrochemical techniques for investigating electrode processes, apart from product identification, were employed. Nevertheless, a mechanism was proposed which was an initial $2e$ oxidation of 1-methylpyrrole (I, Fig. 8-7) to a dication (II, Fig. 8-7), which was attacked by methanol to give 1-methyl-2,5-dimethoxypyrroline (III, Fig. 8-7). Further $2e-2H^+$ oxidation of III (Fig. 8-7) followed by attack of methanol was proposed to give the final product, 1-methyl-2,2,5,5-tetramethoxypyrroline (IV, Fig. 8-7).

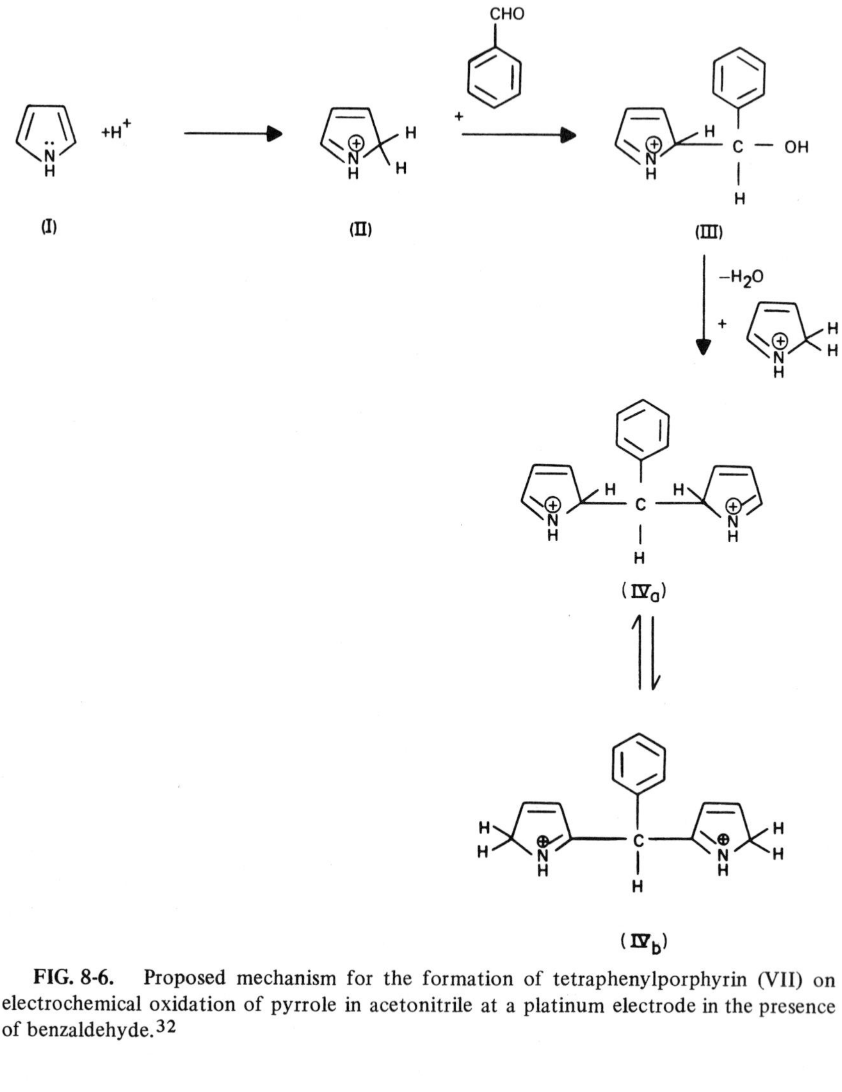

FIG. 8-6. Proposed mechanism for the formation of tetraphenylporphyrin (VII) on electrochemical oxidation of pyrrole in acetonitrile at a platinum electrode in the presence of benzaldehyde.[32]

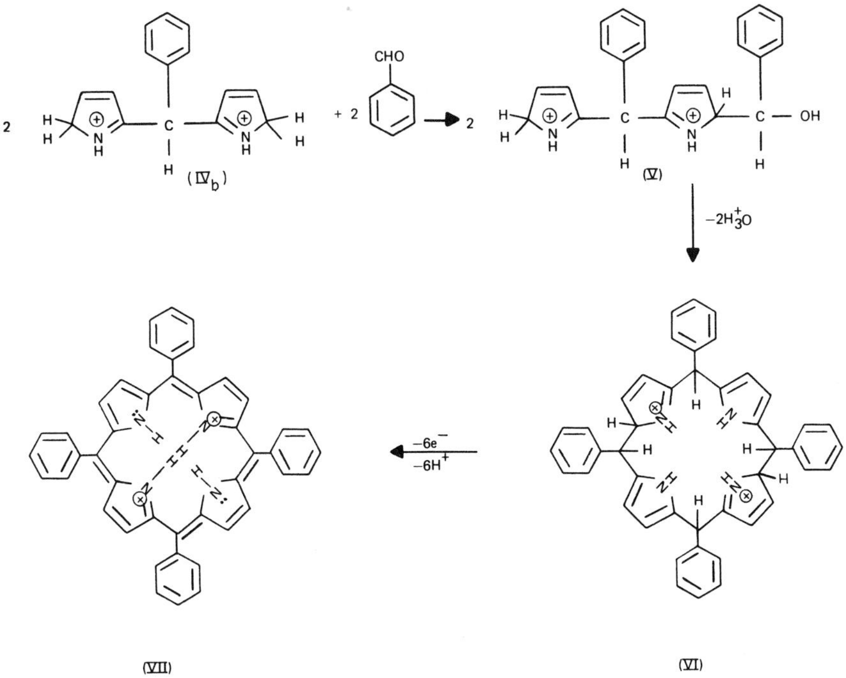

FIG. 8-6. *Continued.*

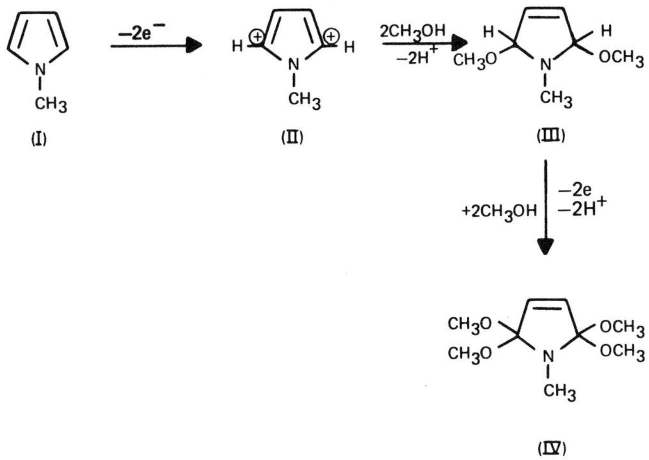

FIG. 8-7. Proposed mechanism of oxidation of 1-methylpyrrole in methanolic solution at a platinum electrode.[35]

Cauquis and Geniés[36] studied the electrochemical oxidation of several pentaphenylpyrroles (I) in acetonitrile at a platinum electrode. In the case of the

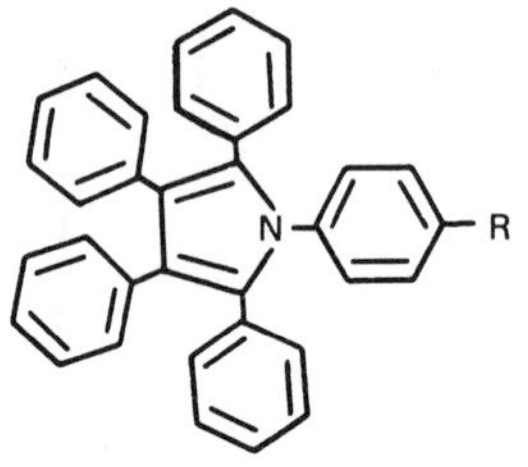

where R = H, CH$_3$,

OCH$_3$ or N(CH$_3$)$_2$

(I)

pentaphenylpyrroles where R was H, CH$_3$, or OCH$_3$, two oxidation waves were observed at the rotating platinum electrode, the half-wave potentials being very close for all three compounds (Table 8-2). In the case of the N(CH$_3$)$_2$-substituted pentaphenylpyrrole three waves were observed, the first two occurring at considerably more negative potential than those of the other three compounds. Coulometry and other data supported the view that the first wave of the pentaphenylpyrroles substituted with H, CH$_3$, or OCH$_3$ was a one-electron oxidation giving a relatively stable cation radical. This was formed by removal of an electron from the π orbitals of the pyrrole ring. The cation radical was sufficiently stable so that its UV, visible, and ESR spectra could be readily obtained. In the case of the N(CH$_3$)$_2$ derivative the protonated species was electrochemically oxidized. The first wave was also a $1e$ process, but the cation radical was proposed to be formed by loss of an electron from the p-dimethylaminophenyl group, not from the pyrrole moiety.

TABLE 8-2   Half-Wave   Potentials   for   Electrochemical Oxidation of Some Pentaphenylpyrroles in Acetonitrile at a Rotating Platinum Electrode[a]

Substitution, R[b]	Half-wave potential (V versus Ag/10^{-2} M Ag$^+$ in MeCN)		
	Wave I	Wave II	Wave III
H	0.755	1.30	
CH$_3$	0.745	1.30	
OCH$_3$	0.735	1.30	
N(CH$_3$)$_2$	0.520	0.93	1.3

[a] Data from Cauquis and Geniés.[36]
[b] See structure I above.

V. ELECTROCHEMICAL REDUCTION OF BILE PIGMENTS

Cuvelier and co-workers[37] have used a polarographic method for the determination of certain components of bile and gastric juices, but useful mechanistic data were not presented. Tachi[38,39] found that bilirubin (II, Fig. 8-4) gave a 2e polarographic wave at the DME. The wave appeared to be pH independent at pH greater than 9 or 10 (Table 8-3). It was also found that solutions of bilirubin underwent air oxidation and that the oxidized product(s) gave rise to two additional polarographic reduction waves.

TABLE 8-3   Half-Wave Potential
Data for Bilirubin[a]

pH of buffer	$E_{1/2}$ (V)[b]
7.0	−1.33
9.0	−1.42
11.0	−1.45

[a] From Tachi.[38,39]
[b] Presumably versus SCE.

VI. CYTOCHROMES

Many years ago Theorell[40] indicated that cytochrome *c* was polarographically reducible, although the treatment he administered to the crude material would seem to suggest that considerable alteration of the cytochrome *c* itself had taken place. More recently, Griggio and Pinamonti[41] studied bovine heart cytochrome *c* polarographically. A 0.5 m*M* solution gave an irreversible polarographic reduction wave in the pH range 6.5–9.6. The half-wave potential was found to be constant to pH 8.7, after which it shifted linearly to more negative values at higher pH. Since cytochrome *c* normally exists as an iron (III) complex, Griggio and Pinamonti[41] felt that the polarographic behavior was associated with the iron moiety.

Cytochrome *c* in 0.001 *M* hexaminocobalt(II) chloride, 1 *M* NH$_4$OH, and 1 *M* NH$_4$Cl gives rise to a catalytic wave or peak at about −1.5 V (reference electrode not known).[42−44] The height of the catalytic peak at low cytochrome *c* concentrations could be used for analytical purposes. At high concentrations the peak current versus concentration curve reached a limiting value. It can be observed (Fig. 8-3) that cytochrome c possesses two sulfur moieties. The origin of the catalytic process was considered to reside in these sulfur groups. Brezina and Zuman[45] have reviewed much of the analytical methodology associated with the polarographic determination of cytochrome c and biochemical processes involving this compound.

VII. CHLOROPHYLL

Chlorophyll is the general name of the green pigments of plants that are capable of photosynthesis. In the past 10–20 years there have been some spectacular advances in the elucidation of the structure and chemical synthesis of chlorophyll and more recently in the understanding of its biological mode of action.[3,46,47]

The nomenclature applied in chlorophyll chemistry is somewhat complex, and, accordingly, in order to facilitate reading of the subsequent sections of this chapter, some common structures and terms will first be outlined. The structures of chlorophyll a and b are shown in Fig. 8-3. The tetrahedral carbon atoms at positions 7 and 8 result in chlorophyll being assigned to a class of compounds called chlorins. Woodward[48] has pointed out that the reduction of a porphyrin (I, Fig. 8-8) by addition of two electrons and two protons can result in the formation of two types of isomeric dihydroporphyrins: the chlorins (II, Fig. 8-8), of which chlorophyll is a member, and the phlorins (III, Fig. 8-8). The chlorins and phlorins have quite different chemical properties. In particular, the chlorins (and hence chlorophyll) are stable to air oxidation while the phlorins are not. Two-electron, two-proton reductions of the latter dihydroporphyrins can give rise to three types of isomeric tetrahydroporphyrins: the chlorin–phlorins (IV, Fig. 8-8), the bacteriochlorins (V, Fig. 8-8), which were mentioned earlier, and the porphomethenes (VI, Fig. 8-8). Of the tetrahydroporphyrins, the bacteriochlorins are the most stable to air oxidation. A number of isomeric structures are possible within the various classes of di- and tetrahydroporphyrins. It is interesting that the stable types of dihydro- and tetrahydroporphyrins, i.e., the chlorins and bacteriochlorins and their biologically important derivatives chlorophyll and bacteriochlorophyll, respectively, are saturated *only* in peripheral pyrrole ring positions, whereas the more readily air oxidizable isomers possess methene bridge saturations.

Other chlorophyll derivatives are also important in the following discussion. Among these are phaeophytin or (pheophytin), which is the magnesium-free derivative of chlorophyll. A chlorophyllide is the derivative resulting from the

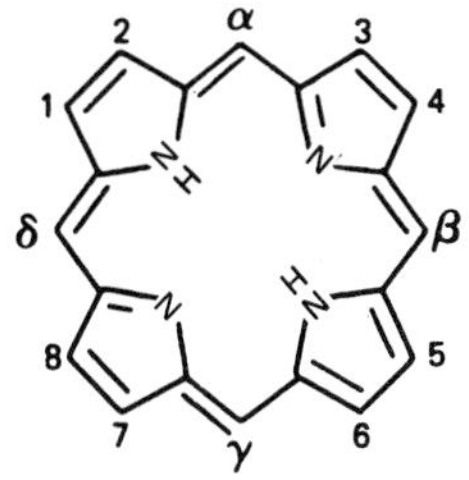

I. PORPHYRIN

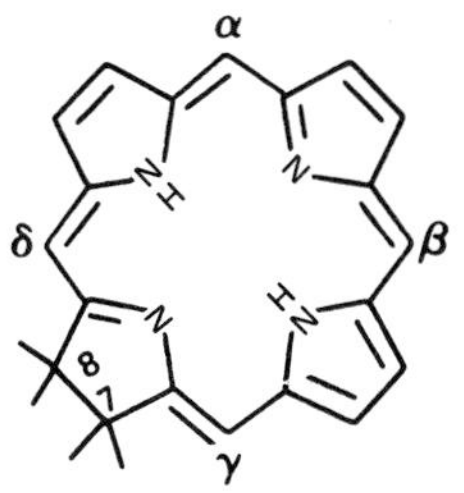

II. CHLORIN

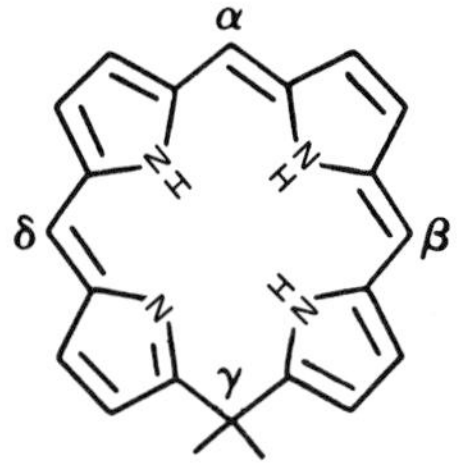

III. PHLORIN

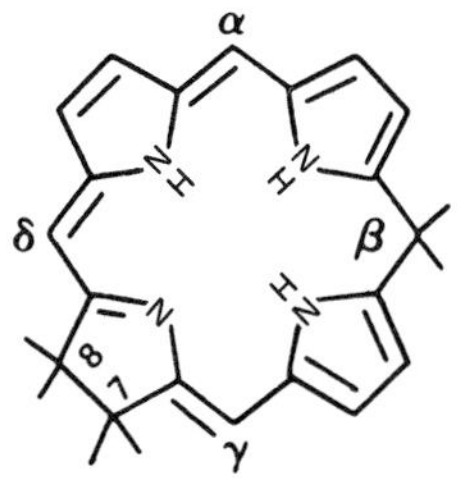

IV. CHLORIN-PHLORIN

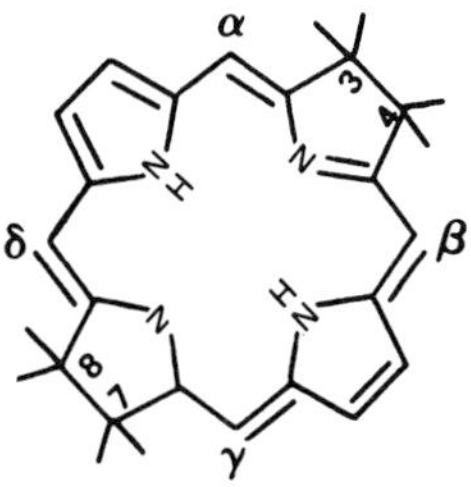

V. BACTERIOCHLORIN

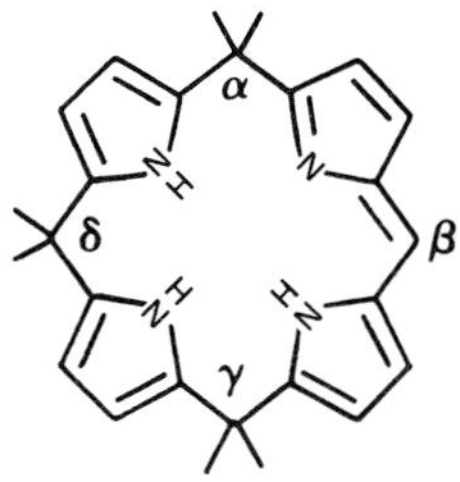

VI. PORPHOMETHENE

removal of the alcohol esterified to the C-7 propionic acid group of a chlorophyll. Thus, chlorophyllide *a* is simply chlorophyll *a* with the phytyl group of the latter

$$(-\overset{\text{O}}{\underset{}{\text{C}}}-OC_{20}H_{39})$$

replaced by the simple carboxylate group

$$(-\overset{\overset{\displaystyle O}{\|}}{C}-OH)$$

When the carboxylate group is present as a methyl or ethyl ester, the compounds are known as methyl or ethyl chlorophyllides. A *meso*-chlorophyll is one in which the C-2 vinyl group has been reduced to an ethyl group.

A. Electrochemical Reduction of Chlorophylls and Related Compounds

Probably the earliest report of the electrochemistry of chlorophyll was that of Van Rysselberghe and co-workers,[49] who showed that extracts of both chlorophyll *a* and *b* and mixtures of the two in ether or dioxane added to water containing tetramethylammonium bromide supporting electrolyte gave rise to polarographic reduction waves (Table 8-4). Saponification of chlorophyll gives rise to phytol, which these workers found gave a polarographic reduction wave

$$CH_3\underset{|}{\overset{\overset{\displaystyle CH_3}{|}}{CH}}\cdot CH_2(CH_2CH_2\overset{\overset{\displaystyle CH_3}{|}}{CH}\cdot CH_2)_2-CH_2CH_2\overset{\overset{\displaystyle CH_3}{|}}{C}=CHCH_2OH$$

Phytol

at about the same $E_{1/2}$ observed for the chlorophylls. Thus, the polarographic reduction wave of chlorophyll was ascribed to hydrogenation of the double bond in the phytol chain. Knobloch,[50] however, apparently found that phytol was polarographically nonreducible.

TABLE 8-4   Half-Wave Potential Data for Chlorophyll *a* and *b*[a]

Compound	$E_{1/2}$ (V versus SCE)	Extract
Chlorophyll *a*	−1.88	Dioxane
Chlorophyll *b*	−1.88	Dioxane
Chlorophyll *a* + *b*	−1.93	Diethyl ether

[a] From Van Rysselberghe *et al.*[49]

In 1964 Felton, Sherman, and Linschitz[51] examined the polarography of a number of chlorophyll derivatives in dimethyl sulfoxide solvent and found that the compounds (I–IV, Fig. 8-9) gave rise to two well-formed, cathodic, diffusion-controlled waves (Table 8-5). Rather surprisingly, however, at the plateau of either the first or second wave, the product was spectrally identical with the so-called *phase test intermediate* (V, Fig. 8-9). Similarly, phase test intermediates were formed when chlorophyll *a* and *b* were electrochemically reduced. By analogy with similar electrochemical studies on metallotetraphenylporphyrins having no reducible side groups,[52] it was concluded[51] that the two reduction waves of chlorophyll in DMSO were due to formation of π-electron ions. Thus, the first wave was due to formation of the π-electron monoanion (a radical anion), and the second wave was due to formation of the π-electron dianion of I, II, III, or IV (Fig. 8-9). This idea was further confirmed by the fact that addition of a proton source to the DMSO, e.g., acetic acid, caused the first wave to shift anodically and to increase in height at the expense of the second wave, in a manner parallel to that observed for aromatic hydrocarbons.[53] It was further found that controlled potential electrolysis of I (Fig. 8-9) on the first polarographic wave resulted in disappearance of the second wave and that the product could not be electrochemically reoxidized back to the initial material at potentials 0.2 V anodic of the first wave. This indicated that, although V (Fig. 8-9) might be the ultimate product of the reaction, it was *not* the primary electrode product. Clearly, in order to convert the radical anion (II, Fig. 8-10), formed in the first wave process, to the phase test intermediate, loss of a hydrogen atom is necessary (see Fig. 8-9). The nature of the hydrogen acceptor in these systems was not known, although it was suggested that it could be the solvent, an impurity, or even another radical anion leading to hydrogen evolution. The actual structure of the phase test intermediate seems to be in some doubt, although it is probably some type of radical species.[54] If the

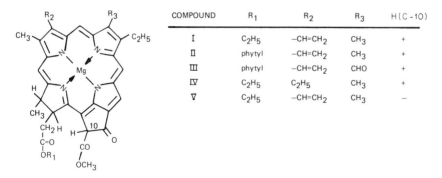

FIG. 8-9. Chlorophyll derivatives examined by Felton, Sherman, and Linschitz.[51]

TABLE 8-5 Half-Wave Potentials and n Values for the Polarographic Reduction of Some Chlorophyll Derivatives in DMSO[a,b]

Compound	Name	Wave I $E_{1/2}$ (V versus SCE)	n	Wave II $E_{1/2}$ (V versus SCE)	n
I	Ethyl chloro- phyllide *a*	−1.11	1.1	−1.54	0.93
II	Chlorophyll *a*	−1.12	0.95	−1.54	0.80
III	Chlorophyll *b*	−1.05	1.0	−1.46	0.87
IV	*meso*-Ethyl chloro- phyllide *a*	−1.17	1.0	−1.64	0.84

[a] Data from Felton *et al.*[51]

[b] Background electrolyte, 0.1 M tetrapropylammonium perchlorate.

interpretation of this electrochemical process is correct, then the chlorophyll radical anion would appear to be unstable with respect to the formation of the phase test intermediate and loss of hydrogen to acceptors. Therefore, it was suggested[51] that such transformations might provide a catalytic pathway by which charge-transfer reactions involving chlorophyll might effect H transfer, by virtue of the high stability of the phase test intermediate.

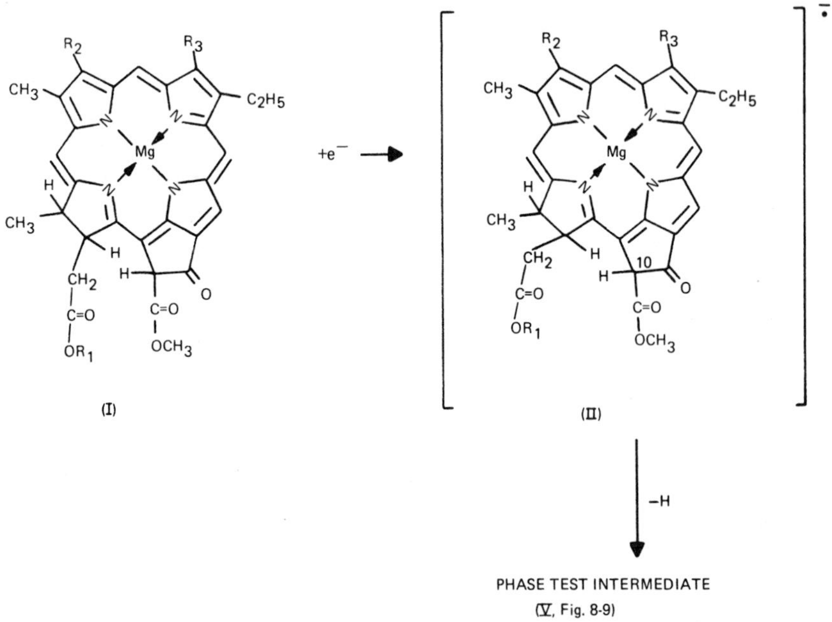

FIG. 8-10. Apparent route of electrochemical reduction of chlorophylls in DMSO.[51] For R substituents, see Fig. 8-9.

Inhoffen and co-workers[55-57] have found that chlorins can be reduced electrochemically at mercury electrodes in weakly acidic solution apparently by way of a single step at ca. −0.5 V. Electrolysis of a number of chlorins at the potential of their first wave led to the formation of chlorophlorins; the identity of the latter species was confirmed by NMR spectroscopy. Thus, chlorin e_6 trimethyl ester (I, Fig. 8-11) gave rise to the β-chlorophlorin (II, Fig. 8-11). The chlorophlorins produced in this manner were found to be unstable toward oxidizing agents (halogens, quinones, chromic acid), allowing them to be oxidized to the starting chlorins. The half-lives of the chlorophlorins under the influence of atmospheric oxygen are given in the original papers with a discussion of the relationship between their stability and structure.[55-57] In most cases of electrochemical reduction of natural 7,8-chlorins, the β-phlorin was produced although in some cases a mixture of β- and α-phlorins was observed.[58]

Kiselev and Kozlov[59] also found that chlorophyll *a* was polarographically reduced in anhydrous dimethylformamide containing 0.1 M LiCl. At chlorophyll *a* concentrations below 0.02 mM two cathodic waves were observed, $E_{1/2} = -1.12$ and -1.56 V versus an aqueous SCE. The slope of the log $i/(i_d - i)$ versus E_{DME} plot for the first wave was 69 mV, which is only slightly greater than one would expect for a reversible 1e reaction. At concentrations of chlorophyll *a* greater than 0.02 mM or when an excess of hydrogen ions was added to the solution, a single 2e wave was observed and $E_{1/2}$ shifted to more positive values. It was concluded[59] that at low concentrations (<0.02 mM) under aprotic conditions the first cathodic wave corresponded to a reversible 1e reduction of chlorophyll *a* (Chl, Eq. 2) to a radical anion. The second wave was

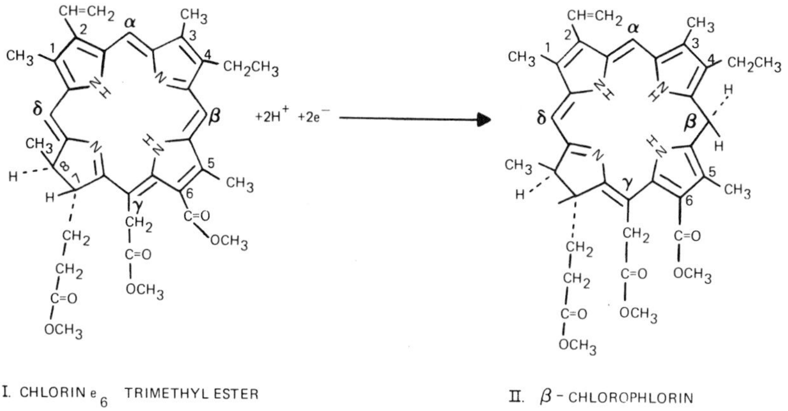

I. CHLORIN e_6 TRIMETHYL ESTER   II. β - CHLOROPHLORIN

FIG. 8-11. Electrochemical reduction of chlorins to chlorophlorins according to Inhoffen *et al.*[55-57]

proposed to be due to a $1e$ reduction of the radical anion to a dianion (Eq. 3). In

$$\text{Chl} + e \;\rightleftharpoons\; \text{Chl}^{\overline{\cdot}} \tag{2}$$

$$\text{Chl}^{\overline{\cdot}} + e \;\longrightarrow\; \text{Chl}^{2-} \tag{3}$$

the presence of excess protons, the single $2e$ wave was thought to result from an ECE type of mechanism (Eq. 4). Thus, addition of the first electron to chlorophyll a gave a radical anion that was rapidly protonated and then reduced in a further $1e$ reaction.

$$\text{Chl} + e \;\underset{E}{\longrightarrow}\; \text{Chl}^{\overline{\cdot}} \;\underset{C}{\overset{H^+}{\longrightarrow}}\; \text{ChlH}\cdot \;\underset{E}{\overset{e}{\longrightarrow}}\; \text{ChlH}^{\overline{\cdot}} \tag{4}$$

Tributsch and Calvin[60] have reported an extremely interesting study on the electrochemistry of excited chlorophylls. This study was based on the fact that semiconductors with a sufficiently large energy gap, in contact with an electrolyte, can be used as electrodes for the study of excited molecules. The behavior of excited chlorophyll molecules at a single crystal ZnO electrode was investigated. When chlorophylls absorb visible light they become excited. When a layer of chlorophyll is deposited onto the ZnO electrode surface these molecules inject electrons from excited levels into the conduction band of the electrode, thus giving rise to an anodic photocurrent. It was also found that, in the presence of electron donors such as hydroquinone or phenylhydrazine in the electrolyte, chlorophyll molecules, absorbing quanta, could mediate the pumping of electrons from the reducing agent into the conduction band of the semiconductor electron acceptor. By applying an adequate potential to the ZnO electrode ($+0.5$ V versus an unspecified reference electrode), the electron capture by the semiconductor electrode was irreversible. Using such techniques it is clear that some properties of excited chlorophyll at the semiconductor electrode are very similar to the expected behavior of chlorophyll in photosynthetic reaction centers. In other words, in both situations chlorophyll exhibits such properties as unidirectional electron transfer and highly efficient charge separation and in effect behaves as an electron pump converting electronic excitation energy into electrical energy.

In a more analytical vein, Serbanescu[61] studied the polarography of many porphyrins, including chlorophyll, obtained from crude oils. In an aqueous ethanolic tetramethylammonium hydroxide or LiOH solution chlorophyll showed two polarographic reduction waves. Similarly, Kramarczyk and Berg[62] employed polarography to follow the photochemical reactions of chlorophyll a and b, pheophytin a and b, bacteriochlorophyll, and sodium chlorophyllin.

B. Electrochemical Oxidation of Chlorophylls

Stanienda[63-65] has found that in propionitrile solution chlorophyll a and b and pheophytin a and b are readily oxidized at a rotating platinum disc electrode in two $1e$ stages (Table 8-6). The two waves are diffusion controlled and are proportional to concentration. In view of the difference in half-wave potentials for the first wave of chlorophyll and that of pheophytin, it appears that magnesium facilitates the electron-removal process. Furthermore, cyclic voltammetry[63] reveals that magnesium stabilizes the oxidation product. Addition of acid to the proprionitrile solution results in the disappearance of the oxidation waves of both chlorophyll and the pheophytins. Accordingly, in view of this effect and certain molecular orbital predictions it was concluded that the oxidative process primarily involves the two lone electron pairs on the ring nitrogen atoms.

Felton and co-workers[66] have shown that ethyl chlorophyllide *a* can be electrochemically oxidized in methylene chloride solution at a platinum anode. Controlled potential coulometry at 0.6 V versus SCE revealed that the process involved a single electron and, in view of the nature of the ESR spectrum of the product, it was concluded that the product was a π cation radical. Further discussion of this finding will be delayed until the end of this chapter.

Artamkina and Kutyurin[67] reported the effect of water on the oxidation of chlorophylls in acetone, at a platinum electrode. In totally anhydrous media chlorophyll *a*, chlorophyll *b*, and bacteriochlorophyll gave no oxidation waves. Oxidizability of the pigments could be restored when 2 moles of water were added to chlorophyll *a*, 0.5 mole to chlorophyll *b*, and 1 mole to bacteriochlorophyll. Addition of water in fact caused the half-wave potentials to shift to more negative values. In acetone containing 5–10% water the chlorophylls gave waves at about 0.48 and 0.58 V versus NCE.

The potentiometric oxidation–reduction potentials of chlorophylls have been reported by Goedheer *et al.*[68] and Kok[69]; the original papers should be consulted for details.

TABLE 8-6   Anodic   Half-Wave   Potentials   for Chlorophylls in Propionitrile Solution[a] at a Rotating Platinum Disc Electrode[b]

| | $E_{1/2}$ (V versus SCE) | |
Compound	Wave I	Wave II
Chlorophyll *a*	0.52	0.77
Chlorophyll *b*	0.65	0.87
Pheophytin *a*	0.86	1.17
Pheophytin *b*	0.99	1.28

[a] Supporting electrolyte, 0.1 M LiClO$_4$.
[b] Data from Stanienda.[64]

VIII. IRON—PORPHYRIN COMPLEXES

Iron—porphyrin complexes are of particular interest because oxygen-transporting hemoglobin is such a complex. Studies involving the electron-transfer reactions of metalloporphyrins in general are of considerable interest because such reactions are apparently involved in the transfer of energy by respiratory enzymes such as the cytochromes in a variety of biological situations. Knowledge of the reduction of hemin (iron—protoporphyrin, Fig. 8-12) appeared until recently to be rather confused. Essentially, it was not clear whether the reduction of hemin (to heme) involved one or two electrons. Conant *et al.*[70] employed a (zero current) potentiometric method and concluded that a dimeric form of hemin reacted to give a monomeric heme product via a two-electron reduction. However, subsequent potentiometric studies by Barron[71] seemed to indicate that monomeric hemin was reduced in a 1*e* process to a monomeric product. Shack and Clark[72] using spectrophotometric data have indicated that both the oxidized and reduced forms of iron—protoporphyrin are dimeric.

In order to resolve this rather serious controversy, Bednarski and Jordan[73,74] studied the reduction of hemin in aqueous solution by a DC polarographic method. It was found generally that the electrochemical behavior of the ferroheme—ferriheme systems exhibited two discrete patterns depending on whether the ionic strength (μ) or the concentration (C^0) of the electroactive species in the bulk of the solution exceeded 0.2 and 10^{-3} *M*, respectively. When the ionic strength of the solution was between 0.1 and 0.2 and the concentration

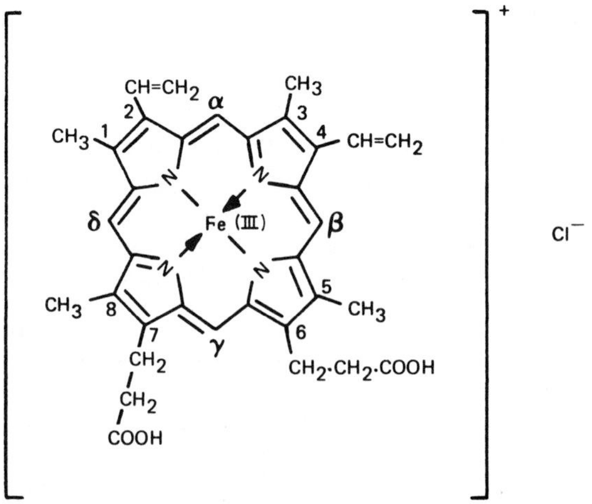

FIG. 8-12.   Iron—protoporphyrin, hemin.

of the electroactive species was below 10^{-3} M, three distinct pH regions were found in which somewhat different reactions occurred. Between pH 7 and 9 and at pH > 12.5 the polarographic wave was asymmetrical about the inflection point, which did not coincide with the half-wave potential (Fig. 8-13). The wave conformed to the theoretical equation expected for the 2e reduction of a dimer to a monomeric species in a reaction occurring with Nernstian reversibility under diffusion-controlled conditions. Between about pH 9 and 12.5, however, the polarograms were symmetrical about the inflection point, which was in turn identical with the half-wave potential. Under these conditions the wave conformed to an equation expected theoretically for the reversible, diffusion-controlled 2e reduction of a ferriheme dimer to a ferroheme dimer. A summary of the polarographic observations for the diffusion-controlled reduction of hemin under the conditions discussed above is presented in Table

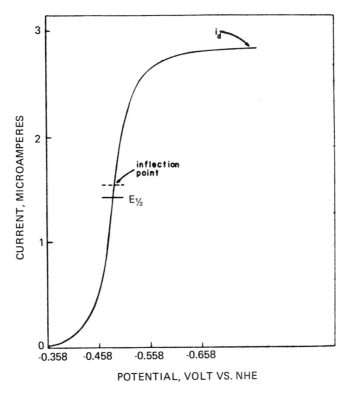

FIG. 8-13. Polarogram of 2.775×10^{-3} M hemin in 0.1 M aqueous KOH at 25°C. Current–voltage curve corrected for residual current and IR drop.[73] [Reprinted with permission from J. Jordan and T. M. Bednarski, *J. Am. Chem. Soc.* **86**, 5690 (1964). Copyright by the American Chemical Society.]

8-7. Some data on half-wave potentials and n values in the three regions delineated above are presented in Tables 8-8 and 8-9.

Calculation of the diffusion coefficient for the ferriheme dimer at pH ≈ 13 (0.1 M KOH) gave a value of 1.64×10^{-6} cm^2 sec^{-1} at 25°C. However, the diffusion coefficient for the dimeric species involved over the range $7 < \mathrm{pH} < 12.5$ was 1.21×10^{-6} cm^2 sec^{-1} at 25°C. On this basis Jordan and Bednarski[73,74] concluded that two discrete ferriheme dimers were involved in the electrode processes. Thus, over the pH range 7–12.5 the ferriheme dimer reduced has a larger Stokes radius and hence a smaller diffusion coefficient. The ferriheme dimer reducible above pH 12.5 has a smaller Stokes radius.

Although it is normally expected that a controlled potential electrolysis of a reversible cathodic wave should give rise to an anodic wave having the same half-wave potential and the same, or very similar, height as the original cathodic wave, in fact this was not observed in the case of reduction of ferriheme. It appeared from electrocapillary and other studies that both the ferriheme and ferroheme species were adsorbed at mercury, but the apparent reason for the failure to observe a normal anodic wave of ferroheme after controlled potential reduction of ferriheme was thought to be due to aggregation of ferroheme. A mechanism was proposed for the three pH regions for the diffusion-controlled reduction of ferriheme (Fig. 8-14). This mechanism indicated that over the entire pH range between 7 and 13 the electroreducible ferriheme was a dimeric species with two electron-acceptor centers. Dimeric ferriheme species are known to coordinate two hydroxyl groups.[75] Since the tetradentate porphyrin ligand forms an Fe(III) chelate which is very stable, the two monomeric ferriheme units had to necessarily be linked through octahedral positions perpendicular to the almost planar tetrapyrrole network. Bednarski and Jordan[74] considered that dimeric ferriheme A$_1$ (Fig. 8-14) was dimerized via a double propionate type of

TABLE 8-7   Summary of Polarographic Observations for the Diffusion-Controlled Reduction of Ferriheme[a,b]

pH Interval	Geometry of wave	Corresponding electrode process
7–9	Asymmetrical	Fe(III) dimer—Fe(II) monomer
9–12.5	Symmetrical	Fe(III) dimer—Fe(II) dimer
>12.5	Asymmetrical	Fe(III) dimer—Fe(II) monomer

[a] Data from Bednarski and Jordan.[74]

[b] Ionic strength solution between 0.1 and 0.2; concentration of electroactive species less than 10^{-3} F.

TABLE 8-8 Data Obtained under Experimental Conditions Where Diffusion-Controlled Dimer–Monomer Electrode Processes Prevailed[a]

Concentration (F)[b]	i_d/C^0 $(\mu A/10^{-3}\,F)$	No. of electrons involved[c]	$E_{1/2}$ (V versus NHE[d])
THAM–HCl buffer, pH 8.89, $\mu = 0.1$[e]			
3.083×10^{-4}	1.46	2.2	-0.192
5.582×10^{-4}	1.54	2.1	-0.195
8.000×10^{-4}	1.57	2.1	-0.205
1.156×10^{-3}	1.57	2.0	-0.209
1.499×10^{-3}	1.49	1.9	-0.212
0.1 M KOH buffer, pH 12.86, $\mu = 0.1$			
2.104×10^{-4}	1.71	2.1	-0.473
2.408×10^{-4}	1.82	2.0	-0.476
3.642×10^{-4}	1.82	2.1	-0.480
4.722×10^{-4}	1.70	2.2	-0.483
5.194×10^{-4}	1.73	2.0	-0.488

[a] Data from Bednarski and Jordan.[74]

[b] Bulk concentration of electroactive species, expressed in terms of formula weight of monomeric ferriheme units per liter of solution.

[c] Determined by a log-plot analysis of the polarographic wave; see original paper for details.

[d] Normal hydrogen electrode.

[e] THAM, tris(hydroxymethyl)aminomethane.

TABLE 8-9 Data Obtained under Experimental Conditions Where Diffusion-Controlled Dimer–Dimer Electrode Processes Prevailed[a]

Concentration (F)[b] $(\times 10^{-4})$	i_d/C^0 $(\mu A/10^{-3}\,F)$	No. of electrons involved[c]	$E_{1/2}$ (V versus NHE)
KHCO$_3$–KOH buffer, pH 9.91, $\mu = 0.18$			
1.499	1.55	2.2	-0.304
2.519	1.50	2.0	-0.301
3.802	1.44	1.9	-0.299
5.776	1.50	1.9	-0.302
7.534	1.40	2.0	-0.303

[a] Data from Bednarski and Jordan.[74]

[b] Bulk concentration of electroactive species, expressed in terms of formula weight of monomeric ferriheme units.

[c] Determined by a log-plot analysis of the polarographic wave; see original paper for details.

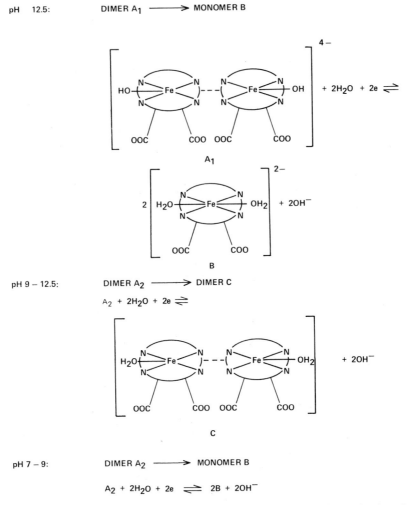

pH 12.5:   DIMER A_1 ⟶ MONOMER B

A_1

$+ 2H_2O + 2e \rightleftharpoons$

$2 \left[H_2O \underset{N}{\overset{N}{\Big(}} Fe \Big) OH_2 \right]^{2-} + 2OH^-$

B

pH 9 – 12.5:   DIMER A_2 ⟶ DIMER C

$A_2 + 2H_2O + 2e \rightleftharpoons$

$+ 2OH^-$

C

pH 7 – 9:   DIMER A_2 ⟶ MONOMER B

$A_2 + 2H_2O + 2e \rightleftharpoons 2B + 2OH^-$

FIG. 8-14.   Electrochemical reduction of hemin in aqueous solution at the dropping mercury electrode.[74] [Reprinted with permission from T. M. Bednarski and J. Jordan, *J. Am. Chem. Soc.* **89**, 1552 (1967). Copyright by the American Chemical Society.]

linkage as proposed by Lemberg and Legge[7] since this best accounted for the small Stokes radius of this species as evidenced by its larger diffusion coefficient. The ferriheme A_2 (Fig. 8-14) was considered to contain a water bridge[76] since ferrohemes coordinate with water more readily than with hydroxyl.[3,7] The dimeric ferroheme C and the monomeric ferroheme B both have their available coordination positions occupied by water (Fig. 8-14).

Under conditions in which the ionic strength of the solution was greater than 0.2 and the concentration of the electroactive species was greater than 10^{-3} F, the polarographic waves of ferriheme contained appreciable kinetic contributions reflecting rate-determining effects of chemical processes preceding the electrode reaction proper. These effects were noted in all the buffer systems and pH regions examined (pH 7–13). Polymerization phenomena were thought to account for the kinetic current observed under these conditions. Thus, the current observed was due to the electroreduction of the appropriate ferriheme dimer (A_1 or A_2, Fig. 8-14) generated *in situ* by depolymerization at the electrode surface of higher molecular weight ferriheme aggregates present in the bulk of the solution.

In addition to the wave corresponding to the electroreduction of the heme iron from the trivalent to the divalent state, which occurred in the potential range of −0.16 to −0.51 V versus NHE, an ill-defined cathodic wave was also observed at much more negative potential, namely, −1.4 to −1.5 V versus NHE. This wave was a typical catalytic hydrogen reduction wave, presumably catalyzed by the protoporphyrin ring.

Davis and Orleron[77] have studied the cyclic voltammetry of iron–porphyrin complexes, particularly to gain information regarding ligand substitution reactions. Recalling Fig. 8-12, which shows the structure of iron–protoporphyrin or hemin, it is clear that the porphyrin ring itself satisfies only the four $X–Y$ coordination positions of the ferric ion. The two remaining coordination positions are located perpendicularly above and below the almost planar porphyrin ring on the Z axis. When these positions on the Z axis are occupied by π-bonding ligands such as pyridine, imidazole, or cyanide, the complexes designated *hemichromes* result. Davis and Orleron[77] were interested in exchanges between the Z axis ligands of hemichromes and other ligands present in solution. Cyclic voltammograms of pyridine hemichrome [bis(pyridyl)ferriprotoporphyrin] and cyanide hemichrome [bis(cyano)ferriprotoporphyrin) are shown in Figs. 8-15 and 8-16, respectively. Both electrode processes conformed to all the theoretical requirements of an uncomplicated, reversible, diffusion-controlled, charge-transfer process involving just a single electron.[78–80] If sodium cyanide was added to a solution of pyridine hemichrome in excess pyridine, i.e., to the solution of Fig. 8-15, new, more negative, reversible cathodic and anodic peaks appeared (Fig. 8-17). By comparing peak potentials and peak currents for this new solution with those of pyridine and cyanide hemichromes (Figs. 8-15 and 8-16), it was evident that a new species was being formed, namely, cyanopyridine hemichrome. This is a complex in which one of the Z ligand positions (apical positions) is occupied by cyanide and the other is occupied by pyridine. Rather interestingly, it was found that the cyclic voltammograms of the solution described in Fig. 8-17 at slow scan rates gave unequal anodic and cathodic peak currents for either set of

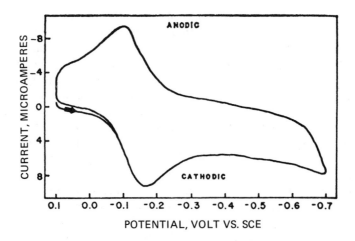

FIG. 8-15. Cyclic voltammogram of pyridine hemichrome at a Pt electrode in 30% aqueous ethanol containing $0.8\,M$ NaNO$_3$.[77] Hemin concentration, 0.81 mM; 2.0 M pyridine. Scan rate, 0.485 V sec^{-1}. [Reprinted with permission from D. G. Davis and D. J. Orleron, *Anal. Chem.* **38**, 179 (1966). Copyright by the American Chemical Society.]

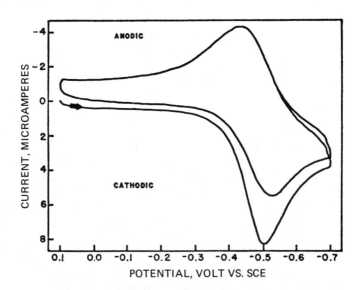

FIG. 8-16. Cyclic voltammogram of cyanide hemichrome at a Pt electrode in 30% aqueous ethanol containing $0.8\,M$ NaNO$_3$.[77] Hemin concentration, 0.81 mM; 0.25 M cyanide ion. Scan rate, 0.246 V sec^{-1}. [Reprinted with permission from D. G. Davis and D. J. Orleron, *Anal. Chem.* **38**, 179 (1966). Copyright by the American Chemical Society.]

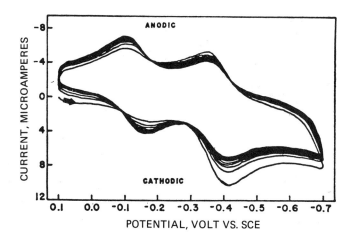

FIG. 8-17. Cyclic voltammogram of pyridine hemichrome at Pt electrode in 30% aqueous ethanol containing 0.8 M NaNO$_3$. Hemin concentration, 0.81 mM; 8.0 x 10^{-4} M cyanide and 2.0 M pyridine.[77] Scan rate, 0.485 V sec^{-1}. [Reprinted with permission from D. G. Davis and D. J. Orleron, *Anal. Chem.* **38**, 179 (1966). Copyright by the American Chemical Society.]

reversible peaks, and the current for each peak changed as the potential was cycled back and forth. However, at fast scan rates (4.7 V sec^{-1}) the peak currents were constant for several cycles, and the anodic and cathodic peak currents for each pair of reversible peaks were constant. These phenomena were taken to indicate that several chemical reactions occurred in the solution subsequent to the reversible electron transfer but that the reactions proceeded at such a rate that they were not noticeable at fast scan rates. A cyclic voltammogram of pure cyanopyridine hemichrome in the presence of excess pyridine (Fig. 8-18) at relatively slow scan rates revealed only a single cathodic peak for reduction of the cyanopyridine hemichrome ($E_p \approx -0.4$ V versus SCE). However, on the reverse sweep both cyanopyridine hemochrome and pyridine hemochrome were oxidized and gave rise to the two observed anodic peaks. As the scan rate was increased, the anodic peak of oxidation of the pyridine hemochrome disappeared ($E_p = -0.1$ V).

The mechanism responsible for these observations is shown in Fig. 8-19. The succeeding chemical reaction, characterized by the rate constant k_{f_1} (0.56 sec^{-1}), was shown to be effectively irreversible. Knowing the nature of the succeeding chemical reaction and the value of its rate constant, Davis and Orleron[77] were able to interpret the cyclic voltammogram shown in Fig. 8-17. The first cycle indicated that both pyridine hemichrome and cyanopyridine hemichrome were reducible and, on the reverse cycle, that cyanopyridine hemochrome and pyridine hemochrome were oxidized. On continued cycling,

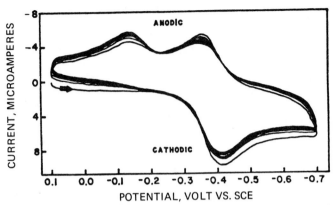

FIG. 8-18. Cyclic voltammogram of cyanopyridine hemichrome at a Pt electrode in 30% aqueous ethanol containing 0.8 *M* NaNO$_3$.[77] Hemin concentration, 0.81 m*M*; 4×10^{-3} *M* cyanide and 2.0 *M* pyridine. Scan rate, 0.485 V sec^{-1}. [Reprinted with permission from D. G. Davis and D. J. Orleron, *Anal. Chem.* 38, 179 (1966). Copyright by the American Chemical Society.]

no new species appeared, but the relative concentrations of the existing species changed, as evidenced by the relative height of the voltammetric peaks. Thus, the cyanopyridine hemochrome produced by electrochemical reduction of cyanopyridine hemichrome reacts with excess pyridine (Fig. 8-19), and hence the concentration of pyridine hemochrome builds up as the experiment continues. The concentration of pyridine hemichrome also increases since it is the product of the oxidation of pyridine hemochrome. However, at fast scan rates these concentration changes do not occur since insufficient time is allowed for the chemical reactions to proceed. The failure of a cathodic peak for pyridine hemichrome to appear on the second and subsequent scans on cyanopyridine hemichrome (Fig. 8-18) was proposed to be due to the fact that at the high concentration of cyanide ion in the solution, the rate of reaction of pyridine hemichrome (formed by oxidation of pyridine hemochrome at the anodic peak, $E_p = -0.1$ V) with cyanide ion to give cyanopyridine hemichrome was very fast.

Cyclic voltammetry of hemin in the presence of pyridine and larger amounts of cyanide (Fig. 8-20) gave rise to only a single cathodic peak due to reduction of the cyanide hemichrome and an anodic peak due to oxidation, *not* of cyanide hemochrome, but of cyanopyridine hemochrome. Accordingly, the reaction of pyridine with cyanide hemochrome to give cyanopyridine hemochrome (Fig. 8-21) was considered to be fast, but reactions of the type shown in Fig. 8-19 were not evident. The value of k_{f_2} for the reaction shown in Fig. 8-21 was about 20 sec^{-1}.

A subsequent report by Davis and Martin[81] using chronopotentiometry,

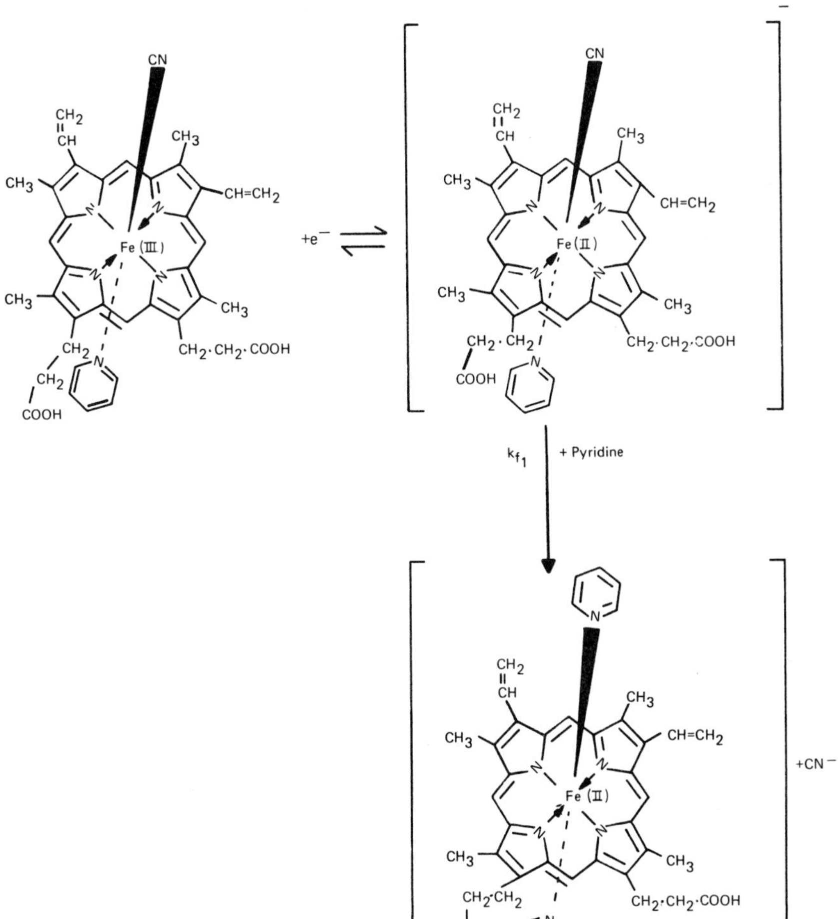

FIG. 8-19. Interpretation of chemical reactions succeeding the electron-transfer process of cyanopyridine hemichrome.[77]

controlled potential coulometry, and polarography confirmed that reduction of cyanide hemichrome was indeed a 1e reversible process, although the fact that chronopotentiometric $i\tau^{1/2}$ increased with decreasing τ indicated that adsorption of the cyanide hemichrome was involved in the overall process.*

---

* In these chronopotentiometric experiments, i was the (constant) current density in amperes per (centimeter)^2, and τ was the transition time in seconds. See Davis[82] for a discussion of chronopotentiometry.

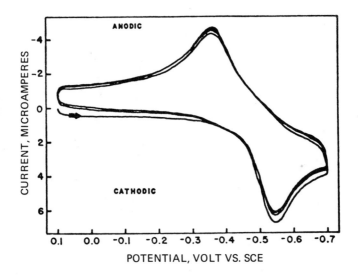

FIG. 8-20.   Cyclic voltammogram of cyanide hemichrome with excess pyridine present at a Pt electrode in 30% aqueous ethanol containing 0.8 M $NaNO_3$.[77] Hemin concentration, 8.1 x 10^{-4} M; pyridine, 2.0 M; cyanide, 0.25 M. Scan rate, 0.246 V sec^{-1}. [Reprinted with permission from D. G. Davis and D. J. Orleron, *Anal. Chem.* **38**, 179 (1966). Copyright by the American Chemical Society.]

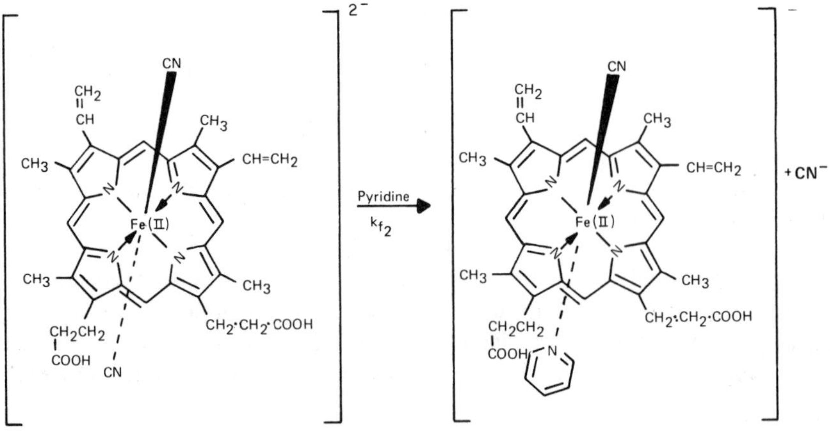

FIG. 8-21.   Interpretation of chemical reaction of cyanide hemochrome with excess pyridine.[77]

Cyanopyridine hemichrome and pyridine hemichrome were also confirmed to undergo $1e$ reversible reductions by similar techniques. Voltammetry at a rotating platinum electrode was used to study reactions 5 and 6.

$$(5)$$

$$(6)$$

Typical voltammograms are shown in Fig. 8-22. The value of log K for reaction 5 was found to be 5.11 and that for reaction 6 was 0.81 using the voltammetric method.

Davis and Martin[81] also found that in alkaline ethanol—water solution the *trans*-hydroxoaquoprotoporphyrin ferrate(III) ion (I, Fig. 8-23) underwent a reversible one-electron reduction at the DME. The polarograms of I (Fig. 8-23)

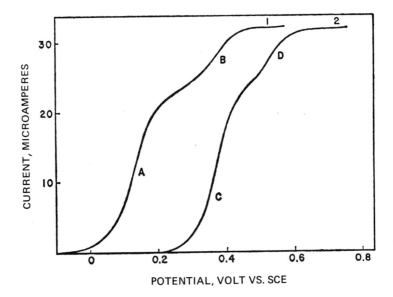

FIG. 8-22. Voltammograms of hemichromes with a rotating platinum electrode. Voltammogram 1: 0.81×10^{-3} *M* hemin, 2.0 *M* pyridine, 3×10^{-4} *M* NaCN, 1 *M* NaNO$_3$, 30% ethanol. Voltammogram 2: 0.81×10^{-3} *M* hemin, 2.0 *M* pyridine, 1×10^{-2} *M* NaCN, 1 *M* NaNO$_3$, 30% ethanol. (A) Reduction of pyridine hemichrome, (B) reduction of cyanopyridine hemichrome, (C) reduction of cyanopyridine hemichrome, (D) reduction of cyanide hemichrome.[81] [Reprinted with permission from D. G. Davis and R. F. Martin, *J. Am. Chem. Soc.* **88**, 1365 (1966). Copyright by the American Chemical Society.]

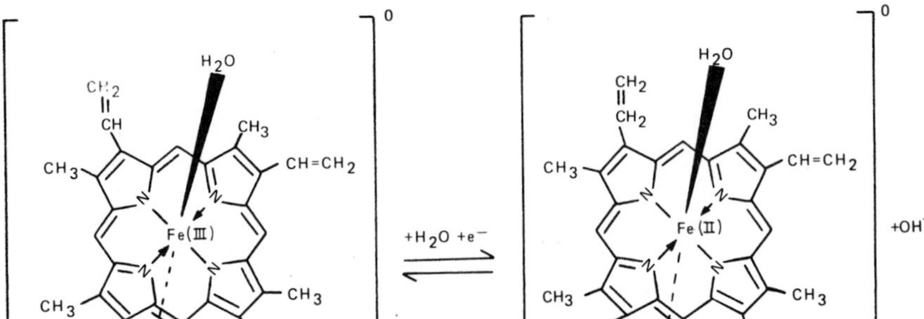

FIG. 8-23. Mechanism of electrochemical reduction of *trans*-hydroxoaquoprotoporphyrin ferrate(III) at the DME in alkaline ethanolic aqueous solution.[81]

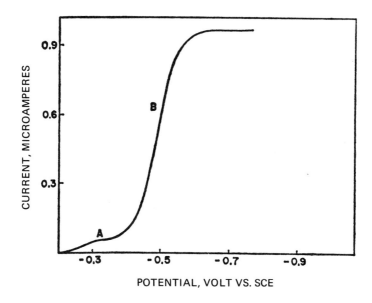

FIG. 8-24. Polarograms of the *trans*-hydroxoaquoprotoporphyrin ferrate(III) ion: $0.90 \times 10^{-3} M$ hemin, $0.025 M$ borax, 30% ethanol, pH 9.5. A: Adsorption wave, B: normal wave.[81] [Reprinted with permission from D. G. Davis and R. F. Martin, *J. Am. Chem. Soc.* **88**, 1365 (1966). Copyright by the American Chemical Society.]

showed a distinct adsorption prewave (Fig. 8-24), indicating that the product of the electrode reaction (II, Fig. 8-23) was adsorbed at the electrode surface. As expected on the basis of the reaction shown in Fig. 8-23, the half-wave potential for the polarographic wave was pH dependent, $dE_{1/2}/d(\text{pH}) = -0.58$ V/pH over the pH range 9.0–12.0. In 30% ethanolic aqueous solution (pH 9.5) the diffusion coefficient of the monomeric hydroxoaquo species was 1.19×10^{-6} cm^2 sec^{-1} at the 1 mM hemin concentration, which was about an order of magnitude greater than that found in aqueous solution without alcohol.[73] This was interpreted as indicating micelle aggregate formation[72] as well as dimer formation, as proposed by Jordan and Bednarski.[73]

The effect of pH on the electrochemical reduction of pyridine hemichrome has also been investigated.[81] Between pH 7.5 and 8.3 it is reversibly reduced at mercury or platinum electrodes at -0.128 V versus SCE (Table 8-10). Above pH 8.3 the polarographic wave shifted to more negative values (Fig. 8-25), although between pH 8.3 and 10.5 the $E_{1/2}$ did not shift linearly with pH. This effect is apparently due to a transition region in which two or possibly more closely related pyridine–hemin species exist. Above pH 10.5 a linear dependence of

TABLE 8-10   Electrochemical Data for Various Hemichromes[a]

	Polarography		Controlled potential coulometry,	Chronopotentiometry	
Complex[b]	$E_{1/2}{}^c$ (V versus SCE)	Slope[d]	n	$E_{T/4}{}^c$ (V versus SCE)	Slope[d]
H(CN)$_2$	−0.499	0.061	1.01	−0.495	0.060
H(CN)$_2$, no ethanol	−0.413	0.059	1.01	−0.420	0.060
H(py)$_2$, pH 7.9	−0.128	0.060	0.98	−0.129	0.059
H(CN)(py)	−0.360	0.061	1.01	−0.358	0.061
H(pic, β or γ), pH 8	−0.147	0.059	0.99	−0.145	0.060

[a] Adapted from Davis and Martin.[81]

[b] The letter H signifies iron(III)−protoporphyrin; py, pyridine; pic, picoline.

[c] Potentials (volt versus SCE) refer to complexes in 30% ethanolic solution except where noted.

[d] A plot of $\log[(i_d − i)/i]$ for polarography or $\log[(\tau^{1/2} − t^{1/2})/t^{1/2}]$ for chronopotentiometry versus potential was used to test Nernstian reversibility. A slope of 0.059 indicates a reversible 1e reduction. Evidence obtained by controlled potential electrolysis supported the 1e change.

$E_{1/2}$ on pH was observed. At pH 9 and above, solutions of pyridine hemichrome exhibit a second well-defined polarographic wave in the range −0.3 to −0.45 V versus SCE (Fig. 8-26). The second wave shifts toward more negative potential with increasing pH $[dE_{1/2}/d(\text{pH}) = −0.060]$ and, as the pH increases, the height

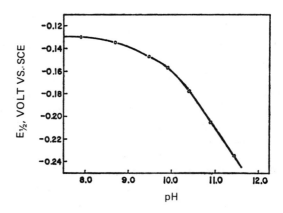

FIG. 8-25. Variation of $E_{1/2}$ with pH for pyridine hemichrome.[81] [Reprinted with permission from D. G. Davis and R. F. Martin, *J. Am. Chem. Soc.* 88, 1365 (1966). Copyright by the American Chemical Society.]

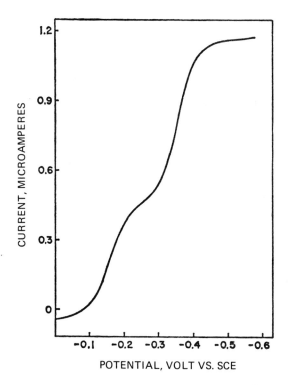

FIG. 8-26.   Polarogram of pyridine hemichrome at pH 10.0: 0.98×10^{-3} M hemin, 0.025 M borax, 2.5 M pyridine, 10% ethanol.[81] [Reprinted with permission from D. G. Davis and R. F. Martin, *J. Am. Chem. Soc.* **88**, 1365 (1966). Copyright by the American Chemical Society.]

of the first wave decreases, while that of the second wave increases. Wave analysis and cyclic voltammetry have revealed that both of the waves are reversible, one-electron reductions and that all iron(III) species have a common product, pyridine hemochrome.

The probable course of electrochemical reduction concluded from the above observations is presented in Table 8-11. Although their conclusions were not entirely consistent with the data, Davis and Martin[81] proposed that the first wave in Fig. 8-26 corresponds to reaction 3 (Table 8-11) and the second wave to reaction 2 (Table 8-11). All species involved in the electrode reactions were thought to be monomeric. Other data on the electrochemical reduction of a variety of hemichrome complexes are presented in Table 8-10.

TABLE 8-11 Electrode Reactions of Iron–Protoporphyrin in Pyridine as a Function of pH[a]

pH Range	Reaction responsible for first wave[b]	Reaction responsible for second wave
7.5–8.3	$$\overset{\displaystyle py}{\underset{\displaystyle py}{|}}\!\!\!Fe(III)P + e \rightleftharpoons \overset{\displaystyle py}{\underset{\displaystyle py}{|}}\!\!\!Fe(II)P \quad (1)$$	None
8.3–9	Transition region involving two or more pyridine—hemin species	None
9–10.5	Transition region involving two or more pyridine—hemin species	$$\overset{\displaystyle OH}{\underset{\displaystyle H_2O}{|}}\!\!\!Fe(III)P + 2py + e \rightleftharpoons \overset{\displaystyle py}{\underset{\displaystyle py}{|}}\!\!\!Fe(II)P + OH^- + H_2O \quad (2)$$
10.5–12.0	$$\overset{\displaystyle OH}{\underset{\displaystyle py}{|}}\!\!\!Fe(III)P + py + e \rightleftharpoons \overset{\displaystyle py}{\underset{\displaystyle py}{|}}\!\!\!Fe(II)P + OH^- \quad (3)$$	$$\overset{\displaystyle OH}{\underset{\displaystyle H_2O}{|}}\!\!\!Fe(III)P + 2py + e \rightleftharpoons \overset{\displaystyle py}{\underset{\displaystyle py}{|}}\!\!\!Fe(II)P + OH^- + H_2O$$

[a] Data from Davis and Martin.[81]

[b] The letter P denotes protoporphyrin ring.

IX. MANGANESE–HEMATOPORPHYRIN IX

Manganese–porphyrins are of some interest because they have been suspected of playing a role in human porphyrin metabolism[83] and in photosynthetic energy transfers.[84] The solution chemistry of the various oxidation states of manganese–porphyrins has been studied by spectrophotometry and potentiometry.[85–87] However, in view of the much greater insight that polarography, cyclic voltammetry, etc., can give into the details of metalloporphyrin systems, Davis and Montalvo[88] studied the Mn(III)– and Mn(II)–hematoporphyrin systems in aqueous ethanol solution. The abbreviated nomenclature system recommended by these authors will be employed here; i.e., Mn(III)Hm and Mn(III)HmdiMe are used for the manganese(III)–hematoporphyrin IX complex and its dimethyl ester, respectively; Hm and HmdiMe are used for the metal-free hematoporphyrin IX and its dimethyl ester, respectively. The structure of hematoporphyrin IX dimethyl ester is shown in Fig. 8-27. In aqueous or aqueous ethanol solution the Z axis coordination positions of manganese are occupied by either water molecules or hydroxyl (OH^-) ions.

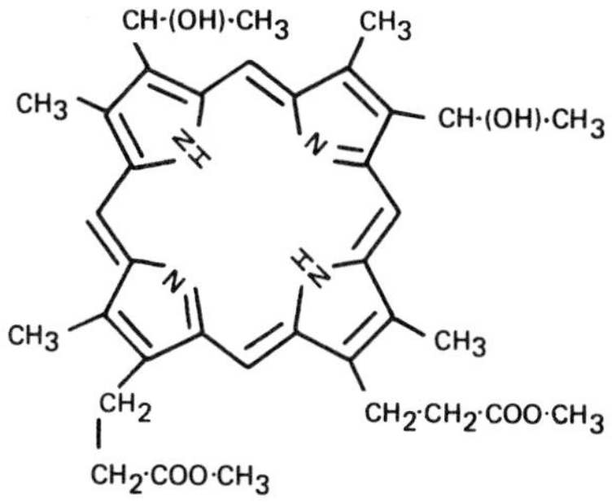

FIG. 8-27.   Structure of hematoporphyrin IX dimethyl ester.

The Mn(III)Hm and Mn(III)HmdiMe give a well-formed polarographic wave (B, Fig. 8-28) in aqueous ethanol solution along with a small adsorption prewave (A, Fig. 8-28) due to adsorption of the product of the electrode reaction [Mn(II)HmdiMe] at the DME surface. The variations of the half-wave potential with pH are shown in Fig. 8-29. The $E_{1/2}$ values agree well with the potentiometric midpoint potentials (E_m) of Loach and Calvin.[87] The shift of the half-wave potential with pH between pH 1 and 4 and between pH 11 and 13 is -49 mV/pH unit, which, since coulometry indicated that the electrode reaction involved one electron, reveals that at low pH one hydrogen ion and one electron are consumed, while at high pH one electron is consumed and one hydroxyl ion is liberated. In the intermediate pH range neither H^+ nor OH^- is produced or consumed since the half-wave potential is essentially pH

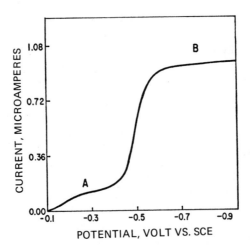

FIG. 8-28.   Polarogram of Mn(III)HmdiMe at pH 9.4, 0.80 m*M* Mn(III)HmdiMe, 47.5% ethanol. Supporting electrolyte, 1.0 *M* NaNO$_3$.[88] [Reprinted with permission from D. G. Davis and J. G. Montalvo, *Anal. Chem.* **41**, 1195 (1969). Copyright by the American Chemical Society.]

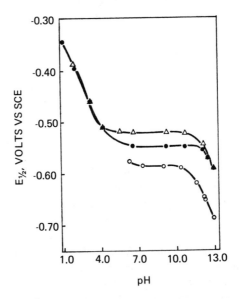

FIG. 8-29.   Variation of half-wave potential with pH.[88] ○, Mn(III)Hm in water; ●, Mn(III)Hm in 47.5% ethanol; △, Mn(III)HmdiMe in 47.5% ethanol; supporting electrolyte 1 *M* NaNO$_3$. [Reprinted with permission from D. G. Davis and J. G. Montalvo, *Anal. Chem.* **41**, 1195 (1969). Copyright by the American Chemical Society.]

independent. At very high pH the half-wave potentials for Mn(III)Hm and Mn(III)HmdiMe coincide because the ester hydrolyzes so rapidly that the polarograms are effectively of Mn(III)Hm. At low pH Mn(II)Hm underwent an irreversible displacement of Mn^{2+} from the porphyrin ring by one or more hydrogen ions. This reaction is sufficiently slow that no appreciable effect on the half-wave potential was expected.

On the basis of the foregoing summary Davis and Montalvo[88] proposed a set of reactions to explain the polarographic data (Table 8-12). These reactions are at variance with those proposed by Loach and Calvin,[87] who suggested that in both the Mn(III) and Mn(II) species one of the Z coordination positions of the metal ion is occupied by an intramolecularly bonded propionic acid carboxylate group.

Thus, in summary, at low pH the electrode reaction is due to a $1e$ reduction of Mn(III) to Mn(II) and addition of a single proton to one of the pyrrole nitrogen atoms. This is an extremely unusual electrode reaction if it does occur, but it appears to be justified because at low pH the Mn^{2+} is displaced rather rapidly by hydrogen ion. Fleischer and Wang[89] have identified such protonated species. The addition of hydroxyl groups to the Z axis coordination positions at pH 6 and above would also appear to be reasonable.

TABLE 8-12 Electrode Reactions of Mn(III)Hm and Mn(III)HmdiMe as a Function of pH[a, b]

pH Range	Electrode reaction	
1.0–4.0	$\overset{\displaystyle H_2O}{\underset{\displaystyle H_2O}{\mid}}\text{Mn(III)}\text{—P} + H^+ + e^- = \overset{\displaystyle H_2O}{\underset{\displaystyle H_2O}{\mid}}\text{Mn(II)}\text{—P—H}$	(1)
4.0–10.6	Transition region of mixed reaction	
6.0–10.6	$\overset{\displaystyle H_2O}{\underset{\displaystyle OH}{\mid}}\text{Mn(III)}\text{—P} + e^- = \overset{\displaystyle H_2O}{\underset{\displaystyle OH}{\mid}}\text{Mn(II)}\text{—P}$	(2)
10.6–12.2	Transition region of mixed reactions	
12.2–13.0	$\overset{\displaystyle OH}{\underset{\displaystyle OH}{\mid}}\text{Mn(III)}\text{—P} + H_2O + e^- = \overset{\displaystyle H_2O}{\underset{\displaystyle OH}{\mid}}\text{Mn(II)}\text{—P} + OH^-$	(3)

[a] Adapted from Davis and Montalvo[88]

[b] The symbol —P is used for either hematoporphyrin or hematoporphyrin dimethyl ester. The charges on the species are neglected for simplicity. The symbol —P—H indicates a hydrogen ion bond to one of the pyrrole nitrogens of the porphyrin.

Over the pH range 5.5–13 in aqueous alcohol or pH 5.5–10.2 in aqueous solution and at concentrations of the Mn(III) species below 1.5 mM (at higher concentrations maxima appeared and dimer formation became appreciable), the normal polarographic wave (B, Fig. 8-28) appeared to conform to the normal criteria of a reversible, diffusion-controlled, one-electron wave, although analysis of wave slope data gave a plot of E_{DME} versus log $i/(i_d - i)$ with a slope greater than 59 mV in many instances. This effect was probably associated with adsorbed reaction product on the electrode surface partially blocking the electrode surface.

Polarograms of the type shown in Fig. 8-28 having a prewave indicate adsorption of the product of an electrode reaction. Cyclic voltammetry confirmed this view but also indicated that the reactant [the Mn(III) species] as well as the product [the Mn(II) species] was adsorbed. Calculation of the number of molecules adsorbed on a mercury drop during its lifetime revealed that the Mn(II)Hm did not form a complete monolayer on the surface of the mercury but that rather incomplete surface coverage was obtained which varied with pH. This result differs from the iron–protoporphyrin case,[81] where an essentially complete monolayer was obtained independent of pH.

The displacement of Mn^{2+} from the Mn(II)Hm or Mn(II)HmdiMe has been studied in some detail and resulted in some rather interesting conclusions. A polarogram of Mn(III)HmdiMe at pH 4.65 is shown in Fig. 8-30. The section labeled B corresponds to the reduction of Mn(III)HmdiMe to Mn(II)HmdiMe according to the previously discussed criteria. The heights of waves C and D increased as the pH became more acidic, while that of B remained constant. At pH values greater than 6 only wave B could be observed. Cyclic voltammetry of Mn(III)Hm at pH 4.65 showed that C and D appeared on the initial cycle at low scan rates but not at high scan rates. In addition, at low scan rates the ratio of the anodic peak current to the cathodic peak current was less than unity for the Mn(III)/Mn(II) hematoporphyrin couple but approached 1 as the scan rate was increased. This behavior was indicative of a reversible electron transfer followed by an irreversible chemical reaction. A polarogram of free Hm gave two polarographic waves that were identical to waves C and D. Thus, waves C and D occur as a result of the irreversible chemical transformation of the Mn(II)–porphyrin complex to the free porphyrin. The mechanism for waves C and D was not elucidated, although these waves have been proposed to result from a multielectron-transfer process because, even though they were of comparable height to wave B, they represented only a small fraction of the decomposition of the Mn(II)–porphyrin species. Controlled potential coulometry of Mn(III)Hm at pH 6.2 revealed that one electron was transformed per molecule of Mn(III)Hm reduced and that the spectrum of the final product was that of Hm. The first-order rate constant for the Mn(II)Hm decomposition to Mn^{2+} and Hm at

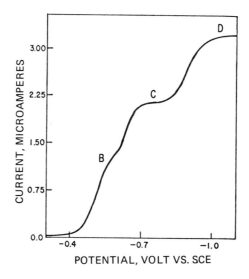

FIG. 8-30.   Direct current polarogram of Mn(III)HmdiMe, pH 4.65, 45% ethanol, 0.99 m*M* Mn(III)HmdiMe.[88] [Reprinted with permission from D. G. Davis and J. G. Montalvo, *Anal. Chem.* **41**, 1195 (1969). Copyright by the American Chemical Society.]

pH 3.79 was found to be $4.7 \pm 0.7 \, \text{sec}^{-1}$ (determined by a chronopotentiometric method.[82] Measurement of the rate constant as a function of pH revealed that at both high and low pH the latter reaction was first order in hydrogen ion but greater than first order at intermediate pH regions. Accordingly, the mechanism for acid displacement of Mn(II)–porphyrin appeared to occur by the initial addition of a proton to a pyrrole nitrogen atom in a rate-determining step, followed by addition of a second proton and a fast decomposition of the resultant transition-state complex. The final transition-state complex leading to rapid decomposition of the Mn(II)–porphyrin complex was therefore considered to have two protons bonded to the pyrrole-type nitrogens and to have a sitting-atop structure of the type shown in Fig. 8-31.

This study is interesting because it reveals that Mn(II)–porphyrin complexes are unstable in an aqueous environment of physiological pH owing to the displacement of Mn^{2+} from the complex. It has been concluded, therefore, that it is unlikely that the Mn(II) state could be important in porphyrin metabolism of cells in man and in photosynthetic processes unless the porphyrin could be otherwise stabilized.[88]

Davis and Montalvo[90] have also examined the electrochemistry of manganese–hematoporphyrin IX complexes in aqueous solutions containing a variety of nitrogenous bases such as picoline, pyridine, and imidazole. The latter

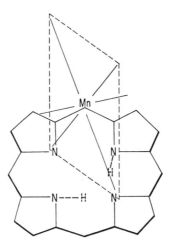

FIG. 8-31.   Final transition-state complex after diprotonation of Mn(II)–porphyrin complexes.[88] [Reprinted with permission from D. G. Davis and J. G. Montalvo, *Anal. Chem.* **41**, 1195 (1969). Copyright by the American Chemical Society.]

species can function as ligands for the Z axis coordination positions of the Mn(III)– and Mn(II)–hematoporphyrin complexes. The most detailed study was carried out with α-picoline as the ligand species. A typical polarogram of Mn(III)Hm at pH 9.4 is shown in Fig. 8-32. The first wave, designated A (Fig. 8-32), corresponded to the one-electron reduction of Mn(III)Hm to Mn(II)Hm. Wave B was thought to be an adsorption prewave preceding the reduction of the porphyrin ring at some potential more negative than −1.0 V. Because of the onset of background discharge it was not possible to observe the latter wave. Wave C was a maximum of the second kind. If the concentration of α-picoline was increased from zero to 1 M, then the half-wave potential of wave A (Fig. 8-32) shifted toward more positive potential to an extent that indicated that two more α-picoline molecules were coordinated with the product of the electrode reaction [Mn(II)Hm] than with the reactant [Mn(III)Hm].[91] Other nitrogenous ligands behaved in a similar fashion. The fact that the dipicoline complex of Mn(II)Hm was easier to form than that of Mn(III)Hm was thought to indicate, not necessarily that Mn(III) does not form strong complexes, but rather that, in the competition with water and hydroxide ion with nitrogenous bases, these bases are at a disadvantage. At concentrations of α-picoline, and presumably other bases, above 1 M the $E_{1/2}$ of wave A was essentially unaffected, indicating that, in the presence of such large excesses of complexing agent, both the Mn(III)Hm and Mn(II)Hm were coordinated at both their Z axis positions with

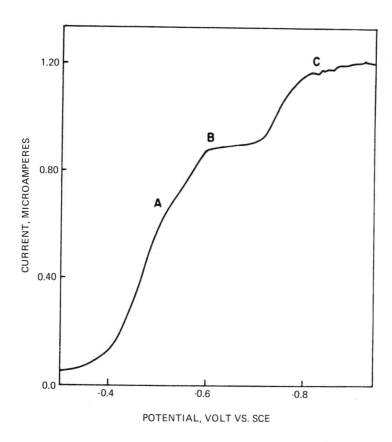

FIG. 8-32. Polarogram for Mn(III)Hm with added α-picoline. Borax buffer, pH 9.4, 0.86 m*M* Mn(III)Hm; 30% α-picoline.[90] [Reprinted with permission from D. G. Davis and J. G. Montalvo, *Anal. Lett.* **1**, 641 (1968). Copyright by Marcel Dekker, Inc., New York.]

α-picoline. A summary of $E_{1/2}$ data for a number of ligands is shown in Table 8-13.

The fact that the $E_{1/2}$ values for the nitrogenous bases differ by only a few millivolts at pH 9 and 13 was thought to indicate that hydroxide ion was not involved in the electrode reaction. Apparently, in the case of the acetate or phosphate ligands hydroxide ion is involved. The wave slope data reported in Table 8-13, if at all significant, deviate slightly from those expected for a reversible 1*e* process. This deviation was thought to be in some way associated with adsorption at the electrode surface.

TABLE 8-13 Polarographic Data for the Reduction of Mn(III)Hm with Added Ligands[a,b]

Ligand	Ligand concentration[c] (M)	(%) Ethanol	$E_{1/2}$ (V versus SCE)[d] at pH 9.0	$E_{1/2}$ (V versus SCE)[d] at pH 13.0	log $i/(i_d - i)$ at pH 9.0	log $i/(i_d - i)$ at pH 13.0
Pyridine	2.48	0	−0.522	−0.525	0.070	0.068
α-Picoline	2.04	0	−0.467	−0.491	0.068	0.060
β-Picoline	2.06	0	−0.496	−0.496	0.068	0.063
γ-Picoline	2.05	0	−0.501	−0.507	0.077	0.064
Imidazole	1.00	20	−0.595	−0.604	0.064	0.060
Nicotine	1.23	0	−0.473	−0.479	0.063	0.059
KOAc	1.00	10	−0.579	−0.660	0.076	0.063
$Na_4P_2O_7$	Saturated solution	10	−0.569	−0.672	0.080	0.060

[a] Data from Davis and Montalvo.[90]

[b] The Mn(III)Hm concentration employed was 0.82 mM.

[c] Sufficiently high to ensure complete complexation of both oxidized and reduced forms of manganese species.

[d] The $E_{1/2}$ for uncomplexed Mn(III)Hm at pH 9 is −0.585 V.

X. OTHER METALLOPORPHYRIN SYSTEMS

Apart from the work described above a number of papers have appeared which describe the electrochemical behavior of a large number of different metals complexed with several different porphyrins. Felton and Linschitz,[92] for example, have reported the polarographic reduction of a number of tetraphenylporphyrin (I, Fig. 8-33, H$_2$TPP), etioporphyrin (II, Fig. 8-33, H$_2$Etio I), and tetraphenylchlorin (III, Fig. 8-33, H$_2$TPC) complexes in dimethylformamide (DMF) and dimethyl sulfoxide (DMSO) solutions. A summary of the polarographic data of these workers is presented in Table 8-14. The polarograms of H$_2$TPP and all the metal–TPP and –TPC complexes showed two well-defined waves of equal height. The slopes of the waves were approximately of the order expected for reversible, 1e processes. The value of the diffusion current constants was about 0.9, which was used to confirm the 1e nature of the waves. The etioporphyrins also gave two main waves, but the second wave was found to be sensitive to concentration, increasing in height and shifting to more negative potential with increasing concentration. The second waves of H$_2$Etio I and CuEtio I were much larger than the first and were assumed to represent complicated reduction processes, unlike the second wave processes for other compounds. The small prewave noted in Table 8-14 for most compounds appears to be associated with some type of adsorption process.

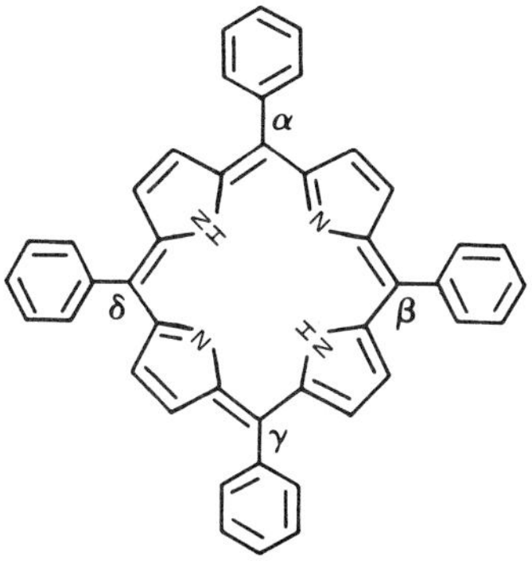

I. $\alpha, \beta, \gamma, \delta$ — TETRAPHENYLPORPHYRIN
(H$_2$TPP)

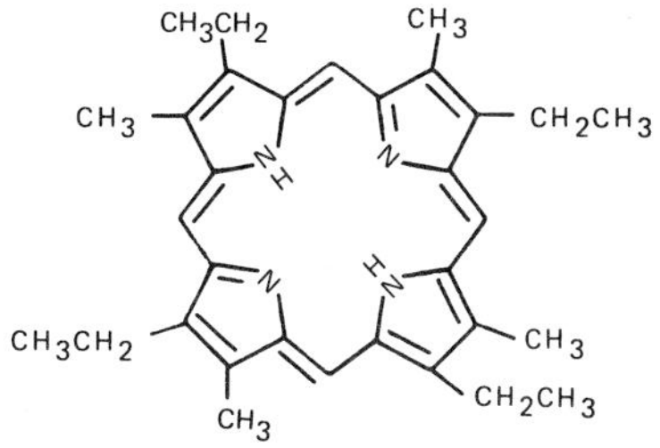

II. ETIOPORPHYRIN
(H$_2$ ETIO I)

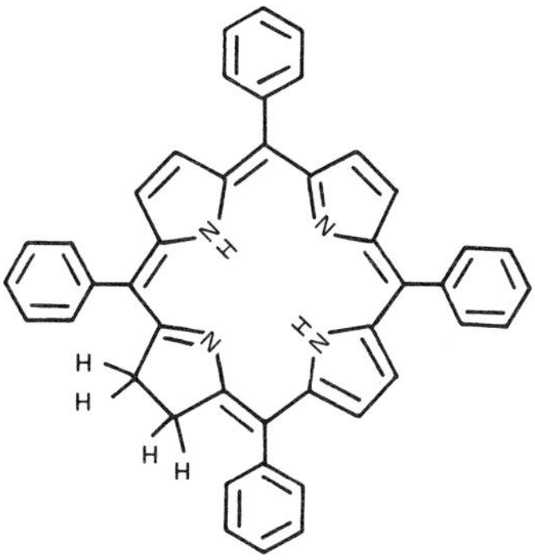

III. TETRAPHENYLCHLORIN
(H$_2$TPC)

FIG. 8-33. Structures of some porphyrins.

Unlike the interpretations of the electrochemistry of metalloporphyrins outlined in earlier sections of this chapter which proposed reduction of the central metal ion, Felton and Linschitz[92] provided some convincing evidence that under the conditions they employed the added electrons enter vacant porphyrin orbitals rather than orbitals centered primarily on the metal ion. Thus, there was a constant difference in the potentials of the first waves between corresponding pairs of TPP and etioporphyrins: (a) H$_2$TPP–H$_2$Etio I, 0.29 V; (b) ZnTPP–ZnEtio I, 0.29 V; (c) CuTPP–CuEtio I, 0.26 V. If the electron were added to an orbital centered primarily on the metal, such a constancy could not be expected for such highly diverse acceptors as protons, filled d-shell ions, and

TABLE 8-14   Polarographic Results on Porphyrins[a]

Compound	$-E_{1/2}$[b] (prewave)	$-E_{1/2}$ (1)[b]	Slope (mV)	$-E_{1/2}$ (2)[b]	Slope (mV)	Δ[c]
MgTPP	0.67	1.35	57	1.80	62	0.45
ZnTPP	0.71	1.31	59	1.72	63	0.41
CdTPP	0.67	1.25	59	1.70	68	0.45
CuTPP[d]	0.72	1.20	64	1.68	62[e]	0.48
NiTPP[f]		1.18		1.75		0.57
PbTPP[g]	0.83	1.10	58	1.52	62	0.42
H$_2$TPP	0.70	1.05	61	1.47	65	0.42
CoTPP	Absent	0.82	65	1.87	68	1.05
SnTPP(OAc)$_2$	0.5	0.81	61	1.26	65[e]	0.45
ZnEtio I[h]	0.89	1.60	57	1.95	85[e]	0.35
CuEtio I[h]	0.95	1.46	62	2.05		0.54
H$_2$Etio I[h]	0.92	1.34	63			
CoEtio I[h]	Absent	1.04	62	1.57	75[e]	0.53
ZnTPC	0.72	1.33	66	1.70	65	0.37
ZnTPP	Over the concentration range 0.37–1.3 mM, diffusion current constants for the first and second waves were constant at 0.89 and 0.90, respectively; plots of wave height versus square root of corrected Hg head (20–60 cm) were linear for both waves, with zero current intercept; higher waves were seen in DMSO solvent at −2.36 and −2.58 V					
H$_2$TPP	Heights of both waves were equal, proportional to concentration (0.2–1.4 mM), and linear in square root of Hg head; an extended wave at −1.87 V was present					
SnTPP	Higher wave at −2.14 V					
PbTPP	Higher wave at −2.34 V					
CoTPP	First wave has a small maximum, 5% of total height					
ZnTPC	Shows small wave at −2.07 V					
CoEtio I	Second wave at −1.57 V was half the height of the first wave; first wave had maximum, independent of dye concentration					
Na$_2$TPP	No wave was observed up to reduction wave of NaClO$_4$ (−1.7 V)					

[a] Unless otherwise specified, the solvent system was 0.1 N tetrapropylammonium perchlorate in DMSO.

[b] Volts versus saturated calomel electrode (aqueous).

[c] $E_{1/2}$ (1) − $E_{1/2}$ (2) = Δ.

[d] 1 : 3 THF–DMF solvent.

[e] Estimated from $E_{3/4} - E_{1/4}$.

[f] Benzene–DMF solvent.

[g] 1 : 3 Tetrahydrofuran–DMSO solvent.

[h] DMF solvent.

transition-metal ions. The positive sign of the $E_{1/2}$ differences also agreed well with the expected decrease in electron affinity of the porphyrin nucleus upon substitution of eight alkyl groups (i.e., etioporphyrin I). Table 8-14 also indicates a constant difference between the potentials of the first and second waves, $\Delta = E_{1/2}(1) - E_{1/2}(2) = 0.44 \pm 0.04$ in the group H_2TPP and the Mg^{2+}-, Zn^{2+}-, Cd^{2+}-, Cu^{2+}-, Pb^{2+}-, and Sn^{4+}TPP complexes. This suggested that in these compounds the second electron was also added to a porphyrin orbital to produce a porphyrin dianion. In the case of CoTPP the value of Δ was 1.05 V, which was quite different from that of other porphyrins. Accordingly, it was suggested that the site of the primary reduction in this case was the metal ion rather than the porphyrin ring.

Further evidence for formation of the porphyrin monoanion at the first wave came from near-infrared spectra of the monoanions of many metalloporphyrins prepared by chemical methods, which were similar to the electrochemical first wave reduction products.

It is also interesting that Zerner and Gouterman[93] found that the sequence Ni $\lesssim$ Cu $<$ Zn $<$ Mg corresponded to the order of increasing negative charge on the ring according to extended Hückel molecular orbital calculations. This order corresponds exactly to that found polarographically by Felton and Linschitz[92] for the first reduction potentials (Table 8-14). The increasing difficulty of reduction in this series is apparently due to increasing coulombic repulsion between the ring charge and the added electron, *not* to changes in the energy of the lowest empty molecular orbital. In fact, the data of Zerner and Gouterman indicate that for the above series of compounds this energy factor remains stationary. However, in the case of the cobalt complex of H_2TPP, the lowest empty molecular orbital is associated with the cobalt ion and lies well below the lowest porphyrin orbital; hence, the preferential reduction of the metal in this complex is reasonable. Closs and Closs[94] have also suggested that in nonaqueous media the negative ions of metalloporphyrin complexes appear to have the excess electrons associated with the π system of the porphyrin ring.

Clack and Hush[95] examined the reduction of a number of porphyrins and metalloporphyrins in dimethylformamide at the DME. Either three or four diffusion-controlled, 1e, polarographic reduction waves were observed (Table 8-15). A typical polarogram, for $\alpha,\beta,\gamma,\delta$-tetraphenylporphyrin in DMF, is shown in Fig. 8-34. The 1e nature of the first wave was inferred by comparison with the course of the reaction with sodium, with which a mononegative ion is formed. Since the heights of the succeeding waves were the same as that of the first wave (after correction for drop time effects), the other waves were also assumed to involve one electron. The slope of the plots of E_{DME} versus log $i/(i_d - i)$ were close to that expected for a reversible, 1e process. In those cases where only three waves were observed (Table 8-15), it was assumed that the fourth wave was masked by background discharge.

TABLE 8-15   Half-Wave Potentials for Polarographic Reduction of Some Porphyrins in Dimethylformamides[a,b]

	$E_{1/2}$ (V versus SCE) for wave[c]			
Compound	I	II	III	IV
Zn–$\alpha,\beta,\gamma,\delta$-Tetraphenylporphyrin	1.32	1.73	2.45	2.67
$\alpha,\beta,\gamma,\delta$-Tetraphenylporphyrin	1.08	1.52	2.38	2.53
[N(n-Pr)$_4$]$_2$–$\alpha,\beta,\gamma,\delta$-Tetraphenylporphyrin	1.45	1.87	2.26	2.44
Etioporphyrin I	1.37	1.80	2.67	
Zn–Etioporphyrin I	1.62	2.00	2.77	
Cu–Etioporphyrin IV	1.48	1.99	2.70	
Zn–Tetrabenzoporphyrin	1.47	1.84	2.49	2.70
Mg–Octaphenyltetraazaporphyrin	0.68	1.11	1.81	2.18

[a] Data from Clack and Hush.[95]
[b] Supporting electrolyte, 0.1 M $(n\text{-Pr})_4NClO_4$.
[c] At 20°C.

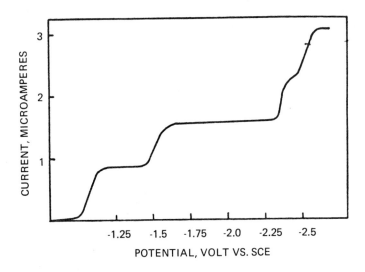

FIG. 8-34.   Current–potential curve for $\alpha,\beta,\gamma,\delta$-tetraphenylporphyrin in dimethylformamide containing 0.1 M N(n-Pr$_4$ClO$_4$).[95] [Reprinted with permission from D. W. Clack and N. S. Hush, *J. Am. Chem. Soc.* **87**, 4238 (1965). Copyright by the American Chemical Society.]

In the case of $\alpha,\beta,\gamma,\delta$-tetraphenylporphyrin the lowest empty molecular orbital is doubly degenerate, and Clack and Hush[95] accordingly assumed, quite reasonably, that the reduction steps corresponded to the stepwise filling of this orbital to its full complement of four electrons. It was also noticed that the intervals between the first and second half-wave potentials, $-^1\Delta^2 E_{1/2}$ in volts, and between subsequent waves were remarkably constant. Averaging overall data, it was found that

$$-^1\Delta^2 E_{1/2} = 0.42 \pm 0.03 \text{ V}$$
$$-^2\Delta^3 E_{1/2} = 0.75 \pm 0.06 \text{ V}$$
$$-^3\Delta^4 E_{1/2} = 0.24 \pm 0.06 \text{ V}$$

Accordingly, it was concluded that substitution of either the central protons or the porphyrin skeleton could be regarded as a $1e$ perturbation, having the effect of shifting the $1e$ orbital levels without appreciably affecting the eigenfunctions, this type of scheme being applicable to metal- and nonmetal-containing porphyrins. This confirmed the view that the level being filled was a porphyrin orbital and that the metal remained in its divalent state. Electronic spectral studies also confirmed this finding.

In support of the mechanisms of Felton and Linschitz[92] and Clack and Hush,[95] Rollmann and Iwamoto[96] have shown that reduction of the tetrasulfonated Ni– and Cu–phthalocyanines (molecules of similar structure to porphyrins) and of the metal-free compounds involved only the ligand species. In the case of the Co species, however, electrochemical reduction involved the Co(III) ion. The latter workers employed ESR spectroscopy and electronic spectra to reach these conclusions.

Lanese and Wilson[97] have examined the electrochemical reduction of Zn–tetraphenylporphyrin (ZnTPP) in dimethylformamide. Direct current polarography (Table 8-16) of ZnTPP gave four diffusion-controlled reduction

TABLE 8-16 Polarographic Data for Reduction of Zn–Tetraphenylporphyrin in Dimethylformamide[a,b]

Wave	$E_{1/2}$ (V versus SCE)	$E_{1/4} - E_{3/4}$ (mV)
I	−1.31	53
II	−1.71	65
III	−2.32	43
IV	−2.53	27

[a] Data from Lanese and Wilson.[97]

[b] Supporting electrolyte, 0.1 M tetraethylammonium perchlorate.

waves at potentials quite similar to those reported by other workers in other nonaqueous solvents (see Tables 8-14 and 8-15).

The polarographic data were confirmed by cyclic voltammetry at a hanging mercury drop electrode (Fig. 8-35). At sweep rates of about $100-200 \, mV \, sec^{-1}$ the first two processes (i.e., equivalent to polarographic waves I and II, Table 8-16) showed cathodic and anodic peak separation indicative of $1e$ reversible processes (Fig. 8-35A). At lower sweep rates, however, the anodic peak

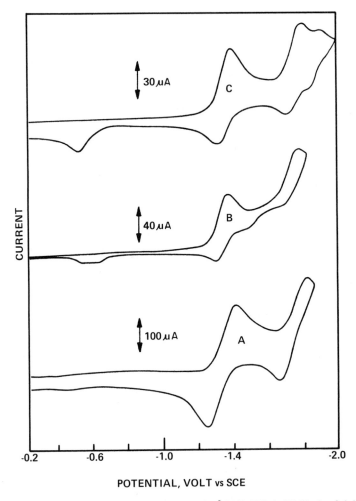

FIG. 8-35.   Cyclic voltammograms of $1.5 \times 10^{-3} \, M$ ZnTPP in DMF plus $0.1 \, M$ TEAP at the HMDE. Sweep rates: A: $178 \, mV \, sec^{-1}$, B: $33 \, mV \, sec^{-1}$, C: $33 \, mV \, sec^{-1}$.[97] (Reprinted with permission of The Electrochemical Society, Inc., New York.)

associated with the second step disappeared and a new peak at about -0.5 V appeared (Fig. 8-35B). On the basis of polarography, cyclic voltammetry, coulometry, and various spectral studies it has been concluded[97] that the first reduction wave or peak is a $1e$ reversible reduction of ZnTPP to a radical anion ZnTPP$^{\cdot-}$ (Eq. 7). The second process at about -1.71 V is a further $1e$ reduction

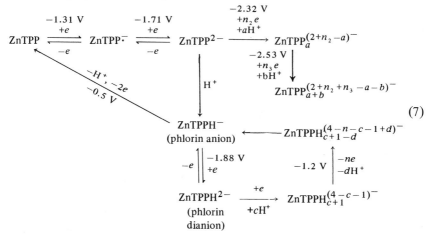

$$(7)$$

of this radical to a dianion, ZnTPP^{2-}. The anodic peak at ca. -0.5 V (see Fig. 8-35B and C) corresponds to the oxidation of a product formed by chemical reaction of the dianion ZnTPP^{2-} (*vide infra*). When the potential was scanned at less than 100 mV sec^{-1} a small reduction peak was observed at about -1.9 V (Fig. 8-35C) with a corresponding anodic peak at ca. -1.82 V. If the scan was maintained at -1.9 V for a few seconds, the oxidation peak at -1.82 V disappeared and a small oxidation peak appeared at -1.2 V, as well as the oxidation peak at -0.5 V mentioned earlier.

The sequence of events occurring when the electrode was polarized at -1.9 V was rationalized[97] as follows: The first step was proposed to involve formation of the porphyrin dianion ZnTPP^{2-} by $1e$ reduction of ZnTPP$^{\cdot-}$ followed by a rapid protonation to form a phlorin anion, ZnTPPH$^-$ (Eq. 7). Then, a second electrochemical step occurs involving a $1e$ reduction of the phlorin anion with formation of the phlorin dianion, ZnTPPH^{2-}. At fast sweep rates protonation of ZnTPP^{2-} to ZnTPPH$^-$ does not occur appreciably so that further reduction of the latter species does not occur. The oxidation peak observed at -0.5 V at slow sweep rates is apparently due to a $2e$–1H$^+$ oxidation of ZnTPPH$^-$ to ZnTPP (Eq. 7). Further $1e$–cH$^+$ reduction of the phlorin dianion (ZnTPPH^{2-}) has been proposed to occur, giving ZnTPPH$_{c+1}^{(4-c-1)-}$. It is this species that gives rise to the anodic peak observed on cyclic voltammetry at -1.2 V and has been proposed to be due to a ne–dH$^+$ process giving ZnTPPH$_{c+1-d}^{(4-n-c-1+d)-}$. The exact identity of the latter species was not known.

The third and fourth electrochemical processes, which occur at more negative potentials (i.e., polarographic waves III and IV, Table 8-16), were proposed to be due to further reduction of $ZnTPP^{2-}$, as is shown in Eq. 7. Clearly, protons are involved in these rather complex processes. The observed behavior on cyclic voltammetry when the potential was scanned through all four waves was very complex owing to many reactions of reduction products with protons.

The electrochemical reduction of cobalt(III)–hematoporphyrin in aqueous solution has been reported by Davis and Truxillo.[98] In the absence of any nitrogenous ligand this molecule is not reducible. However, in the presence of nitrogenous bases such as pyridine and related compounds it does give rise to a well-defined DC polarographic reduction wave (Table 8-17).

In the presence of a sufficient excess of pyridine (the base used most extensively) the DC polarographic wave was found to be diffusion controlled and due to a 1*e* reduction of the central Co(III) to Co(II). The $E_{1/2}$ was independent of pH between pH 7 and 12. By studying the effect of the ligand (pyridine) concentration on the $E_{1/2}$ of the porphyrin it was concluded that at low pyridine concentrations ($<1.5\,M$) at low pH (i.e., ca. pH 6.5) the cobalt(III)–hematoporphyrin complex contained two more pyridine ligands than the cobalt(II)–hematoporphyrin. At higher pyridine concentrations ($>1.5\,M$) at all pH values (pH 6.5–11.5) both cobalt complexes were coordinated with two pyridine ligands. At low pyridine concentrations ($<1.5\,M$) but at high pH (ca. pH 11.5) it was concluded that the cobalt(III) complex has one more pyridine ligand than the cobalt(II) complex. On the basis of these data, the following electrode reactions were proposed:

Low pyridine concentration   ($<1.5\,M$)

$$
\begin{array}{ccc}
\text{py} & & \text{H}_2\text{O} \\
| & & | \\
\text{Co(III)–P} + 2\text{H}_2\text{O} + e \rightarrow \text{Co(II)–P} + 2\text{py} & & \text{low pH} \\
| & & | \\
\text{py} & & \text{H}_2\text{O}
\end{array}
$$

$$
\begin{array}{cccc}
\text{py} & & \text{H}_2\text{O} \\
| & & | \\
\text{Co(III)–P} + \text{H}_2\text{O} + e \rightarrow \text{Co(II)–P} + \text{py} & & \text{high pH} & (8) \\
| & & | \\
\text{py} & & \text{py}
\end{array}
$$

High pyridine concentration   ($>1.5\,M$)

$$
\begin{array}{ccc}
\text{py} & & \text{py} \\
| & & | \\
\text{Co(III)–P} + e \rightarrow \text{Co(II)–P} & & \text{low pH and high pH} \\
| & & | \\
\text{py} & & \text{py}
\end{array}
$$

The product of the reduction was adsorbed, as evidenced by the appearance of an adsorption prewave.

Maricle and Maurer[99] have reported, briefly, on what could potentially be a very interesting area of study. They found that they could study some of the

TABLE 8-17   Direct Current Polarographic Data for Cobalt(III)—
Hematoporphyrin in the Presence of Nitrogenous
Bases[a]

Ligand	Ligand[b] concentration (mole liter^{-1})	$E_{1/2}$ (V versus SCE)
Pyridine	2.46	−0.436
4-Phenylpyridine	0.10	−0.384
4-Picoline	2.05	−0.425
4-Acetylpyridine	1.81	−0.362
Ethyl nicotinate	1.46	−0.291
4,4'-Bipyridine	0.10	−0.300
2-Picoline	2.05	−0.420

[a] Data from Davis and Truxillo.[98]

[b] Cobalt(III)—hematoporphyrin concentration, 1.02 mM.

redox properties of the (photochemically) excited-state molecules of Zn—tetraphenylporphyrin in DMF (containing 0.1 M tetrabutylammonium perchlorate as supporting electrolyte) using an optically transparent SnO_2 electrode. These types of studies have some obvious implications in photosynthetic processes. Similar studies by Tributsch and Calvin[60] have been discussed earlier.

Ricci and co-workers[100] found that hemato-, proto-, pyrro-, meso-, and rhodoporphyrins showed essentially identical polarographic and oscillopolarographic behavior in acidic aqueous solution. Usually two totally irreversible 2e polarographic waves, or two oscillopolarographic indentations, were observed. Occasionally, a third process was observed at very negative potential corresponding to hydrogen ion reduction catalyzed by the porphyrin.

XI. OTHER PORPHYRIN SYSTEMS

Heiling and Wilson[101] have reported the electrochemical reduction of deuteroporphyrin IX dimethyl ester (DPDME) and mesoporphyrin IX dimethyl ester (MPDME) in dimethylformamide solution. The structure of these compounds are shown in Fig. 8-36.

Both deuteroporphyrin IX dimethyl ester and mesoporphyrin IX dimethyl ester give rise to three well-defined polarographic waves at the DME in dimethylformamide solution (Table 8-18). All three waves were found to be diffusion controlled and were linearly dependent on the porphyrin concentration. Analysis of the rising portion of the waves (Table 8-18) revealed that the first two waves were nearly reversible one-electron processes. The third wave was irreversible and was thought to involve more than one electron. Cyclic

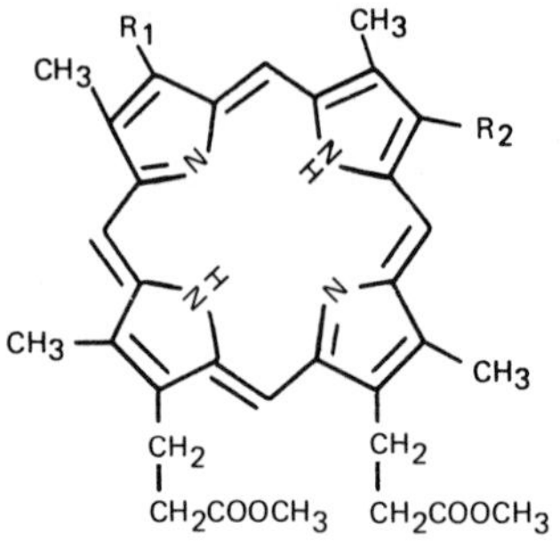

FIG. 8-36.   Structure of deuteroporphyrin IX dimethyl ester (DPDME, $R_1 = R_2 = H$) and mesoporphyrin IX dimethyl ester (MPDME, $R_1 = R_2 = C_2H_5$).

voltammetry at a hanging mercury drop electrode confirmed that the first two waves were nearly reversible, as evidenced by well-formed anodic peaks after the first or first and second cathodic peaks were scanned (Fig. 8-37A). The third peak showed no evidence for reversibility by cyclic voltammetry. If the HMDE was polarized at -1.8 V, i.e., just beyond the second peak (peak 2, Fig. 8-37), followed by an anodic scan, then in addition to the oxidation peaks at $E_p = -1.65$ and -1.27 V (peaks 4 and 5, Fig. 8-37) an additional peak at $E_p = -0.7$ V was observed (peak 6, Fig. 8-37B). Ultraviolet–visible and ESR spectroscopy and thin-layer electrochemistry were extensively employed to evaluate the nature of these processes.

TABLE 8-18   Electrochemical Data for Porphyrin IX Derivatives[a,b]

Compound	Wave	$E_{1/2}$[c] (V versus SCE[d])	Slope (mV)[e]	$(E_p)_c$[f] (V versus SCE[d])	$(E_p)_a$[g] (V versus SCE[d])
DPDME	1	-1.29	60	-1.36	-1.27
	2	-1.68	65	-1.74	-1.65
	3	-2.53	45	-2.62	
MPDME	1	-1.34	58	-1.39	-1.31
	2	-1.73	62	-1.79	-1.70
	3	-2.57	41	-2.64	

[a] Data from Heiling and Wilson.[101]

[b] Data obtained in dimethylformamide containing 0.1 M tetraethylammonium perchlorate.

[c] At the dropping mercury electrode.

[d] Potential versus aqueous SCE.

[e] Slope calculated from $E_{3/4} - E_{1/4}$.

[f] Cathodic peak potential at hanging mercury drop electrode; scan rate for cyclic voltammetry, 100 mV sec^{-1}.

[g] Anodic peak potential at HMDE.

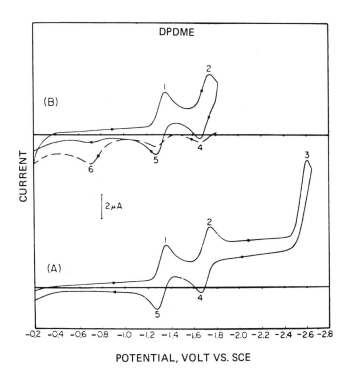

FIG. 8-37. Cyclic voltammetry of DPDME in DMF plus 0.1 *M* TEAP at HMDE. Concentration, 8.0 x 10⁻⁴ *M*. Scan rate, 0.10 V sec⁻¹, single scan. (B) Hold at −1.8 V followed by anodic scan.[101] [Reprinted with permission from G. P. Heiling and G. S. Wilson, *Anal. Chem.* **43**, 545 (1971). Copyright by the American Chemical Society.]

In summary, polarographic wave 1 (peak 1 at HMDE) was proposed to be due to the reduction of the parent porphyrin (P, Eq. 9) in a reversible 1e step to an anion radical (P$\cdot^-$, Eq. 9). This was readily detected by ESR spectroscopy. In the presence of available protons the radical underwent a disproportion reaction to give PH⁻ and P. This type of disproportionation reaction had been noted by earlier workers[94]; in addition, the observed spectrum of PH⁻ was characteristic of a phlorin anion.[48,102] The second polarographic wave was a reversible 1e reduction of P$\cdot^-$ to the dianion P²⁻. Once again, P²⁻ was found to react with protons to give PH⁻. It is this PH⁻ species, formed in a chemical follow-up reaction of P$\cdot^-$ or P²⁻ with protons, that gives rise to the anodic peak in cyclic voltammetry at HMDE at $E_p = -0.7$ V (peak 6, Fig. 8-37B). The reaction responsible for this peak was oxidation of PH⁻, in a 2e−1H⁺ process, back to P. The third irreversible polarographic wave was proposed to be a 2e reduction of P²⁻ to a porphomethene of undefined structure, but represented as PH$_b^{(4-b)^-}$.

A summary of these primary electrochemical processes is presented in Eq. 9. The

$$P + e^- \underset{-1.29\,V}{\overset{}{\rightleftharpoons}} P^{\cdot -} \underset{}{\overset{-1.68\,V \atop +e^-}{\rightleftharpoons}} P^{2-} \xrightarrow[bH^+]{2e^-} Ph_b^{(4-b)-}$$

$$\text{(green)}$$

$$\text{slow} \Big\downarrow {\scriptstyle +\frac{1}{2}H^+} \quad \text{slow} \quad +H^+$$

$$\tfrac{1}{2}PH^- + \tfrac{1}{2}P$$

$$-2e^-, -H^+$$

$$PH^-$$

$$-0.7\,V \qquad \text{(yellow-green)}$$

$$(9)$$

potentials noted in this scheme are those for DPDME. The same scheme applies to MPDME, except that the secondary chemical reactions of $P^{\cdot -}$ and P^{2-} with protons are very slow.

In addition to these fairly straightforward reactions, it was also found that the phlorin anion, PH^-, could be further reduced by a $2e$ step to a porphomethene, which was a bright purple species and was proposed to have a form represented by $PH_{n+1}^{[4-(n+1)]-}$. This process is not observed under normal polarographic or cyclic voltammetric conditions, but it is observed in controlled potential electrolyses at potentials slightly more negative than the second polarographic wave (i.e., at ca. -1.95 V). The $2e$ nature of the process was based on thin-layer current–voltage curves and on spectral data. However, the number of protons associated with reduction to the purple species was not elucidated. In the presence of water the purple porphomethene gave rise to an orange species. The same species could be obtained by electrolysis of P at potentials corresponding to the second polarographic wave in the presence of water. The structure of the orange form was again not elucidated in detail but was proposed to have the form designated $PH_{n+m+1}^{[4-(n+m+1)]-}$. Both the purple $PH_{n+1}^{[4-(n+1)]-}$ and the orange $PH_{n+m+1}^{[4-(n+m+1)]-}$ could be oxidized back to the phlorin anion, PH^-. The potentials associated with these reactions and the proposed reaction routes are presented in Eq. 10.

$$PH^- \underset{-1.15\,V}{\overset{\approx -1.95\,V \atop +2e^- + nH^+}{\underset{-2e^-, -nH^+}{\rightleftharpoons}}} PH_{n+1}^{[4-(n+1)]-} \quad \text{(purple)}$$

$$\text{(yellow-green)} \qquad\qquad mH^+ \;(H_2O)\Big\downarrow$$

$$-2e^-, -(n+m)H^+$$

$$PH_{n+m+1}^{[4-(n+m+1)]-} \quad \text{(orange)}$$

$$-1.15\,V$$

$$(10)$$

Heiling and Wilson also examined the electrochemistry of tetraphenylporphyrin (**II**), tetraphenylchlorin (**III**), and tetraphenylbacteriochlorin (**IV**) in

(Ⅱ)   (Ⅲ)   (Ⅳ)

dimethylformamide solvent.[103] Tetraphenylporphyrin (TPP) gave four well-defined, diffusion-controlled polarographic waves at the DME in dimethylformamide (Table 8-19). Analysis of the rising portion of the first two waves indicated that both processes were reversible one-electron reactions. The latter two waves were irreversible multielectron processes. Cyclic voltammetry at the HMDE of TPP confirmed the polarographic results (Fig. 8-38A). In other words, the first two cathodic steps showed well-formed, reversible anodic peaks on the reverse scan, but the last two steps showed no evidence for reversible anodic peaks even at scan rates as high as 100 V sec^{-1}. An anodic peak at -0.7 V (peak 7) is clearly visible in Fig. 8-38A and B. This peak was not observed until the cathodic scan limit was extended to at least the third cathodic peak (peak 3,

TABLE 8-19   Polarographic Data for Reduction of Tetraphenylporphyrin Derivatives at the DME in Dimethylformamide[a,b]

Compound	Wave	$E_{1/2}$ (V versus SCE[c])	Slope (mV)[d]
Tetraphenylporphyrin (TPP)	1	−1.08	60
	2	−1.45	61
	3	−2.36	35
	4	−2.48	35
Tetraphenylchlorin (TPC)	1	−1.12	59
	2	−1.52	60
	3	−2.43	37
Tetraphenylbacteriochlorin	1	−1.10	58
	2	−1.55	60

[a] Data from Heiling and Wilson.[103]

[b] Supporting electrolyte; 0.1 *M* tetraethylammonium perchlorate.

[c] Aqueous SCE.

[d] $E_{3/4} - E_{1/2}$.

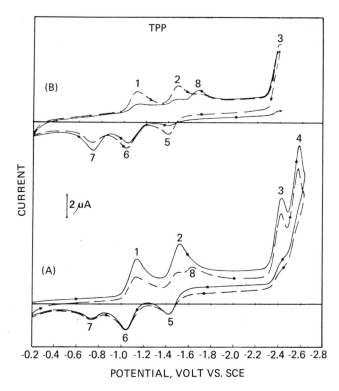

FIG. 8-38. Cyclic voltammetry of TPP in DMF plus 0.1 *M* TEAP at HMDE. Concentration, 6.0×10^{-4} *M*. Scan rate, 0.10 V sec^{-1}. (A) Solid line, first cycle; dashed line, tenth cycle. (B) Electrolysis at -2.4 V followed by anodic scan. Dashed line indicates second cycle.[103] [Reprinted with permission from G. P. Heiling and G. S. Wilson, *Anal. Chem.* **43**, 551 (1971). Copyright by the American Chemical Society.]

Fig. 8-38). The anodic peak at -0.7 V was thought to be associated with the oxidation of the final product produced at peak 3. After electrolysis of TPP at potentials corresponding to the third cathodic peak (peak 3, Fig. 8-38B) followed by an anodic scan, peak 7 is clearly observed. Also associated with this is the cathodic peak 8 at ca. -1.6 V. By observing the spectral properties of the products of TPP by electrolyses in an optically transparent thin-layer cell,[104] it was concluded that the first electrochemical step was a one-electron reduction of TPP to give the radical anion TPP$\overline{\cdot}$; this species gave a strong ESR signal. The product of the second reduction step was a blue dianion designated TPP^{2-}. These processes are summarized in Eq. 11.

$$\text{TPP} + e^- \underset{}{\overset{-1.08 \text{ V}}{\rightleftharpoons}} \text{TPP}\overline{\cdot} \underset{-e^-}{\overset{\overset{\textstyle -1.45 \text{ V}}{+e^-}}{\rightleftharpoons}} \text{TPP}^{2-} \qquad (11)$$

The electrochemical data supported the view that the third step was a $2e$ process. In view of the irreversible nature of the reaction, it was proposed that the final product was protonated. It will be recalled that Clack and Hush[95] had proposed that a stable trinegative anion was produced upon polarographic reduction of several porphyrins and metalloporphyrins (Section X) in dimethyl-formamide. Heiling and Wilson,[103] however, could not observe a stable trinegative ion either by spectrometric or electrochemical techniques.

The scheme proposed by Heiling and Wilson[103] for the third and fourth polarographic and cyclic voltammetric steps of TPP and related reactions is shown in Eq. 12. Thus, the product of the third electrochemical step of TPP was proposed to be a species of the type $TPPH_n^{(4-n)-}$ where n could be equal to 1, 2, or 3. The latter species could be oxidized electrochemically at ca. -1.08 V to a species designated $TPPH_*{}^-$, which in turn slowly decomposed to give a phlorin anion, $TPPH^-$. The oxidation of $TPPH_n^{(4-n)-}$ to $TPPH_*{}^-$ occurs at the same potential as the oxidation of $TPP^{\overline{\cdot}}$ to TPP, i.e., peak 6 (Fig. 8-38A and B). It was therefore not possible to decide this from electrochemical data alone but required additional spectral evidence. The species designated $TPPH_*{}^-$ is the product of the third peak (peak 3, Fig. 8-38A and B) that gives rise to the anodic peak observed at -0.7 V (peak 7). This was particularly so if light of 730 nm was passed through the cell, because under these conditions the conversion of $TPPH_*{}^-$ to $TPPH^-$ described above was very slow. The product of the anodic peak at -0.7 V (peak 7) was a new porphyrin designated TPP^*. It is this species that could be reduced directly to TPP^{2-} at a potential slightly more negative than the normal second cathodic step (i.e., peak 8, Fig. 8-38A and B) observed by cyclic voltammetry. The exact structures of TPP^* and many of the other products and intermediates just described remain in some doubt.

$$
\begin{array}{ccc}
& \xrightarrow{-2.36\ \text{V}} & \\
TPP^{2-} + nH^+ + 2e^- & \longrightarrow & TPPH_n^{(4-n)-} \\
\uparrow & & \downarrow \\
\begin{array}{l} -1.6\ \text{V} \\ +2e^- \end{array} & \quad h\nu \quad & -1.08\ \text{V} \\
TPP^* + H^+ + 2e^- & \xleftarrow{-0.7\ \text{V}} & TPPH_*{}^- + (n-1)H^+ + 2e^- \\
\downarrow & & \downarrow \\
H_2O & & k \\
& \xrightarrow{-0.7\ \text{V}} & \\
TPP + H^+ + 2e^- & \longrightarrow & TPPH^-
\end{array}
\qquad (12)
$$

$$
TPPH_n^{(4-n)-} + mH^+ + 2e^- \xrightarrow{-2.48\ \text{V}} TPPH_{n+m}^{[6-(n+m)]-}
$$

$$
\downarrow {\scriptstyle +1.3\ \text{V}}
$$

$$
TPP
$$

Controlled potential electrolysis of TPP at potentials corresponding to the fourth polarographic wave gave a colorless product. Although little evidence for the nature of this process was presented, it was proposed (Eq. 12) to be a two-electron reduction of the product of the third wave, $TPPH_n^{(4-n)-}$, to a 6e-reduced porphyrin designated $TPPH_{n+m}^{[6-(n+m)]-}$. This could be reoxidized back to TPP in about 80% yield by electrolysis at +1.3 V.

Tetraphenylchlorin (TPC) gave three well-defined, diffusion-controlled polarographic waves in dimethylformamide at the DME (Table 8-19). Analysis of the rising portions of the waves suggested 1e, 1e, and 2e steps, respectively, for each of the waves. Once again cyclic voltammetry indicated that the first two processes were reversible (Fig. 8-39) and that the third process was irreversible. If a controlled potential electrolysis at the most negative peak (peak 3, Fig. 8-39) was carried out for a short time and then an anodic potential sweep was initiated, two additional anodic peaks (peaks 6 and 7) were observed. The electrochemical data and spectral data (UV–visible and ESR) supported the view that the first step was a 1e reduction of TPC to the anion radical $TPC^{\cdot-}$, and the second step was a further 1e reduction to the dianion TPC^{2-} (Eq. 13). Electrolysis at the third step involved a further 2e reduction of TPC^{2-} in a

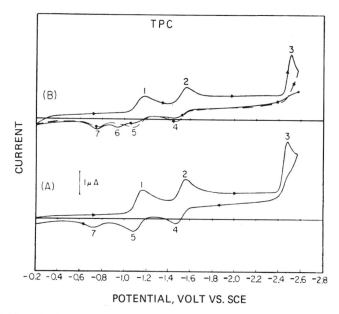

FIG. 8-39. Cyclic voltammetry of TPC in DMF plus $0.1\,M$ TEAP at HMDE. Concentration, $3.0 \times 10^{-4}\,M$. Scan rate, $0.10\ V\ sec^{-1}$. (A) Single cycle. (B) Electrolysis at -2.6 V followed by anodic scan. Dashed line indicates second cycle.[103] [Reprinted with permission from G. P. Heiling and G. S. Wilson, *Anal. Chem.* **43**, 551 (1971). Copyright by the American Chemical Society.]

reaction apparently involving protons to give $TPCH_n^{(4-n)^-}$. It was this species that could be oxidized at -0.9 V (peak 6, Fig. 8-39B). By analogy with the proposed reactions of TPP, the product of the oxidation of $TPCH_n^{(4-n)^-}$ at -0.9 V (peak 6) was assumed to be $TPCH^-$, a chlorin–phlorin anion. Further oxidation of this species at -0.7 V (peak 7, Fig. 8-39B) was shown to give TPC, the starting material.

$$
TPC + e^- \underset{}{\overset{-1.12\ V}{\rightleftharpoons}} TPC^{\cdot} \underset{-e^-}{\overset{\substack{-1.52\ V \\ +e^-}}{\rightleftharpoons}} TPC^{2-}
$$

$$
TPC^{2-} + nH^+ + 2e^- \xrightarrow{-2.43\ V} TPCH_n^{(4-n)^-} \tag{13}
$$

$$
\downarrow -0.9\ V
$$

$$
TPC + H^+ + 2e^- \xleftarrow{-0.7\ V} TPCH^- + (n-1)H^+ + 2e^-
$$

In the case of tetraphenylbacteriochlorin (TPB) only two diffusion-controlled polarographic waves were observed (Table 8-19). The slope of both waves was in accord with that expected for reversible $1e$ processes. This was confirmed by the cyclic voltammetry of the compound at the HMDE (Fig. 8-40). The product of the first wave was paramagnetic (ESR), and, accordingly, the first wave was proposed to be due to the reversible $1e$ reduction of TPB to the anion radical $TPB^{\cdot}$. The second wave was due to further reversible reduction of this anion

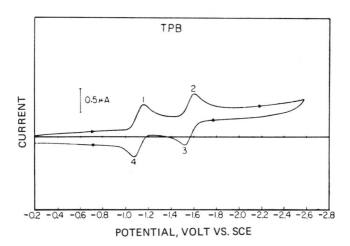

FIG. 8-40. Cyclic voltammetry of TPB in DMF plus 0.1 M TEAP at HMDE. Concentration, $8.0 \times 10^{-5}\ M$. Scan rate, 0.10 V sec^{-1}.[103] [Reprinted with permission from G. P. Heiling and G. S. Wilson, *Anal. Chem.* **43**, 551 (1971). Copyright by the American Chemical Society.]

radical to the dianion TPB^{2-}. It was also found that TPB could be oxidized to TPC at +1.5 V. These reactions are summarized in Eq. 14.

$$TPB + e^- \underset{-1.10\text{ V}}{\overset{}{\rightleftharpoons}} TPB^{\cdot} \underset{-e^-}{\overset{\substack{-1.55\text{ V}\\+e^-}}{\rightleftharpoons}} TPB^{2-}$$

$$\downarrow {\scriptstyle +1.5\text{ V}}$$

$$TPC + 2e^- + 2H^+$$

$$(14)$$

Neri and Wilson[105] have examined the electrochemical reduction of *meso*-tetra(4-*N*-methylpyridyl)porphyrin tetraiodide [(TPyCH$_3$P)I$_4$] in aqueous acid solution. This porphyrin is of interest because it is readily soluble in water.[106] Over the pH range studied (pH 0–6) up to three DC polarographic reduction waves were observed (Table 8-20). The first wave (wave 1, Table 8-20) was diffusion controlled and was a 2e process. Between pH 3 and 6, the $E_{1/2}$ for wave 1 was pH dependent with $dE_{1/2}/d(\text{pH}) \approx 60$ mV, suggesting that two

TABLE 8-20   Direct Current Polarography of *meso*-Tetra(4-*N*-methylpyridyl)-porphin Tetraiodide in Aqueous Solution[a]

Solution conditions	Wave	Limiting current[b]	$E_{1/2}$ (V versus Ag/AgCl)[c]
1 *M* HCl	1	D	−0.041
	2	K	−0.240
	3	D–K	−0.458
0.01 *M* HCl / 0.1 *M* NaCl	1	D	−0.100
	2	D–K	−0.378
	3	D–A	−0.445
0.1 *M* NaOAc/ HOAc / 0.1 *M* NaCl, pH 4.0	1	D	−0.180
	2	D–A	−0.509
0.1 *M* NaOAc/ HOAc / 0.1 *M* NaCl, pH 4.9	1	D	−0.238
	2	D–A	−0.565
0.1 *M* PIPES[d] / 0.1 *M* NaCl, pH 6.0	1	D	−0.303
	2	A	−0.610

[a] Data from Neri and Wilson.[105]
[b] D, diffusion controlled; K, kinetic controlled; A, adsorption controlled.
[c] Ag/AgCl,NaCl$_{(sat)}$; $E^{0'} = 0.202$ V versus standard hydrogen electrode.
[d] Piperazine-*N*,*N'*-bis(2-ethanesulfonic acid).

protons were involved in the reduction step. Below pH 2 the first wave was probably pH independent, although it was partially obscured by anodic mercury dissolution. Below pH 2 the second wave was kinetically controlled, the kinetic step being due to a chemical reaction of the wave 1 product. By use of polarography, coulometry, and thin-layer spectroelectrochemistry it was demonstrated that the wave 1 process below pH 2 is a reversible $2e$ reduction of the porphyrin diacid [$P(0)H_4^{2+}$, Fig. 8-41] to give what may be an isophlorin

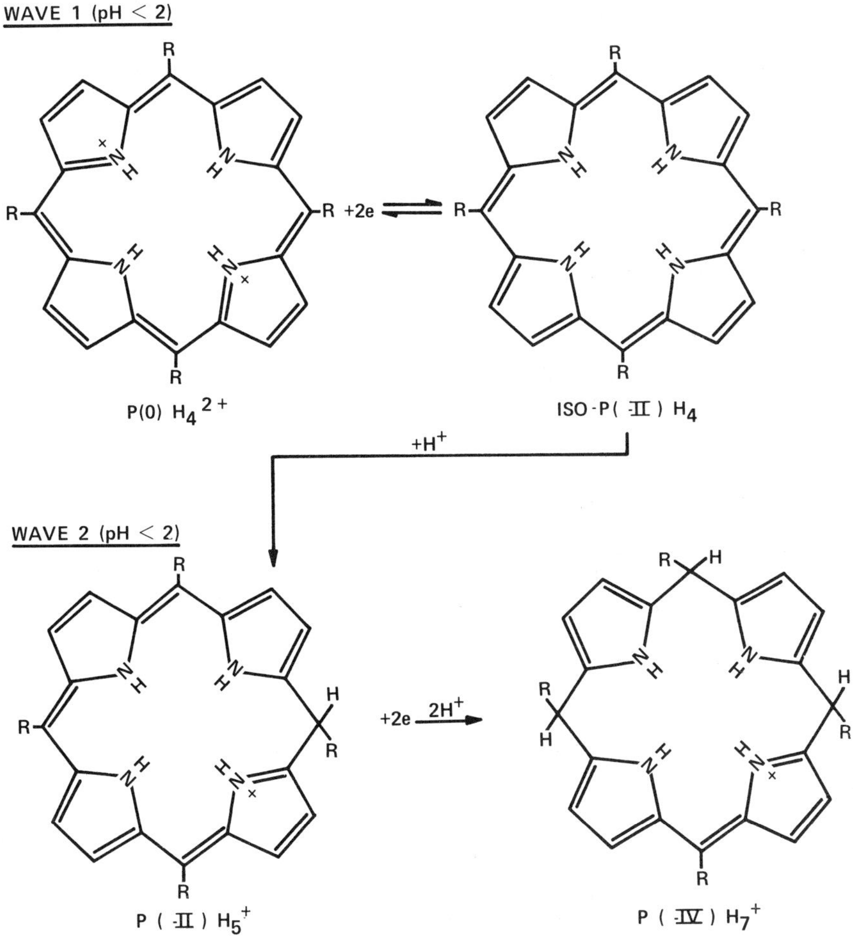

WAVE 1 (pH < 2)

$P(0) H_4^{2+}$ +2e ⇌   ISO-P(II) H_4

+H^+

WAVE 2 (pH < 2)

$P(II) H_5^+$ +2e $\xrightarrow{2H^+}$ $P(IV) H_7^+$

FIG. 8-41. Proposed mechanism for diffusion-controlled polarographic wave 1 and kinetically controlled wave 2 of *meso*-tetra(4-*N*-methylpyridyl)porphin tetraiodide in aqueous solutions at pH ⩽ 2.[105]

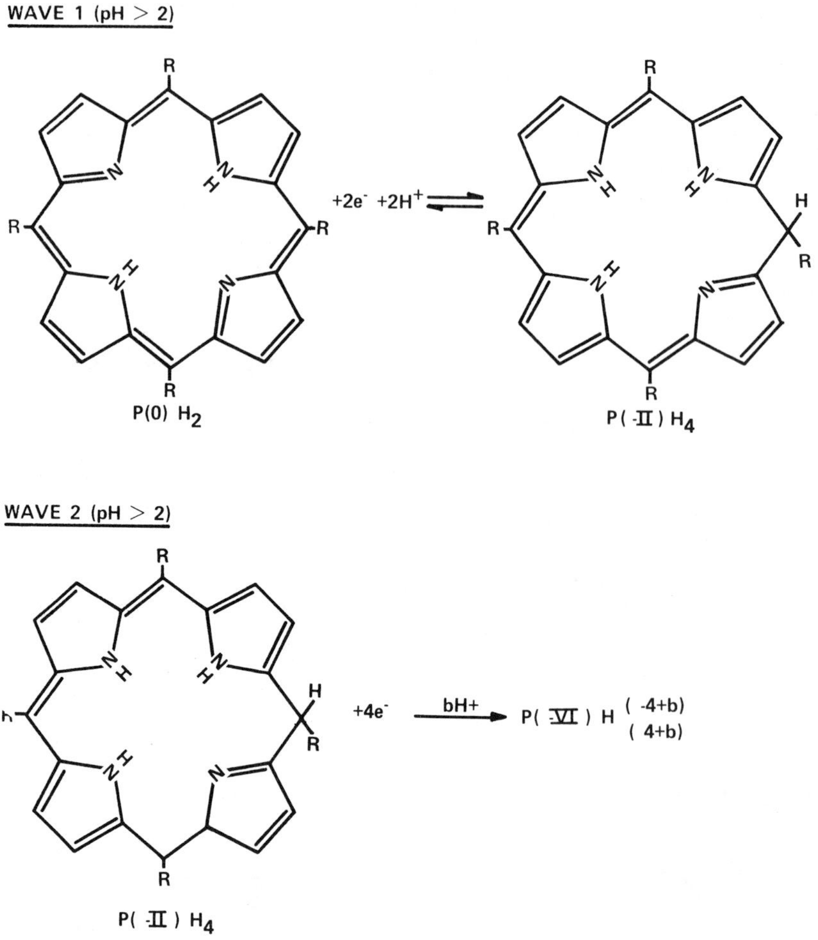

WAVE 1 (pH > 2)

P(0) H$_2$ $+2e^-$ $+2H^+$ ⇌   P(-II) H$_4$

WAVE 2 (pH > 2)

$+4e^-$ $\xrightarrow{bH+}$ P(-VI) H $\begin{pmatrix} -4+b \\ 4+b \end{pmatrix}$

P(-II) H$_4$

FIGURE 8-42

FIG. 8-42. Proposed mechanism for polarographic waves 1 and 2 of *meso*-tetra(4-*N*-methylpyridyl)porphin tetraiodide in aqueous solution between pH 2 and 6.[105]

[iso-P($-$II)H$_4$, Fig. 8-41], a very reactive species postulated by Woodward.[107] Protonation of this intermediate then gives a phlorin monocation [P($-$II)H$_5^+$], which is reduced in the wave 2 process below pH 2 in a 2e reaction to give P($-$IV)H$_7^+$ (Fig. 8-41). The latter process is kinetically controlled. The third wave observed below pH 2 is presumably due to further reduction of P($-$IV)H$_7^+$.

Between pH 2 and 6 the mechanism of the porphyrin reduction changes because dissociation equilibria associated with P(0)H$_4^{2+}$ and P($-$II)H$_5^+$ (see Eqs. 15a and

b) are shifted to the right.

$$P(0)H_4{}^{2+} \; \rightleftharpoons \; 2H^+ + P(0)H_2 \qquad pK_a = 4.01 \qquad (15a)$$

$$P(-II)H_5{}^+ \; \rightleftharpoons \; H^+ + P(-II)H_4 \qquad pK_a = 0.96 \qquad (15b)$$

The first of the two waves observed above pH 2 is due to reduction of the porphyrin free base $[P(0)H_2]$ to the phlorin free base $[P(-II)H_4]$ (Fig. 8-42). It has been suggested[105] that the second wave is a complicated irreversible $4e$ process resulting ultimately in the formation of a porphyrinogen.

XII. ELECTROCHEMICAL OXIDATION OF PORPHYRINS AND METALLOPORPHYRINS

Reports have begun to appear concerning the electrochemical oxidation of porphyrins, usually in nonaqueous media. Stanienda[108] and Stanienda and Biebl[109] found that a very large number of porphyrins and metalloporphyrins are oxidized at a platinum electrode in butyronitrile (Table 8-21). There were normally two voltammetric waves, each of which corresponded to a $1e$ process. Occasionally, the formation of the second step was not observed, apparently as a result of adsorption or decomposition of the oxidation product on the platinum electrode. The data in Table 8-21 indicate that in general the half-wave potentials for the Mg–, Ba–, Zn–, Cd–, Pb–, and Cu–porphyrins occur at less positive potentials than those for the free porphyrins, while $E_{1/2}$ for the Co, Sn, and Fe complexes occur at more positive potential. It is also noticeable that the Co(II)–porphyrins are initially and preferentially oxidized at the metal ion to give Co(III), while in all other cases electrons are removed from porphyrin orbitals. Recalling the structure of the various porphyrin molecules reported in Table 8-21 (Fig. 8-43), it is clear that substituents in the β position of the pyrrole nuclei have only a relatively small effect on $E_{1/2}$, while N-alkylation results in a large shift of the half-wave potentials to more positive potentials. The products of the electrochemical oxidations of these metalloporphyrins were proposed to be cations in which the positive charge is localized on the nitrogen atoms.

The electrochemical oxidation of the tetraphenylporphyrin complexes of first-row transition elements from iron to zinc in benzonitrile solution has been studied by Wolberg and Manassen,[110] using a platinum electrode. It was found that tetraphenylporphyrin and its metal complexes all gave at least two well-resolved one-electron oxidation peaks. Tetraphenylporphyrin itself gave two irreversible one-electron peaks, while it appeared that all the peaks observed for the other compounds were reversible, as evidenced by cyclic voltammetry, except for the second and third peaks observed for the Ni(II) complex, which were irreversible. Typical voltammetric data are presented in Table 8-22. It can be seen that the Fe(II), Co(II), and Ni(II) compounds give three peaks. The first of these is due to the 1e oxidation of the metal ion. All the other peaks are due to 1e oxidations of the porphyrin ring system.

In order to decide whether in fact metal oxidation or ligand oxidation had

TABLE 8-21   Anodic Half-Wave Potentials for the Electrochemical Oxidation of Porphyrins at a Rotating Platinum Electrode in Butyronitrile Containing 0.1 M LiClO$_4$[a,b]

Compound	$E_{1/2}$ (V versus SCE) for wave	
	I	II
$\alpha,\beta,\gamma,\delta$-Tetraphenylporphyrin	0.97	1.12
Mg(II)—Tetraphenylporphyrin	0.54	0.86
Ba(II)—Tetraphenylporphyrin	0.46	0.75
Zn(II)—Tetraphenylporphyrin	0.71	1.03
Cd(II)—Tetraphenylporphyrin	0.63	0.93
Pb(II)—Tetraphenylporphyrin	0.63	0.96
Cu(II)—Tetraphenylporphyrin	0.90	1.16
Ag(II)—Tetraphenylporphyrin	0.54	
Ni(II)—Tetraphenylporphyrin	0.95	
Co(II)—Tetraphenylporphyrin	0.32,[c] 1.06	1.26
Etioporphyrin I	0.77	
Mg(II)—Etioporphyrin I	0.40	0.77
Zn(II)—Etioporphyrin I	0.51	
Pb(II)—Etioporphyrin I	0.50	0.76
Ag(II)—Etioporphyrin I	0.30	0.58
Pt(II)—Etioporphyrin I	0.75	1.37
Pd(II)—Etioporphyrin I	0.70	1.40
Ni(II)—Etioporphyrin I	0.70	1.28
Co(II)—Etioporphyrin I	0.30,[c] 0.87	1.18
Cu(II)—Etioporphyrin I	0.63	
N-Methyletioporphyrin I	1.20	
Zn(II)—Methyletioporphyrin I	1.09	
Protoporphyrin IX dimethyl ester	0.83	
Mg(II)—Protoporphyrin IX dimethyl ester	0.52	0.94
Ca(II)—Protoporphyrin IX dimethyl ester	0.52	0.92
Cd(II)—Protoporphyrin IX dimethyl ester	0.54	
Zn(II)—Protoporphyrin di-n-amyl ester	0.61	
Cu(II)—Protoporphyrin di-n-amyl ester	0.73	
Fe(III)—Protoporphyrin IX dimethyl ester fluoride	1.04	1.32
Fe(III)-Protoporphyrin IX dimethyl ester hydroxide	1.06	1.33

TABLE 8-21 *Continued*

Compound	$E_{1/2}$ (V versus SCE) for wave	
	I	II
Mesoporphyrin IX dimethyl ester	0.78	
Zn(II)—Mesoporphyrin IX dimethyl ester	0.50	0.97
Ni(II)—Mesoporphyrin IX dimethyl ester	0.72	
Deuteroporphyrin IX dimethyl ester	0.76	1.33
Zn(II)—Deuteroporphyrin IX dimethyl ester	0.60	1.04
Ni(II)—Deuteroporphyrin IX dimethyl ester	0.72	1.27
Co(II)—Deuteroporphyrin IX dimethyl ester	0.26,[c] 0.94	
Hematoporphyrin IX dimethyl ester	0.77	
Zn(II)—Hematoporphyrin IX dimethyl ester	0.53	1.00
Koproporphyrin I tetramethyl ester	0.80	
Zn(II)—Koproporphyrin I tetramethyl ester	0.57	1.05
Cu(II)—Koproporphyrin I tetramethyl ester	0.70	1.20

[a] Data from Stanienda[108] and Stanienda and Biebl.[109]

[b] At 20°C.

[c] Wave for Co(II) → Co(III).

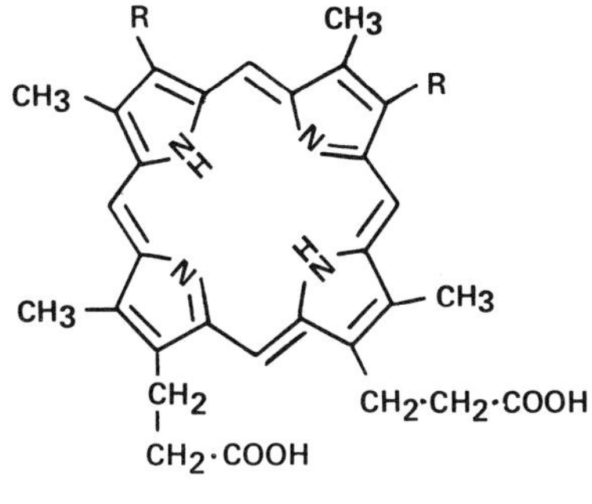

Protoporphyrin: R = -CH=CH$_2$
Mesoporphyrin: R = -CH$_2$CH$_3$
Deuteroporphyrin: R = -H
Hematoporphyrin: R = CH$_3$$\overset{|}{\underset{|}{C}}$(OH)
                                              H
Koproporphyrin: R = -CH$_2$$\underset{|}{CH_2}$
                                          COOCH$_3$

FIG. 8-43. Structure of some porphyrins.

TABLE 8-22 **Voltammetric Oxidation Potentials for Some Tetraphenyl-porphyrins at a Platinum Electrode in Benzonitrile**[a,b]

Compound	Oxidation potential (V versus SCE) metal oxidation[c] $(M^{2+} \rightarrow M^{3+})$	Ligand oxidation (V versus SCE)	
		I	II
Tetraphenylporphyrin		1.00^d	1.20^d
Fe—Tetraphenylporphyrin	−0.32	1.18	1.50^e
Co—Tetraphenylporphyrin	0.52	1.19	1.42^e
Ni—Tetraphenylporphyrin	1.00^d	1.10^d	1.40^d
Cu—Tetraphenylporphyrin		0.99	1.33
Zn—Tetraphenylporphyrin		0.79	1.10

[a] Data from Wohlberg and Manassen.[110]
[b] A 0.002 M solution of porphyrin in a 0.1 M solution of n-Bu$_4$NClO$_4$ in benzonitrile.
[c] Oxidation potentials were determined at 85% of the maximum anodic peak height.
[d] Accuracy, ±0.02 V.
[e] Measured in 0.1 M n-Bu$_4$NBF$_4$ in benzonitrile solution.

occurred, Wolberg and Manassen[110] carried out controlled potential coulometry to first determine the electron number, and then the products of such oxidations were examined by visible spectrophotometry, by ESR spectroscopy, and by magnetic susceptibility measurements. The type of oxidation that had taken place could usually be unequivocably determined from the information derived from all these techniques. In the case of the Cu(II) and Zn(II) complexes the electrode mechanism could be represented as

$$M(II)TPP \xrightarrow[\substack{\text{ligand} \\ \text{peak I}}]{-e} [M(II)TPP]^{+\cdot} \xrightarrow[\substack{\text{ligand} \\ \text{peak II}}]{-e} [M(II)TPP]^{2+} \qquad (16)$$

In the case of the Zn(II) complex the monocation radical, the product of the first oxidation peak, gave, as might be expected, an ESR signal expected for ligand oxidation. However, the Cu(II) complex monocation radical did not give an ESR signal, although magnetic susceptibility measurements unequivocally demonstrated that ligand oxidation had occurred. The lack of an ESR signal was due to broadening of the triplet species $[Cu(II)TPP]^{+\cdot}$. In the case of the Ni(II)−, Co(II)−, and Fe(II)−tetraphenylporphyrin complexes the electrode mechanism could be represented as

$$M(II)TPP \xrightarrow[\substack{\text{metal} \\ \text{oxidation}}]{-e} [M(III)TPP]^{+} \xrightarrow[\substack{\text{ligand} \\ \text{peak I}}]{-e} [M(III)TPP]^{2+\cdot} \qquad (17)$$

$$\downarrow \substack{\text{ligand} \\ \text{peak II}} {-e}$$

$$[M(III)TPP]^{3+}$$

Tetraphenylporphyrin could be regarded as being oxidized according to the mechanism

$$H_2TPP \xrightarrow[\substack{ligand \\ peak\ I}]{-e} [H_2TPP]^{\ddot{+}} \xrightarrow[\substack{ligand \\ peak\ II}]{-e} [H_2TPP]^{2+} \tag{18}$$

The monocation radical gave an ESR signal, but, rather interestingly, an estimation of the lifetime of the species from the decay of the ESR signal gave a value of 16 min. Cyclic voltammetry, however, indicated that the cation radical was much less stable, having a lifetime of less than 10^{-2} sec. This indicated either that the decomposition mechanism at the electrode was different from that in solution, or that the ESR spectrum was that of a further decomposition product rather than $[H_2TPP]^{\ddot{+}}$ itself.

Consideration of these and earlier data reveals, first, that the metal ion associated with the porphyrin ligand has quite an appreciable effect on the potential required for ligand oxidation; i.e., the metal ion seems to set or fix the ligand oxidation potential. Second, in the case of the free porphyrin, although it is oxidized, it appears from the cyclic voltammetric evidence that the oxidation products are exceedingly unstable, although it is not quite clear from the various reports to what in fact these oxidation products decompose. Upon complexation with an appropriate metal ion, however, the stability of the products appears to be enhanced somewhat. Wolberg and Manassen[110] have pointed out an interesting consequence of these conclusions to the understanding of the enzymatic activity of chlorophyll. It will be recalled that chlorophyll, a magnesium complex of a porphyrin, is an absorber of radiant energy with the ligand acting as the active site. This behavior appears to correlate quite well with the cases of metal valence change and ligand oxidation observed in the metallotetraphenylporphyrin series of compounds. A free radical (cation) could be formed from the free porphyrin molecule, but, as the cyclic voltammetry of H_2TPP indicated, a complexed metal ion, which could be Mg in the case of the chlorophyll series, serves to stabilize the molecule and also to reduce the oxidation potential. Thus, the Mg–porphyrin complex is a necessary agent for photosynthesis, and the lowered oxidation potential of the complex with respect to the free ligand is apparently an indicator of the photosensitivity of the pigment.

Fajer and co-workers[111] electrochemically oxidized magnesium–octaethyl-porphyrin (Fig. 8-44) at a platinum anode in dichloromethane containing tetrapropylammonium perchlorate. At a potential of 0.6 V versus SCE this compound could be completely oxidized in a reversible one-electron process to a cation radical. The ESR spectrum of the product clearly indicated that the electron was abstracted from the porphyrin ring. By following the visible

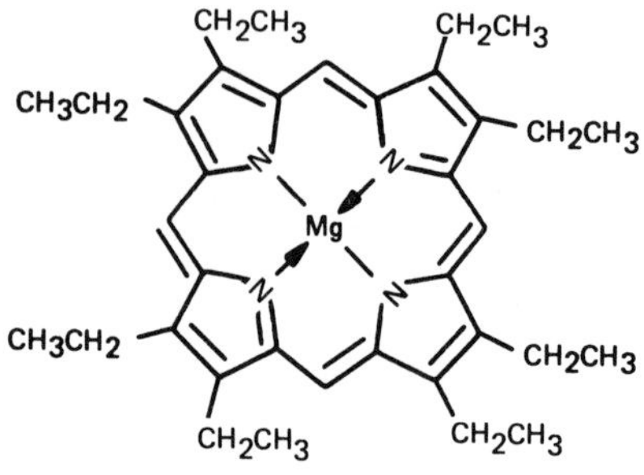

FIG. 8-44.   Magnesium—octaethylporphyrin.

spectrum of solutions of the cation radical as a function of concentration, it was also found that it underwent a dimerization reaction at $0°C$ according to Eq. 19.

$$2[MgOEP]^{\ddot{+}} \rightleftharpoons [MgOEP^{\ddot{+}}]_2 \qquad\qquad (19)$$

Continued electrolysis of $[MgOEP]^{\ddot{+}}$ at 0.9 V versus SCE indicated that another electron could be removed from the cation radical to give a dication $[MgOEP]^{2+}$. The cation radical $[MgOEP]^{\ddot{+}}$ and the dication could be converted back to the original MgOEP simply by electrochemical reduction at 0.0 V. Zinc—tetraphenylporphyrin underwent a reversible one-electron oxidation at 0.8 V versus SCE in the same solvent system to give the expected cation radical $[ZnTPP]^{\ddot{+}}$, which could be isolated as the $ZnTPP^{\ddot{+}}$ ClO_4^- salt and characterized by ESR spectroscopy. The cation radical could be oxidized at 1.1 V in a one-electron process to the corresponding dication. Extensive ESR studies of cation radicals of a number of porphyrins have been carried out by Fajer *et al.*,[112,113] who have found that controlled potential electrochemical oxidation at a platinum electrode in dichloromethane or chloroform is a very convenient way for forming such radicals.

The work reported by Fajer *et al.*[111-113] clearly established the existence of stable π cations of magnesium and zinc porphyrins. Electron spin resonance spectra indicated that the first electron was removed from the ligand and yielded a cation radical, and the fact that the difference between the half-wave potentials for the first and second waves is more or less constant and independent of the metal ion confirmed that the second electrooxidation process also involved removal of an electron from the ligand. The agreement between molecular orbital calculations and the observed electronic spectra also supported the formation of π cations.

Although it has not been definitively proved an increasing body of information suggests that primary photosynthetic processes involve oxidation of chlorophyll[114,115] and bacteriochlorophyll.[116,117] Fajer *et al.*[111] have suggested, considering MgOEP as a model system, that the oxidized chlorophylls might properly be considered π cation radicals. In fact, as mentioned earlier, electrooxidation of ethyl chlorophyllide *a* and chlorophyll *a* converts these molecules into bleached species whose optical and ESR spectra are consistent with such an assignment.[66] It can also be observed from various data quoted in this chapter that the magnesium–porphyrin complexes, in general, are rather easily electrochemically oxidized. This fact in itself might account for the appearance of magnesium–porphyrin complexes (usually chlorins) in the photosynthetic apparatus.

In a somewhat different vein Dolphin *et al.*[118] have utilized the fact that zinc–tetraphenylporphyrin could be oxidized to a π dication as the basis for synthesis of a metalloisoporphyrin. Thus, controlled potential oxidation of ZnTPP (I, Fig. 8-45) at 1.1 V versus SCE in dichloromethane–tetrapropyl-

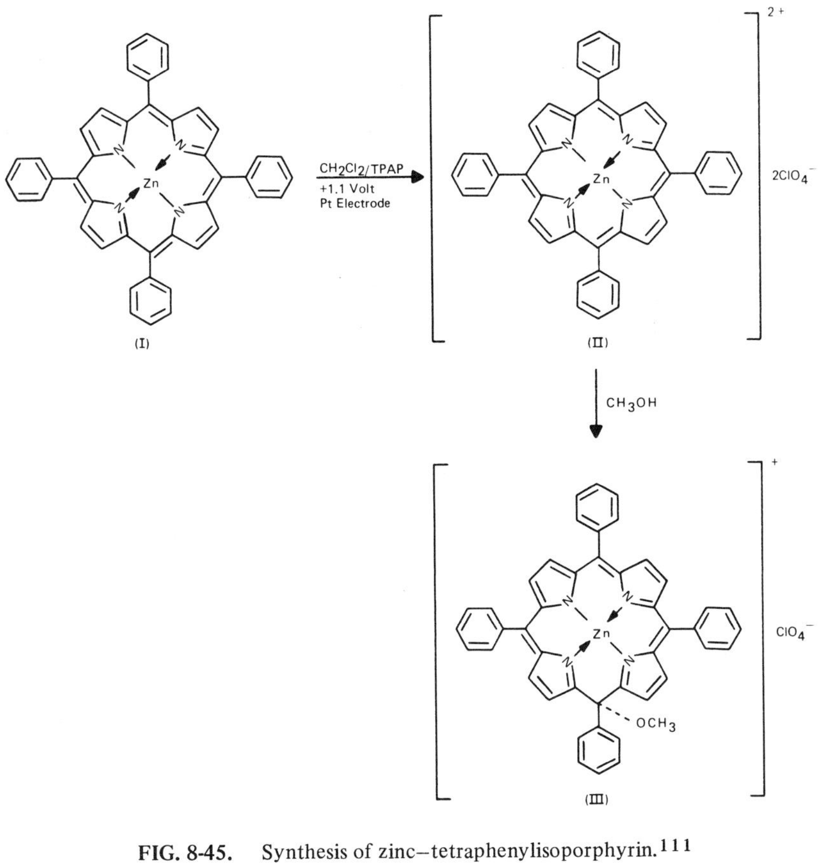

FIG. 8-45. Synthesis of zinc–tetraphenylisoporphyrin.[111]

ammonium perchlorate yields, as indicated earlier, the dication (II, Fig, 8-45). Although this species, a powerful electrophile, is stable in the absence of nucleophiles, the addition of methanol resulted in the formation of the isoporphyrin (III, Fig. 8-45), which could be isolated and appropriately characterized.

XIII. CONCLUSIONS

It is probably fair to say that the electrochemical studies outlined in this chapter have revealed a rather significant amount of information regarding the electron-acceptor and electron-donor properties particularly of the porphyrins. Only a meager amount of mechanistic electrochemistry of the simpler pyrroles and bile pigments has been carried out, although such studies are most important to a complete understanding of the electron-transfer properties of the pyrrole—porphyrin family as a whole.

The situation regarding porphyrins is much healthier. Electrochemical studies are clearly a very powerful tool for studying the effects of the nature of the complexed metal ions of metalloporphyrins and the ease and position of the electron-transfer process. They can also be used to very readily follow and interpret ligand exchange reactions.

The electrochemical studies of the oxidation and reduction of chlorophylls and related molecules in nonaqueous media seem to generate more biologically significant findings than such studies in aqueous media. Many nonbiochemists seem to feel that since most animals and plants are composed primarily of water, the electrochemical processes taking place in aqueous media yield the most biologically significant information. However, it is becoming apparent that many biological electron-transfer and other processes take place in an environment that is essentially free of liquid water. Accordingly, the *in vitro* electrochemical studies in nonaqueous media could in many instances be of considerable biological importance. In particular, it is relatively common to find electron-transfer processes in nonaqueous medium proceeding via sequential one-electron steps, much as such processes are thought to occur in biological systems. In aqueous solution, however, it is often observed that the one-electron steps merge to form multielectron processes. These thoughts do not imply that all biological processes occur in a "nonaqueous" environment. However, it would seem a good policy to study the electrochemical reactions of biologically important molecules under both aqueous and nonaqueous conditions if possible so that as much information as possible about the effect of the environment on the processes can be obtained.

The synthetic utility of the electrochemical reduction of certain porphyrins has again been demonstrated. Thus, under certain conditions electrochemistry can provide simple and straightforward solutions to certain synthetic problems. Although rather little use has been made of the analytical potential of the electrochemical properties of the pyrrole—porphyrin family of molecules, there is little doubt that useful applications can be found.

REFERENCES

1. K. Schofield, "Hetero-Aromatic Nitrogen Compounds: Pyrroles and Pyridines." Plenum, New York, 1967.
2. A. H. Corwin, *Heterocycl. Comp.* **1**, 277 (1950).
3. J. E. Falk, "Porphyrins and Metalloporphyrins." Am. Elsevier, New York, 1964.
4. G. E. W. Wolstenholme and E. C. P. Millar, eds., "Porphyrin Biosynthesis and Metabolism," Ciba Found. Symp. Little, Brown, Boston, Massachusetts, 1955.
5. Runge, *Ann. Phys. Chem.* **31**, 67 (1834).
6. Anderson, *Trans. R. Soc. Edinburgh* **21**, Part IV, 571 (1857); *Ann. Chem. Pharm.* **105**, 349 (1858).
7. R. Lemberg and J. W. Legge, "Haematin Compounds and Bile Pigments." Wiley (Interscience), New York, 1949.
8. C. H. Gray, "The Bile Pigments." Methuen, London, 1953.
9. G. Scaramelli, *Boll. Sci. Fac. Chim. Ind. Bologna* **3**, 205 (1942); *Chem. Zentralbl.* 1943–II **2**, 1180 (1943); *Chem. Abstr.* **38**, 6281 (1944).
10. B. Pullman and A. Pullman, "Quantum Biochemistry," p. 434. Wiley (Interscience), New York, 1963.
11. A. P. Terent'ev, L. A. Yanovskaya, and E. A. Terent'eva, *Dokl. Akad. Nauk SSSR* **70**, 649 (1950).
12. F. Capellina and A. Drusiana, *Gazz. Chim. Ital.* **84**, 939 (1954); *Chem. Abstr.* **49**, 10764 (1955).
13. G. B. Bonino and G. Scaramelli, *Ric. Sci.* **2**, 111 (1935).
14. G. B. Bonino and G. Scaramelli, *Ber. Dtsch. Chem. Ges. B* **75**, 1948 (1942).
15. F. Cappellina and V. Lorenzelli, *Atti Accad. Sci. Ist. Bologna, Cl. Sci. Fis., Rend.* [11] **5**, 1 (1958); *Chem. Abstr.* **53**, 7826h (1959).
16. G. Scaramelli, *Boll. Sci. Fac. Chim. Ind. Bologna* **1**, 239 (1940); *Chem. Zentralbl.* **2**, 152 (1942); *Chem. Abstr.* **37**, 5058 (1943).
17. J. Nakaya, H. Kinoshita, and S. Ono, *Nippon Kagaku Zasshi* **78**, 940 (1957); *Chem. Abstr.* **53**, 21277 (1959).
18. M. Deželic, A. Lacković, and M. Trkovnik, *Croat. Chem. Acta* **32**, 31 (1960); *Chem. Abstr.* **54**, 13902 (1960).
19. F. Capellina and V. Lorenzelli, *Ann. Chim. (Rome)* **48**, 893 (1958); *Chem. Abstr.* **53**, 9857 (1959).
20. N. Tyutyulkov and I. Bakyrdzhiev, *C. R. Acad. Bulg. Sci.* **12**, 133 (1959); *Chem. Abstr.* **54**, 11804 (1960).
21. G. Scaramelli, *Mem. R. Accad. Ital. Cl. Sci. Fis., Mat. Nat.* [7] **1**, 471 (1940).
22. G. Scaramelli, *Mem. R. Accad. Ital. Cl. Sci. Fis., Mat. Nat.* [7] **1**, 575 (1940); *Chem. Abstr.* **37**, 1423 (1943).
23. M. Person, *Bull. Soc. Chim. Fr.* p. 1832 (1966); *Chem. Abstr.* **65**, 13218 (1966).
24. A. P. Terent'ev, L. A. Yanovskaya, and E. A. Terent'eva, *Zh. Obshch. Khim.* **22**, 859 (1952); *Chem. Abstr.* **47**, 3294 (1953).

25. A. P. Terent'ev, L. A. Yanovskaya, and E. A. Terent'eva, *Dokl. Akad. Nauk SSSR* 70, 649 (1950).
26. G. B. Bonino and G. Scaramelli, *Ric. Sci.* 6 (5–6), 1 (1935).
27. F. Cappellina and A. Drusiani, *Ric. Sci.* 30, Suppl. 5, 297 (1960); *Chem. Abstr.* 55, 17306 (1961).
28. J. Tirouflet and P. Fournari, *C. R. Hebd. Seances Acad. Sci.* 248, 1182 (1959).
29. G. Scaramelli, *Boll. Sci. Fac. Chim. Ind. Bologna* 2, 99 (1941); *Chem. Zentralbl.* 2, 269 (1942); *Chem. Abstr.* 37, 4390 (1943).
30. J. Tirouflet and E. Laviron, *Z. Anal. Chem.* 173, 43 (1960); *Chem. Abstr.* 54, 13937 (1960).
31. F. Cappellina and L. Pederzini, *Collect. Czech. Chem. Commun.* 25, 3344 (1960); *Chem. Abstr.* 56, 4515 (1962).
32. A. Stanienda, *Z. Naturforsch., Teil B* 22, 1107 (1967).
33. H. Lund, "Elektrodereaktioner i organisk polarografi og voltammetri." Aarhus Stiftsbogtrykkerie, Aarhus, 1961.
34. H. Lund, *Adv. Heterocycl. Chem.* 12, 261 (1970).
35. N. L. Weinberg and E. A. Brown, *J. Org. Chem.* 31, 4054 (1966).
36. G. Cauquis and M. Geniés, *Bull. Soc. Chim. Fr.* p. 3220 (1967).
37. R. Cuvelier, P. Tronche, J. A. Berger, C. Laroussinie, G. Androucl, and J. Dauphin, *Congr. Soc. Pharm. Fr.* [*C.R.*], *9th, 1957* p. 29 (1957); *Chem. Abstr.* 53, 22178 (1959).
38. I. Tachi, *J. Agric. Chem. Soc. Jpn.* 7, 533 (1931).
39. I. Tachi, *Mem. Coll. Agric. Kyoto Imp. Univ.* 42, 1 (1938).
40. H. Theorell, *Biochem. Z.* 298, 258 (1938).
41. L. Griggio and S. Pinamonti, *Atti Ist. Veneto Sci., Lett. Arti Cl. Sci. Mat. Nat.* 128, 15 (1965–1966); *Chem. Abstr.* 67, 87793r (1967).
42. C. Carruthers, *J. Biol. Chem.* 171, 641 (1947).
43. C. Carruthers, *Fed. Proc., Fed. Am. Soc. Exp. Biol.* 6, 242 (1947).
44. C. Carruthers and V. Suntzeff, *Arch. Biochem.* 17, 261 (1948).
45. M. Brezina and P. Zuman, "Polarography in Medicine, Biochemistry and Pharmacy." Wiley (Interscience), New York, 1958.
46. T. W. Goodwin, ed., "Chemistry and Biochemistry of Plant Pigment." Academic Press, New York, 1965.
47. T. W. Goodwin, ed., "Porphyrins and Related Compounds." Academic Press, New York, 1969.
48. R. B. Woodward, *J. Pure Appl. Chem.* 2, 383 (1961).
49. P. Van Rysselberghe, J. M. E. McGee, A. H. Groppe, and R. W. Lane, *J. Am. Chem. Soc.* 69, 809 (1947).
50. E. Knobloch, see Ref. 45, p. 533.
51. R. Felton, G. M. Sherman, and H. Linschitz, *Nature (London)* 203, 637 (1964).
52. R. Felton, Ph.D. Thesis, Harvard University, Cambridge, Massachusetts, 1964 (quoted in Felton *et al.* [51]).
53. G. Hoijtink, J. van Schooten, E. de Boer, and W. Aalberberg, *Recl. Trav. Chim. Pays-Bas* 73, 355 (1954).
54. E. I. Rabinowitch, "Photosynthesis and Related Processes," p. 1778. Wiley (Interscience), New York, 1956.
55. H. H. Inhoffen and P. Jäger, *Tetrahedron Lett.* p. 3387 (1965).
56. H. H. Inhoffen and R. Mählhop, *Tetrahedron Lett.* p. 4283 (1966).
57. H. H. Inhoffen, P. Jäger, R. Mählhop, and C. D. Mengler, *Justus Liebigs Ann. Chem.* 704, 188 (1967).

58. H. H. Inhoffen, *Pure Appl. Chem.* **17**, 443 (1968).

59. B. A. Kiselev, Y. N. Kozlov, and V. B. Erstigneer, *Biofizika* **15**, 594 (1970); *Chem. Abstr.* **73**, 115,771b (1970).

60. H. Tributsch and M. Calvin, *Photochem. Photobiol.* **14**, 95 (1971).

61. A. Serbanescu, *Petrol. Gaze (Bucharest)* **17**, 135 (1966); *Chem. Abstr.* **65**, 13421 (1966).

62. K. Kramarczyk and H. Berg, *Abh. Dtsch. Akad. Wiss. Berlin. Kl. Med.* p. 363 (1966); *Chem. Abstr.* **66**, 112287t (1966).

63. A. Stanienda, *Z. Phys. Chem. (Leipzig)* **229**, 257 (1965); *Chem. Abstr.* **63**, 10228 (1965).

64. A. Stanienda, *Naturwissenschaften* **24**, 731 (1963).

65. A. Stanienda, *Z. Phys. Chem. (Leipzig)* **229**, 257 (1965).

66. R. H. Felton, D. Dolphin, D. C. Borg, and J. Fajer, *J. Am. Chem. Soc.* **91**, 196 (1969).

67. I. Y. Artamkina and V. M. Kutyurin, *Dokl. Akad. Nauk SSSR* **196**, 980 (1971); *Chem. Abstr.* **74**, 136524j (1971).

68. J. C. Goedheer, G. H. H. De Haas, and P. Schuller, *Biochim. Biophys. Acta* **28**, 278 (1958).

69. B. Kok, *Biochim. Biophys. Acta* **48**, 527 (1961).

70. J. B. Conant, C. A. Alles, and C. O. Tongberg, *J. Biol. Chem.* **79**, 89 (1928).

71. E. S. Barron, *J. Biol. Chem.* **121**, 285 (1937).

72. J. Shack and W. M. Clark, *J. Biol. Chem.* **171**, 143 (1947).

73. J. Jordan and T. M. Bednarski, *J. Am. Chem. Soc.* **86**, 5690 (1964).

74. T. M. Bednarski and J. Jordan, *J. Am. Chem. Soc.* **89**, 1552 (1967).

75. J. N. Phillips, *Rev. Pure Appl. Chem.* **10**, 35 (1960).

76. R. I. Walter, *J. Biol. Chem.* **196**, 151 (1951).

77. D. G. Davis and D. J. Orleron, *Anal. Chem.* **38**, 179 (1966).

78. R. S. Nicholson and I. Shain, *Anal. Chem.* **36**, 706 (1964).

79. J. E. B. Randles, *Trans. Faraday Soc.* **44**, 327 (1948).

80. A. Ševčík, *Collect. Czech. Chem. Commun.* **13**, 349 (1948).

81. D. G. Davis and R. F. Martin, *J. Am. Chem. Soc.* **88**, 1365 (1966).

82. D. G. Davis, *Electroanal. Chem.* **1**, 157 (1966).

83. D. C. Borg and G. C. Catzias, *Nature (London)* **182**, 1677 (1958).

84. M. Calvin, *Rev. Pure Appl. Chem.* **15**, 1 (1965).

85. P. A. Loach and M. Calvin, *Biochemistry* **2**, 361 (1963).

86. J. F. Taylor, *J. Biol. Chem.* **135**, 569 (1940).

87. P. A. Loach and M. Calvin, *Biochim. Biophys. Acta* **79**, 379 (1964).

88. D. G. Davis and J. G. Montalvo, *Anal. Chem.* **41**, 1195 (1969).

89. E. B. Fleischer and J. H. Wang, *J. Am. Chem. Soc.* **52**, 3486 (1960).

90. D. G. Davis and J. G. Montalvo, *Anal. Lett.* **1**, 641 (1968).

91. L. Meites, "Polarographic Techniques," 2nd ed. Wiley (Interscience), New York, 1966.

92. R. H. Felton and H. Linschitz, *J. Am. Chem. Soc.* **88**, 1113 (1966).

93. M. Zerner and M. Gouterman, *Theor. Chim. Acta* **4**, 44 (1966).

94. G. L. Closs and L. E. Closs, *J. Am. Chem. Soc.* **85**, 818 (1963).

95. D. W. Clack and N. S. Hush, *J. Am. Chem. Soc.* **87**, 4238 (1965).

96. L. D. Rollman and R. T. Iwamoto, *J. Am. Chem. Soc.* **90**, 1455 (1968).

97. J. G. Lanese and G. S. Wilson, *J. Electrochem. Soc.* **119**, 1039 (1972).

98. D. G. Davis and L. A. Truxillo, *Anal. Chim. Acta* **64**, 55 (1973).

99. D. L. Maricle and A. H. Maurer, *Chem. Phys. Lett.* **2**, 602 (1968).

100. A. Ricci, S. Pinamonti, and V. Bellavita, *Ric. Sci., Suppl.* **30**, 2497 (1960); *Chem. Abstr.* **60**, 14127 (1964).

101. G. P. Heiling and G. S. Wilson, *Anal. Chem.* **43**, 545 (1971).
102. D. Mauzerall, *J. Am. Chem. Soc.* **84**, 2437 (1962).
103. G. P. Heiling and G. S. Wilson, *Anal. Chem.* **43**, 551 (1971).
104. See, for example, A. T. Hubbard and F. C. Anson, *Electroanal. Chem.* **4**, 129 (1970).
105. B. P. Neri and G. S. Wilson, *Anal. Chem.* **44**, 1002 (1972).
106. P. Hambright and E. B. Fleischer, *Inorg. Chem.* **9**, 1757 (1970).
107. R. B. Woodward, *J. Pure Appl. Chem.* **2**, 383 (1961).
108. A. Stanienda, *Naturwissenschaften* **52**, 105 (1965).
109. A. Stanienda and G. Biebl, *Z. Phys. Chem. (Frankfurt am Main)* **52**, 254 (1967).
110. A. Wolberg and J. Manassen, *J. Am. Chem. Soc.* **92**, 2982 (1970).
111. J. Fajer, D. C. Borg, A. Forman, D. Dolphin, and R. H. Felton, *J. Am. Chem. Soc.* **92**, 3451 (1970).
112. J. Fajer, A. Forman, D. C. Borg, and R. H. Felton, *J. Am. Chem. Soc.* **93**, 2790 (1971).
113. J. Fajer, D. C. Borg, A. Forman, A. D. Adker, and V. Varadi, *J. Am. Chem. Soc.* **96**, 1238 (1974).
114. B. Kok, *Biochim. Biophys. Acta* **48**, 527 (1961).
115. J. Weikard, A. Muller, and H. T. Wett, *Z. Naturforsch., Teil B* **18**, 139 (1963).
116. J. D. McElroy, G. Feher, and D. C. Mauzerall, *Biochim. Biophys. Acta* **172**, 180 (1969).
117. J. R. Bolton, R. K. Clayton, and D. W. Reed, *Photochem. Photobiol.* **9**, 209 (1969).
118. D. Dolphin, R. H. Felton, D. C. Borg, and J. Fajer, *J. Am. Chem. Soc.* **92**, 743 (1970).

9

Pyridines and Pyridine Nucleotides

I. INTRODUCTION, NOMENCLATURE, AND STRUCTURE

Pyridine is a six-membered ring compound containing a single heterocyclic nitrogen atom and a double-bond system analogous to that of benzene. Pyridines are numbered by two systems. The system employed in the older literature uses Greek letters (Fig. 9-1A), while the numbering system employed in the more recent literature is that shown in Fig. 9-1B.

An enormous amount of study has been devoted to pyridine and its derivatives. Much of the chemistry, physical properties, and historical information on pyridines is included in the works of Mosher,[1] Klingsberg,[2] and Schofield[3] and the references quoted therein.

The nomenclature of simple pyridine derivatives is quite straightforward. Most of the biological interest in pyridine derivatives centers around nicotinamide, 3-carbamidopyridine (Fig. 9-2). In nature the important forms of nicotinamide are the two pyridine coenzymes. Coenzyme I is nicotinamide adenine dinucleotide (NAD⁺), which is often referred to as diphosphopyridine nucleotide (DPN⁺). Coenzyme II is nicotinamide adenine dinucleotide

A   **B**

FIG. 9-1. Structure and numbering systems for pyridine.

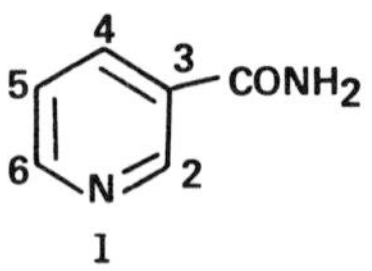

FIG. 9-2.   Structure of nicotinamide (3-carbamidopyridine).

phosphate (NADP$^+$), which is often called triphosphopyridine nucleotide (TPN$^+$). The structures of these and other important pyridine derivatives are shown in Table 9-1.

TABLE 9-1   Name and Structure of Some Important Pyridine Derivatives

Name	Structure
Pyridoxal	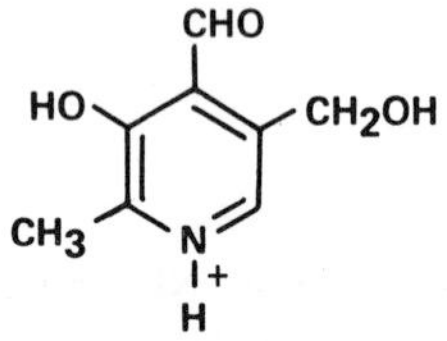
Pyridoxine (pyridoxol)	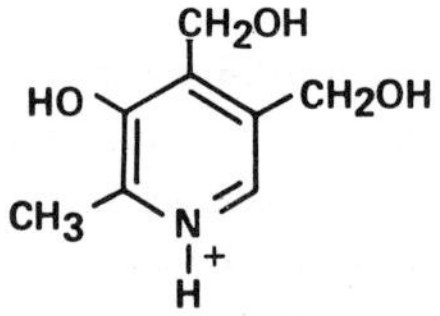
Pyridoxamine	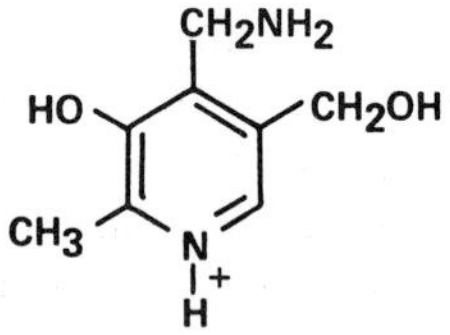
Pyridoxal phosphate	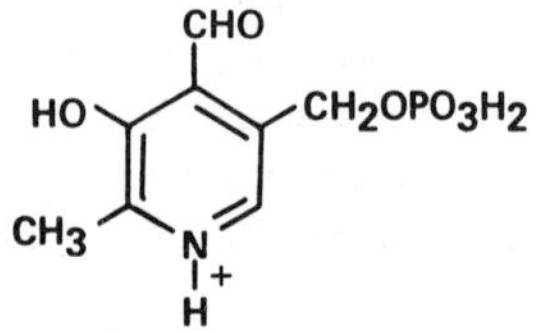
Pyridoxamine phosphate	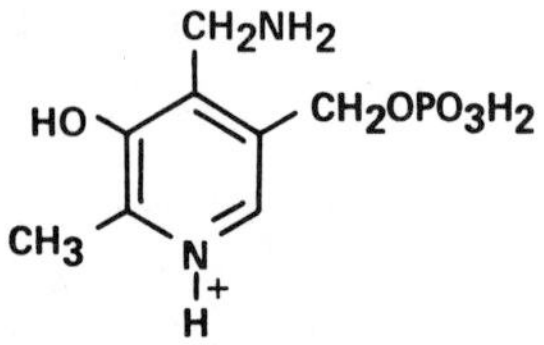

TABLE 9-1 *Continued*

Name	Structure
Nicotinic acid	
Nicotinamide	
Coenzyme I [nicotinamide adenine dinucleotide (NAD^+), or diphospho-pyridine nucleotide (DPN^+)]	
Coenzyme II [nicotinamide adenine dinucleotide phosphate ($NADP^+$), or triphosphopyridine nucleotide (TPN^+)]	

II. OCCURRENCE AND BIOLOGICAL SIGNIFICANCE OF PYRIDINES

Niacin is the official name given to nicotinic acid or nicotinamide. This vitamin is widely distributed in plant and animal tissue. However, the coenzyme forms of the vitamin are the pyridine nucleotides. For the sake of clarity and consistency, and in accord with the 1961 recommendations of the Commission on Enzymes of the International Union of Biochemistry, coenzyme I will be referred to as NAD^+ and coenzyme II as $NADP^+$ throughout this chapter. However, it should be recognized that much of the current literature uses the older designations DPN^+ and TPN^+, the names originally proposed by Warburg.[4]

The pyridine nucleotides are coenzymes for the dehydrogenase group of enzymes that are responsible for catalysis of certain biological oxidation—reduction reactions. Both NAD^+ and $NADP^+$ are reduced in the processes in which they act as coenzymes. The general reductive process and the structures of the reduced forms of NAD^+ or $NADP^+$ are shown in Fig. 9-3. It is not unlikely that the reaction proceeds via a hydride ion ($H:^-$) transfer rather than via the discreet proton and electron route shown (*vide infra*).

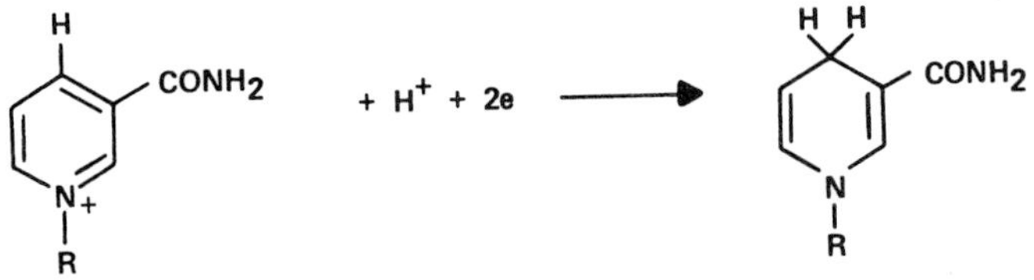

FIG. 9-3. Probable mode of biological reduction of NAD^+ or $NADP^+$ (I) to NADH or NADPH (II) (R = remainder of NAD^+ or $NADP^+$ molecules).

As examples of some redox reactions in which the pyridine nucleotides participate, one might well consider the action of the enzyme alcohol dehydrogenase, which oxidizes ethanol to acetaldehyde according to Eq. 1.

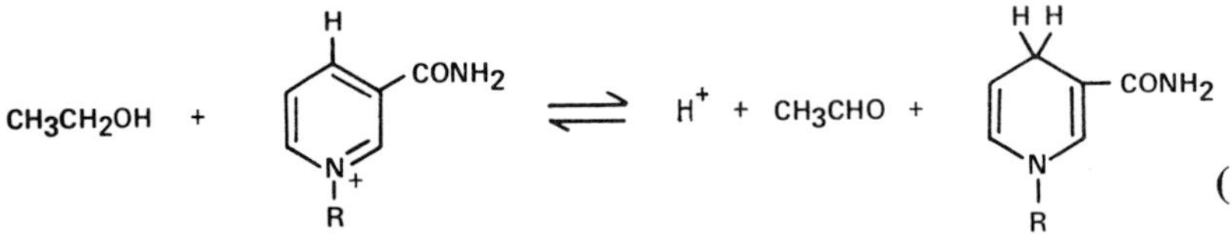

$$(1)$$

Often, the reduced forms of the pyridine nucleotides function as reducing agents, particularly with the flavin coenzymes. Thus, in order to reduce the disulfide of glutathione, the glutathione reductase enzymes are employed. These enzymes contain flavin adenine dinucleotide (FAD) as their prosthetic group. In the presence of NADH the FAD is first reduced to $FADH_2$, which then accomplishes the reduction of the disulfide moiety (Eq. 2).

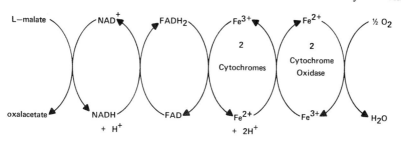

In many biological oxidation reactions, for example, the oxidation of malate, the pyridine nucleotides are the immediate oxidizing agents. The supply of pyridine nucleotides in a cell obviously is limited and, unless means for regeneration are provided, the supply of oxidized nucleotides becomes exhausted. Nature has provided the means and often utilizes molecular oxygen as the ultimate oxidant. The series of oxidation–reduction reactions employed to reoxidize the pyridine (and flavin) nucleotides is often called the *electron-transport chain*, a simplified version of which is shown in Fig. 9-4.[5] The curved arrows in Fig. 9-4 merely indicate the flow of electrons. Thus, L-malate is oxidized by NAD^+ to give oxalacetate and NADH. The latter is then oxidized by FAD to give NAD^+ again and $FADH_2$. Cytochromes, which were briefly discussed in Chapter 8, Section VI, then enter the chain until one cytochrome

FIG. 9-4. Simplified electron-transport chain for oxidation of pyridine and flavin nucleotides ultimately by molecular oxygen.

oxidase enzyme reduces oxygen to water. The net result is that L-malate has been oxidized to oxalacetate and oxygen has been reduced to water. The coenzymes and enzymes are all back in their oxidized form ready for the subsequent electron-transfer sequence.

There is another pyridine nucleotide, nicotinamide mononucleotide, NMN^+, which is less well known and which is an important intermediate in the biosynthesis of NAD^+ and $NADP^+$.

The vitamin B_6 group of compounds is also comprised of pyridine derivatives, namely, pyridoxal, pyridoxine, and pyridoxamine (Table 9-1), and are widely distributed in plants and animals. Pyridoxal and pyridoxamine also occur as their phosphate derivatives, which in fact are the coenzyme forms of vitamin B_6 (Table 9-1). Deficiencies of vitamin B_6, if extreme, can lead to serious disorders of the central nervous system. These compounds are not involved in electron-transport or oxidation–reduction processes as such, but rather appear to function in amino acid metabolism, such as transamination, decarboxylation, and racemization processes.

Because of the involvement of the pyridine nucleotides in electron-transport processes, electrochemists have examined them and simpler pyridine derivatives in some detail. Although pyridine itself and its simple derivatives do not appear to be biologically active, at least from the viewpoint of normal or beneficial effects on organisms, it is possible that a knowledge of their behavior might have considerable bearing on the behavior of the pyridine nucleotides. Accordingly, this chapter is divided into three broad sections, the first and second dealing with the electrochemistry of pyridines and pyridine derivatives in general, and the third with that of biologically important derivatives.

III. ELECTROCHEMISTRY OF PYRIDINES

A. Electrochemical Reduction

1. Pyridine

It appears that the first study of the electrochemical reduction of pyridine was that of Emmert,[6] who employed a lead cathode and electrolyzed pyridine in a 40% sulfuric acid solution. The work, of course, was not done under controlled potential conditions. The products identified were γ,γ'- and α,α'-dipiperidyls and possibly some polypiperidyls.

Shikata and Tachi[7-9] studied the polarographic reduction of pyridine in unbuffered solution. Using a background electrolyte of 0.1 M KCl containing 1 mM HCl, they apparently found that a 12.5 mM solution of pyridine gave two waves, with half-wave potentials of -1.5 and -1.7 V (presumably versus SCE).

These workers concluded that the first wave was due to reduction of pyridinium ion and the second to reduction of the undissociated pyridine molecule. The ultimate product of the reaction was claimed to be piperidine. Tompkins and Schmidt[10] examined the polarography of pyridine more closely and found that, in buffered solutions below about pH 6 containing 1–5 mM pyridine, no waves were produced. However, at higher pH in a 0.1 M potassium phosphate buffer a poorly defined wave could be observed ($E_{1/2}$ = −1.7 V versus SCE at pH 6.1 and 7.4). In the unbuffered medium employed by Shikata and Tachi,[7−9] they obtained effectively the same data as the earlier workers.

Because of the confusion in the literature regarding the polarography of pyridine in unbuffered medium, it is worthwhile restating some of the observations of Tompkins and Schmidt.[10] A solution of 1 mM HCl and 0.1 M KCl (pH 2.9) gave a diffusion-controlled wave, the current for which was 16.9 μA, $E_{1/2}$ = −1.49 V versus SCE. This wave was due to reduction of hydrogen ions. On adding pyridine to the same solution sufficient to make it 10 mM, the pH changed to 5.6 and the wave at −1.49 V decreased to 6.48 μA, but neither its $E_{1/2}$ nor slope changed. When less HCl was employed so that the pH of the pyridine-containing solution was 6.2, two waves were formed. The first had a limiting current of 1.09 μA, $E_{1/2}$ = −1.51 V, and the second had a limiting current of 5.28 μA, $E_{1/2}$ = −1.78 V. In 0.1 M KCl in the absence of added acid or alkali (pH 7–8), the half-wave potential was −1.8 V and was independent of the pyridine concentration. The limiting current was proportional to the pyridine concentration (±10%). The $E_{1/2}$ of this wave shifted toward more negative potentials as the pH increased, which suggested that hydrogen ions were involved in the electrode reaction. Tompkins and Schmidt considered that in the light of their data the mechanism of Shikata and Tachi was doubtful. Accordingly, the wave at −1.49 V was proposed to be a straightforward reduction of hydrogen ions. In unbuffered KCl solution of pH 7–8 the single wave produced gave an n value, calculated from the Ilkovič equation, of 0.6. When the KCl was replaced by 0.1 M phosphate buffer, pH 6.1 or 7.4, the apparent n value was 6. In other words, the current for the wave associated with pyridine was dependent on the supply of hydrogen ions (in effect the buffer capacity) and on the concentration of pyridine. It was therefore concluded that the mechanism of the electrochemical reaction for the process at $E_{1/2}$ of ca. −1.7 to −1.8 V was a primary reduction of hydrogen ions to atomic hydrogen, presumably catalyzed by pyridine (Eq. 3A). The pyridine molecules were then proposed to be reduced to piperidine (Eq. 3B) by the hydrogen so liberated.

Shchennikova and Korshunov[11] also found that at pH < 6 pyridine gave a diffusion wave at −1.55 V versus SCE and another at −1.76 V at pH > 6. Knobloch[12,13] has proposed that pyridine, and indeed many other related derivatives, in buffered solutions at the DME give rise to catalytic hydrogen

$$H^+ + e \xrightarrow{\text{PYRIDINE}} H \qquad A$$

$$\text{pyridine} + 6H \longrightarrow \text{piperidine} \qquad B \tag{3}$$

PYRIDINE   **PIPERIDINE**

waves. The catalytic activity was ascribed to the pyridinium type of cation. As might be expected, the height of the limiting current for the catalytic wave was proportional to the concentration of hydrogen ions, the capacity of the buffer system, and the concentration of the catalyst through a relationship given by a Langmuir isotherm. In view of this effect, and by examining current versus time curves, Knobloch[12,13] concluded that, at potentials at which the catalytic effect starts, adsorption of pyridine takes place. This effect was also confirmed by electrocapillary studies. The pH and buffer-capacity effects were explained by Knobloch as being due to a regeneration of the active cationic form of the catalyst by protonation of pyridine. In other words, the electrochemically active form of the hydrogen ion abstracted from water is the pyridinium cation (Eq. 4). As shown earlier, it is probable that either the hydrogen atoms so

$$\text{pyridine} + H_2O \rightleftharpoons \text{pyridinium} + OH^-$$

$$\text{pyridinium} + e^- \longrightarrow \text{pyridinyl} + H \tag{4}$$

produced can reduce pyridine, or a true faradaic electrochemical reduction of pyridine occurs, or both.

Kaye and Stonehill[14,15] have studied the polarographic wave of pyridine at -1.5 V in the Shikata and Tachi supporting electrolyte (1 mM HCl plus 0.1 M KCl) in the hope of gaining evidence for formation of a stable free radical of pyridine. They were not able to obtain such evidence by analysis of the wave slope, which is not surprising since the wave that these workers were observing was due to reduction of hydrogen ion.

Reynolds and Lindsay[16,17] also examined the polarography of pyridine in unbuffered 0.1 M KCl solution containing added amounts of KOH or HCl. A single polarographic wave was observed for pyridine in 0.1 M KCl containing up to 2×10^{-4} M KOH, $E_{1/2} \approx -1.7$ V versus SCE. The height of the single wave was proportional to the pyridine concentration. In 0.1 M KCl containing $2 \times 10^{-3} - 1.23 \times 10^{-3}$ M HCl two waves were produced which appeared to be at $E_{1/2} \approx -1.6$ V and -2.0 V versus SCE. As noted by earlier workers, an increase of pyridine concentration decreased the height of the first (more positive) wave and increased the height of the second wave. These workers, however, considered that the first wave, in acidic pyridine solutions, was the reduction of hydrogen ion catalyzed by pyridine. The second wave was also thought to be a catalytic hydrogen ion reduction but superimposed on the reduction of pyridine. In neutral or alkaline solution, or in the presence of below 1.3×10^{-3} M HCl, the single wave observed was proposed to be a straight-forward reduction of pyridine, although no electron values were determined nor was any substantial evidence presented to support this view. The apparent reduction of pyridine in these processes was proposed to be irreversible. Kůta and Drabek[18,19] also found that solutions of pyridine in 0.1 M KCl and 10^{-3} M HCl (or a weak nonreducible acid of $pK_a = 2$–5) gave two polaro-graphic waves at $E_{1/2} = -1.53$ and -1.8 V versus NCE. The more positive wave was diffusion controlled and was apparently due to hydrogen ion reduction. However, they found that in the presence of pyridine this wave was shifted toward more positive potential and its slope changed owing to the catalytic effect of pyridine. These data are, of course, at variance with the report of Tompkins and Schmidt.[10] The height of the more negative wave increased with decreasing height of the mercury column and with increasing acid concentration. It also increased with increasing pyridine concentration but tended toward a limiting value. Again, the wave at more negative potential was thought to be due to catalytic reduction of hydrogen ion. In solutions of 0.1 M LiCl or KCl only the more negative wave was observed ($E_{1/2} = -1.8$ V versus NCE). As some additional confirmation of the catalytic nature of this wave, it was found that its height decreased with increasing concentration of ethanol added to the solution. Pozdeeva and Gepshtein[20] had earlier concluded that the single wave in LiCl solutions was diffusion controlled, but Kůta and Drabek[18,19] rejected this idea.

Kalvoda[21] utilized oscillopolarography to study the electrochemical reduc-tion of pyridine and concluded that pyridine was reduced at very negative potentials, as evidenced by the formation of cathodic peaks. Anodic peaks were also observed due to reoxidation of some of the reduction products, which were shown to include dipyridyl, polypiperidyls, certain nitrogen-free aldehydes (e.g., $OHC \cdot CH{=}CH \cdot CH_2 CHO$), and various pyran derivatives.

The general idea that adsorption plays a role in the overall mechanism of the polarographic wave observed for pyridine in unbuffered, nearly neutral solutions is supported by the data of Knobloch,[13] who thought that pyridine was adsorbed (*vide supra*), and by Mairanovskii and Barashkova,[22] who found that, in 0.5 M KCl, pyridine gave a small adsorption prewave at potentials slightly more positive than the main catalytic hydrogen wave. The prewave was thought to be caused by adsorption of the radicals that were the products of the reversible reduction of pyridinium cations. Mairanovskii and co-workers have also studied the electrochemical behavior of pyridine and pyridinium salts with special emphasis on the effect of formation of a surface-active polymeric product on the DME surface,[23,24] on the rate of the protonation of pyridine[25-30] at an electrode surface, and on the effect of experimental parameters on the rate of the latter reactions.[31-35] The references listed here and those in the monograph of Mairanovskii[36] should be consulted for details of these processes.

A number of reports have appeared on the polarography of pyridine in nonaqueous solution. Generally, these studies have been far from exhaustive and at best give the half-wave potential and an *n* value. Given[37] found that pyridine in 90% dioxane–water solution containing tetra-*n*-butylammonium iodide as supporting electrolyte gave a 2*e* (Ilkovič equation) reduction wave. Similarly, Anthoine and co-workers[38] found a single 2*e* wave for pyridine in anhydrous dimethylformamide containing 0.1 *M* tetraethylammonium iodide. Other workers have confirmed this result.[39,40] Wiberg and Lewis,[41] using the same solvent system, have reported that pyridine and many other azines are reduced at the DME via an ECE type of mechanism. The first step in this mechanism was proposed to be a 1*e* reduction of the azine to a radical anion. In many instances fast sweep cyclic voltammetry at a hanging mercury drop electrode indicated that this radical anion could be reoxidized, giving rise to a well-formed anodic peak. The cathodic peak for the 1*e* radical anion formation and the anodic peak for its reoxidation were typical of an almost reversible redox couple. Pyridine did not give an anodic peak on cyclic voltammetry, apparently because of a very fast follow-up chemical reaction of the radical anion. Both polarography and cyclic voltammetry in DMF supported the view that pyridine is adsorbed at mercury electrodes.[41] A summary of the principal data on the polarographic half-wave potentials of pyridine is presented in Table 9-2.[7-10,16-20,37-43]

Pyridine is undoubtedly adsorbed at mercury,[44-58] platinum,[59,60] and gold electrodes,[61,62] as shown by extensive studies using alternating current polarography and oscillopolarography and by studies of the differential capacity of electrodes. A discussion of these phenomena is beyond the scope of this chapter, and the original references should be consulted for details.

2. Pyridinium Ion

The electrochemical activity of pyridine and its ability to function as an agent for catalytic hydrogen reduction appear to be associated with the pyridinium cation.[63,64] The first step is protonation of the pyridine molecule to give the pyridinium cation (Eq. 5A), which is then reduced to a free-radical species

TABLE 9-2 Polarographic Half-Wave Potential Data for Pyridine at the Dropping Mercury Electrode

Background electrolyte	pH[a]	Pyridine conc (mM)[b]	$E_{1/2}$ (V versus SCE[c])	Reference
0.1 M KCl, 1 mM HCl		12.5	I,[d] −1.5	7–9
			II, −1.7	
0.1 M K phosphate	6.1–7.4	1–10	I, −1.7	10
0.1 M Na borate	8.7		I, −1.8	10
Phosphate	7.4		I, −1.69	10
0.1 M Tetraethyl-ammonium phosphate	7.0		I, −1.7	10
0.1 M Na borate	8.6		I, −1.8	10
0.1 M KCl	7.0		I, −1.8	10
0.1 M Tetraethyl-ammonium bromide	7.0		I, −1.8	10
0.1 M KCl + up to 2 x 10⁻⁴ M KOH			I, −1.7	16,17
0.1 M KCl + 2 x 10⁻³ M HCl			I, −1.6	16.17
			II, −2.0	
0.1 M KCl + 10⁻³ M HCl			I, −1.53 (versus NCE)	18,19
			II, −1.8 (versus NCE)	
0.1 M KCl or 0.1 M LiCl			I, −1.8 (versus NCE)	18,19
0.1 M LiCl in			I, −1.73	20
50% ethanol	6.18		I, −1.70	43
Tetra-n-butyl-ammonium iodide in 90% dioxane–water		0.7–1.3	I, −2.01 (versus Hg pool)	37
0.1 M Tetraethyl-ammonium iodide in anhydrous dimethyl-formamide		0.15–1.9	I, −2.10 (versus Hg pool)	38–40
0.1 M Tetra-n-butyl-ammonium iodide in anhydrous dimethyl-formamide		0.5–3	I, −2.76 (versus aq. Ag/AgCl)	42
0.1 M Tetraethyl-ammonium iodide in anhydrous dimethyformamide			I, −2.15 (versus Hg pool)	41

[a] If applicable or if stated in original paper.

[b] If known.

[c] Except where otherwise indicated.

[d] Larger Roman numeral signifies more negative half-wave potential.

(Eq. 5B). This in turn dimerizes and liberates hydrogen gas (Eq. 5C), although the mechanism operative in Eq. 4 probably occurs as well.

$$C_5H_5N + H_2O \rightleftharpoons C_5H_5NH^+ + OH^- \tag{5A}$$

$$C_5H_5NH^+ + e^- \rightleftharpoons C_5H_5NH\cdot \tag{5B}$$

$$2C_5H_5N\cdot \longrightarrow (C_5H_5N)_2 + H_2 \tag{5C}$$

Tompkins and Schmidt[65,66] were probably the first investigators to observe that *N*-alkylpyridinium derivatives were reduced polarographically in a one-electron process. On the basis of the slope of the waves observed, it appeared that the waves might be due to a reversible electrode process. The primary product of the one-electron process was thought to be a free radical that rapidly dimerized (Fig. 9-5).

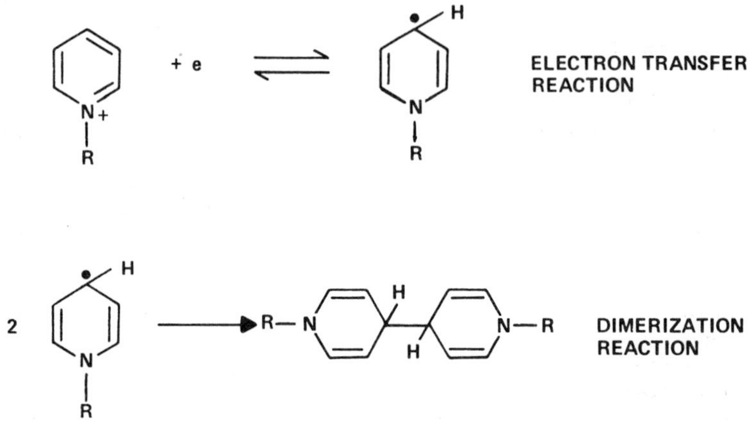

FIG. 9-5.   Proposed mechanism of polarographic reduction of *N*-alkylpyridinium salts.[65,66]

Colichman and O'Donovan[67] investigated a series of β- (**I**) and γ-substituted (**II**) propylpyridinium halides polarographically and proposed, purely by virtue

(**I**)   (**II**)

of the slope analysis of the waves $[E_{DME}$ versus $\log(i_d - i)/i = 0.059 \pm 0.002$ V$]$, that all pyridinium salts were reduced at the DME by a one-electron reversible reaction in aqueous solution. If this mechanism is correct, then a mechanism similar to that show in Fig. 9-5 is probably effective. Wave slope data in dioxane solutions [E_{DME} versus $\log(i_d - i)/i \approx 0.10$ V] were interpreted as indicating that the pyridinium salts were irreversibly reduced in this medium. Typical half-wave potential data are shown in Table 9-3. It is interesting that in unbuffered neutral solution many of the compounds examined by Colichman and O'Donovan[67] gave rise to catalytic hydrogen reduction waves. Mairanovskii[68,69] has developed equations to explain the characteristics of the polarographic waves of N-alkylpyridinium salts on the basis of the reversible electrochemical step and the fast dimerization of the electrode products (radicals) formed.

Ochiai and Kataoka[70] found that electrochemical reduction of 1-methylpyridinium sulfate gave 1-methylpiperidine and 1,1′-dimethyl-4,4′-bipiperidine, while 1-benzylpyridinium chloride gave 1-benzylpiperidine and 1,1′-dibenzyl-4,4′-bipiperidine. It is unfortunate that the exact conditions and electrode materials for this electrolysis are unavailable. However, it is indicative that under certain conditions extensive reduction of the pyridine nucleus can occur under electrolytic conditions. Data on half-wave potentials of some other pyridinium derivatives are presented in Table 9-4.[71,72] The electrode reactions associated with these processes are not known.

A series of papers by Elving and co-workers has appeared concerning the electrochemistry of pyridinium ion, with pyridine as the solvent. Pyridine is a Lewis base that acts by donating the available electron pair from its heterocyclic nitrogen atom to form a coordinate bond with the hydrogen ion of an acid, thus promoting the dissociation of the acid and forming a pyridinium ion (Eq. 6).

$$\text{HA} \quad + \quad \underset{\ddot{N}}{\bigcirc} \quad \rightleftharpoons \quad \underset{\underset{H}{\overset{|}{N^+}}}{\bigcirc} \quad + \quad \text{A}^- \tag{6}$$

Spritzer, Costa, and Elving[73] found that a number of Brönsted acids of aqueous pK_a less than 9, a pyridinium salt, an alkylpyridinium salt, and a Lewis acid such as an alkyl halide (which forms a quaternary salt with pyridine) all gave essentially the same half-wave potential and diffusion current constant in pyridine containing lithium perchlorate as supporting electrolyte. The value of the polarographic diffusion current constant and controlled potential coulometry indicated that the diffusion-controlled electrode process involved a single electron. The half-wave potential for the pyridinium ion reduction wave was -1.34 ± 0.05 V versus NAgE (N AgNO$_3$/Ag electrode in pyridine). Although no attempt was made to identify reaction products, these workers thought that two reaction sequences could occur, giving rise to current flow: (a) formation of a free radical with subsequent dimerization to a tetrahydrobipyridine (Eq. 7A) or

TABLE 9-3   Half-Wave Potentials for Substituted Propylpyridinium Salts[a]

Salt[b]	$E_{1/2}$ (V versus SCE) in aqueous solution				$E_{1/2}$ (V versus SCE) in nonaqueous solution	
	pH 7	pH 8	0.05 M KOH	0.05 M KNO$_3$	0.05 M LiCl in EtOH[c]	0.02 M TEAB[d] in 90.5% dioxane
n-Propyl bromide	−1.38	−1.36	−1.37	−1.37, −1.43[e]	−1.29	−1.15
Isopropyl bromide		−1.40	−1.38	−1.37, −1.43[e]		−1.20
β-OH bromide	−1.36	−1.36	−1.37	−1.42		−1.23
β-Cl bromide	−1.22	−1.24	−1.23	−1.24, −1.40[e] −1.72[f]		−1.10
β-Br bromide	−1.34	−1.33	−1.33	−1.33, −1.43[e] −1.66[f]		−1.15
β-I bromide		−1.40	−1.39	−1.42	−0.19[f] −1.35	
γ-OH bromide	−1.28	−1.28	−1.30	−1.28, −1.42[e]		−1.16
γ-Cl chloride	−1.12	−1.12	−1.13	−1.13		−1.03
γ-Br bromide	−1.18	−1.17	−1.18	−1.18, −1.38[e] −1.70[f]		−1.10
γ-I iodide	−1.20	−1.19	−1.20	−1.23, −1.60[f]		−0.96

[a] Data from Colichman and O'Donavon.[67]
[b] For structures of salts see text (I and II).
[c] Ethanol.
[d] Tetraethylammonium bromide.
[e] Catalytic waves.
[f] These waves were assumed to be halogen reduction waves.

TABLE 9-4 Polarographic Half-Wave Potentials for the Reduction of Some Pyridinium Compounds at the DME

Compound	Solvent and background electrolyte	pH[a]	$E_{1/2}$[b] (V versus SCE)	Reference
N-Methylpyridinium	Aqueous 0.1 N NaOH		−1.47	71
hydroxide	50% Aqueous ethanol	7	−1.73	71
Trigonelline	Aqueous 0.1 N NaOH		−1.38	71
(N-methylnicotinic	50% Aqueous ethanol	7	−1.33	71
acid betaine)				
Homarine	Aqueous 0.1 N NaOH		−1.34	71
(N-methylpyridine-	50% Aqueous ethanol	7	−1.29	71
2-carboxylic acid)				
N-Methylnicotinamide	Aqueous 0.1 N NaOH		−1.38	71
	50% Aqueous ethanol	7	−1.05	71
Cetylpyridinium	20% Ethanolic		I, −0.38[c]	72
bromide	0.1 M NaOH		II, −1.10[c]	

[a] If known or if stated in original paper.
[b] Except where otherwise indicated.
[c] Reference electrode not known.

(b) a cyclic process involving catalytic hydrogen evolution (Eq. 7B). The latter is, of course, essentially the process thought to be responsible for the catalytic hydrogen wave of pyridine in aqueous solution.

$$(7A)$$

OR

$$(7B)$$

Since large-scale electrolyses of *n*-butyl bromide in pyridine at −30°C followed by evaporation of the solvent gave a blue-colored product, it was concluded that the reduction of pyridinium ion follows the free-radical scheme (Eq. 7A) with attack on the pyridine nucleus, rather than the cyclic catalytic hydrogen reaction (Eq. 7B).

Elving and co-workers[74] compared tetraethylammonium perchlorate (TEAP) with lithium perchlorate as supporting electrolyte for the polarographic reduction of pyridinium ion in pyridine. Although the pyridinium ion reduction wave appeared at more negative potential in the presence of TEAP, the process still appeared to be a one-electron, diffusion-controlled reaction. Some comparative data for the two background solutions are presented in Table 9-5.

Since the height of the pyridinium ion reduction wave can be used for analytical determination of the total acidity (of pK_a less than ca. 9) of a sample that could well contain considerable water, Elving and co-workers[75] examined the effect of water on the polarographic behavior of pyridinium ion in pyridine containing various amounts of water. The addition of water caused the $E_{1/2}$ of the pyridinium ion to shift to more negative values. More recently, Tsuji and Elving[76] examined in more detail the reduction of pyridinium ion formed by addition of a Brönsted acid to pyridine and found that when the supporting electrolyte consisted of large univalent ions, i.e., tetraethylammonium perchlorate, there were systematic relations between the $E_{1/2}$ of the pyridinium ion and the aqueous pK_a and structural type of the acid. When the background electrolyte contained a small cation that was a strong Lewis acid, e.g., Li^+, a pronounced leveling effect was observed, and acids of aqueous pK_a less than ca. 9 produced polarographic waves that had an identical half-wave potential, which was also that observed for acids of aqueous $pK_a <$ ca. 0 in TEAP solution.

A related observation was made by Cisak and Elving,[77] who found that the cathodic polarographic wave of $AlCl_3$ in pyridine was not due to reduction of Al^{3+}, but rather to reduction of the solvent. The Al^{3+} apparently functions by

TABLE 9-5   Half-Wave Potentials for the Polarographic Reduction of Pyridinium Ion Derived from Various Precursor Species in Pyridine Containing TEAP and LiClO$_4$ Background[a]

Pyridinium ion precursor	Background electrolyte	$E_{1/2}$ (V versus 1 N AgNO$_3$/Ag in pyridine)	I^b
Benzoic acid (0.1–10 mM)	TEAP	−1.62	1.40
Benzoic acid (1 mM)	LiClO$_4$	−1.31[c]	2.01
Acetic acid	LiClO$_4$	−1.36[c]	2.16
	TEAP	−1.72	1.82
Pyridinium nitrate	LiClO$_4$	−1.39[c]	2.01
	TEAP	−1.31	1.92

[a] Data from Hickey *et al.*[74]

[b] Diffusion current constant $I = i_1/Cm^{2/3}t^{1/6}$.

[c] In LiClO$_4$ a small prewave usually appeared which was apparently associated with some impurity present in the pyridine solvent.

polarizing the pyridine through formation of a Lewis acid–base adduct with the pyridine nitrogen (I, Fig. 9-6). Large-scale electrolyses of pyridine solutions of $AlCl_3$ at a mercury cathode resulted in the reduction of pyridine in a $2e$ process (i.e., one electron each to two pyridine molecules in the Lewis acid–base adduct, I, Fig. 9-6) to give a complex, possibly of the type shown in II, Fig. 9-6. On subsequent contact with water and air the latter was thought to give a complex mixture or copolymer of unsaturated polyamines and polyamides (III_a, III_b, IV_a, IV_b, Fig. 9-6). A significant point in the proposed reaction scheme is that a C–N bond in pyridine is disrupted.

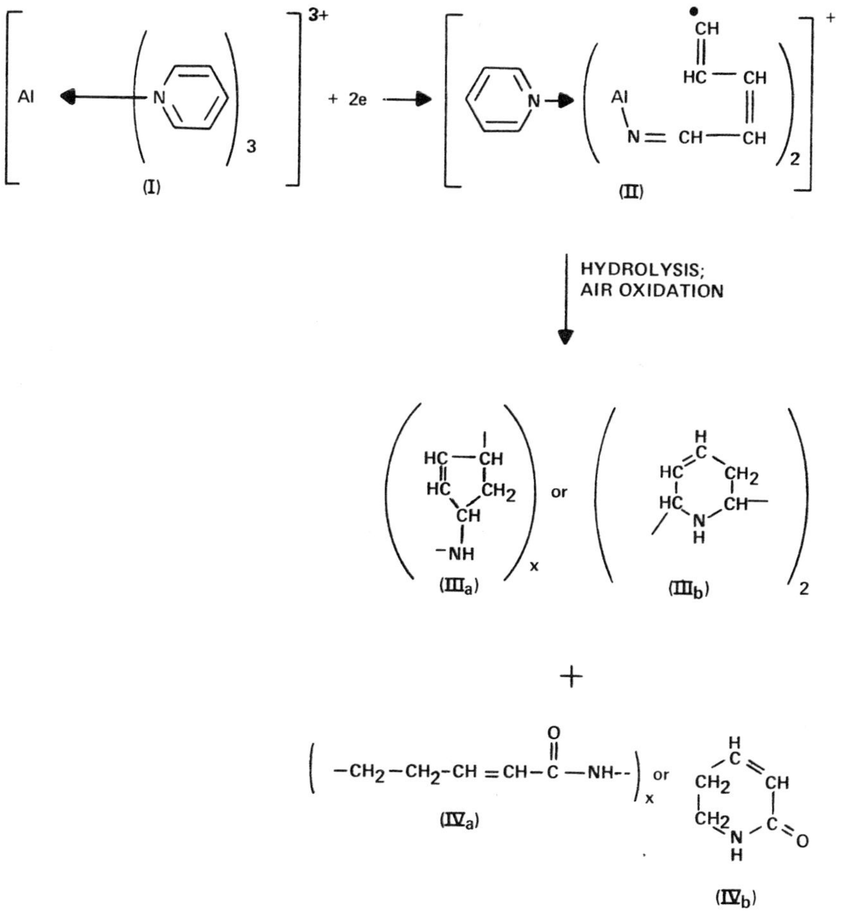

FIG. 9-6. Possible mechanism of electrochemical reduction of pyridine in the presence of $AlCl_3$.[77]

3. Summary

It is unfortunate that even with the considerable amount of data on the polarography of pyridine and pyridinium ion no clear-cut, unequivocal mechanisms can be presently written. The preponderance of evidence from studies on species in aqueous solution, however, does suggest that in buffered and unbuffered solutions at pH values close to 7, pyridine gives rise to a catalytic hydrogen reduction wave at $E_{1/2} \approx -1.7$ V versus SCE. It also appears that pyridine is adsorbed at the electrode surface at potentials where the latter wave occurs. The catalytic mechanism seems to be best explained by the reaction shown in Eq. 4. The wave that occurs in, e.g., 0.1 M KCl plus 1 mM HCl, at ca. -1.5 V versus SCE is undoubtedly primarily due to simple reduction of hydrogen ions, whether pyridine is present or not. However, in the presence of pyridine it may be that the presence of the organic material somehow affects the electrochemical process. At potentials of the more negative catalytic wave there is also a fair body of evidence to support the view that pyridine itself is reduced, either to a piperidine, a dimeric species, or a polymeric species. Whether this is due completely or in part to an electrochemical or chemical reduction does not appear to be certain.

In pyridine solution, pyridinium ion is undoubtedly reduced, yet again there is no clear-cut evidence to confirm whether the overall process is a catalytic reduction of hydrogen ion, or faradaic reduction of the pyridine nucleus, or both.

B. Electrochemical Oxidation

1. Pyridine

Only a few reports of the electrochemical oxidation of pyridine have appeared. Turner and Elving[78] found that a solution of pyridine, 0.1 M in tetraethylammonium perchlorate or tetra-n-butylammonium nitrate, showed a large oxidation peak ($E_p = +1.4$ V versus 1 M AgNO$_3$/Ag in pyridine) at a pyrolytic graphite electrode. The same peak was observed with pyridine in acetonitrile, the peak height increasing with concentration of pyridine. During a large-scale electrolysis of pyridine, the electrolysis solution turned yellowish-brown which, since the same sort of color change had been observed for the persulfate oxidation of pyridine,[79,80] indicated that a similar type of reaction might be occurring electrolytically. The mechanism of the electrochemical oxidation could therefore be that shown in Fig. 9-7, i.e., formation of a

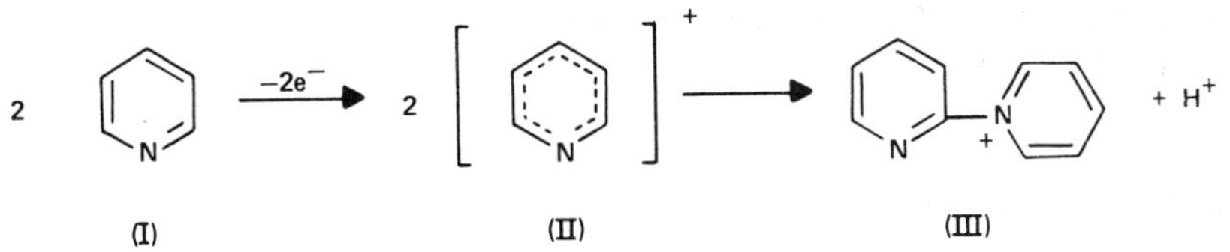

FIG. 9-7. Possible mechanism of electrochemical oxidation of pyridine at the pyrolytic graphite electrode.[78]

carbonium ion that stabilizes itself by forming a pyridinium adduct. There was little real evidence to support such a process.

2. Pyridine Alcohols

Pozdeeva and Novikov[81] have reported that a number of pyridine alcohols are oxidizable at a platinum electrode in aqueous solution at pH 7 (Table 9-6). The mechanism responsible for these processes is not known.

3. Piperidine

There are two reports on the electrochemical oxidation of piperidine. Kaganovich and Damaskin[82] found that at -0.51 V (versus SCE) piperidine, in KCl and Na_2SO_4 solutions, was oxidized at the DME. Analysis of the wave according to the Ilkovič equation led to the conclusion that a $6e$ oxidation of piperidine had occurred and pyridine was the product. Barradas and co-workers[83] have disagreed with this conclusion. Their data suggest that piperidine is oxidized by a $2e$ process to give piperidine N-oxide. Piperidine is very strongly adsorbed at both mercury and platinum electrodes.[82,83]

TABLE 9-6 Peak Potentials for the Oxidation of Pyridine Alcohols at a Pt Electrode in Britton–Robinson Buffer, pH 7[a]

Pyridine derivatives	E_p (V versus SCE)
2-CH_2OH	0.47
3-CH_2OH	0.47
4-CH_2OH	0.47
2-CH_2CH_2OH	0.46
4-CH_3-2-CH_2CH_2OH	0.48
6-CH_3-2-CH_2CH_2OH	0.48
4,6-$(CH_3)_2$-2-CH_2CH_2OH	0.47

[a] Data from Pozdeeva and Novikov.[81]

IV. ELECTROCHEMISTRY OF PYRIDINE DERIVATIVES

A. Pyridine *N*-Oxides

Foffani and Fornasari[84] found that in aqueous solution pyridine *N*-oxide shows a single two-electron polarographic reduction wave. Up to about pH 5–6 the wave was diffusion controlled, but it decreased in height and became kinetically controlled at higher pH; it disappeared at ca. pH 9.5. Accordingly, a mechanism was proposed in which the protonated compound (II, Fig. 9-8) was the polarographically reducible form of pyridine *N*-oxide (I, Fig. 9-8). The ultimate product of the reaction was pyridine (III, Fig. 9-8). The kinetic reaction at higher pH was the protonation reaction (Fig. 9-8a). A number of other workers have confirmed this mechanism.[85–88] Some typical half-wave potential data are presented in Table 9-7.[81,84–89]

Miroslav and Miloslav[90] electrochemically reduced pyridine *N*-oxide. 4-methylpyridine *N*-oxide, and 3-methylpyridine *N*-oxide and their quaternary methylated derivatives at an activated lead electrode in 20% aqueous sulfuric acid and obtained mixtures containing piperidine, 3-piperidine, and pyridine bases.

Kubota and co-workers[91] reduced pyridine *N*-oxide in dimethylformamide and concluded that the reaction involved two electrons. Electron spin resonance spectra of a solution undergoing electrolysis gave no evidence for a radical species. Typical half-wave potential data for the reduction of pyridine *N*-oxide in various nonaqueous systems are presented in Table 9-8.[81,91–93]

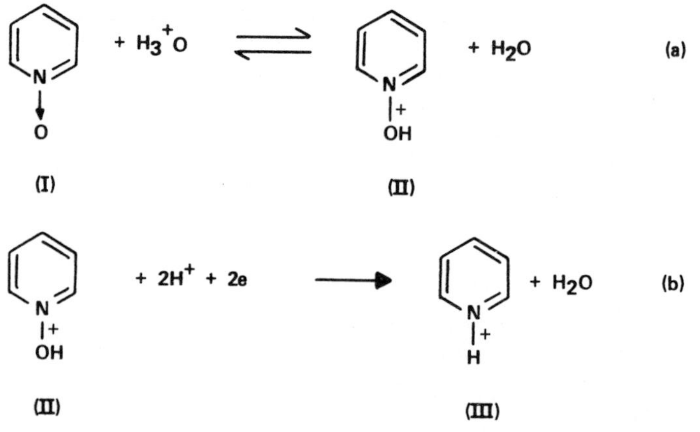

FIG. 9-8.   Mechanism of polarographic reduction of pyridine *N*-oxide.[84]

TABLE 9-7   Half-Wave   Potentials   for   the   Polarographic Reduction of Pyridine *N*-Oxide at the DME

pH	$E_{1/2}$ (V versus SCE)	Reference
0.1 *N* H_2SO_4	−0.902	87,88
2.35	−1.065	84
3.5	−1.279	89
5	−1.281	86
1.5−6.7	−0.849 − 0.0865 pH	85
7	−1.89	81

Anthoine and co-workers[92] found that pyridine *N*-oxide in DMF gave two reduction waves, both of which involved two electrons. The first (more positive) wave was reversible. It was suggested that the reaction responsible for the first wave proceeded through a series of rapid electron–proton transfers (I → V, Fig. 9-9) to give an intermediate having the structure of an unsaturated hydroxylamine derivative (V, Fig. 9-9). The height of the second wave was often less than that of the first wave. Hence, it was assumed that V (Fig. 9-9) could dehydrate to give pyridine (VI, Fig. 9-9) or undergo further electrochemical reduction to products whose nature was not investigated.

Kubota and Miyazaki[85,86] examined a large number of substituted pyridine *N*-oxides in aqueous solution (Table 9-9) and found that they all underwent essentially the same reduction reaction as the parent compound with respect to the *N*-oxide group. Thus, the protonated form of the *N*-oxide group was reduced directly to the substituted pyridine molecule. Above ca. pH 6 the wave height decreased and became kinetically controlled.

TABLE 9-8   Half-Wave Potentials for the Polarographic Reduction of Pyridine *N*-Oxide in Nonaqueous Media

Solvent	Supporting electrolyte	$E_{1/2}$ (V)	Reference electrode	Reference
Dimethylformamide	Tetra-*n*-propyl ammonium perchlorate	−2.297	SCE	91
90% Ethanol−water	Tetraethylammonium iodide	−2.12	SCE	81
75% Dioxane−water	Tetrabutylammonium iodide	−2.52	SCE	81
Dimethylformamide	Tetraethylammonium bromide	I, −1.80 II, −2.24	Hg pool	92
Ethanol	Acetate buffer	−1.364	SCE	93

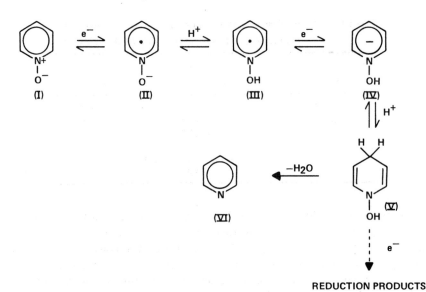

REDUCTION PRODUCTS

FIG. 9-9.   Proposed mechanism for the polarographic reduction of pyridine *N*-oxide in dimethylformamide.[92]

A plot of the $E_{1/2}$ value against the Hammett σ value of the substituent gave a reasonably good linear relation, and, in view of the positive slope of the plot, the reaction was considered to be nucleophilic. A linear relation was also obtained between the $E_{1/2}$ value and lowest empty molecular orbital energy.

A similar study[91] was carried out in dimethylformamide solution. Some substituted pyridine *N*-oxides were reduced by a one-electron process to give a free radical which could be detected by ESR spectroscopy. Others appeared to undergo a two-electron reduction (on the basis of the diffusion current constant) to give a product that did not give an ESR signal. Some typical $E_{1/2}$ data, along with the diffusion current constant and an indication as to the ESR response upon controlled potential electrolysis, are presented in Table 9-10. The peak height of AC polarograms and the half-wave potential of the first reduction waves were found to be approximately linearly related to the radical stability; i.e., generally those compounds that gave an ESR signal were characterized by a large alternating current in AC polarography and a relatively positive half-wave potential. The half-wave potentials were in general linearly related to the calculated energy of the lowest empty molecular orbital and also with the Hammett σ value. Miyazaki and Kubota[94] subsequently reported that 4-nitro-pyridine *N*-oxide gave two polarographic reduction waves in dimethylformamide containing 0.1 M tetra-*n*-propylammonium perchlorate at $E_{1/2} = -0.80$ and

TABLE 9-9 Half-Wave Potentials for the Polarographic Reduction of Substituted Pyridine *N*-Oxides in Aqueous Solution, pH 5.0[a]

Substituent	$E_{1/2}$ (V versus SCE)
4-NH$_2$	-1.465
4-NHOH	-1.434
4-OCH$_3$	-1.416
4-OC$_2$H$_5$	-1.396
4-CH3	-1.357 ($E_{1/2} = -0.883 - 0.0948$ pH)
4-C$_2$H$_5$	-1.331
4-NHCOCH$_3$	-1.189
4-Cl	-1.174
4-Br	-1.094
4-COOC$_2$H$_5$	-0.862
4-CN	-0.857
4-NO$_2$	$-0.204, -1.436$[b]
3-NH$_2$	-1.280
3-OCH$_3$	-1.168
3-CH$_3$	-1.272
3-C$_2$H$_5$	-1.246
3-NHCOCH$_3$	-1.129
3-F	-1.144
3-Cl	-1.071
3-Br	-1.040
3-COOC$_2$H$_5$	-0.992
3-CN	-0.943

[a] Data from Kubota and Miyazaki.[86]

[b] The wave at more positive potential was attributed to the reduction of the NO$_2$ group.

-1.66 V versus aqueous SCE. By use of ESR techniques it was clearly demonstrated that the first polarographic wave involved one electron and formed a radical anion that was quite stable.[95] The second, more negative reduction wave was due to further reduction of the radical anion to a dianion.

Fornasari and Foffani[96] examined the polarographic reduction of the *N*-oxides of picolinic, nicotinic, and isonicotinic acids in aqueous solution. Below pH 3 two well-defined reduction waves were observed (waves I and II). With increasing pH both of these waves decreased in height and reached a minimum value at pH 7–8. Above this pH a third wave (wave III) appeared whose height increased with pH until it reached a value approximately equal to that of the waves in acid solution. Both wave I (acid solution) and wave III (alkaline solution) were proposed to involve two electrons and two protons. In view of the variations of the wave heights with pH, i.e., maximum in acid and

TABLE 9-10   Polarographic Data for the Reduction of Substituted Pyridine
N-Oxides in Dimethylformamide[a,b]

Compound	$E_{1/2}$ (V versus SCE)	I^c	Electrolysis potential[d,e] (V versus SCE)
Pyridine *N*-oxide	−2.297	4.94	−2.60 (no)
4-Methylpyridine *N*-oxide	−2.375	4.77	−2.65 (no)
4-Ethylpyridine *N*-oxide	−2.370	3.79	−2.58 (no)
4-Chloropyridine *N*-oxide	−1.889	4.24	−2.10 (no)
4-Methoxypyridine *N*-oxide	−2.402	3.51	−2.60 (no)
4-Ethoxypyridine *N*-oxide	−2.445	3.50	−2.65 (no)
4-Carbethoxypyridine *N*-oxide	−1.606	2.05	−1.85 (yes)
4-Cyanopyridine *N*-oxide	−1.557	1.99	−1.75 (yes)
4-Nitropyridine *N*-oxide	−0.768	2.04	−1.30 (yes)
3-Methoxypyridine *N*-oxide	−2.315	3.48	−2.52 (no)
3-Carbethoxypyridine *N*-oxide	−1.670	2.07	−1.90 (no)
3-Cyanopyridine *N*-oxide	−1.667	2.05	−1.90 (no)
4-Nitropyridine nitro-^{15}N-*N*-oxide	−0.792	2.04	−1.30 (yes)
4-Cyanopyridine 2,6-d_2-*N*-oxide	−1.595	2.13	−1.75 (yes)
4-Nitropyridine 2,6-d_2-*N*-oxide	−0793	2.04	−1.10 (yes)

[a] Data from Kubota *et al.*[91]

[b] Supporting electrolyte, 0.1 *M* tetra-*n*-propylammonium perchlorate.

[c] $I = i_1/Cm^{2/3}t^{1/6}$.

[d] For controlled potential electrolysis.

[e] Yes indicates that a positive ESR signal was obtained; no indicates a negative ESR signal was obtained.

alkaline solutions and minimum in neutral medium, it was proposed that only the cationic (I, Fig. 9-10) and anionic (III, Fig. 9-10) forms of the acids were electroactive. Thus, wave I was proposed to involve reduction of the cation (I, Fig. 9-10) to the protonated form of the pyridinecarboxylic acid (IV, Fig. 9-10). Although not conclusively proved, wave II was thought to be due to a further

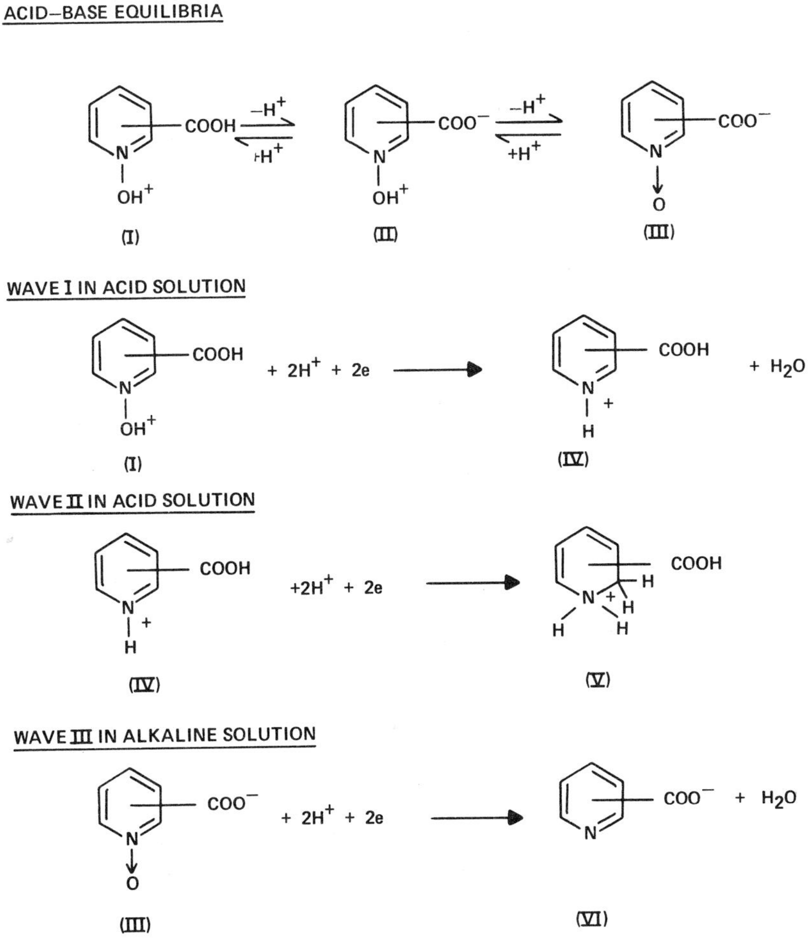

ACID–BASE EQUILIBRIA

(I)   (II)   (III)

WAVE I IN ACID SOLUTION

(I)   (IV)

WAVE II IN ACID SOLUTION

(IV)   (V)

WAVE III IN ALKALINE SOLUTION

(III)   (VI)

FIG. 9-10. Mechanism of polarographic reduction of the *N*-oxides of picolinic, nicotinic, and isonicotinic acids in aqueous solutions.[96]

$2e–2H^+$ reduction of IV (Fig. 9-10) to a dihydropyridinium ion (V, Fig. 9-10). Wave III, observed only in alkaline solution, was thought to be reduction of the anion (III, Fig. 9-10) to the pyridine carboxylate ion (VI, Fig. 9-10). Half-wave potential data are presented in Table 9-11.

Laviron and co-workers[97] studied the polarographic reduction of 2-, 3-, and 4-formylpyridine *N*-oxides in aqueous solution. 3-Formylpyridine *N*-oxide (I, Fig. 9-11) gave rise to two principal reduction processes. The first consisted of

TABLE 9-11   Half-Wave Potentials for the Polarographic Reduction of Pyridinecarboxylic Acid *N*-Oxides[a]

	$E_{1/2}$ (V versus SCE)								
	Picolinic acid wave			Nicotinic acid wave			Isonicotinic acid wave		
pH	I	II	III	I	I	III	I	II	III
1.70	−0.702	−0.962		−0.835	−1.08		−0.67	−0.89	
4.00	−0.861	−1.108		−1.100			−0.91	−1.03	
6.85	−1.199	−1.373		−1.414			−1.30		
7.60	−1.237	−1.419							−1.51
9.65			−1.729			−1.739			−1.54

[a] Data from Fornasari and Foffani.[96]

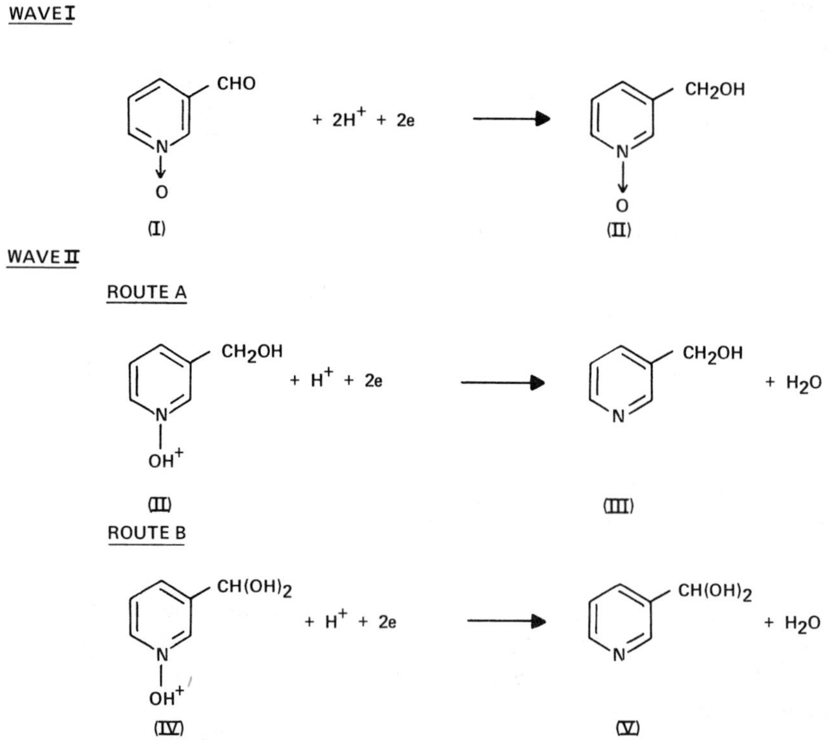

WAVE I

WAVE II

ROUTE A

ROUTE B

FIG. 9-11.   Overall mechanism for polarographic reduction of 3-formylpyridine *N*-oxide.[97]

the reduction of the carbonyl group to 3-hydroxymethylpyridine N-oxide (II, Fig. 9-11). The second process consisted of reduction of the protonated N-oxide group via two routes. One route was by reduction of 3-hydroxymethylpyridine N-oxide (II, Fig. 9-11) to give 3-hydroxymethylpyridine (III, Fig. 9-11); the other was by reduction of the hydrated form of 3-formylpyridine N-oxide (IV, Fig. 9-11) to give 3-formylpyridine (V, Fig. 9-11).

In the case of the 2- and 4-formylpyridine N-oxides in acid solution, the reduction of the N-oxide and carbonyl functions proceeded simultaneously in a four-electron process to give the appropriate hydroxymethylpyridine derivative (II, Fig. 9-12). In basic solution the precise mechanism for the wave was not elucidated. The second process that occurred was apparently due to reduction of the hydrated aldehyde with a protonated N-oxide group (III, Fig. 9-12) to give the appropriately hydrated pyridine aldehyde (IV, Fig. 9-12). Owing to the involvement of aldehyde hydration equilibria and protonation equilibria, the detailed behavior of each of the waves was very complex. The original paper should be consulted for the variations of $E_{1/2}$ and limiting currents with pH and for a complete interpretation of the protonation, hydration, and other phenomena.

Laviron, Gavasso, and Pay[98] subsequently examined the polarography of the acetyl- and benzoylpyridine N-oxides, which showed very similar behavior to that of the formylpyridine N-oxides. Thus, when the substituent was in the 2 or

WAVE I (ACID MEDIUM)

WAVE II

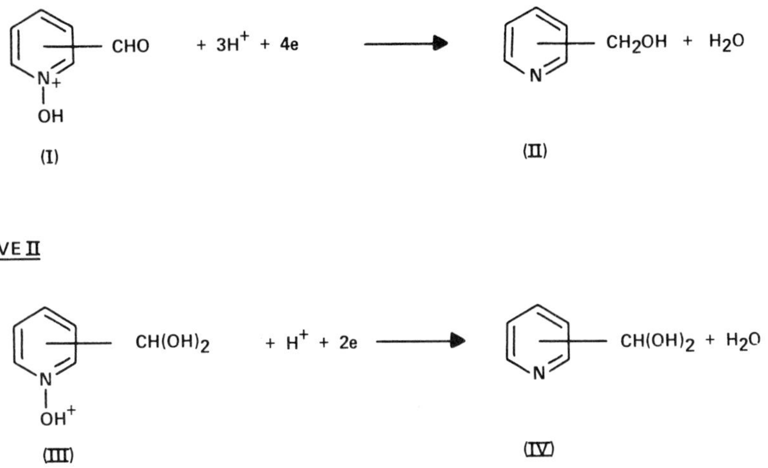

FIG. 9-12. Overall mechanism for polarographic reduction of 2- and 4-formylpyridine N-oxides.[97]

4 position, the *N*-oxide and carbonyl groups were reduced simultaneously, while the 3-isomers were reduced first at the carbonyl and then at the *N*-oxide functions. Again, protonation and hydration phenomena made the overall polarographic behavior complex.

El-Khiami and Johnson[99] found that bis(2-pyridyl) disulfide di-*N*-oxide (I, Fig. 9-13) was irreversibly reduced at the DME to give 2-thiopyridine *N*-oxide (III, Fig. 9-13). The rate-controlling step was thought to involve a single electron and proton to give an unstable dimeric free-radical intermediate (II, Fig. 9-13) that very rapidly underwent a further one electron, one proton reduction to give III (Fig. 9-13). The values of the electron-transfer coefficient, α, and the formal heterogeneous rate constant, $k_{f,h}^0$,[100] were measured at various pH values. Some additional half-wave potential data on substituted pyridine *N*-oxides are presented in Table 9-12.

B. Pyridine Aldehydes and Ketones

Pyridine aldehydes have been studied polarographically by a number of workers.[101−112] The general observations of most workers seem to agree quite well and are probably best summarized in the report of Laviron and Tirouflet.[112] They found that for the 2-, 3-, and 4-pyridine aldehydes there was a considerable variation in the limiting current with pH (Fig. 9-14). In actuality

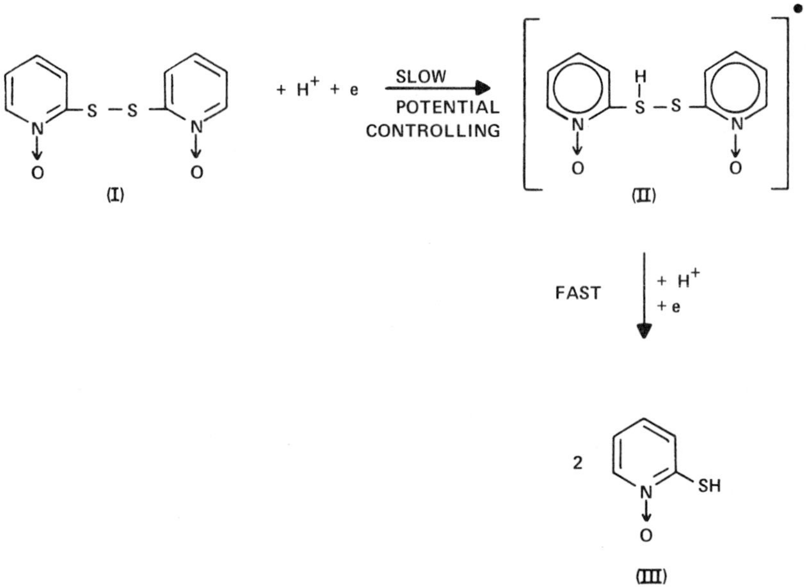

FIG. 9-13.   Proposed mechanism for the polarographic reduction of bis(2-pyridyl)-disulfide di-*N*-oxide.[99]

TABLE 9-12 Half-Wave Potential Data for Polarographic Reduction of Some Substituted Pyridine *N*-Oxides

Compound	pH or Solvent and supporting electrolyte system	$E_{1/2}$ (V)	Reference
Bis(2-pyridyl)	2.0	-0.17^a	99
disulfide di-*N*-oxide	3.2	-0.23^a	
	4.8	-0.33^a	
	5.5	-0.38^a	
	7.3	-0.42^a	
	9.3	-0.42^a	
4-Phenoxypyridine *N*-oxide	3.5	-1.282^b	89
4-Nitropyridine *N*-oxide	3.5	$-0.289,^c$ -1.662	89
2-Methylpyridine *N*-oxide	7	$-^b$	81
	90% EtOH–H$_2$O; 0.1 *M* TEAI^d	-2.07^b	81
	75% Dioxane–H$_2$O; 0.1 *M* TBAI^e	-2.48^b	81
4-Methylpyridine *N*-oxide	90% EtOH–H$_2$O; 0.1 *M* TEAI	-2.07^b	81
	75% Dioxane–H$_2$O; 0.1 *M* TBAI	-2.57^b	81
4-Ethylpyridine *N*-oxide	90% EtOH–H$_2$O; 0.1 *M* TEAI	-2.11^b	81
	75% Dioxane–H$_2$O; 0.1 *M* TBAI	-2.71^b	81
2,3-Dimethylpyridine *N*-oxide	90% EtOH–H$_2$O; 0.1 *M* TEAI	-1.98^b	81
	75% Dioxane–H$_2$O; 0.1 *M* TBAI	-2.58^b	81
2,4-Dimethylpyridine *N*-oxide	90% EtOH–H$_2$O; 0.1 *M* TEAI	-2.13^b	81
	75% Dioxane–H$_2$O; 0.1 *M* TBAI	-2.33^b	81

[a] Versus normal hydrogen electrode (add -0.2412 V at 25°C to convert to SCE).
[b] Versus saturated calomel electrode.
[c] Most positive wave probably due to reduction of the $-NO_2$ group.
[d] Tetraethylammonium iodide.
[e] Tetra-*n*-butylammonium iodide.

the reduction took place in a single 2*e* wave for pyridine 3-aldehyde but as two very close 1*e* waves for the 2 and 4 derivatives. The fundamental electron-transfer processes appear to be a straightforward reduction of the aldehyde function to the corresponding alcohol.

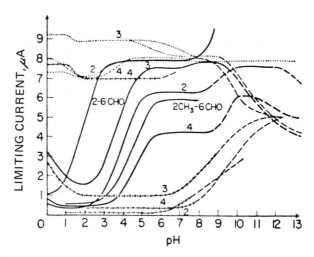

FIG. 9-14. Variation of the limiting current with pH for carbonyl derivatives of pyridine.[112] (———) 2-, 3-, and 4-Pyridine aldehydes, pyridine 2,6-dialdehyde, and 2 methylpyridine 6-aldehyde; (+) 2-, 3-, and 4-*N*-methylpyridine aldehydes and *N*-methyl-pyridine 4-ketone; (•) 2-, 3-, and 4-pyridine ketones. (Reprinted with permission of Pergamon Press, New York.)

The overall behavior of the limiting current for the aldehydes is more clearly shown in Fig. 9-15, where four principal zones have been described. In zone I (pH < 1) the current has a normal value in very strongly acidic solution but decreases as pH 1 is approached. This behavior was interpreted as an acid-catalyzed dehydration. In other words, an equilibrium exists in solution between the hydrated aldehyde and the free aldehyde (Eq. 8) where the dehydration

$$\text{(pyridine)}-CH(OH)_2 \underset{}{\overset{H^+}{\rightleftharpoons}} \text{(pyridine)}-CHO + H_2O \qquad (8)$$

reaction is catalyzed by hydrogen ions. Only the free aldehyde species was thought to be electrochemically reducible. Zone II (1 < pH < 7 or 8) is divided into three subzones: II_a, II_b, and II_c. The current increased as the pH increased from II_a to II_b and attained a plateau in zone II_c (ca. pH 6–8). Comparison of quaternary aldehydes (N-methylated) over the zone II pH region indicated that they gave only very small currents (Fig. 9-14) until about pH 6–7. This indicated that the quaternary aldehydes were almost completely hydrated and that the extent of hydration was due to the presence of the

$$-N^+CH_3$$

group. It was therefore concluded that, when pyridine aldehydes are not

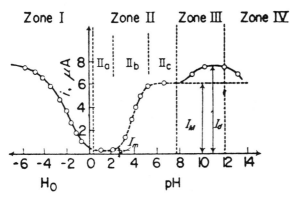

FIG. 9-15.   Variation of the limiting current for pyridine 2-aldehyde as a function of pH and acidity H_0. The dashed line represents a theoretical curve based upon an acid–base equilibrium in zone II. The circles represent experimental points.[112] (Reprinted with permission of Pergamon Press, New York.)

protonated, they are only slightly hydrated but that the pyridinium ion resulted in very extensive hydration. The difference in behavior of the quaternary and nonquaternary aldehydes in the zone II region was therefore proposed to be due to the fact that, for the quaternary compounds, hydration was always extensive and more or less independent of pH, since the carbonyl group remains under the influence of the

$$-N^+CH_3$$

group. However, for the nonquaternary aldehyde some of the aldehyde is in the nonprotonated form and is only slightly hydrated, and a portion is in the pyridinium ion form that is strongly hydrated. Accordingly, the relative amount of free aldehyde and therefore the magnitude of the limiting current depends on the pH. At low pH the pyridinium ion is predominant, and hence the hydrated, nonelectroactive aldehyde is the principal solution species, which results in a small limiting current. As the pH increases the amount of pyridinium ion decreases and the current increases. In zone III (pH > 7) the current increased further owing to a base-catalyzed dehydration. In zone IV in very alkaline solution the current decreased owing to ionization of the free aldehyde.

Volke[101–105,109] in particular has reported a similar type of explanation but in addition measured the hydration equilibrium constant by use of an oscillopolarographic method.[104] A graphical summary of the variation of half-wave potentials for reduction of a number of pyridine aldehydes is presented in Fig. 9-16.

Nakaya[113] has also examined the polarography of pyridine aldehydes and related compounds. Controlled potential electrolysis of pyridine 4-aldehyde (I,

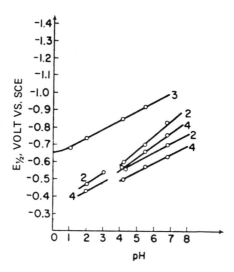

FIG. 9-16.   Variation of half-wave potential with pH for some pyridine aldehydes. 2, 3, and 4: Pyridine 2-, 3-, and 4-aldehdyes, respectively.[112] (Reprinted with permission of Pergamon Press, New York.)

Eq. 9) at pH 7 and 3 on the first wave confirmed its 1e nature, and a dimer (II, Eq. 9) was assumed to be the product.

Fornasari and co-workers[106,107] found that the two 1e reduction waves observed for the pyridine 2- and 4-aldehydes in aqueous solutions were not observed in 50% aqueous ethanol solution, but rather all the pyridine aldehydes gave a single 2e wave. A summary of half-wave potential data is presented in Table 9-13. It can be seen from these data that the half-wave potentials shift linearly more negative with pH $[dE_{1/2}/d(\text{pH}) \approx 65\text{--}75 \text{ mV}]$ up to about pH 8, when there is a sudden rise. The $E_{1/2}$ then remains practically constant with increasing pH. Although no definitive explanation of this effect was advanced, it was thought to be associated with the charge (or in reality lack of charge) associated with the heterocyclic nitrogen at high pH. Attempts to correlate the half-wave potential for the pyridine aldehydes in acid solution with the quantum

TABLE 9-13 Half-Wave Potentials[a] of Pyridine Aldehydes in 50% Ethanol Solution[b]

Pyridine 2-aldehyde		Pyridine 3-aldehyde		Pyridine 4-aldehyde	
pH	$-E_{1/2}$	pH	$-E_{1/2}$	pH	$-E_{1/2}$
0.0	0.287	0.0	0.558	0.0	0.267
2.45	0.471	2.45	0.726	2.45	0.423
2.90	0.496	2.80	0.737	2.70	0.435
3.05	0.536	2.90	0.746	3.00	0.450
3.55	0.550	3.10	0.757	3.65	0.511
3.90	0.582	3.60	0.814	4.65	0.557
5.10	0.665	3.90	0.825	5.60	0.615
5.20	0.682	5.10	0.891	6.15	0.635
6.00	0.745	6.60	1.023	6.35	0.688
6.80	0.820	7.55	1.130	7.75	0.810
7.40	0.890	9.10	1.270	9.20	0.994
9.10	1.087	10.60	1.328	10.25	1.040
10.60	1.142	11.30	1.343	10.45	1.045
11.15	1.150	11.90	1.331	11.05	1.054
11.90	1.162	12.50	1.342	11.63	1.056
12.50	1.171			12.50	1.067

[a] In volts versus the SCE.
[b] Data from Fornasari *et al.*[106]

mechanically calculated π-energy difference between the initial and final state of the primary reduction were not successful. The reasons for this were thought to be the irreversibility of the electrode process and in some way the positive charge carried by the heterocyclic nitrogen atom. At high pH, where the nitrogen does not carry a positive charge, a reasonably good correlation was obtained between the theoretically calculated $E_{1/2}$ and the experimental value.

Tirouflet and co-workers[110] also found that the pyridine aldoximes were polarographically reducible. The pyridine 2-, 3-, and 4-aldoximes all gave a $4e$ wave, the half-wave potentials for which were sufficiently different (Table 9-14) to allow analytical determination of mixtures of the three compounds. Pyridine 4-aldoxime also gave a second wave at more negative potentials which was probably due to a $2e$ reduction. Although the mechanism was not elucidated in detail, the first $4e$ wave involved reduction of the aldoxime to the primary amine as proposed by Volke and co-workers.[114]

The system 2,2'-pyridoin (**III**) − 2,2'-pyridil (**IV**) has been studied polarographically by Holubek and Volke.[115,116] Compound **III** is rapidly air oxidizable in water and so the compounds were accordingly examined in 50% aqueous ethanol solutions buffered to the desired pH. The reduction of pyridoin

TABLE 9-14   Half-Wave   Potentials   for   the   Polarographic Reduction of Pyridine Aldoximes[a]

Compound	pH	$E_{1/2}$ (V versus SCE)
Pyridine 2-aldoxime	2	−0.69
	4.2	−0.73
Pyridine 3-aldoxime	2	−0.82
	4.2	−0.92
Pyridine 4-aldoxime	2	−0.64[b]
	4.2	−0.65[b]

[a] Data from Tirouflet *et al.*[110]

[b] A second wave at more negative potential also observed.

            (III)   **(IV)**

proceeded by a single 2e wave (wave II, Fig. 9-17), which decreased in height at pH below 3 and greater than 8.7 in the shape of a dissociation curve. In alkaline solution wave II was slowly replaced by wave III (Fig. 9-17), which occurred at more negative potential, the sum of the heights of both waves remaining almost constant. A large catalytic hydrogen wave was observed at even more negative potential. A small adsorption prewave was also observed at the foot of wave II. Pyridoin also gave rise to an anodic wave (wave I, Fig. 9-17). The reduction of 2,2′-pyridil proceeded in two main stages. The height of the more positive wave (wave A, Fig. 9-17) was constant at pH above 3.5 but decreased at lower pH. The half-wave potential for wave A was pH dependent up to pH 7.2, when it became almost pH independent. Although coulometry was not carried out, comparison of the wave height for wave A with compounds having known n values revealed that two electrons were consumed in the process. The second wave of 2,2′-pyridil corresponded to wave II of 2,2′-pyridoin with respect to the pH dependence of $E_{1/2}$ and wave height and the formation of wave III in alkaline solution. The mechanism proposed for these processes was that wave I of 2,2′-pyridoin (I, Fig. 9-18) corresponds to its oxidation to 2,2′-pyridil (II, Fig. 9-18). The cathodic waves II and III correspond to the 2e reduction of the protonated and nonprotonated species, respectively, to the corresponding glycol (III, Fig. 9-18). Wave A of 2,2′-pyridil corresponds to formation of 2,2′-pyridoin, while the two more negative waves are exactly the same as for

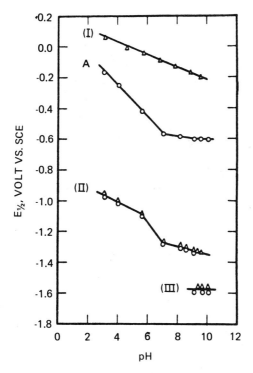

FIG. 9-17. Half-wave potentials of 2,2′-pyridoin (△) and 2,2′-pyridil (○) versus SCE.[116] (Reprinted with permission of Pergamon Press, New York.)

waves II and III of 2,2′-pyridoin. By use of the Kalousek commutator[117] method it was shown that the 2,2′-pyridoin–2,2′-pyridil redox system was irreversible. The decrease in the wave heights for both compounds in acidic solutions was explained by hydration of the ketonic groups.

The carbonyl group present in the isomeric acetylpyridines has been found to be reducible to the corresponding pinacols by controlled potential reduction in 50% ethanolic $1\,M$ KOH at -1.6 V versus SCE at mercury. Using such an electrochemical method, Allen and co-workers[118,119] were able to synthesize 2-methyl-1,2-di-3′-pyridyl-1-propanone (I, Eq. 10) by rearranging the pinacol

$$+ 2H^+ + 2e \longrightarrow \qquad (10)$$

(I)   (III)

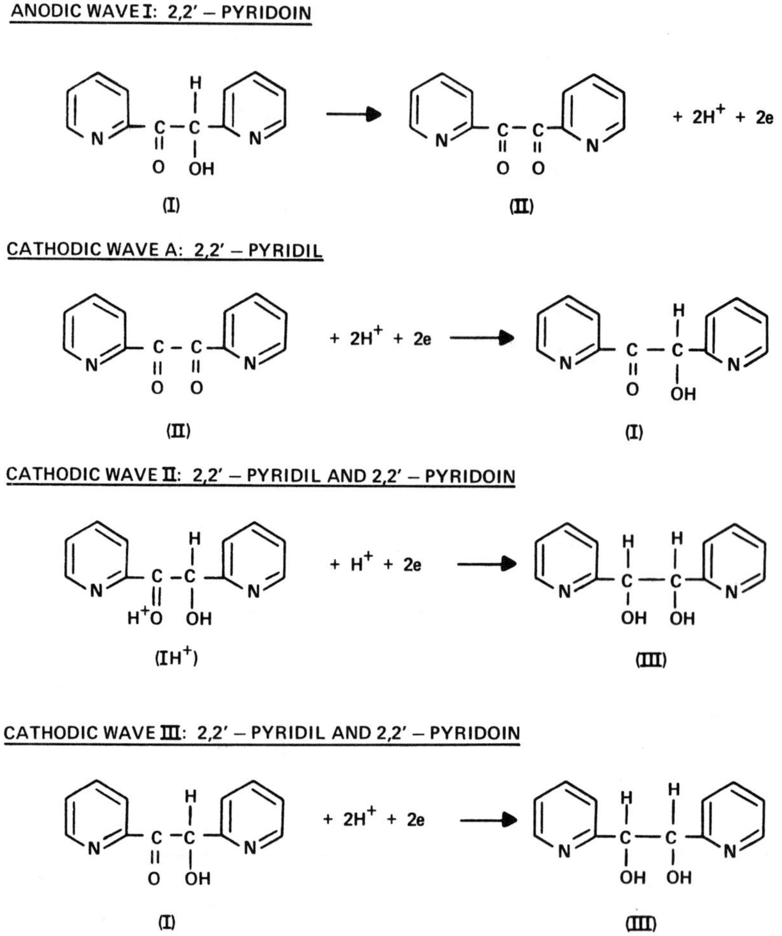

ANODIC WAVE I: 2,2′ – PYRIDOIN

+ 2H$^+$ + 2e

(I)   (II)

CATHODIC WAVE A: 2,2′ – PYRIDIL

+ 2H$^+$ + 2e

(II)   (I)

CATHODIC WAVE II: 2,2′ – PYRIDIL AND 2,2′ – PYRIDOIN

+ H$^+$ + 2e

(IH$^+$)   (III)

CATHODIC WAVE III: 2,2′ – PYRIDIL AND 2,2′ – PYRIDOIN

+ 2H$^+$ + 2e

(I)   (III)

FIG. 9-18.   Proposed mechanism for the polarographic reduction of 2,2′-pyridoin and 2,2′-pyridil. (I) 2,2′-Pyridoin; (II) 2,2′-pyridil, (IH$^+$) protonated form of 2,2′pyridoin; (III) bipyridyl-2,2′-diol.[116]

with concentrated sulfuric acid. Stárka and Buben[120] observed a single polarographic reduction wave for this compound due to reduction of the carbonyl group to give the corresponding alcohol (II, Eq. 10). Allen,[121] however, found that the 2e wave observed by the latter workers was followed by a second, more negative wave, which he proposed to be a 4e process resulting in the reduction of one double bond in each pyridine moiety. Volke[122] has more definitively demonstrated that in fact the second wave is, as might be expected

TABLE 9-15 Half-Wave Potentials for the Polaro-graphic Reduction of 2-Methyl-1,2-di-3'-pyridyl-1-propanone[a]

| pH | $E_{1/2}$ (V versus SCE) for wave | |
	I	II
2.2	−0.759	−1.229
4.0	−0.850	−1.222
6.0	−1.000	−1.303
8.10	−1.151	−1.483

[a] Data from Allen.[121]

for pyridine derivatives, due to a catalytic hydrogen ion reduction. Typical half-wave potential data are presented in Table 9-15.

Volke[123] has also studied the polarography of all three isomeric acetyl-pyridines, which he found were all irreversibly reduced. The least complicated polarographic behavior was observed with 3-acetylpyridine, which gave a single 2e wave in acidic aqueous buffers. Occasionally, an adsorption prewave was noted. As the pH was increased, the wave height decreased until, above pH 9, a wave of about one-half the height of that in acidic solution was obtained. In the case of 2-acetylpyridine, again a single 2e wave was observed in acidic solution which also decreased to a 1e wave at pH > 9. A second, more negative 1e wave was also observed. 4-Acetylpyridine gave a 2e reduction wave in acidic solution which may have been due to two very close 1e waves. At some pH values as many as three polarographic waves were observed for the 4-isomer, and the behavior was complex and probably associated with adsorption phenomena. In acidic solution the overall reduction of acetylpyridines was regarded as a 2e reduction of the acetyl group (I, Fig. 9-19) to the corresponding secondary alcohol (II, Fig. 9-19). In alkaline solution it would appear that the 1e reduction resulted in the formation of a free radical (III, Fig. 9-19) that dimerized to a pinacol (IV, Fig. 9-19). In fact, Ono[124] has reported that in alkaline solution a pinacol could be isolated after controlled potential reduction of acetylpyridines. Half-wave potential data reported by Volke[123] are presented in Fig. 9-20.

C. Nitropyridines

Holubek and Volke[125] found that the isomeric nitropyridines were reduced successively, first in a 4e wave and then in a more negative 2e wave. The height

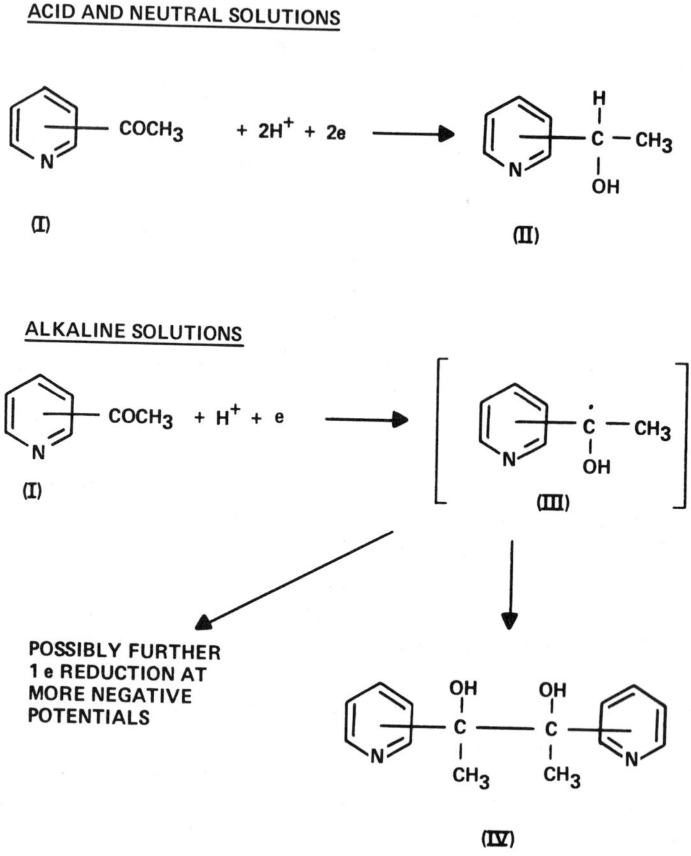

ACID AND NEUTRAL SOLUTIONS

ALKALINE SOLUTIONS

FIG. 9-19.   Proposed mechanism for the polarographic reduction of acetylpyridines in aqueous solution.[123]

of the more negative wave was lowered with increasing pH of the solution. The first wave (wave I) corresponded to reduction of the nitro group to the corresponding hydroxylamine derivative (Fig. 9-21). The second wave was due to reduction of the protonated form of the latter to a primary amine. The decrease in height of wave II at high pH was attributed to dissociation of the reducible cation of the pyridine hydroxylamine. Half-wave potentials in aqueous solution and in solutions containing various amounts of ethanol are presented in Table 9-16.

Tomasik[126] has measured the half-wave potentials for polarographic reduction of a series of 2-substituted 5-nitropyridines in dimethylformamide (Table 9-17). Simply on the basis of the observed polarographic limiting current

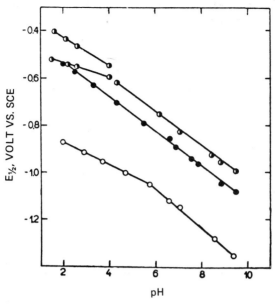

FIG. 9-20. Variation of half-wave potential with pH for acetylpyridines.[123] (●) First wave of 2-acetylpyridine (the second wave at pH > 9 occurs at potentials 150 mV more negative than the first wave); (○) 3-acetylpyridine; (◐) 4-acetylpyridine. (Reprinted with permission of Academia Publishing House, Prague.)

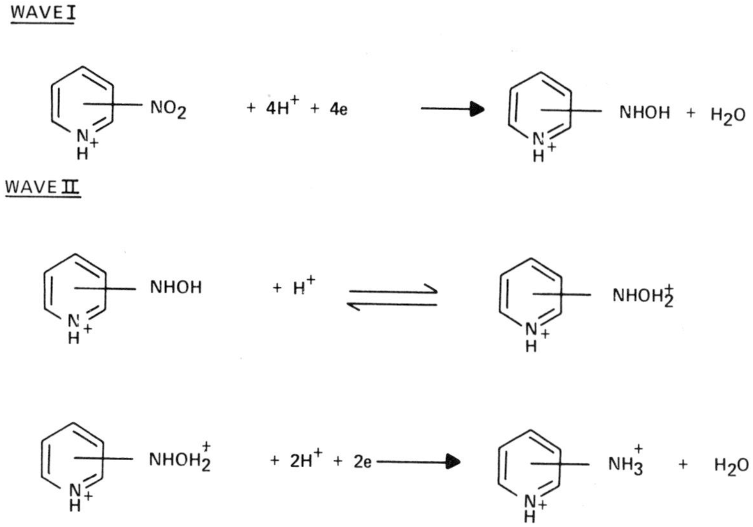

WAVE I

Pyridine—NO_2 + $4H^+$ + 4e ⟶ Pyridine—$NHOH$ + H_2O

WAVE II

Pyridine—$NHOH$ + H^+ ⇌ Pyridine—$NHOH_2^+$

Pyridine—$NHOH_2^+$ + $2H^+$ + 2e ⟶ Pyridine—NH_3^+ + H_2O

FIG. 9-21. Proposed mechanism of polarographic reduction of nitropyridines.[125]

TABLE 9-16   Half-Wave Potentials for the Polarographic Reduction of Nitropyridines[a,b]

Compound	$E_{1/2}$ (V versus SCE) at pH			
	4.0	5.1	7.0	9.6
Aqueous solution				
2-NO$_2$	−0.17	−0.23	−0.32	−0.49
3-NO$_2$	−0.22	−0.28	−0.37	−0.54
4-NO$_2$	−0.12	−0.19	−0.27	−0.40
10% Ethanol solution				
2-NO$_2$	−0.17	−0.25	−0.37	−0.53
3-NO$_2$	−0.23	−0.31	−0.42	−0.59
4-NO$_2$	−0.12	−0.19	−0.28	−0.42
50% Ethanol solution				
2-NO$_2$	−0.32	−0.44	−0.54	−0.60
3-NO$_2$	−0.37	−0.50	−0.59	−0.69
4-NO$_2$	−0.16	−0.28	−0.38	−0.44

[a] Data from Holubek and Volke.[125]
[b] Information available only for the first wave, i.e., for nitro group reduction.

TABLE 9-17   Half-Wave Potentials for the First
Polarographic Waves of 2-Substi-
tuted 5-Nitropyridines in Dimethylformamide[a,b]

2 Substituent	$E_{1/2}$ (V versus SCE)
H	−0.895
F	−0.890
Cl	−0.830
Br	−0.825
I	−0.835
CH$_3$	−0.955
OH	−1.070
OCH$_3$	−1.045
SCH$_3$	−0.950
NHCOCH$_3$	−0.985
NH$_2$	−1.165
NHCH$_3$	−1.215
N(CH$_3$)$_2$	−1.210
NHNH$_2$	−1.110

[a] Data from Tomasik.[126]
[b] Supporting electrolyte, 0.1 M NaNO$_3$

and wave slope it was concluded that the first wave in all cases was essentially a 1e reduction with formation of a radical anion. The reduction site was apparently the nitro group at position 5. Linear correlations between the $E_{1/2}$ values for the first wave of these compounds and Hammett σ values and various UV and IR absorption bands were found. Many compounds apparently gave further waves, but half-wave potential or mechanistic data were not presented.

D. Halogenopyridines

Among the halogenopyridines the 4-chloro-, 2- and 4-bromo-, and all three isomeric iodopyridines are reducible at the DME.[127] In acid solutions the reduction has been proposed to proceed in a single 2e wave (wave I), the height of which decreases with increasing pH. At pH values above 6.5 wave I is followed by a catalytic hydrogen wave. As wave I for the iodopyridines decreases in height with increasing pH, a more negative wave (wave II) appears at more positive potentials than the catalytic wave, the sum of the currents for wave I and wave II being equal to the height of wave I at low pH. The half-wave potential for wave II is independent of pH. The products of wave I for all the halogenopyridines, and of wave II for the iodopyridines, were halide ion and pyridine. The wave I process was therefore assumed to be a 2e–1H^+ reduction of the protonated halogenopyridine (I, Fig. 9-22) to pyridinium ion (II, Fig. 9-22) and halide. Obviously, at pH values above its pK_a the pyridine formed would be uncharged. The pH-independent wave II process of the iodopyridines was considered to be much the same as for benzenoid halogenohydro-carbons,[128] i.e., a 1e rate-controlling reduction of the iodopyridine (III, Fig. 9-22) to a free-radical species (IV, Fig. 9-22) and iodide ion. The free radical is then rapidly reduced to (V, Fig. 9-22), which abstracts a proton from water to give pyridine (VI, Fig. 9-22). The variations of half-wave potential with pH for the halogenopyridines are shown in Fig. 9-23.

Evilia and Diefenderfer[129] repeated the study of Holubek and Volke[127] and found that the DC polarographic $E_{1/2}$–pH relationships were the same between pH 1 and 8. Above pH 9.5, however, a major difference was noted in that the $E_{1/2}$ values for the three isomeric iodopyridines were identical, although the $E_{1/2}$–pH plots (Fig. 9-24) do not exactly appear to support that view. A further discrepancy between the two groups of workers was noted since Evilia and Diefenderfer did not observe that 3- or 4-iodopyridine gave two waves at high pH, although 2-iodopyridine did. By using AC polarography and measuring phase angles Evilia and Diefenderfer[129] concluded that at low pH there are two consecutive electrochemical steps involved in the polarographic reduction of halogenopyridines. Accordingly, a mechanism was proposed in which the halogenopyridine (I, Eq. 11) is first reduced in a rate-controlling 1e–1H^+

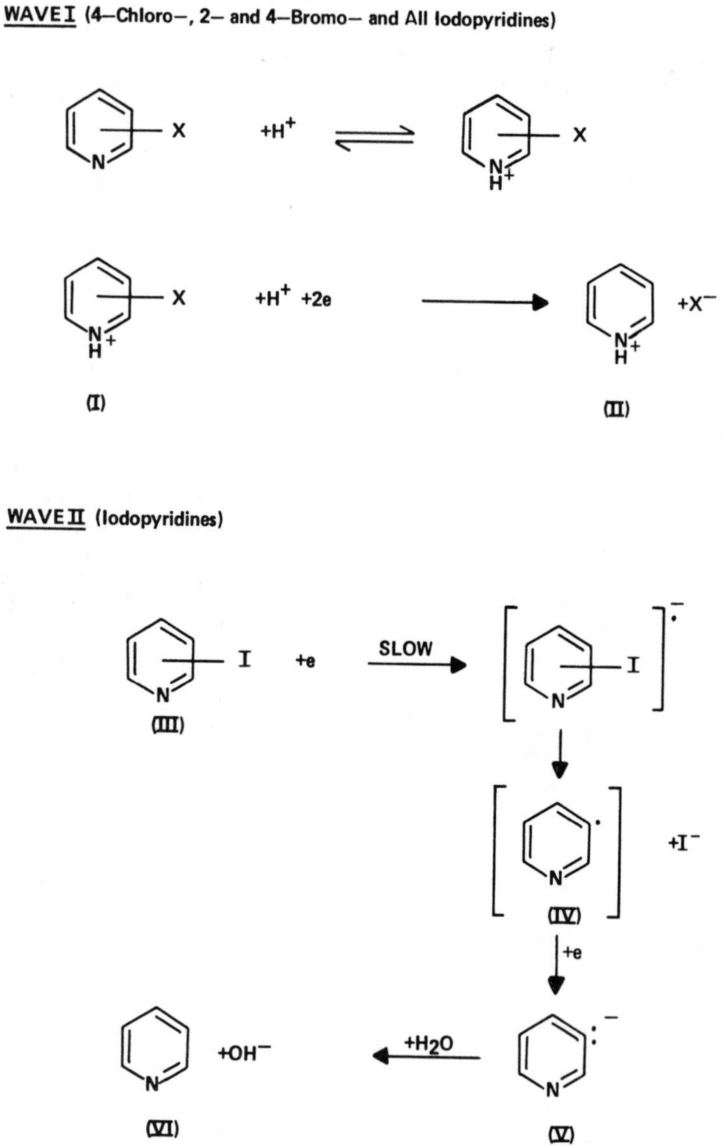
WAVE I (4–Chloro–, 2– and 4–Bromo– and All Iodopyridines)
X
+H+
X
X
+H+ +2e
+X−
(I)
(II)
WAVE II (Iodopyridines)
I
+e
SLOW
I
(III)
+I−
(IV)
+e
+OH−
+H2O
(VI)
(V)

FIG. 9-22. Proposed mechanism for the polarographic reduction of halogeno-pyridines.[127]

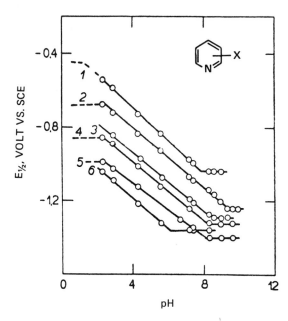

FIG. 9-23. Variation of half-wave potentials with pH for wave I of halogenopyridines according to Holubek and Volke.[127] (1) 2-Iodo, (2) 4-iodo, (3) 3-iodo, (4) 4-bromo, (5) 4-chloro, (6) 2-bromo. Waves II for iodopyridines are pH independent and occur at -1.42 V for 2-iodopyridine, -1.44 V for 3-iodopyridine, and -1.39 V for 4-iodopyridine. (Reprinted with permission of Academia Publishing House, Prague.)

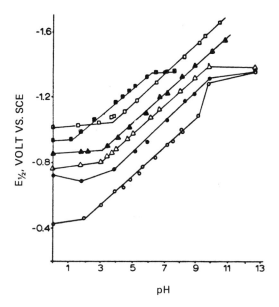

FIG. 9-24. Variation of half-wave potentials with pH for the first waves of (■) 2-bromopyridine, (□) 4-chloropyridine, (▲) 4-bromopyridine, (△) 3-iodopyridine, (●) 4-iodopyridine, and (○) 2-iodopyridine according to Evilia and Diefenderfer.[129] The solvent used was 50% aqueous methanol by volume. (Reprinted with permission of Elsevier Publishing Company, Amsterdam.)

process to give the radical (II, Eq. 11). Further 1e reduction gives the anion (III, Eq. 11), which decomposes to pyridine (IV, Eq. 11) and halide ion.

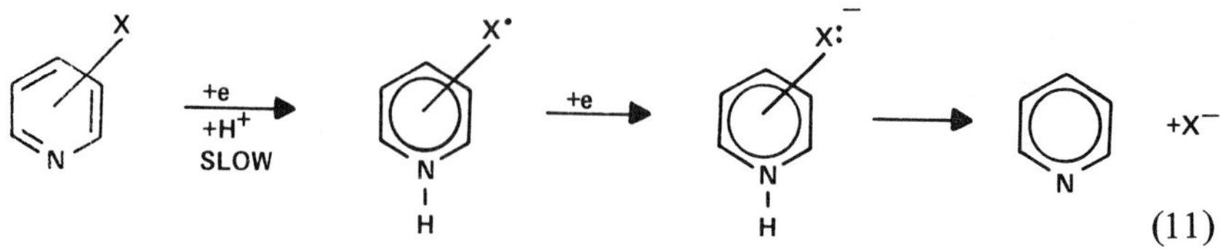

(11)

(I)   (II)   (III)   (IV)

At high pH, where all iodopyridines were claimed to be reduced at identical half-wave potentials,[129] it was proposed that some species must be formed that are energetically similar regardless of the position of the substituent. Since the AC polarographic data were consistent with an EC (electrochemical–chemical) mechanism, the reaction sequence shown in Eq. 12 was proposed.

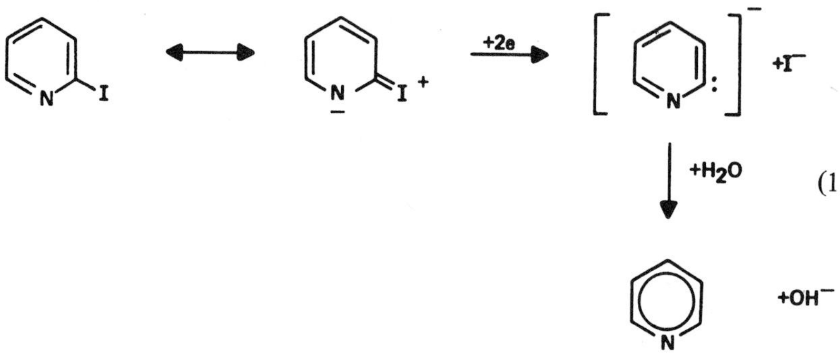

(12)

Mairanovskii and Baisheva[130] have also reported the DC polarographic behavior of 2-chloropyridine. Their data agree closely with those of the previously discussed workers.[127,129]

E. Other Pyridine Derivatives

Phenylpyridylethylenes and phenylpyridylbutadienes are reduced at the DME.[131] Isomers in which the polarographically active ethylene or butadiene group is substituted in the 2 or 4 position in the pyridine ring give a single, two-electron, pH-dependent wave (wave I). With increasing pH the height of this wave decreases, indicating that probably the N-protonated species is electro-active. In the case of the ethylene derivatives (I, Fig. 9-25), the product of the electrode reaction is the appropriate isomer of ethane (II, Fig. 9-25), while in the case of the 1,3-butadiene derivatives (III, Fig. 9-25), the product is the appropriate butene isomer (IV, Fig. 9-25). In the case of the 3-isomers, the same

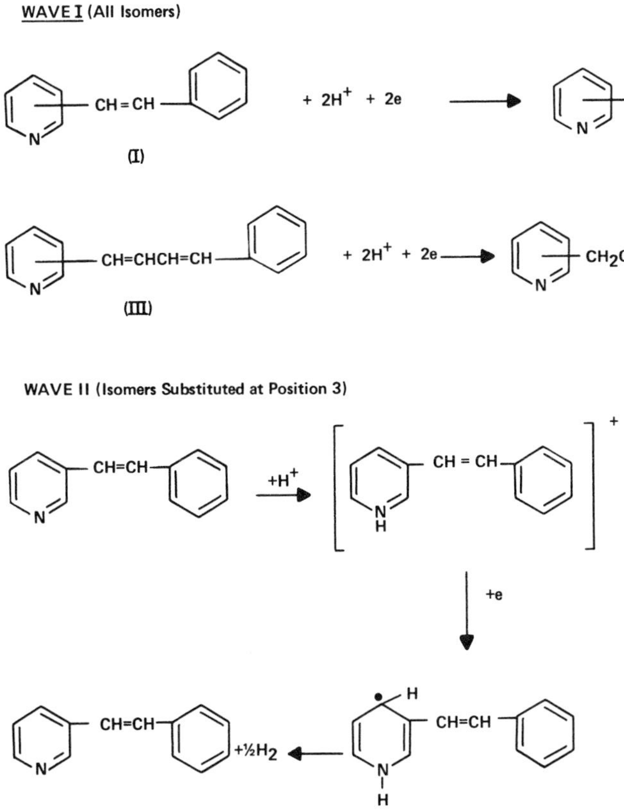

WAVE I (All Isomers)

WAVE II (Isomers Substituted at Position 3)

FIG. 9-25. Proposed mechanism for the polarographic reduction of phenyl-pyridylethylenes and phenylpyridylbutadienes.[131]

type of behavior is observed in acid solution as with the 2- and 4-isomers. However, in neutral or weakly alkaline solutions a second wave (wave II) appears due to catalytic hydrogen ion reduction. The mechanism proposed for the latter process is presented in Fig. 9-25. Some half-wave potential data are presented in Table 9-18.

Isomeric pyridylvinylquinolines and pyridylvinylisoquinolines are also irreversibly reduced at the DME.[132] The first wave observed polarographically was a two-electron reduction of the vinyl group to derivatives of ethane. The second wave was due to reduction of the quinoline or isoquinoline rings, and occasionally a third wave was observed due to catalytic hydrogen ion reduction. Half-wave potential data are shown in Table 9-19.

Zahlan and Linnell[133] have reported that bipyridines are electrochemically reducible (Table 9-20). Although the details of the mechanism were not

TABLE 9-18 Half-Wave Potentials for the Polarographic Reduction of Phenylpyridylethylenes and Phenylpyridylbutadienes in 40% Ethanol Solution[a]

Compound	pH	Supporting electrolyte	$E_{1/2}$ (V versus SCE)
1-Phenyl-4-(4′-pyridyl)-1,3-butadiene		0.1 M KCl	−1.45
		0.1 M HCl	−0.95
		0.1 M NaOH	−1.48
	2.5		−0.96
	4.0		−1.02
	5.5		−1.08
	7.0		−1.17
	8.5		−1.30
	9.5		−1.39
	10.5		−1.46
1-Phenyl-4-(3′-pyridyl)-1,3-butadiene		0.1 M KCl	−1.80
1-Phenyl-4-(2′-pyridyl)-1,3-butadiene		0.1 M KCl	−1.48
		0.1 M HCl	−1.08
		0.1 M NaOH	−1.65
	2.5		−1.07
	4.0		−1.10
	5.5		−1.15
	7.0		−1.24
	8.5		−1.34
	9.5		−1.41
	10.5		−1.48
1-Phenyl-2-(4′-pyridyl)ethylene		0.1 M KCl	−1.48
		0.1 M HCl	−1.08
		0.1 M NaOH	−1.63
	2.5		−1.08
	4.0		−1.09
	5.5		−1.13
	7.0		−1.20
	8.5		−1.29
	9.5		−1.36
	10.5		−1.42
1-Phenyl-2-(2′-pyridyl)ethylene		0.1 M KCl	−1.54
		0.1 M HCl	−1.10
		0.1 M NaOH	−1.69
	2.5		−1.10
	4.0		−1.12
	5.5		−1.17
	7.0		−1.26
	8.5		−1.36
	9.5		−1.43
	10.5		−1.49

[a] Data from Chodkowski and Jakubczak.[131]

TABLE 9-19 Half-Wave Potentials for Polarographic Reduction of Pyridylvinylquinolines and Pyridylvinylisoquinolines in 40% Ethanol Solution[a]

Compound	pH	Solvent and/or supporting electrolyte	$E_{1/2}$ (V versus SCE)
2-[β-(4′-Pyridyl)-vinyl]quinoline		0.1 M HCl	−0.43, −0.95
		0.1 M KCl	−1.16, −1.55
		0.1 M NaOH	−1.20, −1.68
	2.5		−0.47, −0.95
	3.8		−0.53, −0.98
	11.4		−1.14, −1.63
		0.175 M Bu$_4$NI[b]	−1.30, −1.90
2-[β-(2′-Pyridyl)-vinyl]quinoline		0.1 M HCl	−0.44, −1.00
		0.1 M KCl	−1.08, −1.45, −1.57
		0.1 M NaOH	−1.23, −1.70
	2.5		−0.48, −0.96
	3.8		−0.56, −1.02
	11.4		−1.20, −1.63
		0.175 M Bu$_4$NI[b]	−1.35, −1.95
2-[β-(3′-Pyridyl)-vinyl]quinoline		0.1 M HCl	−0.48, −0.69
		0.1 M KCl	−0.96, −1.10, −1.30, −1.59
		0.1 M NaOH	−1.42, −1.68
	2.5		−0.55, −0.72
	3.8		−0.66, −0.82
	11.4		−1.40, −1.72
		0.175 M Bu$_4$NI[b]	−1.45, −2.07
1-[β-(2′-Pyridyl)-vinyl]isoquinoline		0.1 M HCl	−0.40
		0.1 M KCl	−1.06, −1.69
		0.1 M NaOH	−1.25
	2.5		−0.47, −1.00
	3.8		−0.56, −1.08
	11.4		−1.12
		0.175 M Bu$_4$NI[b]	−1.35, −2.05
1-[β-(3′-Pyridyl)-vinyl]isoquinoline		0.1 M HCl	−0.50, −0.75
		0.1 M KCl	−0.92, −1.26, −1.68
		0.1 M NaOH	−1.34
	2.5		−0.55, −0.78
	3.8		−0.63, −0.81
	11.4		−1.22, −1.33, −1.80
		0.175 M Bu$_4$NI[b]	−1.38, −2.03
2-[β-(2′-Pyridyl)-ethyl]quinoline	2.5		−0.94
	3.8		−0.99
	11.4		−1.58
		0.175 M Bu$_4$NI[b]	−1.95
2-[β-(3′-Pyridyl)-ethyl]quinoline	2.5		−0.63
	3.8		−0.75
	11.4		−1.68
		0.175 M Bu$_4$NI[b]	−2.09

[a] Data from Chodkowski and Jakubczak.[132]
[b] Tetra-n-butylammonium iodide, in 75% dioxane.

TABLE 9-20 Half-Wave Potentials for the Polarographic Reduction of Bipyridines[a]

Compound	pH	$E_{1/2}$ (V versus Hg pool) for wave	
		I	II
2,2'-Bipyridine	4.2	-1.12^b	-1.24
	4.6	-1.14^b	-1.25
	5.8	-1.24^b	
	8.8	-1.46^b	
2,3'-Bipyridine	4.6	-1.06	$-^c$
	9.1	-1.36	$-^c$
2,4'-Bipyridine	4.6	-1.04	$-^c$
	5.8	-1.10	$-^c$
	6.4	-1.14	$-^c$
	8.5	-1.31	$-^c$
	13.0	-1.38	$-^c$
3,3'-Bipyridine	4.6	-1.55	
	5.0	-2.02	
4,4'-Bipyridine	4.0	-0.88	
	4.2	-0.89	
	4.6	-0.90	-1.06
	6.4	-1.06	-1.16
	8.8	-1.20	
	8.9	-1.24	
	13.0	-1.30	

[a] Data from Zahlan and Linnell[133]
[b] This is the average $E_{1/2}$ of a prewave and the main wave.
[c] A second, very indistinct wave appeared.

investigated, it appeared that most bipyridines gave two waves. In the case of 4,4'-bipyridine, the first wave involved one electron, as probably did the second. 2,2'-Bipyridine also showed what appeared to be two 1e waves at pH below about 5. Silvestroni[134] found similar behavior for 2,2'-bipyridine. The overall 2e reaction product is probably a dihydro derivative.

Dipyridylethylenes are polarographically reducible (Table 9-21). Volke and Holubek[135] found that in buffered 50% ethanol solutions 1,2-di(pyridyl)-ethylene, 1,2-di(2-pyridyl)ethylene, and 1-(2-pyridyl)-2-(4-pyridyl)ethylene gave just a single 2e reduction wave, the product being the appropriate ethane derivative (Eq. 13). Contrary to this, however, 1-(2-pyridyl)-2-(3-pyridyl)-

$$\text{(pyridyl)}-CH=CH-\text{(pyridyl)} + 2H^+ + 2e \longrightarrow \text{(pyridyl)}-CH_2 \cdot CH_2-\text{(pyridyl)} \quad (13)$$

TABLE 9-21 Half-Wave Potentials for the Polarographic Reduction of Dipyridylethylenes[a]

Compound	pH	$E_{1/2}$ (V versus SCE)
1,2-Di(4-pyridyl)ethylene	2.3	−0.60
	9.6	−1.03
1,2-Di(2-pyridyl)ethylene	2.3	−0.64
	9.6	−1.12
1-(2-Pyridyl)-2-(4-pyridyl)ethylene	2.3	−0.60
	9.6	−1.06

[a] Data from Volke and Holubek.[135]

ethylene and 1-(3-pyridyl)-2-(4-pyridyl)ethylene gave rise to two 1e waves in acid solution; at higher pH these two waves merged to give a single 2e wave. Salvatore[136] has investigated the polarographic reduction of *trans*-dipyridylethylenes in acetonitrile. The half-wave potentials so obtained were found to correlate linearly with the energies of lowest empty molecular orbitals.

Tomasik[137] has reported that various 2-substituted 5-phenylazopyridines give well-formed polarographic reduction waves in dimethylformamide. Based on the magnitude of the limiting current and wave slope analysis it was concluded that the first wave of these compounds (Table 9-22) was a reversible 1e process, although the mechanism or nature of products was not studied. The $E_{1/2}$ values were generally satisfactorily correlated with Hammett constants.

TABLE 9-22 Half-Wave Potentials for the Polarographic Reduction Waves of 2-Substituted 5-Phenylazopyridines in Dimethylformamide[a,b]

2 Substituent	$E_{1/2}$ (V versus SCE)
I	−1.02
Br	−1.06
Cl	−1.09
F	−1.15
H	−1.17
NHCOCH$_3$	−1.42
CH$_3$	−1.25
OH	−1.23
	−1.62
NH$_2$	−1.43
	−1.52

[a] Data from Tomasik.[137]
[b] Supporting electrolyte, 0.2 M NaNO$_3$.

In acid solution 4-cyanopyridine gives rise to a pH-dependent $4e$ irreversible polarographic wave that apparently results in reduction of the pyridine ring.[138] With increasing pH the wave height diminishes until at pH 10.5 a normal $2e$ wave is formed. The reduction of 2-cyanopyridine appears to be complex; three waves are formed at pH 2.2, the sum of which corresponds to a $4e$ reduction. The reduction of 3-cyanopyridine proceeds in a single $1e$ wave. The original paper should be consulted for details[138]; half-wave potentials are shown in Fig. 9-26.

Anabasine (**V**) [α-(β-pyridyl)piperidine] appears to undergo a $2e$ polaro-

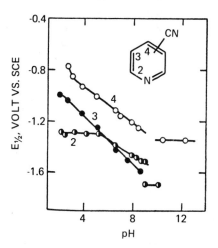

(V)

graphic reduction reaction.[139] In 0.1 M Na$_2$HPO$_4$, 0.1 M KCl buffer, the $E_{1/2}$ is -1.70 V versus SCE. The mechanism of the process is not known.

Sadler and Bard[140] have shown that 4,4'-azopyridine (I, Fig. 9-27) is polarographically reduced in dimethylformamide by a mechanism similar to that observed for aromatic hydrocarbons in aprotic media. Thus, reduction occurs in two $1e$ stages. The product of the first wave was a stable anion radical (II, Fig. 9-27) detected by ESR spectroscopy. The second electron transfer, which gave a dianion (III, Fig. 9-27), was thought to be followed by a chemical

FIG. 9-26. Variation of $E_{1/2}$ with pH for 2-, 3-, and 4-cyanopyridines.[138] (Reprinted with permission of Academia Publishing House, Prague.)

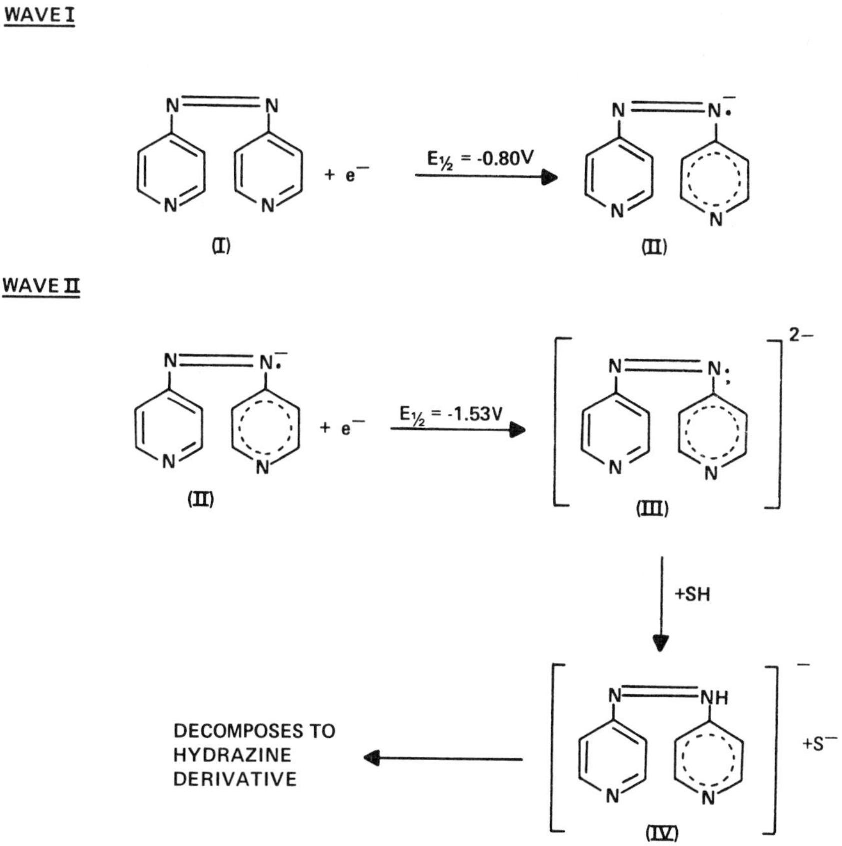

WAVE I

WAVE II

FIG. 9-27. Proposed mechanism for the polarographic reduction of 4,4'-azopyridine in dimethylformamide containing 0.1 M tetrabutylammonium perchlorate. Potentials quoted are versus aqueous SCE.[140]

reaction with the solvent producing a protonated species (IV, Fig. 9-27), which subsequently decomposed to the corresponding arylhydrazine.

Several groups of workers have examined the polarography of isonicotinohydrazide,[141-143] which is of interest because of its antitubercular activity.[144,145] It is generally agreed that in solutions between pH 1 and 8 this compound gives rise to two well-defined polarographic waves. Above pH 8 the two waves merge to give a single wave. The overall reduction appears to involve four electrons and four protons, the $-CONH \cdot NH_2$ grouping being reduced to $-CH_2NH \cdot NH_2$ or to $-CH_2OH$ and $H_2N \cdot NH_2$.[143] The polarographic waves have been utilized for pharmaceutical analysis. At pH 1.5 the half-wave potentials occur at -0.52 and -0.70 V versus SCE.[142]

Mikhailova and co-workers[146] examined the polarographic reduction of a large number of substituted pyridine derivatives (Table 9-23) and concluded, rather surprisingly, that in general the principal reduction wave was due to an irreversible reduction of the $>$C=N$-$ bond of the pyridine ring.

Other pyridine derivatives that have been examined are N-methyl-pyridinium bromide, trigonelline, 1-N-methylnicotinamide bromide, 1-N-methylisonicotin-amide bromide, 1-tetraacetylglucosido-3-carbamyl pyridinium bromide,[147] pyridine-containing carbamates, thiocarbamates and dithiocarbamates,[148] 3,5-diiodo-4(1H)-pyridone, and 1-(2,3-dihydroxypropyl)-3,5-diiodo-4(1H)-pyridone.[149] Other reviews and reports have also appeared; the original literature should be consulted for further information.[150−154]

V. ELECTROCHEMISTRY OF BIOLOGICALLY IMPORTANT PYRIDINES

Basically, this section deals with derivatives of nicotinic acid. It is of interest to compare the electrochemistry of the parent compound and some of its isomers with that of model compounds related to the pyridine nucleotides and with the nucleotides themselves.

A. Electrochemistry of Nicotinic Acid and Related Compounds

1. Nicotinic Acid, Isonicotinic Acid, and Picolinic Acid

Probably the first report of the polarographic reduction of nicotinic acid was that of Shikata and Tachi,[155,156] who observed two waves. It was proposed (incorrectly, as is now known) that the first stage of the reaction was reduction of the carboxyl group to an aldehyde and that the second stage was reduction of the pyridine ring. Quite well-defined waves were observed in 0.1 M NaHCO$_3$ of pH ca. 8.4, but no waves were observed in strongly alkaline solution. Lingane and Davis[157] confirmed this information. At pH below 10 comparison of background polarograms with those of nicotinic acid[10] showed that the presence of nicotinic acid, or its reduction product, shifted the background discharge potentials 0.2–0.3 V more positive. Nakaya[158] found that nicotinic acid gave a two-step reduction process at concentrations above 3 mM but that the diffusion-controlled waves were proportional to concentration only between 0.2 and 1 mM. The two waves reported by Nakaya[158] for nicotinic acid were observed in strongly acidic solution and were claimed to each involve a single electron. In weakly acidic media a single one-electron wave was observed. It is not known whether coulometry or simple polarography was employed to arrive at these electron values. The ultimate product, after transfer of two electrons,

TABLE 9-23 Half-Wave Potentials for the Polarographic Reduction of Substituted Pyridine Derivatives[a]

Compound	$E_{1/2}$ (V versus SCE) at pH			
	1.85	6.18	7.00	11.60
		−1.70		
	−0.90	−1.25	−1.06, −1.80	
	−0.86	−1.23	−0.97, −1.46	−1.50

R_1	R_2	1.85	6.18	7.00	11.60
H	H	−0.86	−1.22	−0.97,	−1.51
Cl	Cl	−0.94	−1.29	−1.47	
OCH_3	OCH_3	−1.07	−1.47	−1.65	−1.62
$(C_2H_5)_2N$	$(C_2H_5)_2N$	−1.17		−2.06	

R_1	R_2	1.85	6.18	7.00	11.60
H	H	−0.80	−1.20	−1.66	
Cl	Cl	−0.95	−1.33	−1.72	−1.83
Cl	OCH_3	−1.09	−1.48	−1.92	
OCH_3	OCH_3	−1.10	−1.51	−2.03	
Cl	$(C_2H_5)_2N$	−1.09	−1.54		
$(C_2H_5)_2N$	$(C_2H_5)_2N$	−1.23	−1.66		

[a] Data from Mikhailova *et al.*[146]

TABLE 9-24 Half-Wave Potentials for the Polarographic Reduction of Pyridinecarboxylic Acids and Their Derivatives

Compound	pH (supporting electrolyte)	$E_{1/2}$ (V versus SCE) for wave I	II	III	Reference
Nicotinic acid (pyridine-3-carboxylic acid)	(0.4 M HCl)	−1.1			10
	1.0 (HCl)	−1.1			10
	3.3 (Cit–Phos)[a]	−0.9	−1.2		10
	6.6 (OAc)[b]	−1.52			10
	7.0 (Phos)[c]	−1.55			10
	7.0 (KCl)	−1.1	−1.3	−1.5	10
	7.0 (OAc)[b]	−1.63			10
	8.0 (Phos)[c]	−1.57			10
	8.1 (NaHCO$_3$)	−1.62			10
	8.6 (Borate)	−1.63			10
	8.7 (Borate)	−1.66			10
	9.1 (Phos)[c]	−1.68			10
	9.0 (TMAB)[d]	−1.60			157
	8.4 (NaHCO$_3$)	−1.6			155
Isonicotinic acid (pyridine-4-carboxylic acid)	1–9	−0.651 − 0.080 pH			161
Picolinic acid (pyridine-2-carboxylic acid)	1–9	−0.741 − 0.081 pH			161

Methyl isonicotinate			
	1	-0.84^e	162
	2.6	-0.885^e	162
	3.4	-0.957^e	162
	4.4	-0.995^e	162
	5.4	-1.055^e	162
	6.4	-1.110^e	162
	7.5	-1.185^e	162
	8.2	-1.230^e	162
	9.0	-1.275^e	162
	10.1	-1.290^e	162
	10.7	-1.290^e	162
Nicotinamide	$0.1\ M\ Na_2CO_3$	-1.61^e	13, 169
	$0.1\ M\ NaOH$	-1.74^e	165
Picolinamide	1 to ca. 11	$-0.597 - 0.087\ pH$	161
	8.26	-1.613	161
	11.68	-1.627	161
Isonicotinamide	1 to ca. 9	$-0.526 - 0.083\ pH$	161
	10.10	-1.32	161
	10.22	-1.282	161
	10.98	-1.30	161
	11.6	-1.33	161
	12.4	-1.36	161

[a] Citrate–phosphate buffer.
[b] Acetate buffer.
[c] Phosphate buffer.
[d] Tetramethylammonium borate.
[e] Versus normal calomel electrode.

was proposed to be a dihydro compound resulting from reduction somewhere in the pyridine nucleus.

Campanella and De Angelis[159] have reported that nicotinic acid exhibits four types of cathodic polarographic waves. The half-wave potentials for these waves were not reported. One wave was an adsorption prewave, and the other three were hydrogen ion reduction waves catalyzed by various ionic forms of nicotinic acid. It would appear that the cationic forms of the acid are principally involved in the catalytic process. It was claimed that under no conditions was nicotinic acid reduced. Some half-wave potential data for the reduction of nicotinic acid are shown in Table 9-24.

Isonicotinic acid has been reported to show a single, well-defined polarographic wave that has a constant height up to about pH 7, when it begins to decrease in height in the form of a dissociation curve and disappears above pH 10.8.[160,161] Picolinic acid shows a rather similar behavior.[160,161] Jellinek and Urwin[161] found that the first wave of picolinamide (*vide infra*) also behaved in a similar fashion and accordingly concluded that the electroactive species was not, as might be expected, the species having an undissociated carboxyl group, but rather the 1-*N*-protonated compounds, VI_a or VI_b, for

(**VI$_a$**)   (**VI$_b$**)

picolinic and isonicotinic acids, respectively. The reduction of picolinic acid in acid solution involves two electrons due to formation of a dihydro compound[158] of some type. Campanella and De Angelis[159] have proposed that three different types of cathodic polarographic waves are formed in the presence of isonicotinic acid. All waves are apparently of a catalytic nature. Similarly, picolinic acid was proposed to give two polarographic waves. As with nicotinic acid, isonicotinic acid and picolinic acid were apparently not reduced upon controlled potential electrolysis at potentials corresponding to their polarographic reduction waves. Gaseous hydrogen formed in the catalytic hydrogen ion reduction was the sole product of the electrolysis.[159] There is obviously a pressing need for careful and complete studies of the polarography of the pyridine monocarboxylic acids in order to properly define the number of waves, the dependence of $E_{1/2}$ on pH, and most importantly the mechanism of the processes responsible for the waves.

The methyl esters of nicotinic acid and isonicotinic acid have also been studied. The methyl ester of isonicotinic acid gives a single pH-dependent reduction wave[162] (possibly involving two electrons) between pH 1 and 11. At pH above about 8 the ester is hydrolyzed somewhat to give nicotinic acid; above pH ca. 11 the hydrolysis is very rapid. Methyl nicotinate gives two 1e waves,[158] the first wave being due to formation of a radical. The methyl ester of picolinic acid has also been briefly examined.[158]

2. Nicotinamide

Nicotinamide does not appear to have been extensively studied. In acid medium it shows two waves[147,157,163-165] similar to nicotinic acid. A catalytic wave at negative potentials has also been observed.[12] It has been suggested, although with minimal supporting evidence, that the waves involve 1e transfers.[166] N-Methylnicotinamide, according to Moret, also gives two 1e waves.[167] The product of the first wave was proposed to be a radical and that of the second a N-methyldihydronicotinamide.

Elving and co-workers[168] have reported the only detailed study of nicotinamide. They have shown that, in acidic media, nicotinamide exhibits two closely adjacent DC polarographic reduction waves (waves I and II, Table 9-25).

TABLE 9-25   Variation in the Polarographic Behavior of Nicotinamide with pH[a]

pH	Wave	$E_{1/2}$[b] (V versus SCE)	I[c]
3.2	I	−1.10	4.2
	II	−1.20	4.0
3.6	I	−1.11	3.0
	II	−1.22	4.3
3.7	I	−1.13	3.3
	II	−1.25	5.1
4.4	I	−1.15	2.3
	II	−1.27	5.6
5.0	I	−1.20	2.2
	II	−1.31	4.8
7.0	I	−1.39	2.0
	II	−1.51	2.5
8.0		−1.57	3.7
9.2		−1.59	3.9
9.4		−1.60	3.8
9.6		−1.61	3.4
9.7		−1.60	3.7
12.0		−1.70	3.8

[a] Data from Schmakel *et al.*[168]

[b] Nicontinamide concentration, 0.35 mM.

[c] $I = i_1/Cm^{2/3}t^{1/6}$.

A catalytic wave was also observed at more negative potentials but was not studied in any detail. The $E_{1/2}$ for wave I was nearly pH independent below pH 4 and above pH 9 (Table 9-25) but varied linearly with pH between pH 4 and 9: $E_{1/2} = -0.73 - 0.079$ pH. Between pH 3 and 7, the height of wave II varied between 1 and 2.5 times the height of wave I, although the close proximity of the more negative catalytic wave made measurement of the height of wave II uncertain. Wave II occurred 0.1–0.2 V more negative than wave I between pH 3.2 and 7. At pH $\geqslant 8$ the two waves were very close together so that individual half-wave potentials for waves I and II could not be readily measured, although addition of tetraethylammonium chloride to the solution allowed the individual waves to be observed. Coulometry revealed that both the wave I and wave II processes involved one electron. After controlled potential electrolysis at wave I at pH 10 (one electron) both cathodic waves I and II were absent but two anodic waves were present. The largest, wave I_a, had an $E_{1/2}$ of -0.45 V versus SCE; the smallest, wave II_a had an $E_{1/2}$ of -0.11 V. Controlled potential electrolysis at wave II (two electrons total) gave only anodic wave II_a. Fast sweep cyclic voltammetry (14–32 V sec^{-1}) at a hanging mercury drop electrode at pH 9 and above showed the presence of a reversible couple corresponding to the polarographic wave I process, suggesting the formation of a highly reactive radical intermediate during the 1e reaction. This reversible couple could not be observed below pH 9, indicating that the reaction of the intermediate radical is exceedingly fast in these pH regions.

On the basis of these results Elving et al.[168] proposed a reaction scheme for the electrochemical reduction of nicotinamide. The wave I process was clearly a 1e reaction to give a radical. Below pH 4 the pH independence of $E_{1/2}$ and the fact that in this pH range nicotinamide is protonated (pK_a = 3.3 at 20°C[169]) indicate that nicotinamide protonated at N-1 (IH$^+$, Fig. 9-28) is reduced to a neutral radical (IIH, Fig. 9-28), which very rapidly dimerizes to an apparent 6,6′ species (III, Fig. 9-28). Between pH 4 and ca. 9 the $E_{1/2}$ for wave I was pH dependent (Table 9-25); hence, in this range a mechanism was proposed whereby neutral nicotinamide (I, Fig. 9-28) is reduced in a 1e–1H$^+$ reaction to a neutral radical (IIH, Fig. 9-28), which again dimerizes. At pH $\geqslant 9$, the $E_{1/2}$ for wave I again becomes pH independent; hence, in these pH regions it has been suggested that nicotinamide is reduced in a 1e reaction to a radical anion (II, Fig. 9-28). This radical anion could be detected by fast sweep cyclic voltammetry, because its dimerization would be slower, and hence its lifetime longer, than for the neutral radical formed at lower pH owing to electrostatic repulsion. Dimerization would give the dianionic form of the 6,6′ species (IV, Fig. 9-28), which upon protonation would give the neutral 6,6′-dimer (III, Fig. 9-28). The rate constant for dimerization of the radical anions in alkaline solution was determined to be 1.8 x 10^6 liters mol^{-1} sec^{-1} at 30°C.

The cathodic wave II process of nicotinamide between ca. pH 3 and 9 was

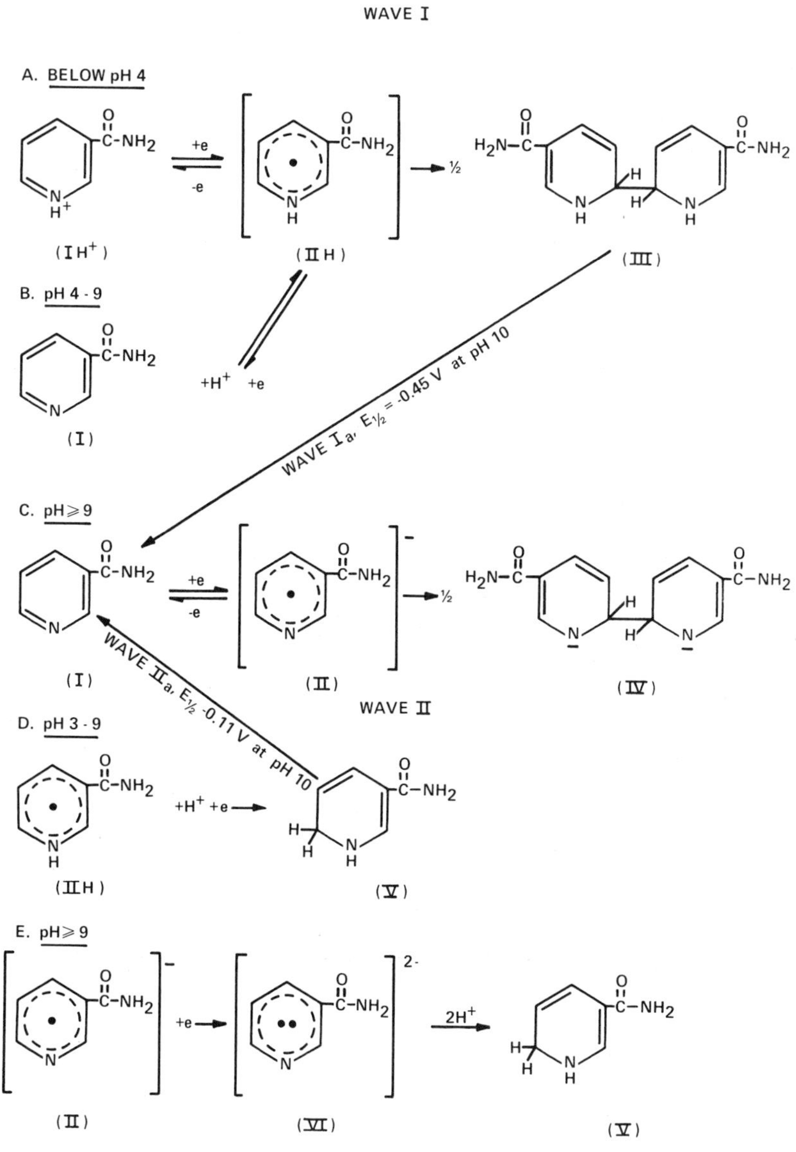

WAVE I

FIG. 9-28. Proposed reaction paths for the electrochemical behavior of nicotinamide and its reduction products.[168]

proposed to be a $1e–1H^+$ reduction of the neutral radical formed in the wave I process (IIH, Fig. 9-28) to 1,6-dihydronicotinamide (V, Fig. 9-28). At pH $\geqslant 9$ the wave II process (which is generally merged with wave I) was proposed to be a $1e$ reduction of the radical anion formed in the wave I reaction (II, Fig. 9-28) to a dianion (VI, Fig. 9-28), which rapidly protonates to give 1,6-dihydronicotin-amide (V, Fig. 9-28). The anodic polarographic wave (wave I_a) observed after controlled potential electrolysis of nicotinamide at wave I in alkaline solution was proposed to be due to oxidation of the 6,6'-dimer (III, Fig. 9-28) to nicotinamide (I, Fig. 9-28). Anodic wave II_a observed after controlled potential electrolysis at cathodic wave II was hence proposed to be due to oxidation of 1,6-dihydronicotinamide (V, Fig. 9-28) to nicotinamide (I, Fig. 9-28).

Alternating current polarography indicated that adsorption of nicotinamide and its reduction products was negligible.[168]

Santhanam and Elving[170] have shown that nicotinamide gives two cathodic polarographic waves in acetonitrile (AN) and dimethyl sulfoxide (DMSO) (Table 9-26). Both waves were diffusion controlled and, by measurements of wave slopes, coulometry, and product analysis, were shown to be $1e$ processes. Cyclic voltammetry of nicotinamide in DMSO at the hanging mercury drop electrode and pyrolytic graphite electrode revealed that at sweep rates below ca. 1 V sec^{-1} neither cathodic peak I_c ($E_p = -2.10$ V) or II_c ($E_p = -2.60$ V) (equivalent to cathodic polarographic waves I and II, respectively) exhibited any reversible complementary anodic peaks, although an anodic peak at -0.58 V (peak I_a) was observed. At sweep rates of 6 V sec^{-1} or greater, however, a reversible anodic peak ($E_p = -2.10$ V) corresponding to cathodic peak I_c ($E_p = -2.16$ V) was observed. Controlled potential electrolysis of nicotinamide in DMSO at wave I potentials (one electron) gave a product having a UV spectrum characteristic of the 6,6'-dimer.

Accordingly, a mechanism has been proposed[170] for electrochemical

TABLE 9-26   Half-Wave Potentials for DC Polarographic Waves of Nicotinamide in Nonaqueous Media^a

Solvent	Wave	$E_{1/2}$ (V versus SCE)
Acetonitrile^b	I	−2.00
	II	−2.45
Dimethyl sulfoxide^b	I	−2.01
	II	−2.50

^{a} Data from Santhanam and Elving.[170]
^{b} Background electrolyte, 0.1M tetraethylammonium perchlorate.

reduction of nicotinamide in AN and DMSO which is essentially the same as that proposed in aqueous solution at pH $\geqslant 9$ and can be represented as shown in Fig. 9-29. A number of analytical methods for the determination of nicotinamide by polarographic techniques have appeared.[165,171]

Isonicotinamide and picolinamide have been studied by Jellinek and Urwin.[161] For picolinamide up to about pH 9, a single reduction wave was observed, but at this pH a second wave appeared at more negative potential which grew in height at the expense of the first wave. The half-wave potential for the first wave was strongly pH dependent, while that for the second wave was pH independent (Table 9-24). Apparently, two waves were involved in the polarographic reduction of isonicotinamide, but their half-wave potentials were so similar that they could not be readily distinguished. Nevertheless, at pH about 9 the first wave had completely disappeared and only the second wave remained. The second wave of isonicotinamide and picolinamide, which in fact are the only waves at high pH, were found to be the most suitable for analytical purposes. Although no mechanistic data were presented, the second waves were thought to involve two electrons on the sole basis of the current observed for the wave.

3. NAD⁺ Model Compounds

It will be recalled that in NAD^+ (nicotinamide adenine dinucleotide, diphosphopyridine nucleotide, DPN^+, or coenzyme I) the heterocyclic nitrogen atom of nicotinamide is quaternized. This has prompted several workers to examine the electrochemistry of NAD^+ model compounds with the heterocyclic nitrogen atom quaternized by various substituents, usually alkyl groups.

Some of the earliest polarographic work was that of Ciusa and co-workers,[172] who studied nicotinamide substituted with alkyl groups at N-1; the chloride, bromide, or iodide salts were normally employed. It was found that nicotinamide hydrohalides gave only a single reduction wave, whereas the halides alkylated at N-1 gave two waves. The larger the alkyl group, the less negative was the half-wave potential for the first wave. The $E_{1/2}$ for the second wave was essentially unaffected by the size of the N-1 substituent (Table 9-27).[147,165,172–175] The mechanism of the polarographic reduction was not outlined in any detail, but the 1-N-alkyl derivatives were proposed to be reduced in two $1e$ processes, the first wave giving a radical. The nature of the second step was not investigated.

Baxendale and co-workers[176] studied several model compounds in order to elucidate a rather interesting biochemical problem. This problem arose from the fact that, although NAD^+ had been reported to have a low formal potential ($E^{0'} = -0.32$ V versus NHE at pH 7),[177] the reduced form of the coenzyme, NADH, shows little tendency to autooxidize in the absence of certain intermediaries; i.e., NADH can be oxidized by flavin nucleotides but not by

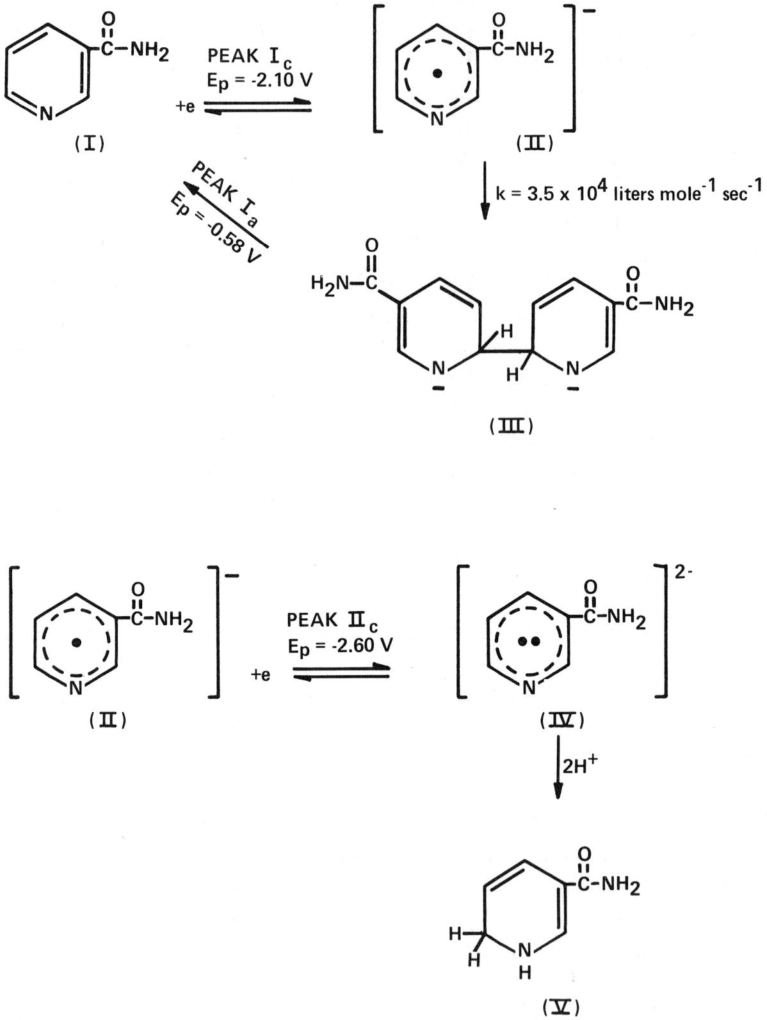

FIG. 9-29. Proposed reaction scheme for the electrochemical behavior of nicotinamide in acetonitrile and DMSO.[170] Potentials shown are those observed on cyclic voltammetry at the HMDE at sweep rates of $\leqslant 1$ V sec^{-1} in DMSO containing 0.1 M tetraethylammonium perchlorate.

TABLE 9-27   Half-Wave Potentials of 1-N-Substituted Nicotinamide Derivatives

| 1-N-Derivative | Anion | pH | $E_{1/2}$ (V versus SCE) | | Reference |
			Wave I	Wave II	
H	Cl	9.65		−1.675	172
	Br	9.65		−1.680	172
	I	9.65		−1.680	172
CH$_3$	Cl	9.65	−1.075	−1.650	172
	Br	9.65	−1.070	−1.645	172
	I	9.65	−1.070	−1.645	172
C$_2$H$_5$	Cl	9.65	−1.065	−1.650	172
	Br	9.65	−1.070	−1.654	172
	I	9.65	−1.060	−1.650	172
C$_3$H$_7$	Cl	9.65	−1.055	−1.650	172
	Br	9.65	−1.055	−1.655	172
	I	9.65	−1.055	−1.653	172
C$_4$H$_9$	Cl	9.65	−1.048	−1.655	172
	Br	9.65	−1.050	−1.655	172
	I	9.65		−1.655	172
C$_6$H$_5$CH$_2$	Cl	9.65	−1.003	−1.673	172
−CH$_2$C$_6$H$_4$SO$_3$$^-$		7–9	−0.96		173, 174
−CH$_2$CH$_2$SO$_3$$^-$		7–9	−1.02		173, 174
D-Glucopyranosidyl	Br		−0.871		165, 175
D-Glucopyranosidyl	Br		−0.991		165, 175
tetraacetate	Br		a		165, 147

a One wave observed in acid solution and two waves in alkaline solution; numerical data not given.[173]

oxygen itself. It was found that 1-*N*-methylnicotinamide chloride could be electrochemically reduced and that the product also showed very slow reactivity toward oxygen, although it could be readily oxidized by certain dyes and by iodine, Fe^{3+}, and Fe(CN)$_6$$^{3-}$, i.e., oxidants having a higher $E^{0'}$. Since the $E^{0'}$ of 1-*N*-methylnicotinamide had an apparent value of −0.36 V versus NHE at pH 9.1, it was clear that the model compound and NAD$^+$ behaved very similarly. The product of reduction of NAD$^+$ in biological situations (i.e., NADH) appears to be the 1,4- (or *para*-) dihydro compound. The product of electrochemical reduction of the model compound, however, could well be the 2,6- or 2,4-dihydro compound. Accordingly, another model compound, *N*-methyl acridan (**VII**), which was produced by reduction of *N*-methylacridinium chloride (**VIII**), was examined since in this case only a *para*-dihydro derivative could be produced upon reduction.

The kinetics of oxidation of **VII** were studied and were shown to proceed through the conjugate acid **IX**. It was therefore considered improbable that the first step in the oxidation of **IX** would involve the removal of an electron since the energy requirements to produce a double positive charge on the molecule

(**VII**)   (**VIII**)   (**IX**)

would be prohibitive. Accordingly, it was assumed that a hydrogen atom (or, more correctly, a hydride ion) transfer step was more probable and that this effect might account for the very slow reaction rate of N-methyl acridan, of the dihydro derivative of 1-N-methylnicotinamide, and of NADH itself with oxygen. In other words, it seems to be generally accepted that oxygen is reduced by an electron-transfer mechanism[178] but that the oxidation–reduction of the NADH/NAD$^+$ system and model compound systems proceeds by a hydrogen atom transfer reaction. Hence, the two redox systems are incompatible. This *might* explain why the flavin coenzymes act as intermediates in the electron-transport chain; i.e., they act as hydrogen acceptors from NADH and pass an electron on to a cytochrome, reducing the Fe^{3+} moiety in the latter to Fe^{2+}. A true electron-transfer reaction between Fe^{2+} and oxygen could then result in oxidation of Fe^{2+} to Fe^{3+} and reduction of oxygen (see Fig. 9-4). It was concluded by Baxendale *et al.*,[176] therefore, that the components of the respiratory or electron-transport chain are graded or react among each other with respect not only to $E^{0\prime}$ values, but also to the chemical mechanism of the redox processes.

In order to further understand the reduction of N-1-substituted nicotinamide derivatives, the latter workers investigated the details of the electrochemical reduction of a model compound, 1-N-methylnicotinamide iodide, which earlier they had understood only qualitatively.[179] 1-N-Methylnicotinamide iodide showed two polarographic waves, the first (more positive) of which (wave I) was pH independent, while the second (wave II) shifted to more negative potential with increasing pH (Table 9-28). The mechanism responsible for the two waves was not studied thoroughly, but it was suspected that the first wave was due to a one-electron reduction of the 1-N-methylnicotinamide cation (I, Fig. 9-30) to a free radical (II, Fig. 9-30) that dimerized under conditions of controlled potential electrolysis at wave I potentials to give III (Fig. 9-30). The second wave appeared to be due to a one-electron, one-proton reduction of the free radical formed in the wave I process to give a dihydro compound (possibly IV, Fig. 9-30).

TABLE 9-28 Half-Wave Potentials for the Polarographic Reduction of 1-*N*-Methylnicotinamide Iodide[a]

Buffer system	pH	$E_{1/2}$ (V versus SCE)	
		Wave I	Wave II
Glyme	9.0	−0.94	−1.57
Phosphate	7.0	−0.94	≈ −1.50
	5.15	−0.94	−1.32
HCl	2.03	≈ −0.9	Obscured

[a] Data from Leach *et al.*[179]

Nakaya[180] examined 1-*N*-methylnicotinamide iodide, 1-*N*-propylnicotinamide iodide, 1-*N*-benzylnicotinamide chloride, and 1-*N*-benzylnicotinamide iodide. All four compounds behaved similarly to the quaternary nicotinamide salts discussed earlier; i.e., they gave two diffusion-controlled polarographic waves. The first was pH independent and involved one electron, giving a free radical. The second wave was pH dependent and involved one electron and one proton to give a dihydro compound of some type. (Half-wave potential and other polarographic data of Nakaya[180] are not presented here but are available in Japanese in the original paper.)

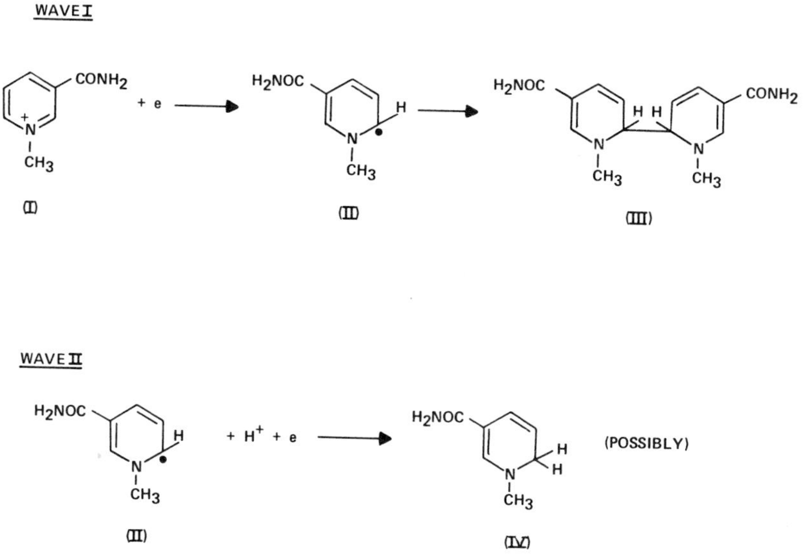

WAVE I

WAVE II

FIG. 9-30. Proposed mechanism for the polarographic reduction of 1-*N*-methyl-nicotinamide iodide.[179]

Paiss and Stein[181] have described electrolysis experiments with 1-*N*-methyl-nicotinamide salts. The electrolyses were carried out at constant current, although the potential of the mercury pool cathode was monitored throughout the electrolysis with respect to a calomel electrode. The electrolysis was terminated when the cathode potential reached about −1.1 V. This potential is about equal to the half-wave potential of the first pH-independent wave observed by other workers. However, Paiss and Stein[181] claimed that a dihydro compound was formed which was postulated to be 1-*N*-methyl-1,2-dihydro-nicotinamide. By electrolysis of 1-*N*-propylnicotinamide under similar conditions, Paiss and Stein obtained a dihydro compound when 2×10^{-3} *M* concentrations of starting materials were employed. However, when 3×10^{-2} *M* concentrations were employed, a mixture of dihydro compound and dimer was claimed to be formed.

Subsequent work by Burnett and Underwood[182] using polarography and electrolysis at controlled potential failed to confirm the Paiss and Stein results. It was proposed that, as a result of the very high currents employed in the latter workers' studies (1 A), considerable hydrogen gas may have been generated at the cathode, leading to formation of dihydro derivatives at potentials where a free radical, and hence a dimer, should have been the expected product. Burnett and Underwood[182] found that 1-*N*-methylnicotinamide (i.e., 1-methyl-3-carbamidopyridinium chloride) exhibited a single, diffusion-controlled, pH-independent polarographic wave below pH 7 (Table 9-29). In basic solution a second, diffusion-controlled, pH-dependent wave appeared at more negative potential (Table 9-29). Controlled potential electrolysis of 1-*N*-methylnicotina-mide (I, Fig. 9-31) at potentials on the first wave involved transfer of one electron, and the product was shown to be a dimer (III, Fig. 9-31) where the two pyridine moieties were linked through the 6 position. The dimer also gave a polarographic reduction wave at close to but not identical with the potentials observed for the second wave of the parent compound. Electrolysis on this dimer wave involved uptake of two additional electrons per molecule of original pyridinium compound to give a dimeric product reduced at positions 4 and 5 in both nicotinamide rings (IV, Fig. 9-31). However, electrolysis of 1-*N*-methyl-nicotinamide directly at potentials corresponding to its second polarographic wave consumed two electrons. The product was fairly conclusively demonstrated to be 1-*N*-methyl-1,4-dihydronicotinamide (V, Fig. 9-31). In other words, the dimer formed on controlled potential electrolysis at the first wave does not function as an intermediate in the formation of the 1,4-dihydromonomer. The polarographic reduction of 1-*N*-methylnicotinamide[170] in AN and DMSO and 1-*N*-benzylnicotinamide in dimethylformamide[183] has been examined and, in the absence of proton donors, only a single 1*e* cathodic wave was observed (Table 9-30). The product of the 1*e* reaction was shown to be a neutral radical that dimerizes in a manner essentially identical to that proposed for wave I of

TABLE 9-29   Half-Wave Potentials for the Polarographic Reduction of 1-*N*-Methylnicotinamide Chloride[a]

| Buffer system | pH | $E_{1/2}$ (V versus SCE) | |
		Wave I	Wave II
Acetate	4.0	−1.11	
	5.0	−1.11	
Citrate	6.0	−1.12	
Phosphate	6.0	−1.10	
	7.0	−1.11	
Tris[b]	7.0	−1.11	
Pyrophosphate	7.5	−1.11	−1.68
Phosphate	8.0	−1.11	−1.70
Pyrophosphate	8.8	−1.11	−1.70
Tris	9	−1.12	
Phosphate	9	−1.10	−1.72
Ammonia	9.9	−1.09	−1.78
NaOH	13	−1.12[c]	−1.84[c]

[a] Data from Burnett and Underwood.[182]

[b] Tris(hydroxymethyl)aminomethane.

[c] Solutions unstable; after 2 hr both waves were replaced by one wave at −1.52 V.

1-*N*-methylnicotinamide in aqueous solution by Burnett and Underwood.[182] Addition of a proton donor to DMSO or AN solutions of 1-*N*-methylnicotinamide resulted in formation of a second, more negative polarographic wave, which has been shown to be due to a further 1e–1H⁺ reduction of the neutral radical formed in the first 1e reaction.[170] However, the dimer formed upon controlled potential electrolysis at the first 1e wave was not reducible in AN or DMSO. Hence, the reports of Burnett and Underwood[182] that the dimer is reduced in aqueous solution have been seriously questioned.[170] The mechanism of Burnett and Underwood[182] agrees quite closely with that of Leach *et al.*[179] However, there is a serious discrepancy between the half-wave potential data of these two groups of workers (cf., Tables 9-28 and 9-29). Comparison with the data of other workers tends to support the values quoted by Burnett and Underwood.[182]

Cunningham and Underwood[184] used fast sweep cyclic voltammetry to prove the existence of an unstable one-electron reduction product of several 1-*N*-alkylnicotinamide salts. Using a stationary hanging mercury drop electrode, it was found that having once scanned to the first peak, E_p = ca. −1.00 V versus SCE (which corresponds to the first one-electron wave observed at the DME), then, provided the reverse scan toward positive potentials was fast enough, a

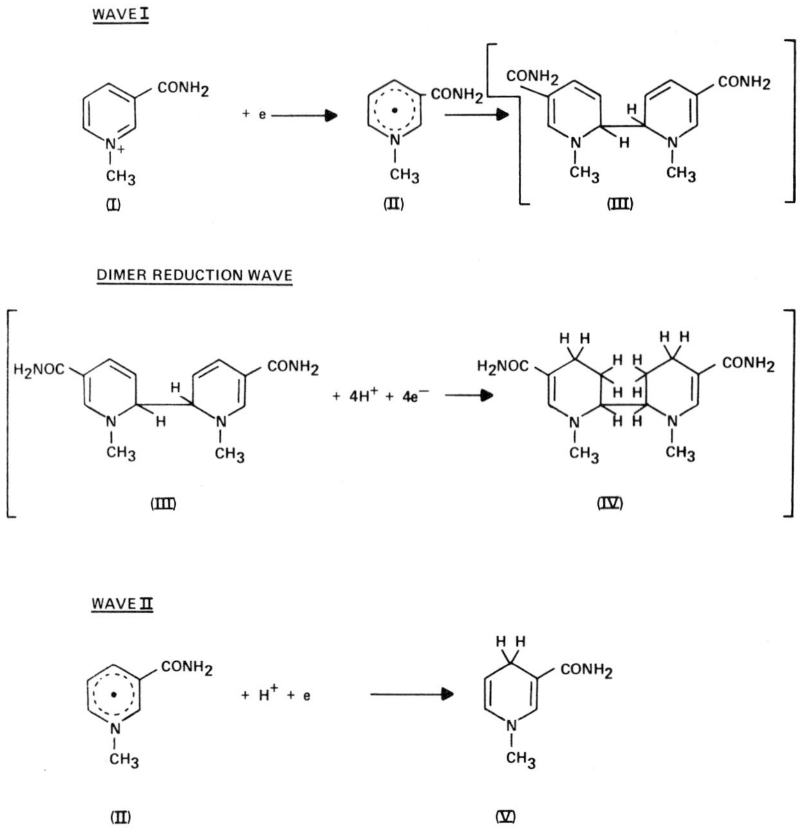

WAVE I

DIMER REDUCTION WAVE

WAVE II

FIG. 9-31.   Proposed mechanism of reduction of 1-*N*-methylnicotinamide according to Burnett and Underwood.[182] Reactions and products shown in brackets are observed only after controlled potential electrolysis at the first wave.

TABLE 9-30   Half-Wave Potentials Observed for Polarographic Reduction of 1-N-Substituted Nicotinamides in Nonaqueous Media

Compound	Solvent	$E_{1/2}$ (V versus SCE)	Reference
1-*N*-Methyl-	Acetonitrile[a]	−1.04	170
nicotinamide	Dimethyl sulfoxide[a]	−1.01	170
1-*N*-Benzyl-	*N,N'*-Dimethylformamide[b]	−1.02	183
nicotinamide			

[a] Supporting electrolyte, 0.1 *M* tetraethylammonium perchlorate.
[b] Supporting electrolyte, potassium perchlorate.

well-formed anodic peak could be observed. These two peaks formed a reversible one-electron couple, and it is entirely reasonable to assume that the reduction product was the free radical of the 1-N-alkylnicotinamide. However, ESR studies by the same workers failed to detect the presence of such a free radical. If the scan rates employed were slow, the compounds exhibited a pH-independent cathodic peak at about -1.0 V versus SCE and an anodic peak at much more positive potential (-0.35 V). This anodic peak observed at slow scan rates appeared to be due to oxidation of the dimer (III, Fig. 9.31), which had had time to form by dimerization of the free radical (II, Fig. 9-31). As might be expected, a second cathodic peak was observed at more negative potential in neutral or alkaline solution due to further one-electron, one-proton reduction of the free radical to a dihydro compound (see Fig. 9-31). Strong adsorption phenomena complicated the electrochemical behavior. Some typical peak potential (E_p) data are presented in Table 9-31.

By varying the sweep rates necessary to observe the anodic peak due to oxidation of the free-radical product of peak I, it was concluded that the lifetime of the free radical was somewhere between 0.2 and 20 msec. Similarly, it was concluded that for 1-N-methylnicotinamide the dimerization rate constant of the free radical was approximately 2×10^{-2} mole^{-1} sec^{-1}. In dimethyl sulfoxide solution the dimerization rate constant for the free radical of 1-N-methylnicotinamide is about 10^6 liters mole^{-1} sec^{-1}.[170]

Underwood and co-workers[185] have also investigated the electrochemical reduction of several 1,1'-polymethylenebis(3-carbamidopyridinium bromides), where n in structure **X** ranged from 0 to 3. All of these compounds exhibited a

(X)

reduction wave at -0.7 to -1.0 V versus SCE (Table 9-32). At pH above 8 an additional poorly defined wave was observed at potentials very close to background discharge. This wave was thought to be due to catalytic hydrogen evolution. Cyclic voltammetry of the compound **X** with $n = 0$ at the hanging mercury drop electrode in the presence of sufficient Triton X-100 to eliminate adsorption peaks gave a single cathodic peak, $E_p = -0.80$ V versus SCE, and a corresponding anodic peak, $E_p = -0.51$ V (sweep rate, 1.0 V sec^{-1}). These peaks and the cathodic polarographic wave were diffusion controlled. The large separation of the anodic and cathodic peak potentials indicated that the

TABLE 9-31   Peak Potentials for the Voltammetric Reduction of Some 1-*N*-Alkylnicotinamides at HMDE[a]

Compound	E_p (V versus SCE)[b]	
	Peak I	Peak II
1-*N*-Methylnicotinamide[c]		−1.140
1-*N*-Ethylnicotinamide	−1.100	−1.140
1-*N*-*n*-Propylnicotinamide	−1.090	−1.160
1-*N*-Benzylnicotinamide	−1.000	−1.150

[a] Data from Cunningham and Underwood.[184]
[b] At pH 7; scan rate, 1.15 V sec^{-1}.
[c] Iodide salts were employed.

electrode process was irreversible. It was not possible to detect formation of an intermediate radical by fast sweep cyclic voltammetry such as was observed with various other NAD$^+$ model compounds.[184] Controlled-potential electrolysis and coulometry of all the compounds gave values of two electrons taken up per molecule of starting material. After electrolysis at a potential on the cathodic wave, the solution of the ethylene compound (**X**, $n = 0$) exhibited an anodic

TABLE 9-32   Half-Wave Potentials for the First Polarographic Reduction Wave of Some 1,1′-Polymethylenebis(3-carbamidopyridinium bromides)[a,b]

Compound with n	pH	Buffer	$E_{1/2}$ (V versus SCE)
0	5.1	Acetate	−0.83
	7.0	Phosphate	−0.82
	9.0	Pyrophosphate	−0.82
1	5.1	Acetate	−0.78
	7.0	Phosphate	−0.78
	9.0	Pyrophosphate	−0.78
2	6.0	Acetate	−1.00
	7.0	Phosphate	−0.98
	9.0	Pyrophosphate	−0.98
3	6.0	Acetate	−1.0[c]
	7.0	Phosphate	−1.0[c]
	9.0	Pyrophosphate	−1.0[c]

[a] Data from McClemens *et al.*[185]
[b] For general structure, **X**, see text.
[c] Approximate value; a prewave at ca. −0.82 V was observed.

polarographic wave, $E_{1/2} = -0.25$ V versus SCE. Coulometric oxidation on this wave regenerated the starting compound and two electrons per molecule were transferred. On the basis of the coulometric data and UV spectral information it was concluded that 1,1'-ethylenebis(3-carbamidopyridinium bromide) (I, Fig. 9-32) is reduced in a 2e reaction to II (Fig. 9-32). Partial oxidation of II (Fig. 9-32) by allowing oxygen into the yellow reduced solution yielded a red product that gave a fairly good ESR spectrum. The red material was proposed to be the radical IV (Fig. 9-32). Complete oxidation of II with oxygen

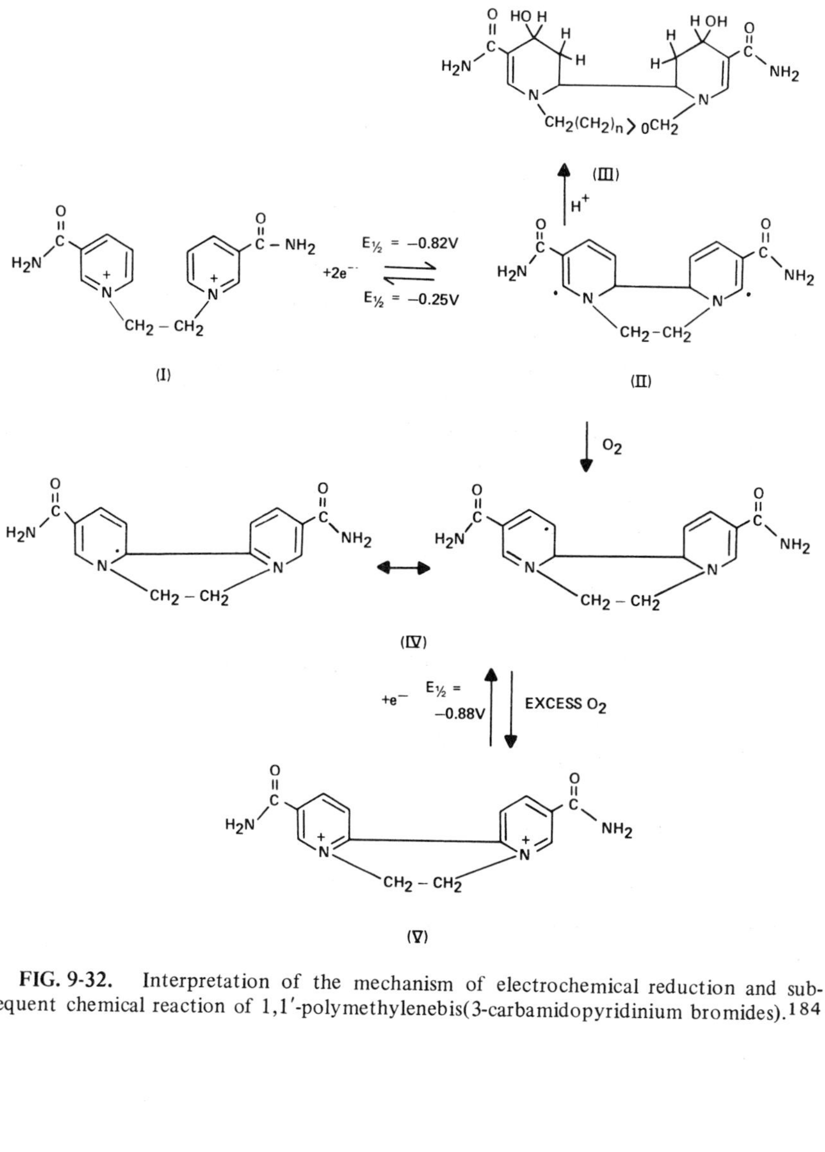

FIG. 9-32.   Interpretation of the mechanism of electrochemical reduction and subsequent chemical reaction of 1,1'-polymethylenebis(3-carbamidopyridinium bromides).[184]

was thought to give V (Fig. 9-32). This compound could be electrochemically reduced ($E_{1/2}$ = −0.88 V versus SCE) in a 1e reaction to the radical IV, (Fig. 9-32). With increasing values of n (see structure X) the products of electrochemical reduction were thought to be of similar structure to II (Fig. 9-32), but protonation of the product and hydration was proposed to give III (Fig. 9-32).

4. Summary

For the sake of clarity it is worthwhile summarizing the electrochemical reduction of 1-*N*-alkylnicotinamide derivatives. All, except the 1,1′-polymethyl-enebis(3-carbamidopyridinium bromides) discussed immediately above, appear to give two 1e polarographic and voltammetric reduction waves in aqueous solution. The first is due to the reversible formation of a free radical, and the second is due to further reduction of this radical to a 1,4-dihydro compound, although the structures of neither product have been unequivocably proven. In the case of 1-*N*-methylnicotinamide controlled potential electrolysis at potentials corresponding to the first wave gives a dimeric product. Dimerization apparently occurs at the 6 position. It appears that other 1-*N*-alkylnicotinamides probably also dimerize at the 6 position, although 1-*N*-propylnicotinamide salts have been claimed to dimerize at the 4 position.[180] The free radicals formed in the first wave of 1-*N*-alkylnicotinamides have been detected as an anodic peak by fast sweep cyclic voltammetry and, in the case of 1-*N*-methylnicotinamide, have a life-time of a few milliseconds. The dimer produced by dimerization of the free radical is electrochemically oxidized with considerably more difficulty than the free radical itself, but it can be further reduced at about the same potential as the free radical, although this reduction has been questioned. The products supposedly obtained on reduction of the free radical and the dimer are, however, quite different.

5. Electrochemical Oxidation of NADH Model Compounds

Blaedel and Haas[186] have examined the electrochemical oxidation of a number of NADH analogs, including 1-*N*-(2,6-dichlorobenzyl)-1,4-dihydronicotinamide and 1-*N*-methyl-1,4-dihydronicotinamide in acetonitrile at platinum and glassy carbon electrodes. In unbuffered acetonitrile at low-voltage scan rates, three anodic peaks were observed at the carbon electrode for all NADH analogs examined (Table 9-33).

Cyclic voltammetry revealed that peak I was irreversible but that peak II was reversible. If a slight excess of a base such as pyridine or *t*-butylamine was added,

TABLE 9-33   Voltammetric Oxidation Behavior of NADH Model Compounds in Acetonitrile at a Glassy Carbon Electrode[a,b]

| Compound | E_p in unbuffered solution (V versus Ag/0.01 M AgClO$_4$ in 0.1 M TEAP plus acetonitrile) | | |
	Peak I	Peak II	Peak III
1-N-(2,6-Dichlorobenzyl)-1,4-dihydronicotinamide	0.30	0.74	1.25
1-N-Methyl-1,4-dihydronicotinamide	0.23	0.73	1.22
1-N-n-Propyl-1,4-dihydronicotinamide	0.57	1.02	1.56
1-N-Benzyl-1,4-dihydronicotinamide	0.30	0.69	1.22

[a] Data from Blaedel and Haas.[186]

[b] Supporting electrolyte, 0.1 M tetraethylammonium perchlorate, voltage scan rate, 2.0 V min^{-1}.

peaks II and III disappeared and a single peak was produced, the height of which was approximately double that of peak I in unbuffered solution. In the presence of base this single oxidation peak occurred at potentials slightly negative of peak I.

In unbuffered solution the height of peak I was close to that expected for a one-electron oxidation. In basic acetonitrile coulometry revealed that, overall, two electrons were transferred upon oxidation of dihydronicotinamides (RNH, Eq. 14), and the product was the corresponding NAD$^+$ analog (RN$^+$) in quantitative yield.

$$+ \; 2e \; + \; HB \qquad (14)$$

(RNH)   (RN$^+$)

Cyclic voltammetry of 1-N-methyl-1,4-dihydronicotinamide (MeNH) in unbuffered and buffered (basic) acetonitrile solution also revealed that after the first anodic peak was scanned at the glassy carbon electrode, two cathodic peaks were observed at ca. $E_p = -1.40$ and -1.65 V. The more positive of the latter peaks corresponded to the one-electron reduction of the oxidation product, 1-*N*-methylnicotinamide (MeN$^+$), to a free radical, as observed by earlier workers.[182,184] The more negative peak, which was considerably small than the latter peak, was proposed to arise from the two-electron reduction of MeN$^+$ to MeNH.

Sweep rate studies of several nicotinamide analogs in unbuffered acetonitrile solutions indicated that anodic peaks II and III were formed as a result of a chemical follow-up reaction on the product of peak I. In other words, at fast voltage scan rates peaks II and III disappeared because under such conditions there was insufficient time for the primary product of peak I to undergo homogeneous chemical reactions to give the products responsible for peaks II and III. Further electrochemical, spectrophotometric, and chromatographic studies on the products obtained on oxidation of 1-*N*-(2,6-dichlorobenzyl)1,4-dihydronicotinamide (ClBzNH) in unbuffered acetonitrile revealed that indeed 1-*N*-(2,6-dichlorobenzyl)nicotinamide (ClBzN$^+$) was a product of the reaction. In addition, at least two additional products were detected, one of which had properties similar to those observed for a dimeric, nonabsorbing product of acid decomposition of MeNH and other dihydronicotinamides[187] and did not absorb in the UV region of the spectrum. A second additional product(s) did absorb UV light.

Coulometry of MeNH and ClBzNH in unbuffered acetonitrile at potentials corresponding to the first anodic peak (peak I) gave n values ranging from 0.63 to 0.72. The observed nonintegral n values of less than unity along with the detection of a product formed upon acid decomposition of dihydro-nicotinamides suggested that a proton liberated in the electrooxidation reaction was reacting with the starting dihydronicotinamides, which then, at least partially, decomposed. On the basis of these and other studies it was considered that the electrochemical oxidation of NADH analogs (RNH) in acetonitrile proceeded via a primary one-electron oxidation (peak I) to give a protonated pyridinyl radical (RNH$^{+\cdot}$):

$$\mathrm{RNH} \rightarrow \mathrm{RNH}^{+\cdot} + e \qquad \text{(peak I)} \qquad (15)$$

The RNH$^{+\cdot}$ was considered to be nonelectroactive at the potentials at which RNH was oxidized. The RNH$^{+\cdot}$ then underwent one or more additional reactions involving proton or electron transfer, depending on the conditions. In the absence of base in aprotic acetonitrile, the RNH$^{+\cdot}$ was thought to disproportion-ate (Eq. 16) to form the pyridinium salt RN$^+$ and the protonated form of the

$$2\mathrm{RNH}^{+\cdot} \rightarrow \mathrm{RN}^+ + \mathrm{H(RNH)}^+ \qquad (16)$$

substrate, H(RNH)$^+$. The latter product was then proposed to undergo further chemical reaction to give at least three products (Eq. 17), one of which absorbed

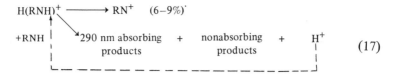

$$H(RNH)_2^+ \longrightarrow RN^+ \quad (6\text{--}9\%)$$

$$+RNH \quad \searrow \quad 290 \text{ nm absorbing} \quad + \quad \text{nonabsorbing} \quad + \quad H^+ \tag{17}$$
$$\text{products} \qquad \text{products}$$

UV light and at least one of which did not absorb UV light. A small amount of RN^+ was also thought to be produced from the decomposition reaction. As a result of the formation of $H(RNH)^+$ by disproportionation of RNH_2^+ and by protonation of RNH with protons liberated in Eq. 17, n values were consistently lower than unity in unbuffered acetonitrile.

In the presence of base the protonated pyridinyl radical (RNH_2^+) was thought to react with base to give the pyridinyl radical ($RN\cdot$),

$$RNH_2^+ + B \rightleftharpoons RN\cdot + HB^+ \tag{18}$$

which underwent immediate electrochemical oxidation to RN^+:

$$RN\cdot \rightleftharpoons RN^+ + e \qquad (-1.0 \text{ to } -1.4 \text{ V}) \tag{19}$$

The disproportionation and acid-decomposition reactions that occurred in unbuffered acetonitrile were blocked; consequently, n values of close to 2 per mole of RNH were observed in basic solution and formation of RN^+ was quantitative.

The processes responsible for anodic peaks II and III in unbuffered acetonitrile were not elucidated. However, the voltage scan rate studies mentioned earlier suggested that the primary product of peak I, RNH_2^+, underwent a chemical follow-up reaction to give a product, C, which could be further oxidized at peak II to an unknown product D (Eq. 20). No doubt, reactions of this type also resulted in a product that gave rise to peak III.

$$RNH \xrightarrow[\text{peak I}]{-e^-} RNH_2^+ \xrightarrow{k} C \xrightarrow[\text{peak II}]{-ne^-} D \tag{20}$$

A number of investigations of the chemical oxidation of dihydronicotinamides have been reported. Wallenfels[188] and Sund[189] have concluded that the chemical oxidations of dihydronicotinamides occur primarily by hydride ion transfer involving the simultaneous loss of a proton and two electrons. These findings do not, of course, coincide with the mechanism of electrochemical oxidation of dihydronicotinamides in acetonitrile. However, several other reports have indicated that dihydronicotinamides are capable of functioning as one-electron donors. Thus, Schellenberg and Hellerman[190] observed that NADH reacted rapidly with several $1e$ transfer agents but only slowly with oxidants that were $2e$ acceptors. Similarly, Westheimer and co-workers[191,192] have proposed that the oxidation of 1-N-alkyl-1,4-dihydronicotinamides might proceed via a radical process. In fact Kosomer[193] has suggested that the species $RNH^{\ddagger}$, which Blaedel and Haas[186] have proposed to be the primary product of electro-chemical oxidation of dihydronicotinamides, might be an important inter-mediate in some of these reactions. Gutman *et al.*[194] carried out a kinetic study of the reduction of iron(III) by NADH in aqueous solution at pH 3–4. In this system they found that NADH behaved as a one-electron donor in the redox reaction. However, in the presence of a large excess of iron(III) or in the presence of flavin adenine dinucleotide (FAD), two-electron behavior was observed. In order to account for their data, Gutman *et al.*[194] invoked a mechanism involving the cation radical $NADH^{\ddagger}$. Thus, the electrochemical studies appear to support the possibility of NADH and related analogs as being potential one-electron donors.

B. Electrochemistry of Pyridine Nucleotides.

1. Electrochemical Reduction

Until relatively recently the electrochemistry of NAD^+ in particular presented an extraordinarily confused picture. The confusion centered around the number of polarographic waves observed (one or two), the nature of the electrochemical product, and the effect of suitable enzyme systems (i.e., no effect, partial oxidation, or complete oxidation) on the reduced product. Even now there is no conclusive evidence to support the actual structure of products proposed to be formed electrochemically.

Many investigators have reported just a single cathodic polarographic wave for NAD^+ in acidic and alkaline buffered solutions, the half-wave potential for which is virtually pH independent (Table 9-34).[15,147,165,174,195–197] In solutions containing tetramethylammonium chloride[198] or tetra-*n*-butyl-ammonium carbonate[199] as supporting electrolyte, two reduction waves have been reported. The second wave is strongly pH dependent.[198–201]

For the purposes of understanding some of the disagreement in the literature with regard to both the electrochemistry of NAD^+ and the biochemistry of its reduction product, it is worthwhile summarizing some of the more important and controversial papers. Carruthers and Suntzeff,[196] having found that nicotinamide could be determined in biological materials by a polarographic method,[202] examined the possibility of determining NAD^+ by a similar method. They found that NAD^+ gave just a single polarographic reduction wave in a sodium citrate–citric acid–dioxane–tetrabutylammonium iodide buffer system, pH 5.1, with an $E_{1/2}$ of about −1.09 V versus SCE. The half-wave potential was practically independent of pH (Table 9-34), and the slope of the wave [E_{DME} versus $\log(i_d - i)/i$] was a straight line of slope 60 mV, which is the value

TABLE 9-34   Half-Wave Potentials for the Polarographic Reduction of NAD$^+$ at the DME in Aqueous Solution

| pH | $E_{1/2}$ (V versus SCE) | | Reference |
	Wave I	Wave II	
3.2–7.1	−0.91 to −0.92		198
7.3–8.7	−0.92 to −0.94		198
9.0	−0.95		198
9.2	−0.96		198
5.0	−0.97	−1.38	198
4.0–8.0	−1.09		165
4.0–9.0	−0.93		199
7–9	−0.93	−1.72	199
7–9	−0.91		174
7.4	−0.93		15
4.2–6.6	−1.09		196
8.3	−1.10		196
10.4	−1.18		196
10.6	−0.98		195
10.3–10.6	−0.98		197
7	−0.995[a]		184

[a] Peak potential at HMDE; scan rate, 1.15 V sec^{-1}.

expected for a 1e reversible process. In view of the fact that the limiting current observed for NAD$^+$ was directly proportional to concentration, an adequate analytical method was developed. Since NADP$^+$ (TPN$^+$ or coenzyme II) had an identical polarographic half-wave potential (*vide infra*), it was not possible to polarographically distinguish between the two common pyridine nucleotides. In a subsequent paper Carruthers and Tech[197] found that NAD$^+$, NADP$^+$, and NMN$^+$ (nicotinamide mononucleotide) were all reducible polarographically by way of a single cathodic wave. However, in a tris buffer of pH 10.3–10.6, all three compounds were reduced at different potentials (Table 9-35). The wave slope data were not in accord with those reported earlier and favored the view that the waves were irreversible. An analytical procedure was developed for the determination of NAD$^+$ in the presence of NADP$^+$. Carruthers and co-workers[196,197] did not give any mechanistic details, except that the single wave that they observed was possibly due to a 1e reaction.

In order to facilitate further understanding of the more recent electro-chemistry, a brief summary of the biochemical and chemical reduction of NAD$^+$ in particular, will be presented. In biological situations and in the presence of suitable enzymes and substrates, NAD$^+$ is reduced to NADH. A typical example is the oxidation of ethanol to acetaldehyde in the presence of the enzyme

TABLE 9-35   Half-Wave Potentials and i_d/C Values for the Polarographic Reduction of Pyridine Nucleotides in Tris Buffer, pH 10.3–10.6[a]

Nucleotide	Concentration range (μg/2 ml)	i_d/C (μA/mmole/liter)	$E_{1/2}$ (V versus SCE)
NMN$^+$	75–302	2.76	−1.14
NAD$^+$	35–526	2.92	−0.98
NADP$^+$	38–570	2.05	−1.23

[a] Data from Carruthers and Tech.[197]

alcohol dehydrogenase (Eq. 21). As shown in Fig. 9-3, NAD$^+$ is reduced at the 4

$$CH_3CH_2OH + NAD^+ \underset{\text{dehydrogenase}}{\overset{\text{alcohol}}{\rightleftharpoons}} CH_3CHO + NADH + H^+ \qquad (21)$$

or *para* position. The NADH can be oxidized by a mechanism shown in Eq. 2 by the flavin coenzymes or simply by adding, e.g., an excess of acetaldehyde, and using the reverse of Eq. 21. It appears to be generally accepted that the reduction of NAD$^+$ proceeds by a hydride ion-transfer reaction rather than by sequential (electron–proton)-transfer processes. Conn and Stumpf[4] have written a simplified mechanism (Eq. 22) for this type of process. This type of

$$(22)$$

mechanism has, from time to time, been questioned, as previously mentioned. Perhaps the most recent and eloquent discussion of the mechanism of reduction of NAD$^+$ is that of Hamilton.[203] This author has proposed that in nearly all biological redox reactions in which hydrogen is transferred, such as reduction of NAD$^+$, the hydrogen is probably transferred as a proton and not as a hydride ion or hydrogen atoms. Further discussion of alternative mechanisms for reduction of NAD$^+$ will be presented later (p. 561).

In order to prepare NADH in the absence of enzymes and organic material, it has been found that dithionite (hydrosulfite) reduces NAD$^+$ and NAD$^+$ model compounds to a product identical spectrally and enzymatically with NADH;[204–206] i.e., enzymatically the product can be completely oxidized to NAD$^+$. A number of reports have appeared on the mechanism of the dithionite reduction process, but that of Colowick and co-workers[207–209] seems to be the most generally accepted. The mechanism proposes an ionic attack of dithionite on the carbon 4 of the nicotinamide ring of NAD^+ (I, Fig. 9-33), giving a sulfinate derivative (II, Fig. 9-33). The formation of NADH (III, Fig. 9-33) is believed to occur by hydrolysis of the sulfinate intermediate.

As mentioned previously, the product of enzymatic reduction and of dithionite reduction is completely oxidizable by appropriate enzyme systems. It has been shown, however, that when NAD^+ is reduced by the reactive intermediates formed from ethanol in γ- or X-irradiated aqueous solutions, the product is not identical with the product of enzymatic or dithionite reduction.[210,211] In fact, the product is totally enzymatically inactive. Mathews and Conn[212] used sodium borohydride as a reducing agent and obtained a NAD^+ reduction product that could be oxidized enzymatically to the extent of about 50%. The mode of reduction in the irradiated system has been assumed to be due to the intermediate formation of free radicals, which carry out the reduction reaction.[213] Stein and Stiassny[214] carried out a comparative study of the reduction of 1-*N*-propylnicotinamide (a NAD^+ model compound) with dithionite by X irradiation of solutions containing ethanol, by borohydride, and by electrochemical methods. The dithionite reduction gave a single dihydro isomer. The X-irradiated solution gave another dihydro isomer, while the borohydride reduction gave a mixture of the latter two isomers. Electrochemically the product was identical to one of the borohydride products and to the isomer

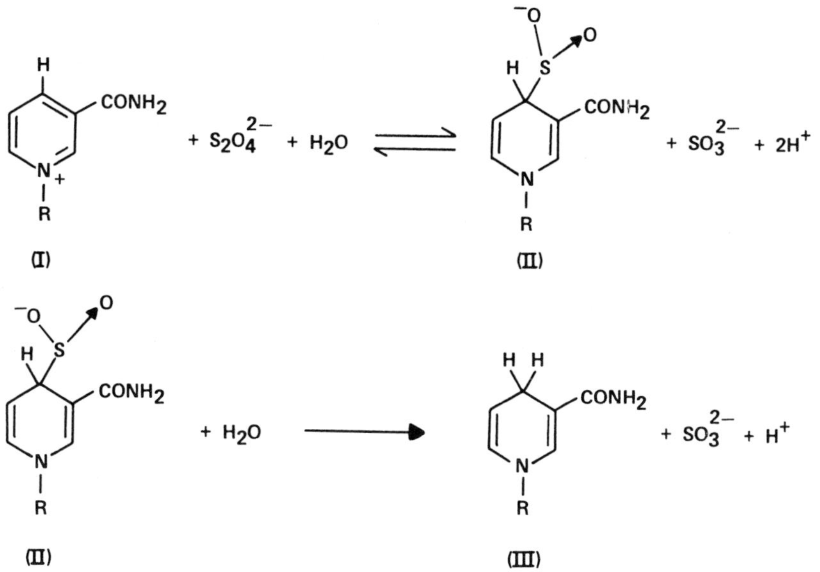

FIG. 9-33. Proposed mechanism for the reduction of NAD^+ to NADH by dithionite.[209]

obtained in the X- and γ-irradiation experiments. Accordingly, it was assumed that the electrolytic and irradiation products were formed by a free-radical mechanism, while the dithionite reduction proceeded by an ionic mechanism. Borohydride reduction was intermediate between these two extremes. The electrode material or the reaction conditions employed by Stein and Stiassny[214] for the electrochemical reduction of the NAD^+ model compounds were not specified.

The site of hydrogen transfer in the enzymatic and dithionite reduction of NAD^+ has been demonstrated by Pullman *et al.*[215] and Loewus *et al.*[216] to occur at the C-4 or *para* position. It was also confirmed that only this reduction product would normally be enzymatically active. Accordingly, it has been suggested[211,212] that the reduction of NAD^+ by borohydride or irradiation might form partially or completely an *ortho-* (2- or 6-) dihydro isomer. Recent evidence does not favor this view (*vide infra*).

Ke[198] examined the polarography of NAD^+ and found that in most solutions it gave a single, pH-independent cathodic wave at $E_{1/2} = -0.93$ V versus SCE (Table 9-34), although in a supporting electrolyte of tetramethylammonium chloride, pH 5, two waves of about equal height were observed at $E_{1/2} = -0.97$ and -1.38 V. The first wave was proposed to be irreversible, although the evidence presented to support this is dubious. The value of the diffusion current constant for the first wave of NAD^+ ($I = i_1/Cm^{2/3}t^{1/6} = 1.47$) was that expected for a one-electron reaction, even though the author[198] thought (quite erroneously) that the use of I was not possible because of the irreversibility of the wave. Coulometry, at an unspecified potential, was claimed to indicate that two electrons were transferred in the electrode reaction. The millicoulometric method applied by Ke has been shown repeatedly to be unreliable,[217–219] and later work has proved this electron number to be incorrect. After electrolysis an anodic wave was observed at approximately 0.6 V more positive than the original cathodic wave. In a subsequent study, Ke[220] electrolyzed NAD^+ in pH 7.6 tris buffer at an unspecified potential and found that upon reduction the electrolyzed solution turned pale yellow. The reduced product had exactly the same paper chromatographic R_f value as NADH but did not possess the well-known fluorescence of NADH. The electrochemically reduced product was able to reduce $AgNO_3$, I_2, 2,6-dichlorophenolindophenol, and methylene blue but was totally inactive with alcohol dehydrogenase and acetaldehyde; i.e., the product was not NADH.

Powning and Kratzing[221] electrochemically reduced NAD^+ at a mercury pool electrode at -1.7 V versus SCE in a number of background electrolytes between pH 7 and 9.2. Under these conditions it was found that about one-half of the reduced coenzyme underwent the same enzyme reactions as the enzymatically reduced NAD^+. No color change was observed in these experiments. The fate of the portion of the NAD^+ that was not enzymatically active was not known, but the possibility of a dimerization reaction or of formation of other dihydro isomers was not ruled out. The activity of the electrochemically reduced product was dependent on the components of the buffer solution employed; i.e., the activities of the reduced product obtained in tris, phosphate, and pyrophosphate buffers were 3, 55, and 50–80% of authentic NADH, respectively. It may also be significant that the activity of electrochemically reduced NADP⁺ was 94% of that expected for authentic NADPH. It should be noted that, at −1.70 V versus SCE, it is likely that considerable evolution of hydrogen gas occurred at the electrode.

Ke,[222] who, it will be recalled, found that NAD⁺ was electrochemically reduced at mercury to a totally enzymatically inactive compound, went on to examine the electrochemical reduction of the coenzyme at a variety of solid metal electrodes. A tris buffer of pH 7.5 was employed for the experiments. Current—voltage curves were determined indirectly by a spectrophotometric method because at the potentials where NAD⁺ was reduced *vigorous hydrogen evolution occurred* at all the metallic electrodes. Under these conditions it was found that at platinum and lead electrodes the reduced form of NAD⁺ showed partial enzymatic activity. Because there was no correlation between the activity of the electrochemical NAD⁺ reduction product and the hydrogen overvoltage or zero charge potential of the various metals examined, it was assumed that the electrolytic reduction proceeded through a free-radical intermediate.

Kono[223] electrolyzed NAD⁺ at mercury in a phosphate buffer, pH 7.2, at a controlled potential of −1.75 V versus SCE. In contradiction to the observations of Ke,[198,220,222] it *was* observed that about one-third of the electrochemically reduced NAD⁺ was active toward alcohol dehydrogenase. Actually, the ratio of enzymatically active component in the reduction product was not affected by changing the controlled cathode potential from −1.75 to −1.40 V. It was also found that the electrolyzed solution, as well as NADH prepared enzymatically, was completely oxidized by a fraction obtained from etiolated mung bean seedlings. The reason for this effect was thought to be that a diaphonase enzyme present in the mung seedling fraction was less specific to the configuration of the reduced form of NAD⁺. It was therefore proposed that the electrochemical product of reduction of NAD⁺ (I, Fig. 9-34) was a mixture of the three isomers shown in Fig. 9-34.[224] The *para*-isomer (II, Fig. 9-34) is, of course, NADH. The 2- (III, Fig. 9-34) and 6- (IV, Fig. 9-34) dihydro isomers were considered to be the isomers inactive toward alcohol dehydrogenase but apparently active toward the mung seedling fraction.

Kono and Nakamura[225] reduced NAD⁺ electrochemically under a variety of conditions and found that the most enzymatically active product was obtained when the reduction was carried out in a sodium tripolyphosphate buffer, pH 7.0, at a platinum wool electrode at −2.0 to −3.0 V versus SCE. Under all conditions employed, at either platinum or mercury electrodes, an appreciable amount of

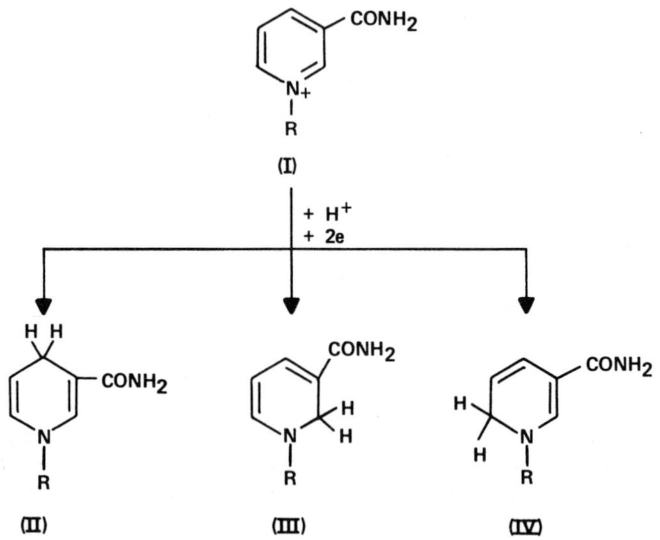

FIG. 9-34. Proposed composition of the product of electrochemical reduction of NAD$^+$ (I) at mercury according to Kono and Suekane.[224]

enzymatically active product was formed. At all controlled potentials employed, vigorous evolution of hydrogen gas must have occurred, and in fact the authors considered that the enzymatically active NADH was probably formed by reduction with hydrogen gas. In support of the findings of Powning and Kratzing,[221] Kono and Nakamura[225] found the NADP$^+$ could be reduced to a product having 95% of the enzymatic activity expected for NADPH.

It was not until 1965 that a definitive study of the electrochemical reduction of NAD$^+$ was reported. Burnett and Underwood[199] found that NAD$^+$ exhibited two cathodic polarographic waves at the DME in aqueous solution (Table 9-34). The second wave was observed distinctly only in tetra-*n*-butylammonium carbonate buffers of pH 7–9. Like other workers, Burnett and Underwood found that the first wave, at $E_{1/2} \simeq -0.93$ V versus SCE, was essentially pH independent. Coulometry revealed that *one electron* was involved in the first wave process. The slope of the wave [E_{DME} versus $\log(i_d - i)/i$] was that expected for a reversible one-electron reaction, i.e., 60 mV. Evidence suggested that the final product of controlled potential electrolysis was a dimer; i.e., the 1e reversible electrode process forming free radicals was followed by an irreversible chemical dimerization step. A 4,4′-dimer (III, Fig. 9-35) was proposed as the product, although this was not confirmed by structure analysis. The existence of a short-lived primary product of the electrode reaction, namely, a NAD$^+$ free radical (II, Fig. 9-35), was confirmed by fast sweep cyclic voltammetry when, the first cathodic wave of NAD$^+$ having once been scanned, an anodic peak was

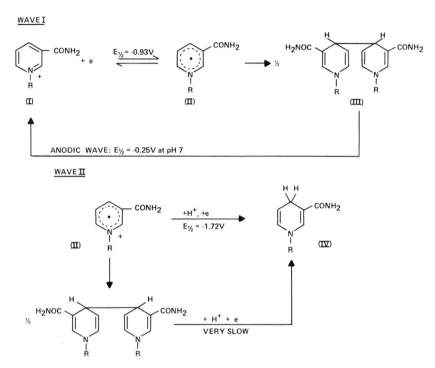

FIG. 9-35. Proposed mechanism for the electrochemical reduction of NAD$^+$ at mercury electrodes.[199]

observed on sweeping toward positive potential. The first cathodic peak and this anodic peak formed an almost reversible couple (Fig. 9-36).[184] The final product of the first wave process, the 4,4'-dimer, was also oxidizable, but at much more positive potentials than the free radical. Thus, at the DME it gave rise to an anodic wave, $E_{1/2} = -0.25$ V versus SCE, at pH 7–9. The anodic wave was due to oxidation of the dimer back to NAD$^+$. The dimer was also oxidizable back to NAD$^+$ by the mung seedling extract employed earlier by Kono and co-workers.[224,225] At potentials where the second wave of NAD$^+$ appeared, the dimeric first wave product was found to be very slowly reducible to NADH (IV, Fig. 9-35) under controlled-potential electrolysis conditions. The mechanism responsible for the second cathodic wave of NAD$^+$ was rather complex and seemed to involve reduction of the NAD$^+$ to a mixture of NADH and dimer. The proposed mechanism is shown schematically in Fig. 9-35.

There is an obvious similarity between this mechanism of reduction of NAD$^+$ and the mechanism observed for 1-N-methylnicotinamide[182] (Fig. 9-31), although the proposed dimerization sites are different. There is also an inconsistency in the mechanism of NAD$^+$ reduction as proposed by Burnett and

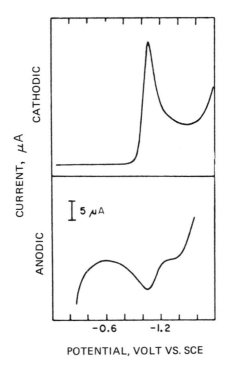

FIG. 9-36.   Cyclic voltammogram of NAD^+ with asymmetric sweep, showing free-radical oxidation. A solution $1.0 \times 10^{-3} M$ in phosphate buffer of pH 7.0; forward (cathodic) sweep rate, 5.7 V sec^{-1}; reverse sweep rate, 115 V sec^{-1}.[199] [Reprinted with permission from J. N. Burnett and A. L. Underwood, *Biochemistry* **4**, 2060 (1965). Copyright by the American Chemical Society.]

Underwood.[199] This arises from the fact that controlled-potential electrolysis of NAD^+ at potentials corresponding to its second polarographic reduction wave is proposed to give small amounts of NADH and a larger amount of dimer. Although the dimer could be transformed to NADH, the process was very slow. If the second wave is due to reduction of the product of the first wave, i.e., the NAD^+ free radical and the dimer, but the dimer is only very slowly reduced and is the major product, then the second wave should be much smaller than the first wave. The data presented in the original papers appear to indicate that the height of the second wave is equal to or greater than that of the first wave.

The studies of Underwood and co-workers[184,199] also revealed that the electrochemical behavior of NAD^+ and related compounds in aqueous solution was complicated by adsorption phenomena, so that attempts to determine the lifetime of the primary free-radical product formed in the first wave process and the rate constant for dimerization of the free radical could not be carried out with any certainty.

It was found that NADH was electrochemically oxidizable back to NAD^+. At a platinum electrode at pH 6 an ill-defined two-electron oxidation peak ($E_p \approx 1.05$ V versus SCE) was observed.

A priori there would appear to be no apparent reason why $NADP^+$ should differ significantly from NAD^+ in its electrochemical behavior. Yet, as noted earlier, much higher yields of NADPH than of NADH have been reported by electrochemical reduction.[221,225] In addition, Carruthers and Tech[197] reported a polarographic $E_{1/2}$ value for the first wave of $NADP^+$ about 250 mV more negative than that of NAD^+ in tris buffer, pH 10.3–10.6 (Table 9-35). However, voltammetry, at a stationary hanging mercury drop electrode, of NAD^+ and $NADP^+$ at pH 7 gave almost identical peak potentials for both compounds ($E_p = -0.995$ V versus SCE for NAD^+ and -1.000 V for $NADP^+$).[184] Cunningham and Underwood[226] therefore made a systematic study of the $NADP^+$ system. Some typical half-wave potential data are presented in Table 9-36.

In acetate, citrate, and phosphate buffers only a single, pH-independent cathodic wave was observed which occurred at almost exactly the same half-wave potential as for NAD^+. The half-wave potential shifted toward more negative potential in tetra-*n*-butylammonium carbonate buffers, but this was considered to be a specific buffer effect rather than protons entering into the potential controlling reaction. At pH values above 7 a second wave of about equal height to the first wave appeared at more negative potential. From the data available it appears that the second wave shifted more negative with increasing pH. The first wave was diffusion controlled, and the second wave appeared to be partially adsorption controlled. The first (more positive) wave

TABLE 9-36   Half-Wave Potentials for the Polarographic Reduction of $NADP^+$

| Buffer system | pH | $E_{1/2}$ (V versus SCE) | | Reference |
		Wave I	Wave II	
Citrate	5–7	−0.92		226
Acetate				
Phosphate				
Tetra-*n*-butyl-ammonium carbonate	5–7	−1.10		226
Pyrophosphate	9	−0.92	−1.70	226
Tetra-*n*-butyl-ammonium carbonate	9	−1.10	−1.75	226
Tris	10.3–10.6	−1.23		197

was essentially reversible, as evidenced by wave slope data and by fast sweep cyclic voltammetry, when a reversible anodic peak could be observed (Fig. 9-37). Controlled potential electrolysis at potentials corresponding to the first wave showed uptake of one electron per molecule of $NADP^+$ reduced. A product was formed which was analogous to that obtained with NAD^+ and was therefore postulated to be a 4,4'-dimer. The product at pH 9 exhibited a small cathodic wave at $E_{1/2} = -1.75$ V versus SCE and a prominent anodic wave at $E_{1/2} = -0.20$ V. The process responsible for the latter wave was oxidation of the dimer to $NADP^+$. Reduction of $NADP^+$ directly at potentials corresponding to its second wave gave a mixture of NADPH and the reduction product of the first wave, i.e., 4,4'-dimer. As was observed with electrolysis of NAD^+ at its second wave, the 4,4'-dimer of $NADP^+$ produced upon electrolysis at potentials of the second wave was very slowly converted to NADPH. The NADPH formed during the reduction could be electrochemically oxidized back to $NADP^+$ at a platinum electrode, $E_p = 0.616$ V versus SCE at pH 9. Thus, apart from certain potentials involved, the electrochemistry of $NADP^+$ can be represented by a mechanism identical to that of NAD^+.

Santhanam and Elving[170] have examined the electrochemical reduction of a number of biologically important nicotinamide nucleotides in dimethyl sulfoxide, including NMN^+, NAD^+, deaminonicotinamide adenine dinucleotide (nicotinamide hypoxanthine dinucleotide, $DNAD^+$), and $NADP^+$. Under DC polarographic conditions all of these compounds give rise to a single cathodic reduction wave (wave I) (Table 9-37). In the presence of proton donors (e.g., benzoic acid), NAD^+ in DMSO exhibits a second, more negative wave (wave II,

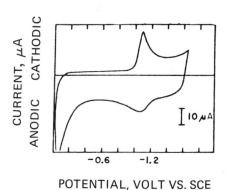

FIG. 9-37.   Cyclic voltammogram for $NADP^+$ with asymmetric sweep, showing free-radical oxidation. A solution $1.0 \times 10^{-3} M$ in phosphate buffer of pH 7.0; forward (cathodic) sweep rate, 5.7 V sec^{-1}; reverse sweep rate, 57.3 V sec^{-1}.[199] [Reprinted with permission from J. N. Burnett and A. R. Underwood, *Biochemistry* **4**, 2060 (1965). Copyright by the American Chemical Society.]

TABLE 9-37   Half-Wave Potentials for the DC Polarographic Reduction of Nicotinamide Nucleotides in Dimethyl sulfoxide[a]

Compound	$E_{1/2}$[b] (V versus SCE)
Nicotinamide mononucleotide (NMN⁺)[c]	−0.99
Nicotinamide adenine dinucleotide (NAD⁺)	−0.98
Deoxynicotinamide adenine dinucleotide (DNAD⁺)[d]	−1.00
Nicotinamide adenine dinucleotide phosphate (NADP⁺)[d]	−1.06

[a] Data from Santhanam and Elving.[170]

[b] Supporting electrolyte, 0.1 M tetraethylammonium perchlorate.

[c] Small adsorption prewave observed at $E_{1/2}$ = −0.38 V.

[d] A second wave at $E_{1/2}$ = −1.96 V observed due to reduction of Na(I) introduced by the use of sodium salts of DNAD⁺ and NADP⁺.

$E_{1/2}$ = −1.99 V) which grows in height with increasing proton-donor concentration and reaches a limiting value equal to the wave I height at a molar ratio of acid to NAD⁺ of about 8. Under cyclic voltammetric conditions a single cathodic peak was observed (peak I_c) at about −1.05 V. Reversal of the sweep at sweep rates up to ca. 2.5 V sec⁻¹ indicated no complementary, reversible anodic peak, although an anodic peak (peak I_a) was observed at much more positive potentials (E_p = −0.23 V for NAD⁺). The other nicotinamide derivatives exhibited similar behavior. At sweep rates ranging from 10 to 80 V sec⁻¹ all nicotinamide nucleotides give a reversible anodic peak corresponding to cathodic peak I_c. Coulometry, polarographic wave analysis, and cyclic voltammetry all supported the view that the polarographic wave I or voltammetric peak I_c was a $1e$ reversible process followed by a fast dimerization step. Examination of the product of controlled potential electrolysis of NAD⁺, NADP⁺, and DNAD⁺ in DMSO by UV spectrophotometry indicated that the product was a 6,6′-dimer. On the basis of these data the polarographic wave I electrode reaction of the cationic nicotinamide nucleotides has been proposed to be an initial $1e$ reaction to give a neutral radical that dimerizes to the 6,6′ derivative (Eq. 23, where R⁺

$$R^+ + e \rightleftharpoons R \cdot \rightarrow \tfrac{1}{2} R{-}R \qquad \text{wave I} \tag{23}$$

represents the nicotinamide nucleotide). The rate constant for dimerization of the free radicals was calculated to be about 10⁶ liters mole⁻¹ sec⁻¹.[170] The 6,6′-dimer was not electrochemically reducible within the available potential range in DMSO. In the presence of proton donors the second wave observed for the nicotinamide nucleotides was proposed to be due to a $1e-1H^+$ reduction of the neutral radical R· (Eq. 24).

$$R· + H^+ + e \rightarrow RH \tag{24}$$

2. Conclusions

Out of all the confusion regarding the electrochemistry of the pyridine nucleotides and their model compounds, it is now reasonably safe to generalize about the gross mechanism.

1. In aqueous solution the pyridine nucleotides can be electrochemically reduced by way of two discrete steps. The first step is an almost reversible $1e$ reduction to give a primary, short-lived free radical that is followed by an irreversible chemical dimerization step. The exact structure of this dimer has not been elucidated.

2. The second wave is complex and appears to involve a competitive reaction. First, some of the free radical produced in the first wave process is further reduced in a $1e-1H^+$ process to the corresponding dihydro compound. Some of the free radicals dimerize before the second electron transfer can occur. The dimer appears to be very slowly reduced to the dihydro compound. That this dimer is reducible, however, may be questionable.

3. The electrode reaction pattern in aprotic nonaqueous media is similar to that observed in aqueous solution except that a second cathodic polarographic wave is not observed unless a proton donor such as benzoic acid is present. In the presence of such a proton donor the second wave in nonaqueous media is a $1e-1H^+$ reduction of the initially formed free radical. The dimer, formed from the free radical produced in the first wave process, is not electrochemically reducible in nonaqueous media.

4. There seems to be only one rational explanation for the varying yields of enzymatically active NADH obtained by various workers using electrolytic methods. This explanation is based on the fact that most reports seem to employ potentials at which vigorous evolution of hydrogen gas would occur. Presumably, the vigorous evolution of hydrogen converts the primary free-radical electrochemical product, or the dimer, or both, to NADH or NADPH. Although at least one report states that hydrogen gas was not visibly evolved in background solution, it is likely that in the presence of NAD^+ or $NADP^+$ or their reduction products some catalytic hydrogen ion reduction would occur. It is also significant that the highest yield of enzymatically active reduction product of NAD^+ was obtained at platinum. Platinum has virtually no hydrogen overpotential, and hence hydrogen evolution at very negative potentials is unavoidable.

5. The reasons for the difference between the activity of the NAD^+ and

NADP$^+$ reduction products formed under similar electrochemical conditions are not clear, and more research is required to resolve this problem.

6. The fact that electrolysis products have never been isolated and subjected to a detailed structure elucidation is unfortunate. Consequently, the structure of the NAD$^+$ and NADP$^+$ dimers are unproven. The potential involvement of *ortho*-dihydro derivatives instead of the enzymatically active *para*-dihydro derivatives also needs investigation.

It is worthwhile considering the potential biological significance of the electrochemical data. First, the standard oxidation–reduction potential (E^0) of the NAD$^+$/NADH system in enzymatic reactions has been calculated to be -0.325 V versus NHE at pH 7.4 by Clarke[227] and -0.318 V at pH 7.0 (30°C) by Rodkey.[228] These potentials correspond to -0.56 to -0.57 V versus SCE. The cyclic voltammetrically reversible peak for the first one-electron reduction of NAD$^+$ occurs at -0.93 V versus SCE (-0.69 V versus NHE). The significance of the difference between these potentials is not clear and, superficially, they would appear to correspond to different processes; i.e., the enzymatic NAD$^+$/NADH system has an apparent E^0 of ca. 0.32 V versus NHE, and the electrochemical reversible $E_{1/2}$ ($E^{0'}$) of ca. -0.69 V corresponds to the

system.

Most biologists and biochemists accept the view that enzymatic reductions of NAD$^+$ and NADP$^+$ to NADH and NADPH, respectively, proceed by hydride ion transfers[229] rather than by a radical- or electron-transfer mechanism. However, it is undisputed that NAD$^+$ and NADP$^+$ participate in biological electron-transport systems in which radicals of other species such as flavins have been recognized. In fact, it has been found by Commoner *et al.*[230] that, during the enzymatic reduction of NAD$^+$ with ethanol and the oxidation of NADH with acetaldehyde, an ESR signal is observed. This finding could not be confirmed by Mahler and Brand,[231] yet they obtained a stable ESR signal in a reaction involving NADH and riboflavin which could not be definitely attributed to the flavin moiety. Schellenberg and Hellerman[190] have found that NADH is oxidized more effectively by reagents that behave as 1e acceptors than by reagents that behave as 2e acceptors, and in fact a free-radical intermediate was postulated for these nonenzymatic reactions. Of course, none of these findings establishes that in a biological situation an electron-transfer process involving free radicals is involved in the oxidation–reduction reactions of the pyridine nucleotides. They do, however, raise the possibility of such intermediates existing and participating in the biological system.

C. Electrochemistry of Pyridoxol and Related Compounds

Pyridoxol (pyridoxine), pyridoxamine, and certain substituted derivatives of these are compounds belonging to the vitamin B_6 group. Although these compounds are not presently thought to be involved in biological electron-transfer reactions in the sense that the pyridine nucleotides are, they have been examined polarographically fairly extensively. A summary of polarographic half-wave potentials is presented in Table 9-38.[101,157,232−234]

Lingane and Davis[157] were the first to observe that pyridoxol (pyridoxine) was electroactive. Volke and co-workers[235] then reported that pyridox-amine gave a 2e reduction wave, but no mechanistic details were presented. The use of polarography for the analytical determination of pyridoxol and related compounds has also been reported by a number of workers.[101,104,105,232,233,236−239] The most comprehensive and systematic study of pyridoxol derivatives is that of Manoušek and Zuman.[234] The bulk of the following discussion is based on their work.

TABLE 9-38 Half-Wave Potentials for the Polarographic Reduction[a] of Pyridoxol and Related Compounds

Compound	Buffer system	pH	$E_{1/2}$ (V versus SCE) Wave I[b]	Wave II	Wave III	Reference
Pyridoxol (pyridoxine)	Tetramethyl-ammonium iodide		−1.8	−2.0		157
		4	−0.55[c]			101
	Veronal	8.95	−0.874			232
	NH_4OH/NH_4Cl	8.7	−1.695			233
	NH_4OH/NH_4Cl	8.7	−1.70			234
		8.9	−1.72			234
		9.3	−1.73			234
		9.8	−1.75			234
		10.1	−1.99			234
4-Methoxymethyl-pyridoxine		8.7	−1.68			234
		8.9	−1.70			234
		9.3	−1.71			234
		9.8	−1.75			234
		10.1	−1.77			234
Pyridoxamine		8.7	−1.53			234
		8.9	−1.58			234
		9.3	−1.61			234
		9.8	−1.67			234
		10.1	−1.69			234

TABLE 9-38 *Continued*

Compound	Buffer system	pH	Wave I[b]	Wave II	Wave III	Reference
				$E_{1/2}$ (V versus SCE)		
Pyridoxamine		8.7	−1.68			234
5-phosphate		8.9	−1.69			234
		9.3	−1.72	−1.82		234
		9.8		−1.86		234
		10.1		−1.88		234
4-Pyridoxthiol		3.9	(−0.21)	−0.99		234
		4.9	(−0.29)	−1.01		234
		5.6	(−0.33)	−1.08		234
		6.6	(−0.39)	−1.14		234
		7.6	(−0.46)	−1.26		234
		8.8	(−0.51)	−1.39		234
		9.7	(−0.54)		−1.55	234
		10.7	(−0.58)		−1.81	234
Bis-4-pyridoxal		3.9	−0.40	−0.99		234
disulfide		4.9	−0.45	−1.01		234
		5.6	−0.48	−1.08		234
		6.6	−0.54	−1.14		234
		7.6	−0.61	−1.26		234
		8.8	−0.67	−1.39		234
		9.7	−0.70		−1.55	234
		10.7	−0.73		−1.81	234
Pyridoxal		8.9	−0.88			234
5-phosphate		10.1	−0.99	−1.13		234
		10.8	−1.03	−1.14		234
		11.2	−1.05	−1.14	−1.39	234
		11.6	−1.05	−1.15	−1.37	234
		11.8		−1.16	−1.35	234
Pyridoxaloxine		9.9	−0.98			234
		10.4	−1.02	−1.22		234
		10.8	−1.06	−1.23	−1.49	234
		11.0	−1.06	−1.24	−1.50	234
		11.2	−1.06	−1.25	−1.50	234
		11.4		−1.26	−1.51	234
Pyridoxaloxine		9.2	−1.04	−1.36		234
5-phosphate		9.8	−1.05	−1.36		234
		10.4	−1.05	−1.36		234
		10.9		−1.36	−1.70	234
		11.5		−1.36	−1.80	234

[a] Except where otherwise stated.

[b] Values in parentheses correspond to anodic wave.

[c] Versus the normal calomel electrode.

1. Pyridoxol (Pyridoxine)

Pyridoxol (I, Fig. 9-38) shows a single diffusion-controlled cathodic polaro-graphic wave in ammoniacal buffers between about pH 8.5 and 10 (Table 9-38). At higher pH the wave was very close to background electrolyte discharge, but the height of the wave appeared to decrease. The process responsible for the polarographic wave is a two-electron reduction of the protonated species at the activated C—OH bond in position 4 to give, ultimately, 4-deoxypyridoxine (II, Fig. 9-38). Since the half-wave potential for the wave is almost pH independent (Table 9-38, Fig. 9-39), it appears likely that protons are not involved in the potential-controlling reaction. The apparent decrease in the height of the wave above pH 10 was assumed to be due to the dissociation of the electroactive pyridinium ion (I, Fig. 9-38).

2. 4-Methoxymethylpyridoxine

4-Methoxymethylpyridoxine (III, Fig. 9-38) shows a single faradaic $2e$ diffusion-controlled wave between about pH 8.5 and 10 in ammonia buffers

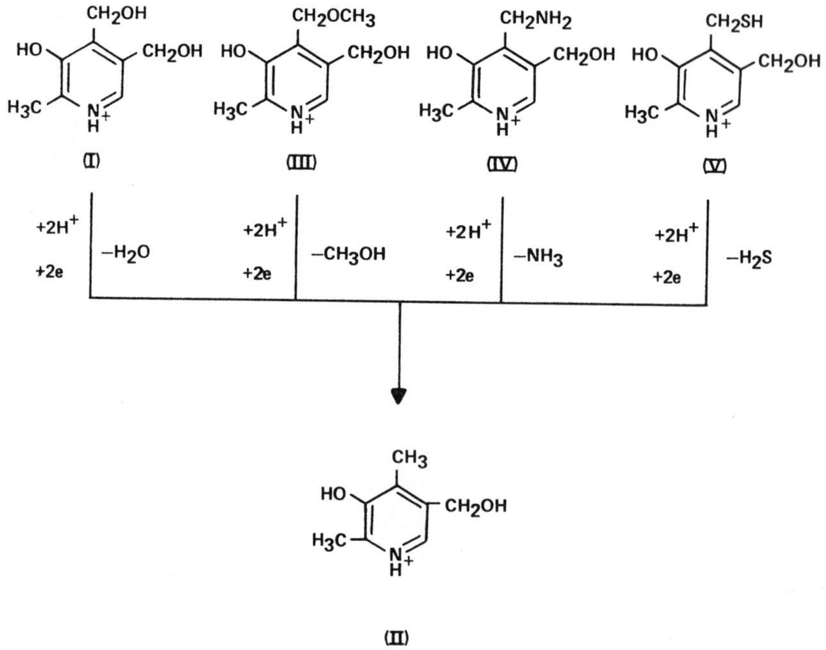

FIG. 9-38. Proposed mechanism of polarographic reduction of pyridoxol (I), 4-methoxymethylpyridoxine (III), pyridoxamine (IV), and 4-pyridoxthiol (V) to 4-deoxypyridoxine (II).[234]

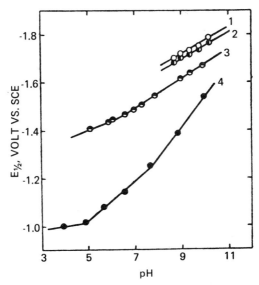

FIG. 9-39. Variation of $E_{1/2}$ with pH for some pyridoxol derivatives. (1) Pyridoxol, (2) 4-methoxymethylpyridoxine, (3) pyridoxamine, (4) 4-pyridoxthiol.[234] (Reprinted with permission of Academia Publishing House, Prague.)

which is very similar to that of pyridoxol (Table 9-38, Fig. 9-39). Methyl alcohol was detected in a solution after controlled potential electrolysis, and the final product was 4-deoxypyridoxine (II, Fig. 9-38). In veronal buffers above pH 8.4 a catalytic wave was also observed at potentials more negative than the normal wave.

3. Pyridoxamine

Pyridoxamine (IV, Fig. 9-38) was first shown to be polarographically reducible by Volke *et al.*[235] Between pH 5 and 11 it shows a single, diffusion-controlled, pH-dependent wave which involves two electrons. At higher pH the limiting current decreases, presumably due to a decrease in the concentration of the protonated electroactive species. Half-wave potential data are presented in Fig. 9-39 and Table 9-38. Controlled-potential electrolysis of pyridoxamine caused the liberation of ammonia, and presumably once again 4-deoxypyridoxine (II, Fig. 9-38) was the product.

4. 4-Pyridoxthiol

4-Pyridoxthiol (V, Fig. 9-38) shows a total of three polarographic waves (Table 9-38). That observed at most positive potential is an anodic wave involving one electron which is ascribed to the formation of a compound with mercury. The first cathodic wave involves two electrons, is pH dependent, and is constant in height and diffusion controlled up to about pH 9, when, with further increase in pH, it decreases. Above pH 9 a second, more negative cathodic wave appears. The first 2e cathodic wave corresponds to reduction of the thioalcohol group at position 4 of the protonated 4-pyridoxthiol since H_2S is evolved on prolonged electrolysis. The process responsible for the second cathodic wave at very negative potentials is not clear.

5. Pyridoxamine 5-Phosphate

In veronal and ammonia buffers pyridoxamine 5-phosphate (I, Fig. 9-40) shows a two-electron reduction process. Below pH 8 just a single wave is observed, but between pH 8 and 10 two waves appear (Table 9-38), the height of the more positive wave decreasing with increasing pH and that of the second wave growing. The sum of the heights of the two waves is practically pH independent. Controlled-potential electrolysis reveals that ammonia is liberated. The fact that the two waves change in height with pH relative to each other indicates the participation of acid—base equilibria in the electrode process, possibly involving the phosphate moiety. It would appear that the ultimate product of electrolysis is the 5-phosphate ester of 4-deoxypyridoxine (II, Fig. 9-40).

6. Bis-4-Pyridoxyl Disulfide

Generally, the polarographic reduction of bis-4-pyridoxyl disulfide (I, Fig. 9-41) proceeds via two waves (Table 9-38). The first, more positive wave involves two electrons and two protons and results in the reduction of the disulfide to 4-pyridoxthiol (II, Fig. 9-41). The second wave is due to further reduction of the two molecules of 4-pyridoxthiol to 4-deoxypyridoxine (III, Fig. 9-41) and H_2S, with the consumption of a further four electrons and four protons. Both of the reduction waves are accompanied by maxima of the first

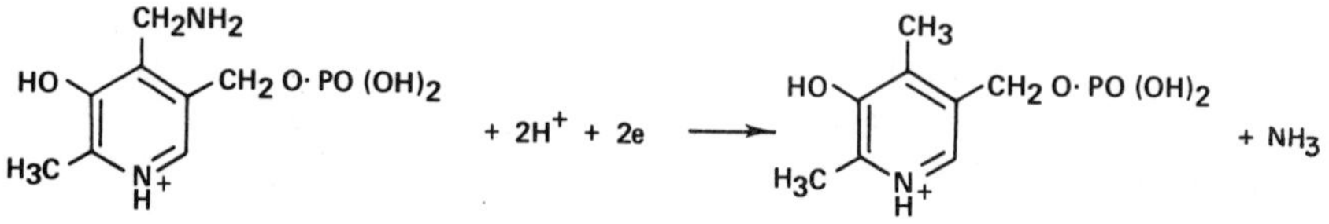

FIG. 9-40.   Possible   mechanism   of   polarographic   reduction   of   pyridoxamine 5-phosphate.

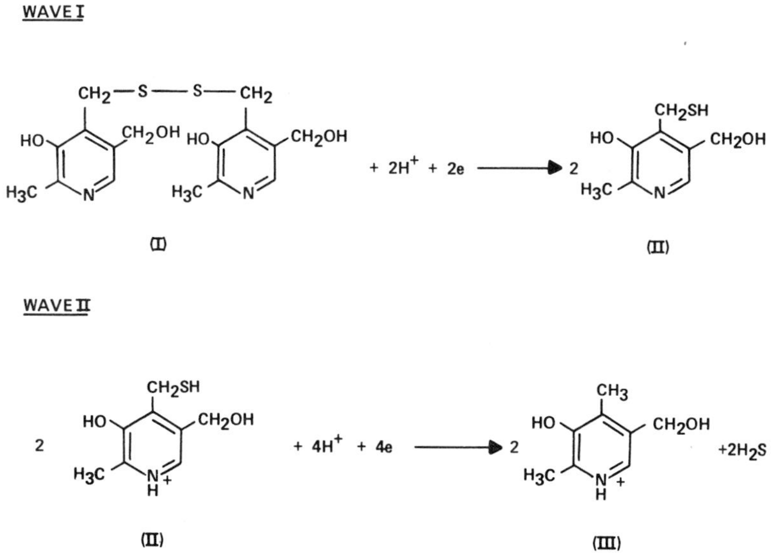

WAVE I

$$\text{(I)} \quad + 2H^+ + 2e \longrightarrow 2 \quad \text{(II)}$$

WAVE II

$$2 \ \text{(II)} \quad + 4H^+ + 4e \longrightarrow 2 \ \text{(III)} \quad + 2H_2S$$

FIG. 9-41. Proposed mechanism for the polarographic reduction of bis-4-pyridoxyl disulfide (I).[234]

kind. At above pH 11 a trough is observed on the limiting current. This decrease in the current has been proposed to be due to repulsion of the dianion (XI) from the negatively charged electrode surface.

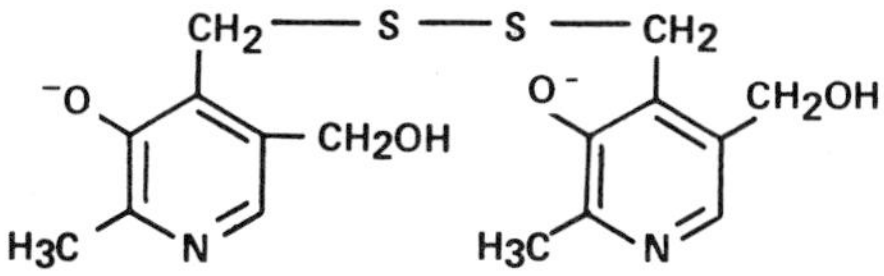

(XI)

7. Pyridoxal

Pyridoxal (XII) appears to behave in a fashion expected for pyridine

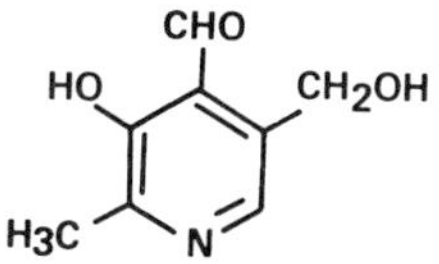

(XII)

aldehydes (see Section IV, B); i.e., the height of the single pH-dependent wave (Fig. 9-42) depends on the acidity of the solution. Volke[101] proposed that the limiting current is governed by the rate of (acid–base)-catalyzed ring opening of

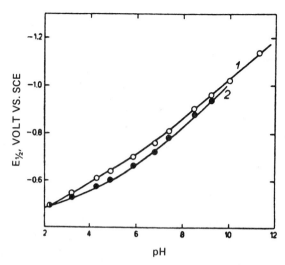

FIG. 9-42.   Variation of $E_{1/2}$ with pH for pyridoxal (1) and pyridoxal 5-phosphate (2).[234] (Reprinted with permission of Academia Publishing House, Prague.)

the hemiacetal form of the aldehyde group. However, Manoušek and Zuman[234] proposed that the rate of dehydration of the hydrated aldehyde group is equally plausible. Nevertheless, in acid solution a single wave is observed which increases in height, as the acidity is made greater, in the form of a dissociation curve until it reaches the height expected for a 2*e* reduction. No doubt the ultimate product is pyridoxol, although, curiously, no wave attributable to this compound was reported.

8. Pyridoxal 5-Phosphate

Below pH 10 pyridoxal 5-phosphate shows only a single, pH-dependent (Fig. 9-42), two-electron, diffusion-controlled wave that, on the basis of oscillopolarographic experiments, is irreversible.[234] Above pH 10 this wave decreases and a second, more negative wave appears (Table 9-38). Above pH 10.6 both the first and second waves decrease in height and a third, even more negative wave appears. The sum of the heights of all three waves is virtually constant. No doubt various acid–base equilibria are responsible for the observed behavior. The polarographic waves are ascribed to reduction of the aldehyde group over the whole pH range.

Manoušek and co-workers utilized the polarographic behavior of pyridoxal and pyridoxal 5-phosphate for an analytical method for both compounds.[232,233,238] Manoušek and Zuman also employed polarography to study the hydrolysis of pyridoxal 5-phosphate.[237,239] Uehara and Nakaya[240] have used polarographic methods to study transamination reactions between amino acids and keto acids involving pyridoxal or pyridoxamine.

9. Pyridoxaloxime

At pH values below 9 pyridoxaloxime (I, Eq. 25) gives a single, four-electron,

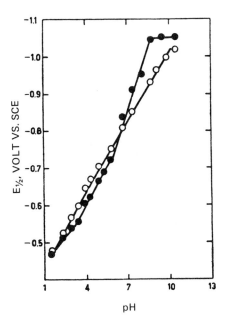

$$+ 4H^+ + 4e \longrightarrow +H_2O \quad (25)$$

(I)   (II)

diffusion-controlled wave (Fig. 9-43).[234] At higher pH two further waves are observed at more negative potentials (Table 9-38). The first wave decreases in height and the second wave grows and then decreases at even higher pH as the third wave appears. The overall process at all pH values appears to be a $4e$

FIG. 9-43. Variation of $E_{1/2}$ with pH for pyridoxaloxime (●) and pyridoxaloxime 5-phosphate (○).[234] (Reprinted with permission of Academia Publishing House, Prague.)

reduction of the oxime group to the corresponding amine (II, Eq. 25), complicated by protonation equilibria. Matsumoto and co-workers[241] have reported similar behavior.

10. Pyridoxaloxime 5-Phosphate

The polarographic behavior of pyridoxaloxime 5-phosphate is very similar to that observed for pyridoxaloxime (Fig. 9-43, Table 9-38), so that a four-electron reduction of the oxime group occurs.

REFERENCES

1. H. S. Mosher, *Heterocycl. Compd.* **1**, 617 (1950).
2. E. Klingsberg, ed. "Pyridine and its Derivatives, Parts 1–4. Wiley (Interscience), New York, 1960, 1961, 1962, 1964.
3. K. Schofield, "Heteroaromatic Nitrogen Compounds: Pyrroles and Pyridines." Plenum, New York, 1967.
4. E. E. Conn and P. K. Stumpf, "Outlines of Biochemistry," 2nd ed. Wiley, New York, 1967.
5. E. E. Conn and P. K. Stumpf, "Outlines of Biochemistry," 2nd ed., pp. 258–259. Wiley, New York, 1967.
6. B. Emmert, *Ber. Dtsch. Chem. Ges.* **46**, 1716 (1913).
7. M. Shikata and I. Tachi, *Mem. Coll. Agric., Kyoto Imp. Univ.* **4**, 19 and 35 (1927).
8. M. Shikata and I. Tachi, *Bull. Agric. Chem. Soc. Jpn.* **3**, 53 (1927); *Chem. Abstr.* **22**, 720 (1928).
9. M. Shikata and I. Tachi, *Chem. News J. Ind. Sci.* **137**, 133 (1928); see also *Chem. Abstr.* **22**, 720, 1894, 2870, 4035, 4036 (1928).
10. P. C. Tompkins and C. L. A. Schmidt, *J. Biol. Chem.* **143**, 643 (1942).
11. M. K. Shchennikova and I. A. Korshunov, *Zh. Fiz. Khim.* **22**, 503 (1948); *Chem. Abstr.* **42**, 7169 (1948).
12. E. Knobloch, *Chem. Listy* **39**, 54 (1945); *Chem. Abstr.* **42**, 1829 (1948).
13. E. Knobloch, *Collect. Czech. Chem. Commun.* **12**, 407 (1947); *Chem. Abstr.* **42**, 1829 (1948).
14. R. C. Kaye and H. I. Stonehill, *J. Chem. Soc.* p. 2638 (1951).
15. R. C. Kaye and H. I. Stonehill, *J. Chem. Soc.* p. 3240 (1952).
16. G. F. Reynolds and A. J. Lindsay, *Z. Phys. Chem. (Leipzig)* **223**, 141 (1963).
17. G. F. Reynolds and A. J. Lindsay, *Abh. Dtsch. Akad. Wiss. Berlin, Kl. Chem., Geol. Biol.* p. 443 (1964); *Chem. Abstr.* **62**, 6378 (1965).
18. J. Kůta and J. Drabek, *Chem. Listy* **49**, 23 (1955).
19. J. Kůta and J. Drabek, *Collect. Czech. Chem. Commun.* **20**, 902 (1955); *Chem. Abstr.* **49**, 5996 (1955).
20. A. G. Pozdeeva and E. M. Gepshtein, *Zh. Obshch. Khim.* **22**, 2065 (1952); *Chem. Abstr.* **47**, 9325 (1953).
21. R. Kalvoda, *Chem. Zvesti* **18**, 347 (1964); *Chem. Abstr.* **62**, 1332 (1965).
22. S. G. Mairanovskii and N. V. Barashkova, *Izv. Akad. Nauk SSSR, Otd. Khim. Nauk* p. 186 (1962); *Chem. Abstr.* **56**, 15286 (1962).

23. G. A. Tedoradze, S. G. Mairanovskii, and L. D. Klyukina, *Izv. Akad. Nauk SSSR, Otd. Khim. Nauk* p. 1352 (1961); *Chem. Abstr.* **55**, 26787 (1961).
24. L. D. Klyukina and S. G. Mairanovskii, *Izv. Akad. Nauk SSSR, Ser. Khim.* p. 1183 (1963); *Chem. Abstr.* **59**, 14901 (1963).
25. M. K. Polievktov and S. G. Mairanovskii, *Elektrokhimiya* **3**, 139 (1967); *Chem. Abstr.* **66**, 115202x (1967).
26. M. K. Polievktov, S. G. Mairanovskii, and M. Gonikberg, *Kinet. Katal.* **8**, 316 (1967); *Chem. Abstr.* **67**, 90316t (1967).
27. S. G. Mairanovskii, *Izv. Akad. Nauk SSSR, Otd. Khim. Nauk* p. 784 (1962); *Chem. Abstr.* **57**, 12252 (1962).
28. S. G. Mairanovskii and M. K. Polievktov, *Collect. Czech. Chem. Commun.* **30**, 4168 (1965); *Chem. Abstr.* **64**, 15718 (1966).
29. S. G. Mairanovskii and V. P. Gul'tyai, *Elektrokhimiya* **1**, 460 (1965); *Chem. Abstr.* **63**, 9778 (1965).
30. S. G. Mairanovskii and R. G. Baisheva, *Elektrokhimiya* **6**, 226 (1970); *Chem. Abstr.* **72**, 128013k (1970).
31. M. K. Polievktov and S. G. Mairanovskii, *Izv. Akad. Nauk SSSR, Ser. Khim.* p. 413 (1965); *Chem. Abstr.* **63**, 9456 (1965).
32. M. K. Polievktov and S. G. Mairanovskii, *Elektrokhimiya* **2**, 1487 (1966); *Chem. Abstr.* **66**, 51660k (1967).
33. S. G. Mairanovskii, J. Koutecký, and V. Hanus, *Zh. Fiz. Khim.* **36**, 2621 (1962); *Chem. Abstr.* **58**, 8646 (1963).
34. S. G. Mairanovskii, V. P. Gul'tyai, and N. K. Lisitsina, *Elektrokhimiya* **6**, 1202 (1970); *Chem. Abstr.* **73**, 126,446a (1970).
35. S. G. Mairanovskii, *Dokl. Akad. Nauk SSSR* **110**, 593 (1956); *Chem. Abstr.* **52**, 3565 (1958).
36. S. G. Mairanovskii, "Catalytic and Kinetic Waves in Polarography." Plenum, New York, 1968.
37. P. H. Given, *J. Chem. Soc.* p. 2684 (1958).
38. G. Anthoine, G. Copens, J. Nasielski, and E. V. Donckt, *Bull. Soc. Chim. Belg.* **73**, 65 (1964).
39. J. Koutecký and R. Zahradnik, *Collect. Czech. Chem. Commun.* **28**, 2089 (1963).
40. C. Parkanyi and R. Zahradnik, *Bull. Soc. Chim. Belg.* **73**, 57 (1964).
41. K. W. Wiberg and T. P. Lewis, *J. Am. Chem. Soc.* **92**, 7154 (1970).
42. B. J. Tabner and Y. R. Yandle, *J. Chem. Soc. A* p. 381 (1968).
43. T. A. Mikhailova, N. I. Kudryashova, and N. V. Khromov-Borisov, *Zh. Obshch. Khim.* **39**, 26 (1969).
44. R. D. Armstrong, *J. Electroanal. Chem.* **20**, 168 (1969).
45. G. H. Aylward and J. W. Hayes, *J. Electroanal. Chem.* **8**, 442 (1964).
46. R. G. Barradas, P. G. Hamilton, and B. E. Conway, *Collect. Czech. Chem. Commun.* **32**, 1790 (1967).
47. B. E. Conway and L. G. M. Gordon, *J. Phys. Chem.* **73**, 3609 (1969).
48. B. B. Damaskin, A. A. Survila, S. Y. Vasina, and A. I. Fedorova, *Elektrokhimiya* **3**, 825 (1967); *Chem. Abstr.* **67**, 87165n (1967).
49. S. L. Gupta and S. K. Sharma, *J. Indian Chem. Soc.* **43**, 53 (1966).
50. R. Kalvoda, *J. Electroanal. Chem.* **1**, 314 (1960).
51. R. Kalvoda and A. Vaškellis, *Collect. Czech. Chem. Commun.* **32**, 2206 (1967); *Chem. Abstr.* **67**, 28732x (1967).
52. B. Kastening and L. Holleck, *Z. Elektrochem.* **64**, 823 (1960); *Chem. Abstr.* **55**, 177 (1961).

53. B. Breyer and S. Hacobian, *Aust. J. Sci. Res., Ser. A* **5**, 500 (1952).
54. H. A. Laitinen and B. Mosier, *J. Am. Chem. Soc.* **80**, 2363 (1938).
55. B. Breyer and H. H. Bauer, "Alternating Current Polarography and Tensammetry," pp. 239–240. Wiley (Interscience), New York, 1963.
56. B. E. Conway and R. G. Barradas, *Electrochim. Acta* **5**, 319 (1961).
57. R. G. Barradas and P. G. Hamilton, *Can. J. Chem.* **43**, 2468 (1965).
58. B. E. Conway, R. G. Barradas, P. G. Hamilton, and J. M. Perry, *J. Electroanal. Chem.* **10**, 485 (1965).
59. Y. Matsuda and H. Tamura, *Electrochim. Acta* **14**, 427 (1959).
60. Y. Matsuda and H. Tamura, *Kogyo Kagaku Zasshi* **70**, 2121 (1967); *Chem. Abstr.* **68**, 65112m (1968).
61. A. Hamelin and G. Valette, *C. R. Hebd. Seances Acad. Sci., Ser. C* **267**, 127 and 211 (1968).
62. M. Petit, N. V. Huong, and J. Clavilier, *C. R. Hebd. Seances Acad. Sci., Ser. C* **266**, 300 (1968).
63. S. G. Mairanovskii, *Zh. Fiz. Khim.* **33**, 691 (1959); *Chem. Abstr.* **54**, 2046 (1960).
64. S. G. Mairanovskii, *Dokl. Akad. Nauk SSSR* **142**, 1327 (1962); *Chem. Abstr.* **57**, 12252 (1962).
65. P. C. Tompkins and C. L. A. Schmidt, *Univ. Calif., Berkeley, Publ. Physiol.* **8**, 221 (1943); S. Wawzonek, *Anal. Chem.* **21**, 64 (1949).
66. P. C. Tompkins and C. L. A. Schmidt, *Univ. Calif., Berkeley, Publ. Physiol.* **8**, 229, 237, and 247 (1943); through Colichman and O'Donovan, Ref. 67.
67. E. L. Colichman and P. A. O'Donovan, *J. Am. Chem. Soc.* **76**, 3588 (1954).
68. S. G. Mairanovskii, *Dokl. Akad. Nauk SSSR* **110**, 593 (1956).
69. S. G. Mairanovskii, "Proceedings of the All-Union Electrochemical Convention, 4th." Moscow, 1956. [Transl.: *Soveshch. Elektrokhim.*, **4**, 223 (1959).]; *Chem Abstr,* **54**, 9558 (1960).
70. E. Ochiai and H. Kataoka, *J. Pharm. Soc. Jpn.* **62**, 241 (1942); *Chem. Abstr.* **45**, 5150e (1951).
71. H. Yasuda and S. Kitagawa, *Jpn. J. Pharm. Chem.* **27**, 779 (1955); *Chem. Abstr.* **51**, 13246 (1957).
72. C. P. Yuan and P. T. Li, *K'o Hsueh T'ung Pao* **17**, 303 (1966); *Chem. Abstr.* **67**, 39958b (1967).
73. M. S. Spritzer, J. M. Costa, and P. J. Elving, *Anal. Chem.* **37**, 211 (1965).
74. J. E. Hickey, M. S. Spritzer, and P. J. Elving, *Anal. Chim. Acta* **35**, 277 (1966).
75. L. Floch, M. S. Spritzer, and P. J. Elving, *Anal. Chem.* **38**, 1074 (1966).
76. K. Tsuji and P. J. Elving, *Anal. Chem.* **41**, 1571 (1969).
77. A. Cisak and P. J. Elving, *Electrochim. Acta* **10**, 935 (1965).
78. W. R. Turner and P. J. Elving, *Anal. Chem.* **37**, 467 (1965).
79. P. Baumgarten, *Ber. Dtsch. Chem. Ges B* **69**, 1938 (1936)
80. P. Baumgarten and E. Dammann, *Ber. Dtsch. Chem. Ges. B* **66**, 1633 (1933).
81. A. G. Pozdeeva and E. G. Novikov, *Zh. Prikl. Khim.* **42**, 2626 (1969); *Chem. Abstr.* **72**, 74009t (1970).
82. R. I. Kaganovich and B. B. Damaskin, *Sov. Electrochem. (Engl. Transl.)* **4**, 221 (1968).
83. R. G. Barradas, M. C. Giordano, and W. H. Sheffield, *Electrochim. Acta* **16**, 1235 (1971).
84. A. Foffani and E. Fornasari, *Gazz. Chim. Ital.* **83**, 1051 (1953); *Chem. Abstr.* **49**, 2901 (1955).
85. T. Kubota and H. Miyazaki, *Bull. Chem. Soc. Jpn.* **35**, 1549 (1962).
86. T. Kubota and H. Miyazaki, *Bull. Chem. Soc. Jpn.* **39**, 2057 (1966).

87. G. Horn, *Monatsber. Dtsch. Akad. Wiss. Berlin* **3**, 386 (1961); *Chem. Abstr.* **57**, 3196 (1962).
88. G. Horn, *Acta Chim. Acad. Sci. Hung.* **27**, 123 (1961).
89. E. Ochiai, *J. Org. Chem.* **18**, 534 (1953).
90. J. Miroslav and F. Miloslav, *Collect. Czech. Chem. Commun.* **35**, 2802 (1970).
91. T. Kubota, K. Nishikida, and H. Miyazaki, *J. Am. Chem. Soc.* **90**, 5080 (1968).
92. G. Anthoine, J. Nasielski, E. V. Donckt, and N. Vanlantem, *Bull. Soc. Chim. Belg.* **76**, 230 (1967).
93. L. V. Varyukhina and Z. V. Pushkareva, *Zh. Obshch. Khim.* **26**, 1740 (1956); *Chem. Abstr.* **51**, 1960 (1957).
94. H. Miyazaki and T. Kubota, *Bull. Chem. Soc. Jpn.* **44**, 279 (1971).
95. K. Ezumi, H. Miyazaki, and T. Kubota, *J. Phys. Chem.* **74**, 2397 (1970).
96. E. Fornasari and A. Foffani, *Gazz. Chim. Ital.* **83**, 1059 (1953); *Chem. Abstr.* **49**, 2901 (1955).
97. E. Laviron, R. Gavasso, and M. Pay, *Talanta* **16**, 293 (1969).
98. E. Laviron, R. Gavasso, and M. Pay, *Talanta* **17**, 747 (1970).
99. I. El-Khiami and R. M. Johnson, *Talanta* **14**, 745 (1967).
100. See L. Meites, "Polarographic Techniques," 2nd ed., Chapter 4 [Wiley (Interscience), New York, 1965], for an introduction to the kinetics of polarographic electron-transfer reactions.
101. J. Volke, *Z. Phys. Chem. (Leipzig)* p. 268 (1958).
102. J. Volke, *Chem. Listy* **52**, 16 (1958).
103. J. Volke, *Collect. Czech. Chem. Commun.* **23**, 1486 (1958); *Chem. Abstr.* **52**, 13482 (1958).
104. J. Volke and P. Valenta, *Collect. Czech. Chem. Commun.* **25**, 1580 (1960); *Chem. Abstr.* **58**, 13448 (1963).
105. J. Volke, *Chem. Listy* **55**, 26 (1961); *Chem. Abstr.* **55**, 6480 (1961).
106. E. Fornasari, G. Giacometti, and G. Rigatti, *Adv. Polarogr., Proc. Int. Congr., 2nd, 1959* **3**, 895 (1960).
107. E. Fornasari, G. Giacometti, and G. Rigatti, *Ric. Sci.* **30**, Suppl. 5, 261 (1961); *Chem. Abstr.* **55**, 16221 (1961).
108. G. Giacometti, *Chim. Ind. (Milan)* **39**, 123 (1957).
109. J. Volke, *Experientia* **13**, 274 (1957).
110. J. Tirouflet, P. Fornasari, and J. P. Chané, *C. R. Hebd. Seances Acad. Sci.* **242**, 1799 (1956).
111. J. Tirouflet and E. Laviron, *C. R. Hebd. Seances Acad. Sci.* **247**, 217 (1958).
112. E. Laviron and J. Tirouflet, *Adv. Polarogr. Proc. Int. Congr., 2nd, 1959* Vol. 2, p. 727 (1960).
113. J. Nakaya, *Nippon Kagaku Zasshi* **81**, 1731 (1960); *Chem. Abstr.* **56**, 2275 (1962).
114. J. Volke, R. Kubíček, and F. Šantavý, *Collect. Czech. Chem. Commun.* **25**, 871 (1960); *Chem. Abstr.* **58**, 13448 (1963).
115. J. Holubek and J. Volke, *Collect. Czech. Chem. Commun.* **25**, 3292 (1960); *Chem. Abstr.* **53**, 5920 (1959).
116. J. Holubek and J. Volke, *Adv. Polarogr., Proc. Int. Congr., 2nd, 1959* **3**, 847 (1960).
117. M. Kalousek, *Collect. Czech. Chem. Commun.* **13**, 105 (1948).
118. W. L. Bencze and M. J. Allen, *J. Am. Chem. Soc.* **81**, 4015 (1959).
119. M. J. Allen and H. Cohen, *J. Electrochem. Soc.* **106**, 451 (1959).
120. L. Stárka and J. Buben, *J. Pharm. Pharmacol.* **12**, 175 (1966).
121. M. J. Allen, *J. Chem. Soc.* p. 1420 (1961).
122. J. Volke, *Collect. Czech. Chem. Commun.* **27**, 483 (1962).

123. J. Volke, *Collect. Czech. Chem. Commun.* **25**, 3397 (1960).

124. S. Ono, *Rev. Polarogr.* **7**, 102 (1959).

125. J. Holubek and J. Volke, *Collect. Czech. Chem. Commun.* **25**, 3286 (1960).

126. P. Tomasik, *Rocz. Chem.* **44**, 341 (1970).

127. J. Holubek and J. Volke, *Collect. Czech. Chem. Commun.* **27**, 680 (1962).

128. L. W. Marple, L. E. J. Hummelstedt, and L. B. Rogers, *J. Electrochem. Soc.* **107**, 437 (1960).

129. R. F. Evilia and A. J. Diefenderfer, *J. Electroanal. Chem.* **22**, 407 (1969).

130. S. G. Mairanovskii and R. G. Baisheva, *Elektrokhimiya* **5**, 893 (1969); *Chem. Abstr.* **71**, 76762 (1969).

131. J. Chodkowski and T. G. Jakubczak, *Rocz. Chem.* **41**, 373 (1967).

132. J. Chodkowski and T. G. Jakubczak, *Rocz. Chem.* **43**, 1037 (1969).

133. A. B. Zahlan and R. H. Linnell, *J. Am. Chem. Soc.* **77**, 6207 (1955).

134. P. Silvestroni, *Ric. Sci.* **24**, 1695 (1954); *Chem. Abstr.* **49**, 5996 (1955).

135. J. Volke and J. Holubek, *Collect. Czech. Chem. Commun.* **27**, 1777 (1962).

136. M. Salvatore, *Ric. Sci.* **39**, 620 (1969).

137. P. Tomasik, *Rocz. Chem.* **44**, 1211 (1970).

138. J. Volke, R. Kubíček, and F. Šantavý, *Collect. Czech. Chem. Commun.* **25**, 1510 (1960); *Chem. Abstr.* **58**, 13448 (1963).

139. R. H. Linnell, *J. Am. Chem. Soc.* **76**, 1391 (1954).

140. J. L. Sadler and A. J. Bard, *J. Am. Chem. Soc.* **90**, 1979 (1968).

141. P. G. W. Scott, *J. Pharm. Pharmacol.* **4**, 681 (1952).

142. J. D. Neuss, W. J. Seagers, and W. J. Mader, *J. Am. Pharm. Assoc.* **41**, 670 (1952).

143. A. Anastasi, E. Mecarelli, and L. Novacic, *Mikrochem. Ver. Mikrochim. Acta* **40**, 113 (1952); *Chem. Abstr.* **47**, 3185 (1953).

144. H. A. Offe, W. Siefken, and G. Domagk, *Naturwissenschaften* **39**, 118 (1952).

145. H. H. Fox, U.S. Patent 2,596,069 (1952).

146. T. A. Mikhailova, N. I. Kudryashova, and N. V. K. Borisov, *Zh. Obshch. Khim.* **39**, 26 (1969).

147. F. Šorm and Z. Šormová, *Chem. Listy* **42**, 82 (1948).

148. E. G. Novikov and A. G. Pozdeeva, *Zh. Prikl. Khim. (Leningrad)* **42**, 2377 (1969); *Chem. Abstr.* **72**, 38234 (1970).

149. A. Sugii, Y. Kabasawa, Y. Niki, and Y. Yamazaki, *Yakugaku Zasshi* **89**, 1066 (1969); *Chem. Abstr.* **71**, 105251 (1969).

150. J. Volke, *Acta. Chim. Acad. Sci. Hung.* **9**, 223 (1956); *Chem. Abstr.* **51**, 13620 (1957).

151. J. Tirouflet and E. Laviron, *Ric. Sci., Suppl.* **4**, 189 (1959); *Chem. Abstr.* **54**, 3041 (1960).

152. C. Parkanyi and R. Zahradnik, *Abh. Dtsch. Akad. Wiss. Berlin, Kl. Chem., Geol. Biol.* p. 363 (1964); *Chem. Abstr.* **62**, 12750 (1965).

153. J. Tirouflet and E. Laviron, *Z. Anal. Chem.* **173**, 43 (1960); *Chem. Abstr.* **54**, 13937 (1960).

154. J. Volke and V. Volkova, *Chem. Listy* **48**, 1031 (1954); *Chem. Abstr.* **48**, 13547 (1954).

155. M. Shikata and I. Tachi, *Bull. Agric. Chem. Soc. Jpn.* **3**, 95 (1927); *Chem. Abstr.* **22**, 1894 (1928).

156. M. Shikata and I. Tachi, *Mem. Coll. Agric., Kyoto Imp. Univ.* **4**, 35 (1927).

157. J. H. Lingane and O. L. Davis, *J. Biol. Chem.* **137**, 567 (1941).

158. T. Nakaya, *Nippon Kagaku Zasshi* **81**, 778 (1960); *Chem. Abstr.* **56**, 270 (1962).

159. L. Campanella and G. De Angelis, *Rev. Roum. Chim.* **16**, 545 (1971).

160. Y. Nagata and I. Tachi, *Bull. Chem. Soc. Jpn.* **27**, 290 (1954).
161. H. H. G. Jellinek and J. R. Urwin, *J. Phys. Chem.* **58**, 168 (1954).
162. Y. Nagata and I. Tachi, *Bull. Chem. Soc. Jpn.* **28**, 113 (1955).
163. P. C. Tompkins and C. L. Schmidt, *Univ. Calif., Berkeley, Publ. Physiol.* **8**, 237 (1943); through Tirouflet and Laviron.[153]
164. P. C. Tompkins and C. L. Schmidt, *Univ. Calif., Berkeley, Publ. Physiol.* **8**, 247 (1943); through Tirouflet and Laviron.[153]
165. M. Brezina and P. Zuman, "Polarography in Medicine, Biochemistry and Pharmacy." Wiley (Interscience), New York, 1958.
166. C. Carruthers and V. Suntzeff, *Arch. Biochem. Biophys.* **45**, 140 (1953).
167. V. Moret, *G. Biochim.* **4**, 192 (1955).
168. C. O. Schmakel, K. S. V. Santhanam, and P. J. Elving, *J. Electrochem. Soc.* **121**, 345 (1974).
169. D. D. Perrin, "Dissociation Constants of Organic Bases in Aqueous Solution." Butterworth, London, 1965.
170. K. S. V. Santhanam and P. J. Elving, *J. Am. Chem. Soc.* **95**, 5482 (1973).
171. W. Kemula and J. Chadkowski, *Rocz. Chem.* **29**, 839 (1955); *Chem. Abstr.* **50**, 6256 (1956).
172. W. Ciusa, P. M. Strocchi, and G. Adamo, *Gass. Chim. Ital.* **80**, 604 (1950); *Chem. Abstr.* **45**, 9059 (1951).
173. B. Janik and P. J. Elving, *Chem. Rev.* **68**, 295 (1951).
174. I. Bergmann, *in* "Polarography 1964" (G. J. Hills, ed.), p. 985. Wiley (Interscience), New York, 1966; through Campanella and De Angelis, Ref. 159.
175. C. Carruthers and V. Suntzeff, *Arch. Biochem. Biophys.* **45**, 140 (1953).
176. J. H. Baxendale, M. G. Evans, and S. J. Leach, *Biochim. Biophys. Acta* **11**, 597 (1953).
177. K. Burton, *Biochim. Biophys. Acta* **8**, 114 (1952).
178. M. G. Evans and N. Uri, *Trans. Faraday Soc.* **45**, 224 (1949).
179. S. J. Leach, J. H. Baxendale, and M. G. Evans, *Aust. J. Chem.* **6**, 395 (1953).
180. J. Nakaya, *Nippon Kagaku Zasshi* **81**, 1459 (1960); *Chem. Abstr.* **56**, 4514 (1962).
181. Y. Paiss and G. Stein, *J. Chem. Soc.* p. 2905 (1958).
182. J. N. Burnett and A. L. Underwood, *J. Org. Chem.* **30**, 1154 (1965).
183. S. Kato, J. Nakaya, and E. Imoto, *Rev. Polarogr.* **18**, 29 (1972).
184. A. J. Cunningham and A. L. Underwood, *Biochemistry* **6**, 266 (1967).
185. D. J. McClemens, A. K. Garrison, and A. L. Underwood, *J. Org. Chem.* **34**, 1867 (1969).
186. W. J. Blaedel and R. G. Haas, *Anal. Chem.* **42**, 918 (1970).
187. A. G. Anderson and G. Berkelhammer, *J. Am. Chem. Soc.* **80**, 992 (1958).
188. K. Wallenfels, *Ciba Found. Study Group* **2**, 10 (1959).
189. H. Sund, *in* "Biological Oxidations" (T. P. Singer, ed.), p. 603. Wiley (Interscience), New York, 1968.
190. K. A. Schellenberg and L. Hellerman, *J. Biol. Chem.* **231**, 547 (1958).
191. F. H. Westheimer, *Adv. Enzymol.* **24**, 469–482 (1962).
192. J. L. Kurz, R. Hutton, and F. H. Westheimer, *J. Am. Chem. Soc.* **83**, 584 (1961).
193. E. M. Kosomer, *Prog. Phys. Org. Chem.* **3**, 136 (1965).
194. M. Gutman, R. Margalit, and A. Schejter, *Biochemistry* **7**, 2778 (1968).
195. V. P. Skulachev and L. I. Denisovich, *Biokhimiya* **31**, 132 (1966).
196. C. Carruthers and V. Suntzeff, *Arch. Biochem. Biophys.* **45**, 140 (1953).
197. C. Carruthers and J. Tech, *Arch. Biochem. Biophys.* **56**, 140 (1955).
198. B. Ke, *Biochim. Biophys. Acta* **20**, 547 (1956).

199. J. N. Burnett and A. L. Underwood, *Biochemistry* **4**, 2060 (1965).
200. V. Moret, *G. Biochim.* **4**, 192 (1955).
201. V. Moret, *G. Biochim.* **5**, 318 (1956); *Chem. Abstr.* **51**, 3710 (1957).
202. C. Carruthers and V. Suntzeff, *Cancer Res.* **12**, 879 (1952).
203. G. A. Hamilton, *Prog. Bioorg. Chem.* **1**, 83 (1971).
204. S. J. Leach, *Adv. Enzymol.* **15**, 1 (1954).
205. P. Karrer, G. Schwarzenbach, F. Benz, and U. Solmssen, *Helv. Chim. Acta* **19**, 811 (1936).
206. P. Karrer and F. J. Stare, *Helv. Chim. Acta* **20**, 418 (1937).
207. M. B. Yarmolinsky and S. P. Colowick, *Fed. Proc., Fed. Am. Soc. Exp. Biol.* **13**, 327 (1954).
208. S. P. Colowick, *in* "The Mechanism of Enzyme Action" (W. D. McElroy and B. Glass, eds.), p. 353. Johns Hopkins Press, Baltimore, Maryland, 1954.
209. M. B. Yarmolinsky and S. P. Colowick, *Biochim. Biophys. Acta* **20**, 177 (1956).
210. A. J. Swallow, *Biochem. J.* **54**, 253 (1953).
211. G. Stein and A. J. Swallow, *Nature (London)* **173**, 937 (1954).
212. M. D. Mathews and E. E. Conn, *J. Am. Chem. Soc.* **75**, 5428 (1953).
213. G. Stein, *Discuss. Faraday Soc.* **12**, 227 (1952).
214. G. Stein and G. Stiassny, *Nature (London)* **176**, 734 (1955).
215. M. E. Pullman, A. San Pietro, and S. P. Colowick, *J. Biol. Chem.* **206**, 129 (1954).
216. F. A. Loewus, B. Vennesland, and D. L. Harris, *J. Am. Chem. Soc.* **77**, 3391 (1955).
217. D. N. Hume, *Anal. Chem.* **28**, 629 (1956).
218. R. D. Weaver and G. C. Whitnack, *Anal. Chim. Acta* **18**, 51 (1958).
219. R. E. Cover and L. Meites, *Anal. Chim. Acta* **25**, 93 (1961).
220. B. Ke, *Arch. Biochem. Biophys.* **60**, 505 (1956).
221. R. F. Powning and C. C. Kratzing, *Arch. Biochem. Biophys.* **66**, 249 (1957).
222. B. Ke, *J. Am. Chem. Soc.* **78**, 3649 (1956).
223. T. Kono, *Bull. Agric, Chem. Soc. Jpn.* **21**, 115 (1957).
224. T. Kono and M. Suekane, *Bull. Agric. Soc. Jpn.* **22**, 404 (1958).
225. T. Kono and S. Nakamura, *Bull. Agric. Soc. Jpn.* **22**, 399 (1958).
226. A. J. Cunningham and A. L. Underwood, *Arch. Biochem. Biophys.* **177**, 88 (1966).
227. W. M. Clarke, *J. Appl. Phys.* **9**, 97 (1938).
228. L. Rodkey, *J. Biol. Chem.* **213**, 777 (1955).
229. F. H. Westheimer, *Enzymes* **1**, 259 (1959).
230. B. Commoner, J. J. Heise, B. B. Lippincott, R. E. Norberg, J. V. Passonneau, and J. Townsend, *Science* **126**, 57 (1957).
231. H. R. Mahler and L. Brand, *in* "Free Radicals in Biological Systems" (M. S. Blois, Jr. *et al.*, eds.), p. 157. Academic Press, New York, 1961.
232. O. Manoušek and P. Zuman, *J. Electroanal. Chem.* **1**, 324 (1960).
233. O. Manoušek and P. Kocova, *Mikrochim. Acta* 754 (1961); *Chem. Abstr.* **56**, 8848 (1962).
234. O. Manoušek and P. Zuman, *Collect. Czech. Chem. Commun.* **29**, 1432 (1964).
235. J. Volke, P. Kubíček, and F. Šantavý, *Collect. Czech. Chem. Commun.* **25**, 871 (1960).
236. O. Manoušek, *Naturwissenschaften* **46**, 323 (1959); *Chem. Abstr.* **54**, 8814 (1960).
237. O. Manoušek and P. Zuman, *Biochim. Biophys. Acta* **44**, 393 (1960); *Chem. Abstr.* **55**, 12498 (1961).
238. O. Manoušek and P. Zuman, *Collect. Czech. Chem. Commun.* **27**, 486 (1962); *Chem. Abstr.* **57**, 1215 (1962).
239. P. Zuman and O. Manoušek, *Collect. Czech. Chem. Commun.* **26**, 2134 (1962).

240. M. Uehara and J. Nakaya, *Nippon Kagaku Zasshi* **90**, 930 (1969); *Chem. Abstr.* **72**, 54 414t (1970).
241. M. Matsumoto, M. Miyazaki, and M. Ishii, *Yakugaku Zasshi* **88**, 1093 (1968); *Chem. Abstr.* **70**, 16525 (1969).

Subject Index

A

Acetyladenine, *see also* 6-(*N*-Acetyl-amino)purine, 121

6-(*N*-Acetylamino)purine, polarographic reduction, 94

AC polarograph, operational amplifier type, 61–63

AC polarography, 20
adsorption, 25
of uncharged substance, 26
alternating current for irreversible reaction, 23
for reversible reaction, 22
base current, 24
suppression, 25
diagnostic criteria for faradaic and tensammetric peaks, 28
peak current and concentration, 25
phase relationships between alternating voltage, faradaic and base currents, 24
relationship to DC polarography, 23
summit potential, 22
tensammetry, 26, 27
test for reversible reaction, 24
tuned amplifier, 62–63

Acetylpyridine(s), electrochemical reduction, 507, 508, 509–511

Acetylpyridine *N*-oxides, polarographic reduction, 499–500

Adenine, 88–92
AC polarography, 91–92
adsorption at DME, 91–92, 210, 274, 275
at pyrolytic graphite electrode, 163–164
analysis in presence of adenosine, 276
in angustmycin A, 78–79
in angustmycin C, 79
anomalous polarographic wave, 99–100
association at DME, 98–99
biochemical oxidation, 165
biological role, 78
in coenzyme(s), 78
in coenzyme A (pantothenic acid), 78
complex with citrate and phosphate, 91
cyclic voltammetry, 91, 163
DC polarography, 88–92
deoxynucleoside, *see* Deoxyadenosine
diffusion current constant, 91
effect of on differential capacitance of DME, 98–99
of high energy radiation, 165–166
on hydrogen overpotential, 91
electrocapillary studies, 92
electrochemical oxidation, pyrolytic graphite electrode, 162–163
electrochemical reduction mechanism, 90
energy of highest occupied molecular orbital, 175
in flavin adenine dinucleotide, 78
formation of mercury compounds, 91